ORGANOCHROMIUM COMPOUNDS

ORGANOMETALLIC CHEMISTRY

A Series of Monographs

BRIAN G. RAMSEY: Electronic Transitions in Organometalloids, 1969.
R. C. POLLER: The Chemistry of Organotin Compounds, 1970.
RUSSELL N. GRIMES: Carboranes, 1970.
PETER M. MAITLIS: The Organic Chemistry of Palladium, Volumes I and II, 1971.
DONALD S. MATTESON: Organometallic Reaction Mechanisms of the Nontransition Elements, 1974.
RICHARD F. HECK: Organotransition Metal Chemistry: A Mechanistic Approach, 1974.
P. W. JOLLY AND G. WILKE: The Organic Chemistry of Nickel, Volume I, Organonickel Complexes, 1974. Volume II, Organic Synthesis, 1975.
P. C. WAILES, R. S. P. COUTTS, AND H. WEIGOLD: Organometallic Chemistry of Titanium, Zirconium, and Hafnium, 1974.
U. BELLUCO: Organometallic and Coordination Chemistry of Platinum, 1974.
P. S. BRATERMAN: Metal Carbonyl Spectra, 1974.
L. MALATESTA AND S. CENINI: Zerovalent Compounds of Metals, 1974.
THOMAS ONAK: Organoborane Chemistry, 1975.
R. P. A. SNEEDEN: Organochromium Compounds, 1975.

ORGANOCHROMIUM COMPOUNDS

R. P. A. Sneeden

Monsanto Research S. A.
Zurich, Switzerland

Academic Press

New York San Francisco London 1975

A Subsidiary of Harcourt Brace Jovanovich, Publishers

ACADEMIC PRESS, INC.
111 Fifth Avenue, New York, New York 10003

United Kingdom Edition published by
ACADEMIC PRESS, INC. (LONDON) LTD.
24/28 Oval Road, London NW1

Library of Congress Cataloging in Publication Data

Sneeden, R P A
Organochromium compounds.

(Organometallic chemistry series)
Bibliography: p.
Includes index.
1. Organochromium compounds. I. Title.
QD412.C7S6 547'.05'532 74-27791
ISBN 0-12-653850-6

PRINTED IN THE UNITED STATES OF AMERICA

CONTENTS

Chapter 3

REACTIONS

PREFACE

In the past twenty years interest in the field of organochromium chemistry and in the much broader field of organotransition metal chemistry has increased. This is a result of the early observations that organotransition metal compounds not only possessed many new and unusual properties but also that they were capable of promoting a multitude of novel, specific, and, at times, commercially significant processes. These factors plus the characteristic curiosity of the researchers have prompted more and more scientists from widely different disciplines to study organotransition metal compounds in a continuing effort to elucidate the fundamental principles associated with the chemistry of these compounds and to emulate the efficiency and selectivity of natural biometallic processes.

The aim of this series of monographs on organometallic chemistry is to provide scientists with a timely appraisal of the state of the art in this vast domain, and, as the chemistry of organochromium compounds is in itself an extensive field, some selection of the material to be discussed was found to be inevitable. This work deals with the major developments in the preparation, characterization, and reaction of the five main classes of organochromium compounds: chromium π-complexes, carbenoid-chromium compounds, solvated σ-bonded organochromium(III) and (II) compounds, solvated sodium and lithium poly(organo)chromate(III) and (II) complexes, and unsolvated σ-bonded tetra(organo)chromium(IV) compounds. The chemistry of chromium hexacarbonyl and simply substituted chromium carbonyls, e.g., $Cr(CO)_{6-n}L_n$; L, amines, phosphines, and the use of ^{13}C-NMR spectroscopy in the characterization of diamagnetic chromium compounds have been the subject of several reviews and therefore are not included.

This monograph is addressed to researchers interested in organotransition metal chemistry. It is hoped that the material presented will stimulate further developments in this challenging, stimulating, and rewarding field.

I extend my sincere thanks to Dr. H. H. Zeiss, President of Monsanto Research S.A., Zurich, and Professor P. M. Maitlis, for their continued interest and encouragement, and to Dr. J. B. O. Sneeden, F.R.S.E., for his invaluable help and advice in the preparation and correction of the manuscript.

Raymond P. A. Sneeden

Chapter 1

PREPARATION

A. Introduction

The development of the chemistry of σ-bonded organochromium compounds can be traced back to two early observations. The first is the reduction of acetylene by chromous salts [Eq. (1-1)], discovered by Berthelot [1]

$$CH{\equiv}CH + Cr(II)/NH_4OH \longrightarrow CH_2{=}CH_2 + Cr^{3+} \qquad (1\text{-}1)$$

in 1866, the second is the interaction of arylmagnesium halides and chromium trichloride in diethyl ether to give biaryls [Eq. (1-2)], discovered by Bennett and Turner in 1914 [2].

$$C_6H_5MgBr + CrCl_3 \xrightarrow{Et_2O} C_6H_5C_6H_5 \qquad (1\text{-}2)$$

Both these discoveries had the effect of sidetracking the development of organochromium chemistry. Thus chromous salts were subsequently used for the reduction of a large number of organic substrates, the emphasis being placed on the reduction of the substrate. It was only in 1957 that Anet and Leblanc [3] successfully prepared solutions containing a pure σ-bonded organochromium compound, Eq. (1-3). They thereby opened up

$$C_6H_5CH_2Cl + 2\,Cr(ClO_4)_2 \xrightarrow{HClO_4} [C_6H_5CH_2Cr(H_2O)_5]^{2+}(ClO_4)_2 + [Cr^{3+}] \qquad (1\text{-}3)$$

a good synthetic route to a large variety of organochromium(III) aquocations [4].

The reaction of aryl Grignards with $CrCl_3$ was, in turn, developed as a method of synthesis of biaryls and led to the discovery of the transition metal-catalyzed synthesis of biaryls from aryl halides [5]. At the same time the reaction of alkyl Grignards with chromium trichloride was reported to effect the reduction of the chromium salt to chromium metal with the concomitant formation of alkane and alkene [6,7]. All these reactions were interpreted in terms of free radical processes. The concept that organochromium compounds are only a source of free radicals and that their fragmentation involves solely the homolysis of the carbon–metal bond dominated the scene until very recently.

During this period, however, Hein and his colleagues studied the interaction of aryl Grignards with $CrCl_3$ in diethyl ether in more detail. In 1919 [8], they found that by controlling the temperature of the reaction it was possible to isolate a water-soluble "phenylchromium compound" (Scheme 1-1, Reaction a), which was subsequently shown [9,10] to be the bis(benzene)chromium π-complex.

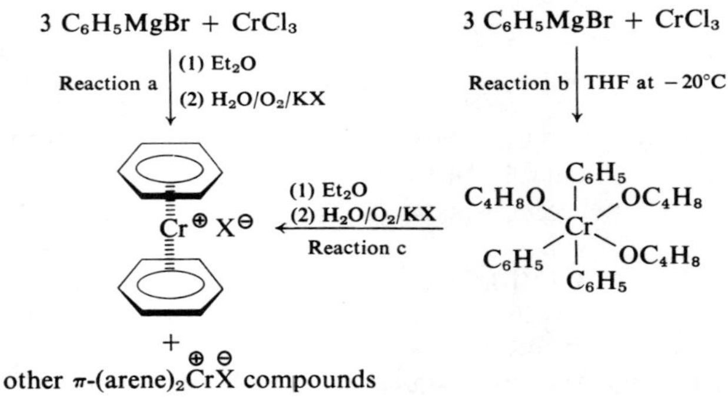

Scheme 1-1

Herwig and Zeiss [11,12] discovered that stoichiometric ratios (3:1) of phenylmagnesium bromide and $CrCl_3$ interact at $-70°$ to $-20°C$ in tetrahydrofuran to give the crystalline solvated σ-bonded tris(phenyl)chromium, Scheme 1-1, Reaction b.

This discovery opened the way for the logical syntheses of many other σ-bonded organochromium(II) and (III) compounds. Zeiss and his colleagues [12,13] also showed that the σ-tris(organo)chromium(III) compounds could be rearranged to bis(arene)chromium π-complexes, thereby establishing a link between the two species, Scheme 1-1, Reaction c. Around the same time

Fischer and his colleagues reported another synthesis of bis(arene)chromium π-complexes [14], Eq. (1-4).

$$3\,CrCl_3 + 2\,Al + AlCl_3 + 6\,C_6H_6 \xrightarrow[(2)\ Na_2S_2O_4]{(1)\ \Delta} \qquad (1\text{-}4)$$

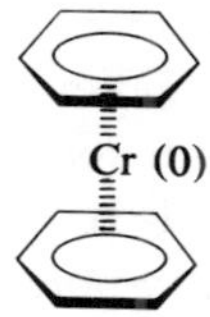

Hein and his colleagues continued their studies on the interaction of organometallic compounds with $CrCl_3$ and found that reaction with phenyllithium led to a new class of σ-bonded organochromium complexes [15], Eq. (1-5). The reaction was subsequently extended to the synthesis of a

$$6\,LiC_6H_5 + CrCl_3 \xrightarrow{(C_2H_5)_2O} [Li_3Cr(C_6H_5)_6][(C_2H_5)_2O]_4 + 3\,LiCl \qquad (1\text{-}5)$$

number of lithium and sodium organochromium(III) and chromium(II) compounds, e.g., $[Li_3Cr(R)_6]$ and $[Li_2Cr(R)_4]$. They were later able to convert a σ-bonded tris(organo)chromium(III) compound into a hexa-(organo)chromium lithium complex [16], Eq. (1-6).

$$3\,LiC_6H_5 + (3\text{-}ClC_6H_4)_3Cr(THF)_3 \xrightarrow{C_2H_5)_2O} [Li_3Cr(C_6H_5)_3(3\text{-}ClC_6H_4)_3][(C_2H_5)_2O]_{2.5} \qquad (1\text{-}6)$$

Fischer and Maasböl, while studying the reactions of organolithium compounds with transition metal carbonyls, found that phenyllithium and $Cr(CO)_6$ interact to give yet another class of organochromium compounds, namely, the "carbenoid chromium carbonyls" [17], Eq. (1-7).

$$C_6H_5Li + Cr(CO)_6 \xrightarrow[(2)\ CH_2N_2]{(1)\ Et_2O} C_6H_5\overset{\overset{\displaystyle OCH_3}{|}}{C}\!-\!Cr(CO)_5 \qquad (1\text{-}7)$$

Although monomeric quadricovalent chromium compounds of the type $Cr(OR)_4$ and $Cr(NR_2)_4$ had been isolated as early as 1959 and 1963 by Hagihara [18] and Bradley [19], respectively, it was only recently that Wilkinson and his colleagues [20], Bower and Tennent [21], and Kruse [22], prepared monomeric σ-bonded tetra(organo)chromium(IV) compounds by the interaction of organolithium compounds and $CrCl_3(THF)_3$ or $Cr(OR)_4$, e.g., Eqs. (1-8) and (1-9):

$$CrCl_3(THF)_3 + 4\,LiCH_2Si(CH_3)_3 \longrightarrow Cr[CH_2Si(CH_3)_3]_4 \qquad (1\text{-}8)$$

$$(t\text{-}BuO)_4Cr + 4\,Li(t\text{-}C_4H_9) \longrightarrow Cr(t\text{-}C_4H_9)_4 \qquad (1\text{-}9)$$

Thus there are to date five known main classes of organochromium compounds:

(1) The chromium π-complexes [e.g., π-bis(benzene)chromium, π-bis(cyclopentadienyl)chromium, π-(benzene)chromium tricarbonyl, and π-(norbornadiene)chromium tetracarbonyl].
(2) Carbenoid-chromium compounds [e.g., $RC(OR')Cr(CO)_5$].
(3) Solvated σ-bonded organochromium(II) and (III) compounds [e.g., $R_3Cr(THF)_n$].
(4) Solvated σ-bonded poly(organo)chromium(II) and (III) lithium and sodium complexes [e.g., $Li_3CrR_6(S)_x$].
(5) Unsolvated σ-bonded tetra(organo)chromium(IV) compounds [e.g., $Cr(t\text{-}C_4H_9)_4$].

The general methods available for the preparation of these five classes of compounds will be presented in the ensuing pages and, for convenience, the preparation of each type of chromium π-complex (i.e., Class 1) will be discussed individually under the headings: bis(arene)chromium π-complexes, π-(cyclopentadienyl)chromium compounds and carbonylchromium π-complexes.

B. Bis(arene)chromium π-Complexes

The preparation of bis(arene)chromium π-complexes has been extensively reviewed and discussed [13,23–28]. The following general methods are available for the preparation of certain bis(arene)chromium π-complexes.

1. Hein Grignard Synthesis

This synthesis, originating in the work of Hein [8], was shown by Zeiss and Herwig [11,12] to involve essentially the rearrangement of a σ-bonded arylchromium species to a "black paramagnetic intermediate," which upon hydrolysis gives either one or a mixture of bis(arene)chromium π-complexes, Scheme 1-2. Thus the preformed tris(aryl)chromium compound may be rearranged to the black intermediate **1** (Scheme 1-2) by simple treatment with diethyl ether [11, 12, 26–29], e.g., Scheme 1-2, Reaction a. Alternatively, the "black intermediate" may be prepared directly by the interaction of stoichiometric ratios (3:1) of aryl Grignard and $CrCl_3$ in diethyl ether [8,12,28], Scheme 1-2, Reaction b. Hydrolysis of the black intermediate **1** and subsequent air oxidation leads to a mixture of bis(arene)chromium(I) π-complexes, e.g., Scheme 1-2, compounds **2**, **3**, **4**. The latter may be isolated as their iodide or tetraphenylborate salts, Scheme 1-2, Reaction c.

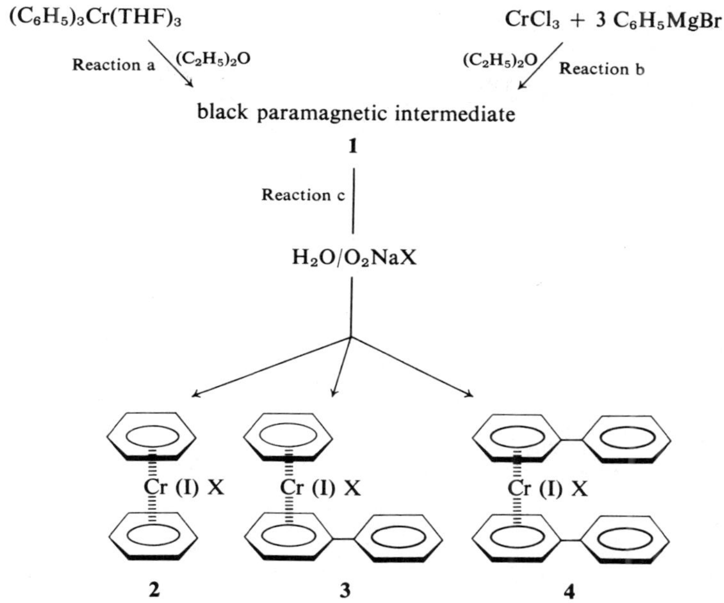

Scheme 1-2. Hein synthesis of π-bis(benzene)Cr(I), π-(benzene)-π-(biphenyl)Cr(I), and π-bis(biphenyl)Cr(I) salts. X = $I^{\ominus}$ or $B(C_6H_5)_4^{\ominus}$.

The main disadvantages to these syntheses are first, that a mixture of bis-(arene)chromium π-complexes is obtained, e.g., compounds **2** to **4**, Scheme 1-2, and second, that it is not possible to predict the overall course of the reactions. The former difficulty may be overcome either by fractional crystallization [10,30–32] or by chromatography [27,33,34] of a suitable salt of the mixture of bis(arene)chromium π-complex cations. The latter difficulty, however, remains an obstacle insofar as not all tris(aryl)Cr(III) compounds undergo rearrangement to bis(arene)chromium π-complexes. Thus, whereas tris(phenyl)- and tris(alkyl- or aryl-substituted-phenyl)Cr(III) compounds rearrange readily to the respective π-complexes [11,12,27–29], tris(3- and 4-ClC_6H_4)-, tris(4-F and 4-BrC_6H_4)-, tris(mesityl)- and tris(2-thienyl)Cr(III) do not [27,35]. Furthermore, whereas aryl Grignards give the same mixture of bis(arene)chromium π-complexes irrespective of whether $CrCl_3$ or $CrCl_2$ [35] is used, the benzyl Grignard gives different products. An ether solution of benzylmagnesium chloride reacts with $CrCl_3(THF)_3$ to give π-(toluene)-π-(2-benzyltoluene)chromium [36] and with $CrCl_2$ to give π-bis(toluene)-chromium [37]. Clearly, these syntheses cannot be regarded as being general ones for the preparation of specific bis(arene)chromium π-complexes.

2. Fischer–Hafner Aluminum Synthesis

In general terms this synthesis involves the interaction of an aromatic hydrocarbon with a metal halide (here $CrCl_3$) in the presence of a reducing system ($AlCl_3/Al$), Eq. (1-10) [13,14,23–25,38]. The crude product (the

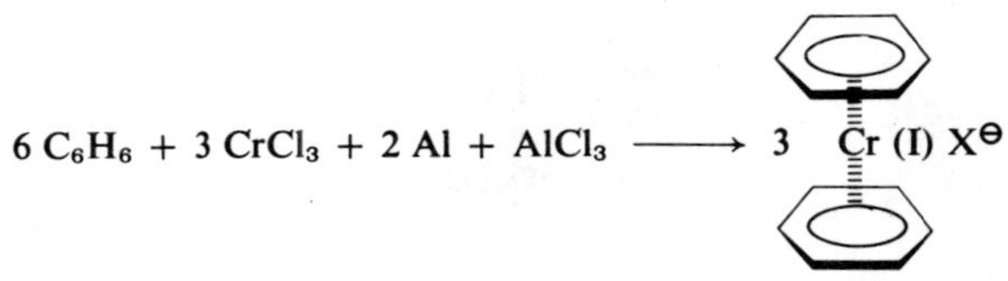

5 (1-10)

cation **5**, X = $AlCl_4$) is hydrolyzed and subsequently reduced by sodium dithionite to the benzene-soluble chromium(0) complex **6**. The latter, on air oxidation, in the presence of KI, furnishes the relatively insoluble π-bis-(arene)chromium(I) iodide **5**, Eq. (1-11), X = I. The combination of these

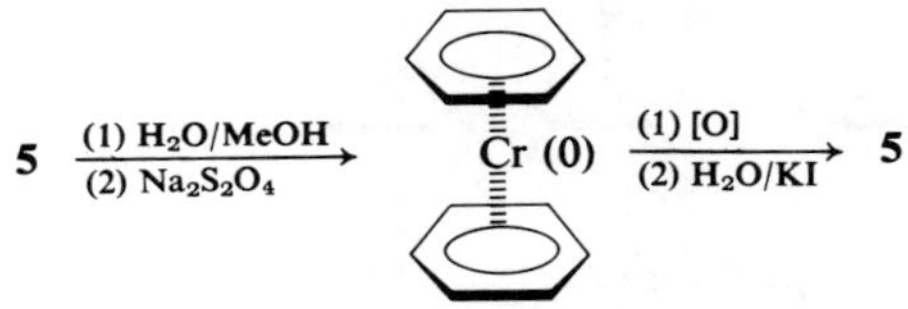

6 (1-11)

reduction and oxidation reactions [i.e., Eq. (1-11)] provides a convenient method for the isolation of pure material from a complex reaction mixture.

The conditions originally used in the Fischer–Hafner synthesis involved rather high temperatures (i.e., bomb-tube experiments); however, as the chemistry of this field developed, it became evident that alkyl-substituted benzenes react more readily than benzene itself and that $AlCl_3$ acts as a catalyst in the exchange of arene groups in bis(arene)chromium π-complexes [39,40].

On the basis of these observations, Fischer and Seeholzer [41] were able to lower the original reaction temperature by the use of catalytic amounts of mesitylene. Presumably, the mesitylene reacts rapidly to give π-bis(mesitylene)chromium; this subsequently undergoes $AlCl_3$-catalyzed exchange (with the arene present in excess) to give the desired π-bis(arene)chromium complex.

It has also been reported that triethylaluminum may replace $AlCl_3/Al$ as reducing species [23,25,42,43] and that a mixture of arenechromium π-complexes results from the interaction of phenyldichloroaluminum and $CrCl_3$, Eq. (1-12) [44].

$$CrCl_3 + n\text{-}AlCl_2C_6H_5 \xrightarrow{H_2O} \pi\text{-}(C_6H_6)_2Cr + \pi\text{-}(C_6H_6)\text{-}\pi\text{-}(C_6H_5C_6H_5)Cr + \pi\text{-}(C_6H_5C_6H_5)_2Cr \quad (1\text{-}12)$$

The Fischer–Hafner synthesis affords a convenient route to certain simple bis(arene)chromium π-complexes. The yields are good, and the products contain only one bis(arene)chromium π-complex. The reaction is of limited application, however, since it requires that the future arene components be inert toward aluminum chloride (at 100°–200°C). Thus, when halogen-substituted arenes (e.g., chlorobenzene or 4,4′-dichlorobiphenyl) are used in the Fischer–Hafner synthesis, the final products consist of the desired bis(halogenoarene)chromium π-complexes together with appreciable quantities of the corresponding halogen-free bis(arene)chromium π-complexes [45, 46]. Similarly it is known that aluminum trichloride will promote the transalkylation and isomerization of alkylbenzenes. Thus, when alkylbenzenes are used in the Fischer–Hafner synthesis, the final products are mixtures of π-complexes. The major component (90%) is the expected π-bis(alkylarene)-chromium compound, contaminated with other compounds in which the arene portions are transformation products of the arene originally used: with mesitylene, the by-products are xylene and tetramethylbenzene; with ethylbenzene, the by-products are benzene, di-, tri-, and tetraethylbenzene; with isopropylbenzene, the by-products are benzene, ethylbenzene, di- and triisopropylbenzene [47–49]. When 1,4-diphenylbutane is used as the arene component in the Fischer–Hafner synthesis, a mixture of four π-bis-(arene) chromium(0) compounds is obtained (in ca. 40% yield); the major constituents are π-(benzene)-π-(tetralin)chromium(0) and π-bis(tetralin)-chromium(0) [49a]. It has been reported that the transalkylation reaction may be avoided by using $CrBr_3$, $AlBr_3$, and Al [47].

3. Fischer, Müller, and Kuzel Alkyl Grignard Synthesis

The interaction of an alkylmagnesium halide and $CrCl_3$, in diethyl ether solution in the presence of 1,3-cyclohexadiene leads to the formation of π-bis(benzene)chromium in modest yields, Eq. (1-13) (R = CH_3CH_2, $CH_3CH_2CH_2$, $(CH_3)_2CH$, or C_4H_9, but not CH_3) [50–52]. This reaction,

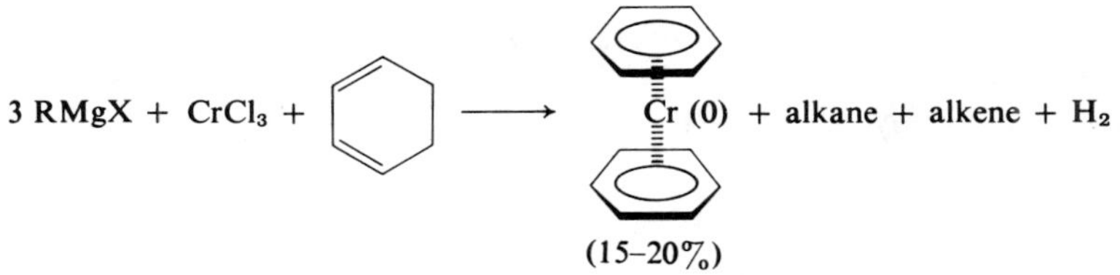

(1-13)

which is catalyzed by UV light, illustrates the capacity of low-valent organochromium species to abstract hydrogen from complexed substrate. The detailed mechanism of the hydrogen abstraction and of the formation of the bis(arene)chromium π-complex is not clear. It has been suggested [50] that the overall reaction involves the initial formation of a tris(alkyl)chromium complex [7, Eq. (1-14)]. This undergoes homolytic fragmentation to give a

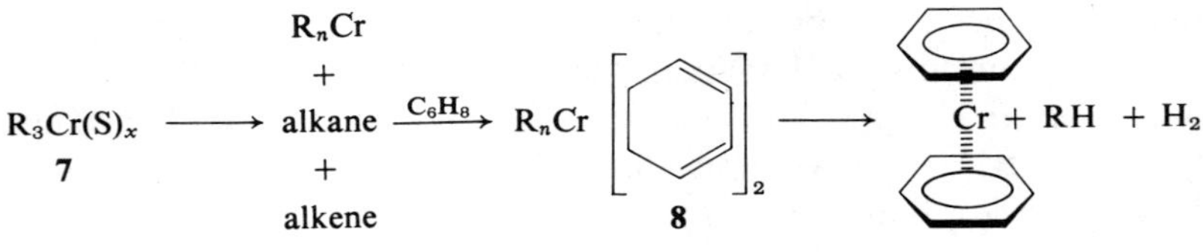

(1-14)

low-valent "chromium species," which forms a complex (**8**) with the diene, which then undergoes fragmentation and hydrogen abstraction to give, finally, π-bis(benzene)chromium, Eq. (1-14). This synthesis of π-bis(arene)-chromium compounds is a priori restricted to the available 1,3-cyclohexadienes, and since these are best prepared by the Birch reduction of the appropriate arenes, it is doubtful whether the method offers any advantage over the direct aluminum/$AlCl_3$ route (see Section B,2).

The alkyl Grignard method has also been used for the preparation of a variety of chromium π-complexes. Thus the reaction of isopropylmagnesium bromide with $CrCl_3$ in the presence of azulene gives π-(azulenium)chromium-(0)-π-azuleniate, **9**, Eq. (1-15), in 6% yield [53]. When isopropylmagnesium

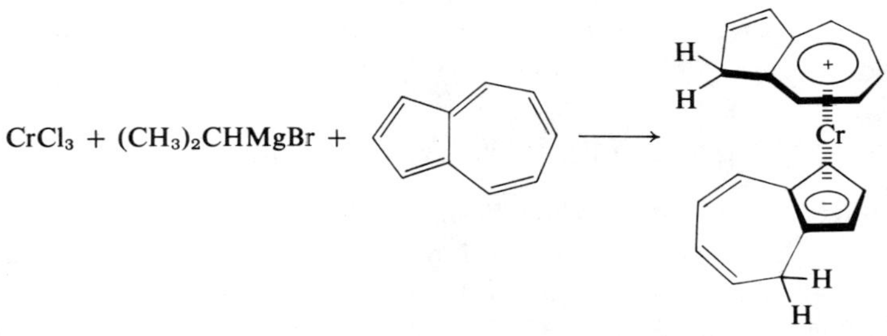

(1-15)

9

bromide and $CrCl_3$ interact in the presence of a mixture of cyclopentadiene and cycloheptatriene, 3% of the mixed complex π-(cyclopentadienyl)-π-(cycloheptatrienyl)Cr(0) **10**, Eq. (1-16), may be isolated [54]. When cyclo-

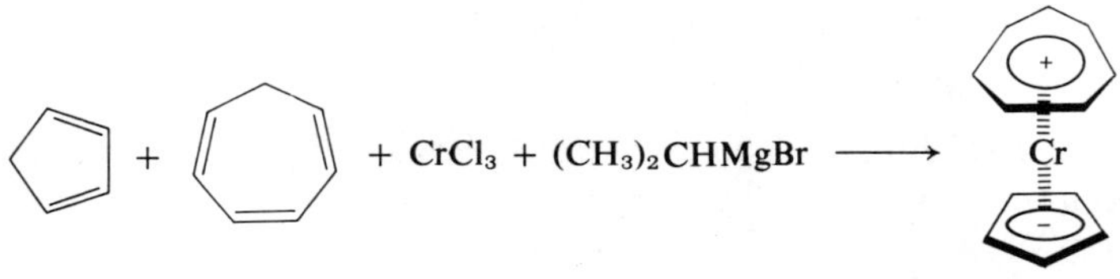

(1-16)

10

heptatriene is used alone, the product, isolated in 5–10% yield, is π-(cycloheptatrienyl)-π-(1,3-cycloheptadiene)Cr(−I), **11**, Eq. (1-17) [55,55a].

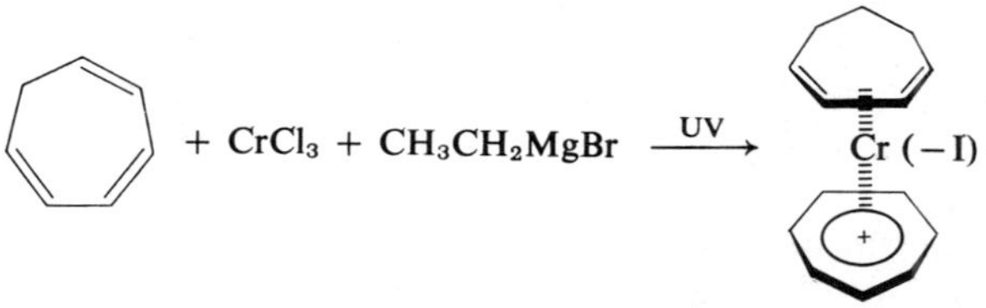

(1-17)

11

All the foregoing reactions with $CrCl_3$, Eqs. (1-13)–(1-17), involve the transfer of hydrogen either to or from the olefinic substrate; in contrast, no such transfers would seem to occur in analogous reactions with π-(cyclopentadienyl)chromium dichloride. Thus, reaction of π-$(C_5H_5)CrCl_2(THF)$ and isopropylmagnesium bromide in the presence of either cycloheptatriene or 1,3,5-cyclooctatriene leads, in 60% yield, to the corresponding paramagnetic π-(cyclopentadienyl)chromium olefin complexes, **12** and **13**, Eqs. (1-18) and (1-19) [56].

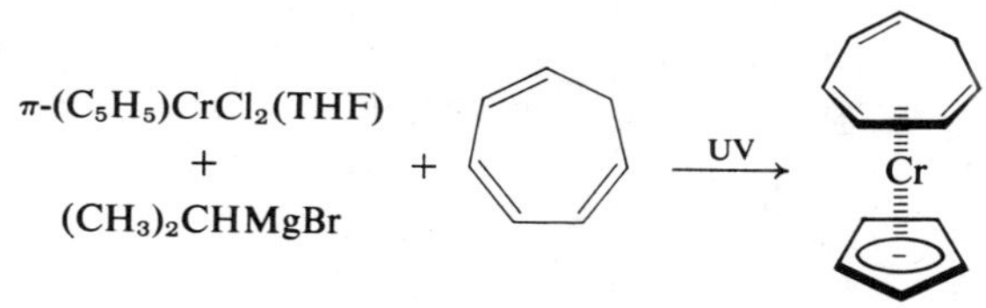

(1-18)

12

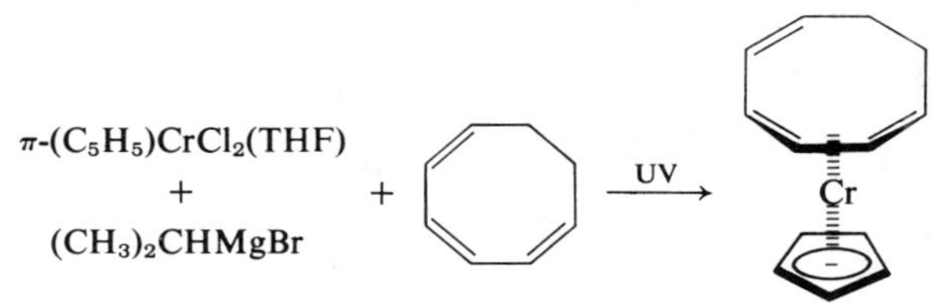

(1-19)

13

The π-(cycloheptatriene)chromium complex (**12**) may be converted readily to the corresponding neutral and cationic π-(cyclopentadienyl)-π-(cycloheptatrienyl)chromium complexes **14** and **15**, respectively, as shown in Eq. (1-20) [56].

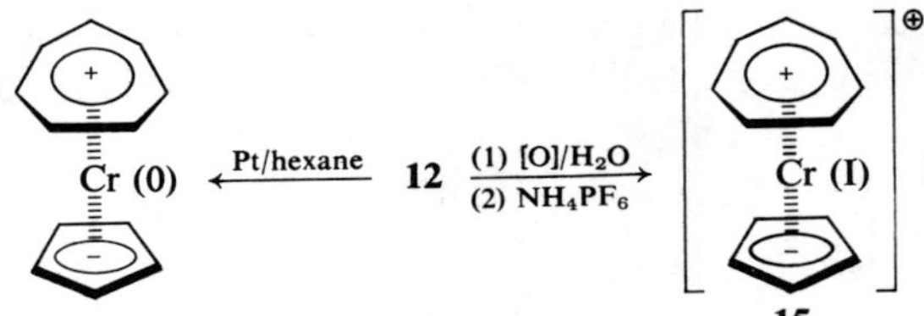

14 15 (1-20)

The mechanism of this "alkyl Grignard" synthesis, in particular that of the hydrogen transfer processes involved in the reactions with $CrCl_3$, has not been elucidated [Eqs. (1-13)–(1-17)].

4. Herwig–Zeiss Acetylene Trimerization

Certain σ-bonded organochromium(III) and (II) systems react with disubstituted acetylenes to give a bis(arene)chromium π-complex together with a variety of organic products derived from the acetylene [e.g., Eq. (1-21)]

$$(C_6H_5)_3Cr(THF)_3 + CH_3C{\equiv}CCH_3 \xrightarrow[(2)\ H_2O/O_2/X^{\ominus}]{(1)\ \text{THF at } 20^\circ\text{C}} \text{organic products (see Chapter 3, Section F)} + [(CH_3)_6C_6]_2CrX \tag{1-21}$$

[23,57–61]. The yields of π-complex in these reactions are usually modest, and depend upon the presence of a substantial excess of acetylene in the reaction mixture [23]. The mechanism of this reaction has not been studied specifically: however, the available data suggest that the bis(arene)chromium π-complex is an end product and not an intermediate one in the formation of the sundry organic products [23]. The future arene part is found as one of the "organic products," and could be formed by the interaction of a chromocyclopentadiene and 2-butyne, e.g., Eq. (1-22) (see Chapter 3, Section F,3,a).

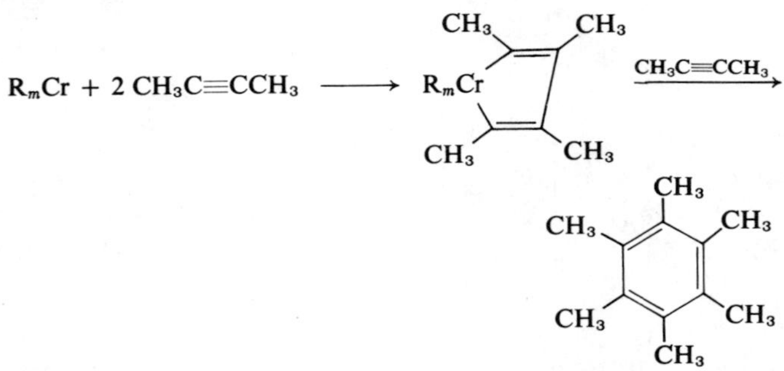

(1-22)

The bis(arene)chromium π-complex might then originate in the direct interaction of low-valent chromium species [Cr(0) or Cr(I)] and the arene. The formation of π-bis(arene)chromium(0) compounds by the reaction of chromium vapor and various arenes has been reported [62,62a,63]. The recent observation [63] that chromium vapor and 2-butyne interact to give hexamethylbenzene but no π-bis(hexamethylbenzene)chromium(0) species shows, however, that the low-valent chromium species cannot be simple active chromium centers. An alternative route to the π-complex could involve the direct interaction of a spirocyclic Cr(IV) compound (**16**) and two molecules of 2-butyne, e.g., Eq. (1-23).

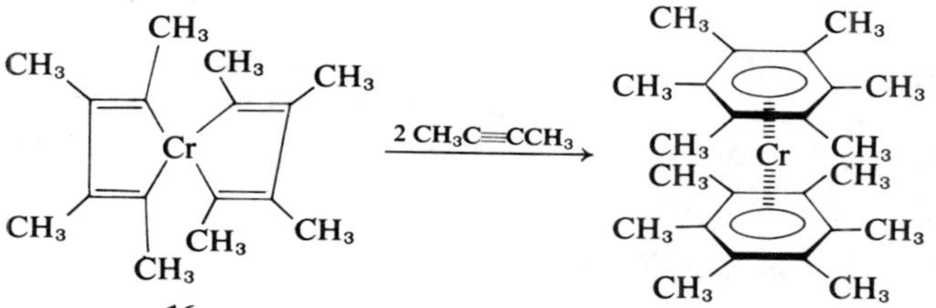

(1-23)

16

A closer study of the acetylenic trimerization route to bis(arene)chromium π-complexes is urgently required, not only with the object of clarifying the mechanism of π-complex formation, but also for the valuable information it would yield on the formation of the organic products.

There are clearly no direct routes known for the synthesis of specific bis(arene)chromium π-complexes, in particular, those bearing functional groups (e.g., COOR, CHO, OR, etc.,). Also, the mechanisms of the available synthetic methods (i.e., Section B, 1–4) are not clearly understood—thus, does the bis(arene)chromium π-complex originate in the direct interaction of a low-valent chromium species and the arene, or in a σ-bonded organochromium precursor? Furthermore, it would be of interest to determine the real scope of the methods and in particular whether any may be modified or adapted to the preparation of π-bis(arene)chromium compounds bearing functional groups.

C. π-(Cyclopentadienyl)chromium Compounds

There are three major classes of π-(cyclopentadienyl)chromium complexes: π-mono(cyclopentadienyl)chromium derivatives, π-bis(cyclopentadienyl)chromium(II) and the salts derived from the corresponding chromium(III) cation, π-complexes involving one cyclopentadienyl unit and another system containing delocalized electrons (e.g., benzene, 1,3,5-cyclooctatriene, or cycloheptatriene). The key compound in the preparation of many of these substances is the highly reactive π-bis(cyclopentadienyl)chromium. These preparations really illustrate the reactions of π-$(C_5H_5)_2$Cr but they are included here since they represent valid preparative methods.

1. π-Mono(cyclopentadienyl) Derivatives

a. Unsolvated Compounds, π-$(C_5H_5)CrX_2$

Unsolvated π-(cyclopentadienyl)dichlorochromium can be prepared in good yield by the interaction of bis-π-(cyclopentadienyl)chromium and hexachlorocyclopentadiene [64], Eq. (1-24). There are no details available

$$\pi\text{-}(C_5H_5)_2Cr + C_5Cl_6 \xrightarrow{160°C} \pi\text{-}(C_5H_5)CrCl_2 \qquad (1\text{-}24)$$

concerning the mechanism of the reaction. The unsolvated dibromo and dichloro compounds can be obtained [65] by removal, in high vacuum, of the solvate from the corresponding tetrahydrofuranates [for their preparation, see Eqs. (1-27a), (1-28), and (1-29)].

b. Solvated Compounds π-$(C_5H_5)CrX_2L$

The pyridine-solvated dibromo and diodo compounds (**17**, X = Br or I) can be prepared by the reaction of the pyridinium trihalide on π-bis-(cyclopentadienyl)chromium, Eq. (1-25) [65] (X = Br or I, L = pyridine).

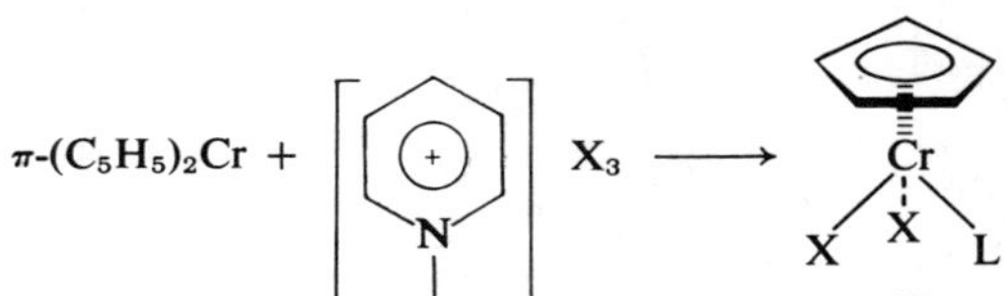

17 (1-25)

The solvated dichloro compound, **17** (X = Cl and L = tetrahydrofuran), can be obtained indirectly from π-bis(cyclopentadienyl)chromium. The latter reacts with CCl_4, in tetrahydrofuran solution, to give an ionic species [**18**, Eq. (1-26)]. The composition of **18** was established by its reaction with

$$\pi\text{-}(C_5H_5)_2Cr + CCl_4 \xrightarrow{THF} \underset{\mathbf{18}}{[\pi\text{-}(C_5H_5)_2Cr][\pi\text{-}(C_5H_5)CrCl_3]} \qquad (1\text{-}26)$$

$NaB(C_6H_5)_4$, to give the π-bis(cyclopentadienyl)chromium(III) salt [π-$(C_5H_5)_2\overset{\oplus}{Cr}\cdot\overset{\ominus}{B}(C_6H_5)_4$] [65]. Extraction of **18** with tetrahydrofuran or dioxane gives the corresponding solvated dichloro-π-(cyclopentadienyl)chromium compounds [**19** and **20**, Eqs. (1-27a) and (1-27b)] [65].

$$\underset{\mathbf{20}}{[\pi\text{-}(C_5H_5)CrCl_2]_2(O_2C_4H_8)} \xleftarrow[\text{(b)}]{\text{dioxane}} \mathbf{18} \xrightarrow[\text{(a)}]{\text{THF}} \underset{\mathbf{19}}{\pi\text{-}(C_5H_5)CrCl_2(OC_4H_8)} \qquad (1\text{-}27)$$

Alternative syntheses involve the interaction of allylic halides with π-(cyclopentadienyl)chromium tricarbonyl dimer or π-bis(cyclopentadienyl)-

chromium. Thus, reaction of the tricarbonyl dimer with allyl iodide or bromide (but not chloride) gives the dihalogeno-π-(cyclopentadienyl)-chromium as initial product; extraction of the latter with tetrahydrofuran or pyridine gives the corresponding solvated compounds in good yields, Eq. (1-28), X = Br or I, L = tetrahydrofuran or pyridine [65]. Similarly π-bis-

$$[\pi\text{-}(C_5H_5)Cr(CO)_3]_2 \xrightarrow{CH_2{=}CHCH_2X} \pi\text{-}(C_5H_5)CrX_2 \xrightarrow{L} \pi\text{-}(C_5H_5)CrX_2L \quad (1\text{-}28)$$

(cyclopentadienyl)chromium reacts with allyl bromide (but not allyl iodide or chloride) to give the unsolvated dibromo compound that is readily converted to the corresponding solvated species by reaction with tetrahydrofuran, Eq. (1-29), L = tetrahydrofuran [65]. The allyl unit formed in these reactions is converted probably into 1,5-hexadiene.

$$\pi\text{-}(C_5H_5)_2Cr \xrightarrow{CH_2{=}CHCH_2Br} \pi\text{-}(C_5H_5)CrBr_2 \xrightarrow{L} \pi\text{-}(C_5H_5)CrBr_2L \quad (1\text{-}29)$$

The tetrahydrofuran of solvation in these π-(cyclopentadienyl)chromium dihalides may be readily exchanged for pyridine or triphenyl phosphine [65].

c. Ionic Species Derived from $[\pi\text{-}(C_5H_5)CrX_3]^{\ominus}$

These compounds are formed in the interaction of π-bis(cyclopentadienyl)-chromium and CCl_4 (X = Cl), Eq. (1-26), or with hydrochloric or hydrobromic acids. In the latter cases the ionic species were isolated as their ammonium salts [65], Eq. (1-30) (X = F, Cl, Br, or I).

$$\pi\text{-}(C_5H_5)_2Cr \xrightarrow{3\,HX} H[\pi\text{-}(C_5H_5)CrX_3] + C_5H_6 + \tfrac{1}{2}\,H_2 \xrightarrow{NH_4Cl} NH_4[\pi\text{-}(C_5H_5)CrX_3] \quad (1\text{-}30)$$

d. Miscellaneous π-(Cyclopentadienyl)chromium Compounds

The interaction of π-bis(cyclopentadienyl)chromium and carbon monoxide, alone or with added hydrogen, leads to a variety of π-(cyclopentadienyl)-chromium-carbonyl derivatives [66], Eqs. (1-31)–(1-33). The dimeric com-

$$\pi\text{-}(C_5H_5)_2Cr + CO \xrightarrow[150^\circ\text{–}170^\circ C]{100\text{ atm}} \underset{\mathbf{21}}{[\pi\text{-}(C_5H_5)Cr(CO)_3]_2} \quad (1\text{-}31)$$

$$\pi\text{-}(C_5H_5)_2Cr + CO \xrightarrow[100^\circ\text{–}110^\circ C]{100\text{ atm}} [\pi\text{-}(C_5H_5)_2Cr][\pi\text{-}(C_5H_5)Cr(CO)_3] \quad (1\text{-}32)$$

$$\pi\text{-}(C_5H_5)_2Cr + CO + H_2 \longrightarrow \underset{\mathbf{22}}{\pi\text{-}(C_5H_5)Cr(CO)_3H} \quad (1\text{-}33)$$

pound **21** and the hydridochromium compound **22** are interconvertible by the addition, or elimination, of hydrogen, Eq. (1-34).

$$\underset{\mathbf{21}}{[\pi\text{-}(C_5H_5)Cr(CO)_3]_2} \underset{\Delta - H_2}{\overset{H_2}{\rightleftharpoons}} 2\ \underset{\mathbf{22}}{\pi\text{-}(C_5H_5)Cr(CO)_3H} \tag{1-34}$$

The hydridochromium compound **22** may be prepared more conveniently either by the interaction of $Cr(CO)_6$ and a metal cyclopentadienide [67], Eq. (1-35) [M = K or Na], or by the interaction of tris(acetonitrile)$Cr(CO)_3$

$$Cr(CO)_6 + MC_5H_5 \xrightarrow{130°C} M[\pi\text{-}(C_5H_5)Cr(CO)_3] + 3\ CO \xrightarrow{CH_3COOH/H_2O} \underset{(60\%)}{\pi\text{-}(C_5H_5)Cr(CO)_3H} \tag{1-35}$$

with cyclopentadiene [68], Eq. (1-36).

$$(CH_3CN)_3Cr(CO)_3 +$$  $$\xrightarrow{THF/20°C} \underset{(20\%)}{\pi\text{-}(C_5H_5)Cr(CO)_3H} + 3\ CH_3CN \tag{1-36}$$

The latter reaction, which may be regarded as the oxidative addition of an alkane to a low-valent transition metal compound, has also been used to prepare the tungsten and molybdenum analogs [69].

Acetyl-pentamethylcyclopentadiene, on the other hand, reacts with $(CH_3CN)_3Cr(CO)_3$ or $Cr(CO)_6$ to give, by loss of CH_3CO, the *dimeric dicarbonylchromium* compound **23**, Eq. (1-37) [70].

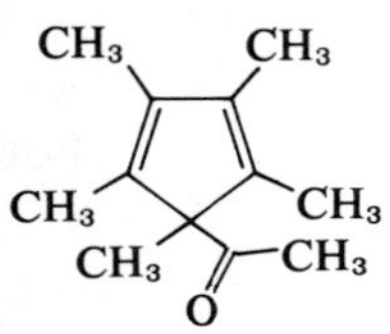

$$+ \begin{matrix} (CH_3CN)_3Cr(CO)_3 \\ \text{or} \\ Cr(CO)_6 \end{matrix} \longrightarrow \underset{\mathbf{23}}{[\pi\text{-}[(CH_3)_5C_5]Cr(CO)_2]_2} \tag{1-37}$$

A single crystal X-ray structure analysis [71] confirms the dimeric nature of the molecule. The Cr–Cr distance 2.276(2) Å is shorter than that found in bridged dimers (2.67 to 2.95 Å) (Table 2.3, Chapter 2) but longer than that found in nonbridged σ-bonded organochromium(II) dimers (1.97 to 1.98 Å) (Table 2.7, Chapter 2).

The dimer $[\pi\text{-}(C_5H_5)Cr(CO)_3]_2$ (**21**) may be prepared from sodium cyclopentadienide and $Cr(CO)_6$. The initial product is the sodium salt, **24**, Eq. (1-38), and this may be converted to the dimer **21** by reaction with either tropylium bromide, Eq. (1-39), or allylic halides [72].

$$NaC_5H_5 + Cr(CO)_6 \xrightarrow[\text{diglyme}]{\Delta} \underset{\mathbf{24}}{\pi\text{-}(C_5H_5)Cr(CO)_3Na} \tag{1-38}$$

$$2\ \pi\text{-}(C_5H_5)Cr(CO)_3Na + 2\ C_7H_7Br \longrightarrow [\pi\text{-}(C_5H_5)Cr(CO)_3]_2 + C_{14}H_{14} \quad (1\text{-}39)$$

24 **21**

π-Bis(cyclopentadienyl)chromium reacts with oxygen, under controlled conditions, to give π-(cyclopentadienyl)chromium oxide tetramer $[(C_5H_5)CrO]_4$ [73].

The π-(cyclopentadienyl)-di(nitrosyl)chromium halides are formed in 65% yield by the successive actions of the hydrogen halide (HCl and HBr) and nitric oxide on π-bis(cyclopentadienyl)chromium [74], Eq. (1-40) (X = Cl or Br). An alternative synthesis, which avoids the isolation of

$$\pi\text{-}(C_5H_5)_2Cr \xrightarrow[(2)\ NO]{(1)\ HX/(C_2H_5)_2O} \pi\text{-}(C_5H_5)Cr(NO)_2X \quad (1\text{-}40)$$

π-bis(cyclopentadienyl)chromium, involves the use of π-(cyclopentadienyl)-chromium dichloride (**25**). Thus, reaction of $CrCl_3$ with sodium [75,76] or thallium cyclopentadienide [77,78] gives **25**, Eq. (1-41) and reaction of the latter with nitric oxide yields π-(cyclopentadienyl)-di(nitrosyl)chromium chloride [**26**, Eq. (1-41)].

$$\underset{\text{or } TlC_5H_5}{CrCl_3 + NaC_5H_5} \xrightarrow{THF} \underset{\mathbf{25}}{\pi\text{-}(C_5H_5)CrCl_2} \xrightarrow{NO} \underset{\mathbf{26}}{\pi\text{-}(C_5H_5)Cr(NO)_2Cl} \quad (1\text{-}41)$$

The di(nitrosyl)chromium compound **26** has been used as starting material in the preparation of a variety of σ-bonded (organo)chromium compounds, the yields of products in these reactions are, however, low. Reaction of **26** with alkyl- or arylmagnesium halides gives the corresponding alkyl- or arylchromium compounds (**27**, R = CH_3, CH_3CH_2, C_6H_5), Eq. (1-42) [74,79], of these, the methyl compound has been fully characterized [74,80].

$$\pi\text{-}(C_5H_5)Cr(NO)_2Cl \xrightarrow[\text{in } Et_2O]{RMgX}$$

27 (1-42)

The interaction of NaC_5H_5 and π-$(C_5H_5)Cr(NO)_2Br$ gives a crystalline bis(cyclopentadienyl)chromium compound, Eq. (1-43). The analysis of the

$$\pi\text{-}(C_5H_5)Cr(NO)_2Br \xrightarrow[\text{in THF}]{NaC_5H_5} \underset{\mathbf{28}}{\sigma\text{-}(C_5H_5)Cr(NO)_2\text{-}\pi\text{-}(C_5H_5)} \quad (1\text{-}43)$$

compound together with its IR and NMR spectra and its reaction with maleic anhydride, led to its formulation as a σ-bonded cyclopentadienyl-π-(cyclopentadienyl)chromium compound [**28**, Eq. (1-43)] [74].

π-(Cyclopentadienyl)-di(nitrosyl)chlorochromium reacts with diazomethane to give, in 3% yield, a green solid formulated as the σ-bonded chloromethylchromium compound **29**, Eq. (1–44) [74,79].

$$\pi\text{-}(C_5H_5)Cr(NO)_2Cl \xrightarrow[(C_2H_5)_2O/Cu]{CH_2N_2} \underset{\mathbf{29}}{\pi\text{-}(C_5H_5)Cr(NO)_2CH_2Cl} \qquad (1\text{-}44)$$

2. π-Bis(cyclopentadienyl)chromium and Its Derivatives

a. π-Bis(cyclopentadienyl)chromium: Chromocene

The general methods available for the synthesis of π-bis(cyclopentadienyl)-chromium(II) have been reviewed and discussed [81]; however, these will be briefly outlined.

The tris(cyclopentadienyl)chromium hexamine complex **30** [prepared [82,83] as outlined in (1) in Eq. (1-45)] gives, on pyrolysis in high vacuum, the volatile π-bis(cyclopentadienyl)chromium in low yield [83] [(2) in Eq. (1-45)].

$$[Cr(NH_3)_6][NO_3]_3 \xrightarrow[LiC_5H_5/NH_3]{(1)} \underset{\mathbf{30}}{[Cr(NH_3)_6][C_5H_5]_3} \xrightarrow[80°\text{–}120°C]{(2)}$$ 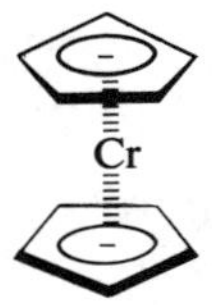 (1-45)

π-Bis(cyclopentadienyl)chromium may be prepared by the interaction of cyclopentadiene and $Cr(CO)_6$ vapors, in an inert atmosphere, at 280°–350°C [84,85]. The crude product, which is contaminated with metallic Cr and unreacted $Cr(CO)_6$, has to be purified by subsequent fractional sublimation in vacuum. The most convenient laboratory method of preparation is by the interaction of "an ionic cyclopentadienide" with a chromic or chromous halide, Eqs. (1-46) and (1-47) [86,67]. Excess of the cyclopenta-

$$NaC_5H_5 + CrCl_3 \xrightarrow{THF} \pi\text{-}(C_5H_5)_2Cr + NaCl \qquad (1\text{-}46)$$

$$C_5H_5MgBr + CrCl_3 \xrightarrow{(CH_3CH_2)_2O} \pi\text{-}(C_5H_5)_2Cr + MgBrCl \qquad (1\text{-}47)$$

dienide reagent may be removed by reaction with CO_2. The chromocene is subsequently purified and isolated by sublimation in high vacuum. The reaction of chromic chloride with molten bis(cyclopentadienyl)magnesium is reported to give a modest yield (22%) of chromocene [87].

b. π-Bis(cyclopentadienyl)chromium Halides

π-Bis(cyclopentadienyl)chromium(II) reacts smoothly with iodine or allyl iodide to give the chromium(III) salt $[\pi\text{-}(C_5H_5)_2Cr]^{\oplus}I^{\ominus}$ [65,88]; in this reaction the organic product is 1,5-hexadiene, Eq. (1-48) [88]. There is also a report that π-bis(cyclopentadienyl)chromium(III) bromide is the product of the direct interaction of lithium cyclopentadienide and $CrBr_3$ [89].

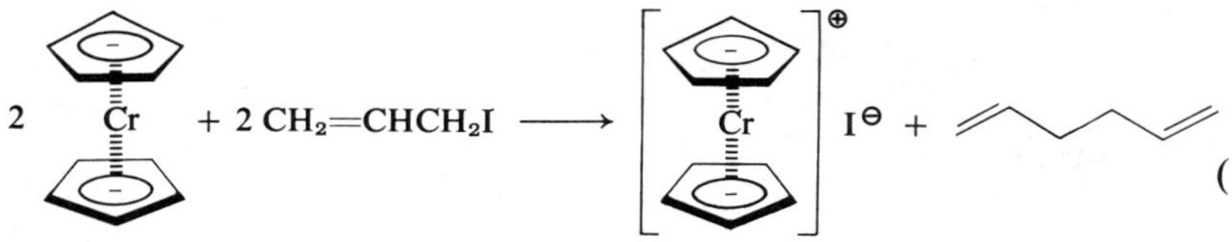

(1-48)

3. π-Complexes with One Cyclopentadiene Ring

There are several mixed π-complexes in which the chromium is bonded to one cyclopentadienyl unit and to a second organic unit containing delocalized electrons. The preparations of these compounds, though highly specific, are illustrative of some unusual reactions and are therefore briefly discussed here.

a. π-(Cyclopentadienyl)-π-(benzene)chromium

Compound **31** results from the interaction of a mixture of phenyl- and cyclopentadienylmagnesium halides and $CrCl_3$ in tetrahydrofuran [90], Eq. (1-49). The same mixed π-complex may also be prepared (in 20% yield) by

$$(C_5H_5)MgBr + C_6H_5MgBr + CrCl_3 \xrightarrow[(2)\ H_2O]{(1)\ THF} \quad (1\text{-}49)$$

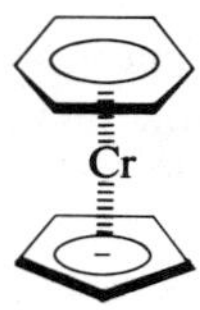

31

the reaction of phenylmagnesium bromide with a preformed cyclopentadienylchromium compound, Eq. (1-50) [91]. Hydrolysis and vacuum sublimation

$$CrCl_3 + NaC_5H_5 \xrightarrow{THF} \pi\text{-}(C_5H_5)CrCl_2 \xrightarrow[\substack{(2)\ \Delta,\ THF \\ (3)\ C_6H_6/H_2O}]{(1)\ C_6H_5MgBr} \mathbf{31} \quad (1\text{-}50)$$

are subsequently necessary, in both the foregoing syntheses, for the final isolation of the pure mixed π-complex **31**. The mechanisms of these reactions have not been elucidated; however, the formation of the mixed π-complex probably proceeds via a σ- to π-rearrangement in a σ-(phenyl)-π-(cyclopentadienyl)chromium compound.

b. π-(Cyclopentadienyl)-π-(cycloheptatrienyl)chromium

Compound **32** can be prepared either by the isopropyl Grignard synthesis [Eq. (1-16) and Eqs. (1-18) and (1-20)] or by a ligand exchange promoted by $AlCl_3$, Eq. (1-51) [92]. Substituted π-(cycloheptatrienyl)-π-(cyclopenta-

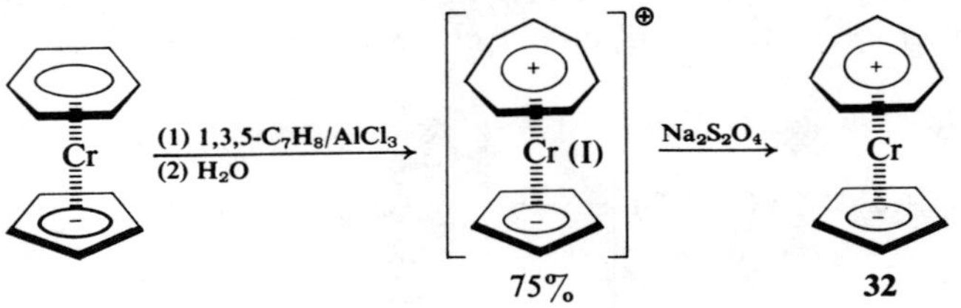

(1-51)

dienyl)chromium(I) complexes may be obtained, in modest yields (6–10%), by a most unusual ring-expansion reaction [91]. In this reaction π-(cyclopentadienyl)-π-(benzene)chromium is subjected to the action of an acyl halide in the presence of aluminum bromide. The product, however, is not the anticipated acylarene complex but is a substituted π-(cycloheptatrienyl)-π-(cyclopentadienyl)chromium(I) cation, Eq. (1-52) ($R = CH_3$ or C_6H_5) [91].

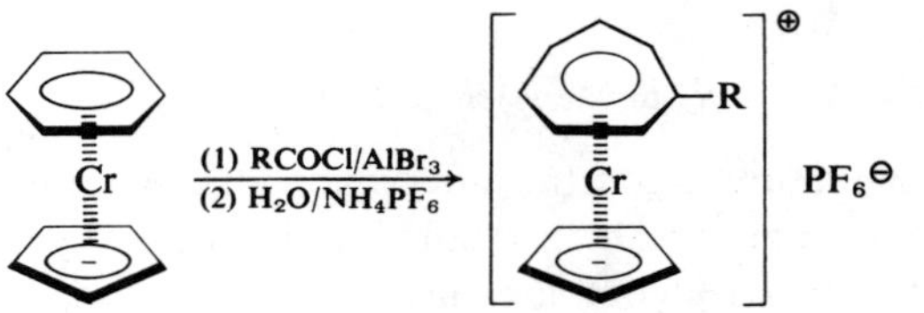

(1-52)

The transformation of the benzene ring into a cycloheptatrienyl ring could proceed as outlined in Scheme 1-3. The role of the chromium may be either

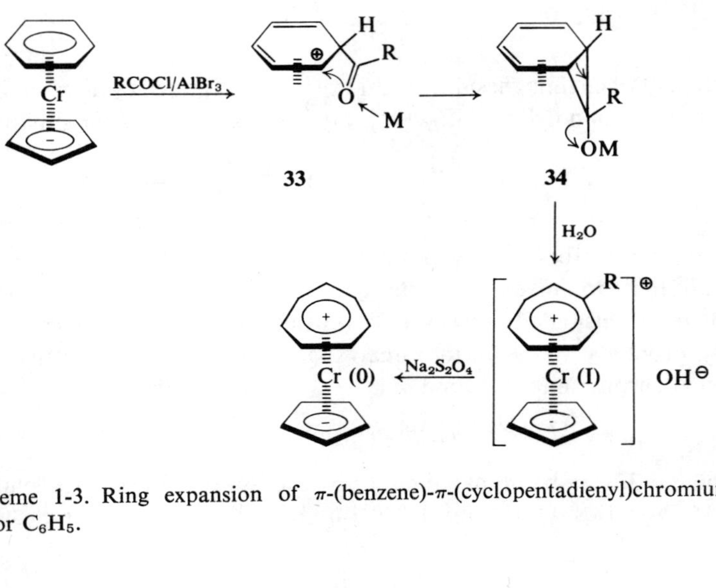

Scheme 1-3. Ring expansion of π-(benzene)-π-(cyclopentadienyl)chromium, $R = CH_3$ or C_6H_5.

to protect the carbonium ion (**33**) or to assist the cyclization and oxygen extrusion in **34** by bonding to oxygen. Attempts to effect the analogous ring expansion on π-(benzene)chromium tricarbonyl were unsuccessful, whereas ring contraction of π-(cycloheptatrienyl)chromium tricarbonyl [to π-(benzene)-chromium tricarbonyl] has been effected [Chapter 3, Section D,2,b,ii].

D. Carbonylchromium π-Complexes

There are two main classes of chromium carbonyl π-complexes: (1) the *tricarbonylchromium π-complexes* in which an arene or triene replaces three of the carbonyl groups of $Cr(CO)_6$ [e.g., π-(arene)$Cr(CO)_3$]; and (2) the *tetracarbonylchromium π-complexes* in which a diene replaces two of the carbonyl groups of $Cr(CO)_6$ [e.g., π-(norbornadiene)$Cr(CO)_4$]. The preparation of π-(arene)chromium tricarbonyls has been reviewed [23] and the methods available for the preparation of both these classes of carbonylchromium π-complexes will now be discussed.

1. π-(Arene)- or (Triene)chromium$(CO)_3$ Compounds

a. Synthesis from $Cr(CO)_6$ *and Arene or Triene*

Nicholls and Whiting [93] found that a large variety of aromatic compounds when heated with $Cr(CO)_6$, under anaerobic conditions, either alone or in an inert solvent (e.g., decalin or diethylene glycol dimethyl ether) led to the formation, in good yield, of π-(arene)chromium tricarbonyl compounds, e.g., Eq. (1-53) (R = H, CH_3, OCH_3, NH_2, $NHCH_3$, $N(CH_3)_2$, $COOCH_3$, F, Cl, CH_2OH, $COCH_3$, etc.). The reaction was subsequently extended to the

$RC_6H_5 + Cr(CO)_6 \xrightarrow{\Delta}$ 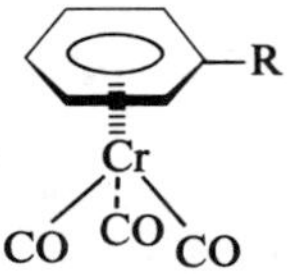 $+ 3\,CO$ (1-53)

preparation of a large variety of π-(arene)chromium$(CO)_3$ compounds [4], of 1,3,5-(cycloheptatriene)chromium$(CO)_3$ [94,95] and of π-(diene)chromium tetracarbonyls (see page 21). The inconvenience of the method is that owing to the volatility of the $Cr(CO)_6$, the latter has to be continuously returned, mechanically, into the reaction vessel. Strohmeier solved this difficulty by designing a "closed-cycle" apparatus in which the condensed solvent vapors return the volatilized $Cr(CO)_6$ into the reaction vessel [96]. Limitations to the reaction were occasioned by the presence of certain substituents in the arene (e.g., COOH, NO_2, $CH{=}CH_2$). Thus, with certain substrates, the reaction conditions are too drastic to permit the isolation of pure π-(arene) or (olefin) compound. In these instances the decomposition temperature of the product is close to, or lower than, that required for the original reaction, Eq. (1-53). This difficulty could, in many cases, be circumvented by the ligand exchange reaction outlined below (Section b).

b. Ligand Replacement Reaction

The arene, triene, or diene ligand of certain π-(arene)- or (triene)chromium-$(CO)_3$ and π-(diene)chromium$(CO)_4$ compounds may be replaced by interaction with certain aromatic compounds, e.g., Eq. (1-54) [97], Eq. (1-55) [95], and Eq. (1-56) [98]. These reactions again require relatively high tempera-

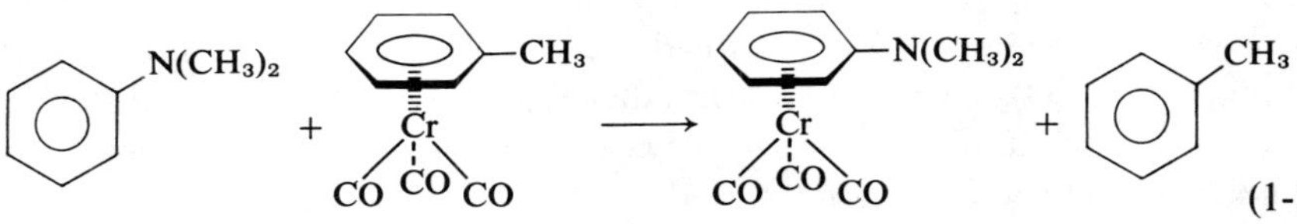

(1-54)

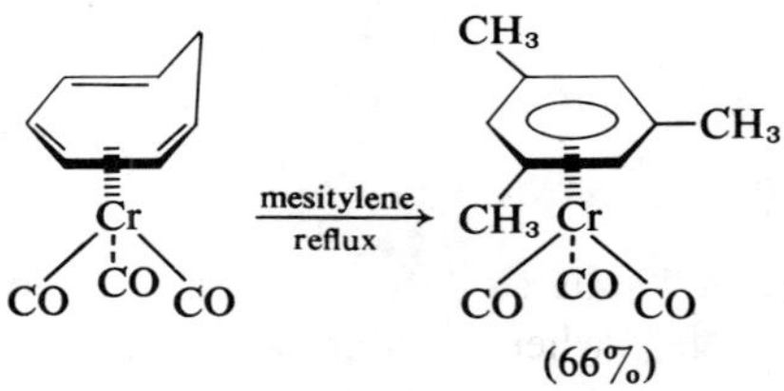

(1-55)

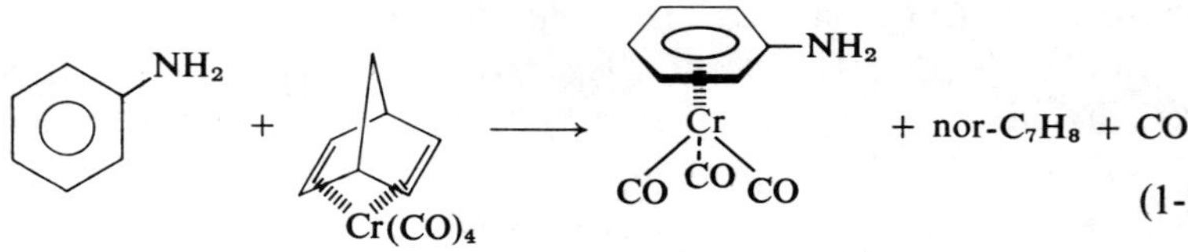

(1-56)

tures. The ligands of tris(pyridine)chromium$(CO)_3$ [99], tris(acetonitrile)-chromium$(CO)_3$ [100,101], and tris(ammonia)chromium$(CO)_3$ [102], on the other hand, are readily replaced by arenes, trienes, and some dienes at relatively low temperatures. This has permitted the preparation of hitherto inaccessible π-(arene)- or (triene)chromium tricarbonyls, e.g., Eqs. (1-57–1-59) [102–105].

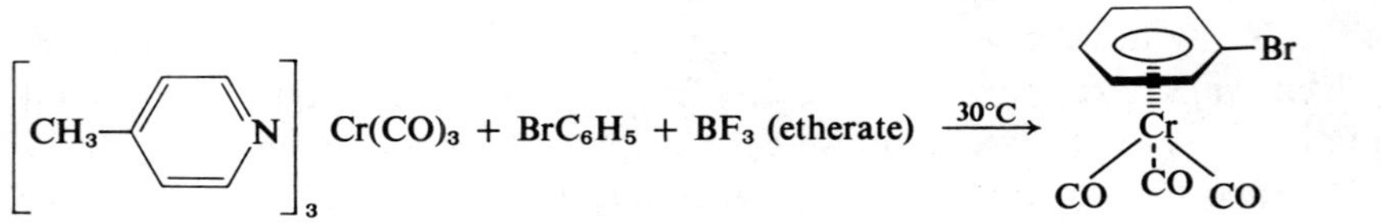

(1-57)

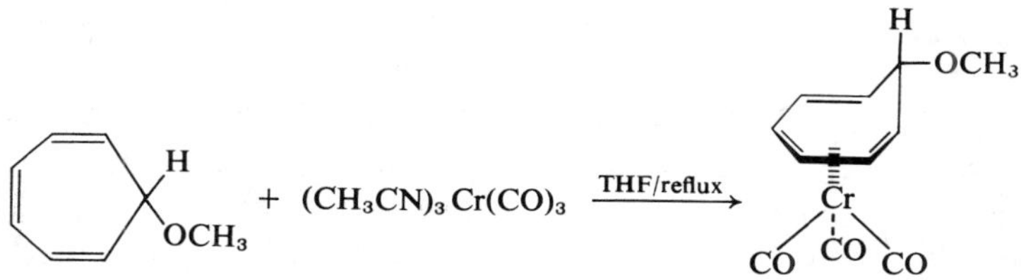

(1-58)

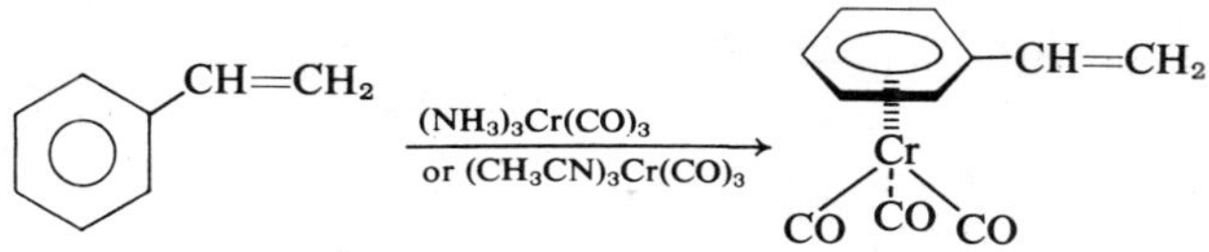

(1-59)

Another type of ligand-exchange reaction is involved in the original synthesis of π-(benzene)chromium$(CO)_3$, Eq. (1-60) [106]. The scope of this

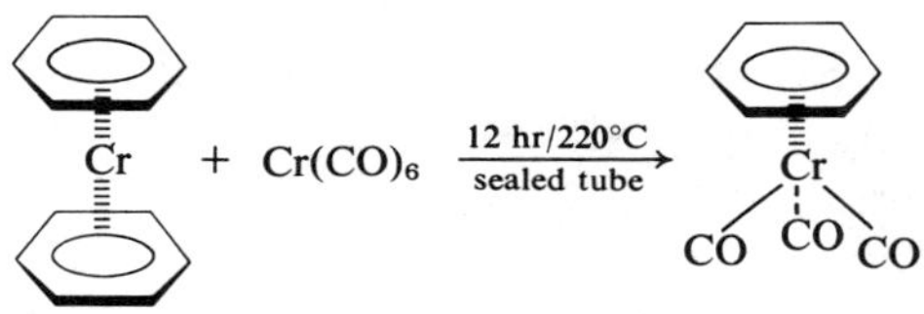

(1-60)

synthesis is restricted both by the rather vigorous reaction conditions required and by the limited number of available bis(arene)chromium π-complexes.

2. π-(Olefin)chromium$(CO)_4$ Compounds

The products formed from the interaction of cyclic dienes and $Cr(CO)_6$ depend upon the nature of both the diene and the solvent used. Thus norbornadiene reacts with $Cr(CO)_6$, in high-boiling petroleum ether (b.p. 120°–140°C) to give an 83% yield of the corresponding tetracarbonyl π-complex **35** [98]. 1,5-Cyclooctadiene, on the other hand, reacts with $Cr(CO)_6$, in refluxing heptane to give only a 2% yield of the tetracarbonylchromium π-complex **36** [107]. However, when the same reaction is carried out in di-*n*-butyl ether, π-(*o*-xylene)chromium$(CO)_3$ (**37**) (2%) and free *o*-xylene are formed [108]. The mechanism of the transformation of 1,5-cyclooctadiene

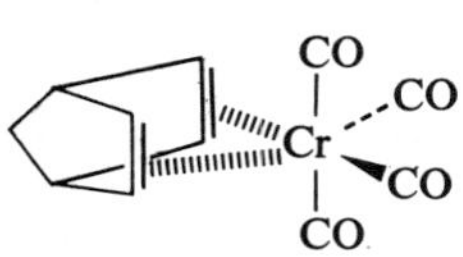

35

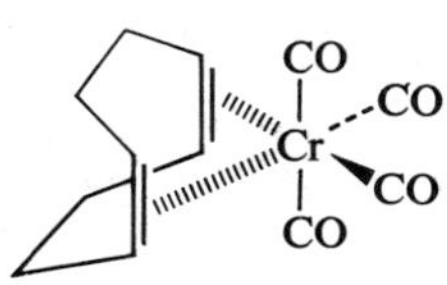

36

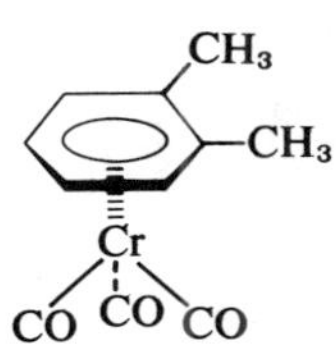

37

(1,5-COD) to *o*-xylene is not known. However, since both 1,5-COD and the 1,5-(COD) π-complex **36** are stable in hot di-*n*-butyl ether, it has been suggested that the transformation of 1,5-COD to *o*-xylene could occur during π-complex formation [108].

Methoxycyclohexa-1,4-diene and related compounds react with $Cr(CO)_6$, in refluxing nonane, to give, by methoxyl-extrusion and loss of hydrogen, the corresponding π-(arene)chromium tricarbonyls, e.g., (b) in Eq. (1-61) [109] which may be photolyzed to give the free arene [e.g., (c) in Eq. (1-61)].

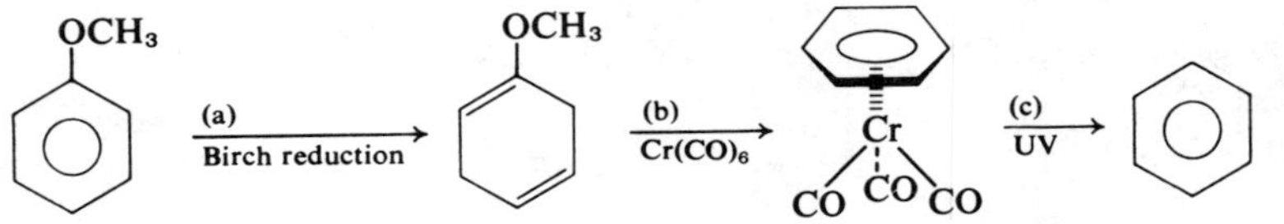

(1-61)

This reaction has been applied to a variety of substituted methoxycyclohexadienes (Table 1.1). The "methoxycyclohexa-1,4-dienes" are readily formed by the Birch reduction ($Li/NH_3/EtOH$) of the appropriate aromatic compounds [e.g., (a) in Eq. (1-61)]. A combination of these three processes, Eq. (1-61), therefore, provides a convenient method for the cleavage of aryl-OCH_3 bond in methyl aryl ethers.

π(Hexamethyl-Dewar-benzene)chromium tetracarbonyl (**38**) (R = CH_3) is formed in low yield (2%) by the interaction of the hydrocarbon and $(CH_3CN)_3Cr(CO)_3$, Eq. (1-62) [110]. Cyclohexa-1,3-diene, on the other

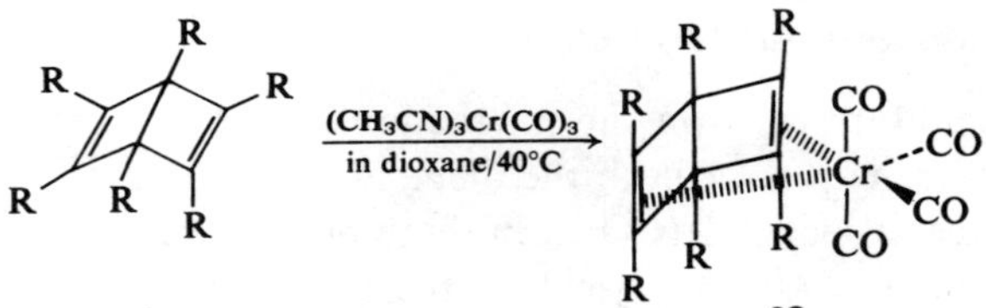

38

(1-62)

hand, reacts with $(CH_3CN)_3Cr(CO)_3$, in refluxing hexane to give a low yield (7%) of π-(benzene)chromium$(CO)_3$ [111].

Three novel tetracarbonylchromium complexes (**39** to **41**) have been prepared, albeit in low yield (1–10%) by the irradiation of an ether solution of $Cr(CO)_6$ and the appropiate dihalogenoalkenes [1,1-bis(chloromethyl)-ethylene, 3,4-dichlorocyclobutene and 3,4-dibromobenzocyclobutene, respectively] [112].

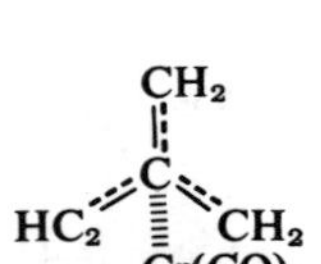

39

40

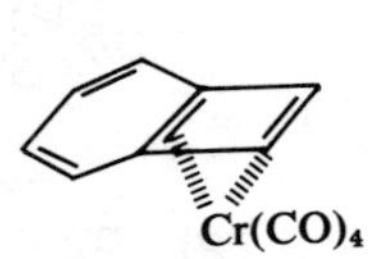

41

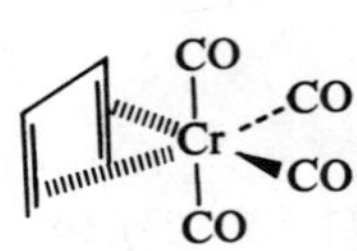

42

TABLE 1.1

Reactions of Methoxycyclohexa-1,4-dienes with $Cr(CO)_6$ [Eq. (1-61b)] [109]

DIENE	ARENE UNIT IN π-(ARENE)$Cr(CO)_3$
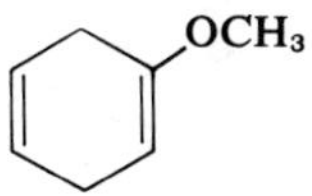	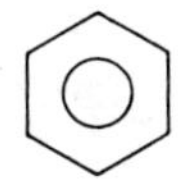
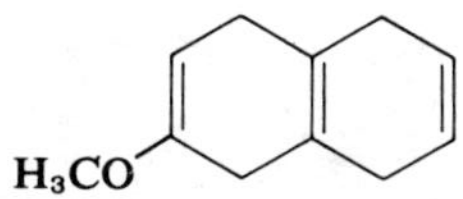	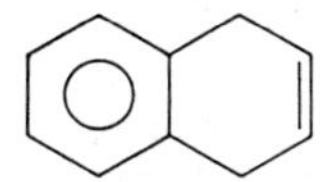
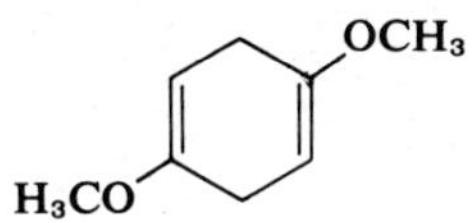 or 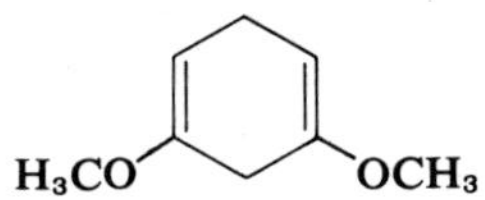	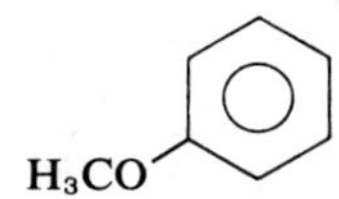
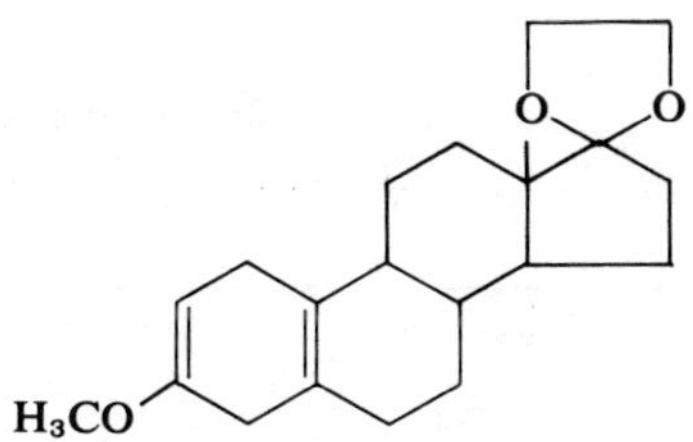	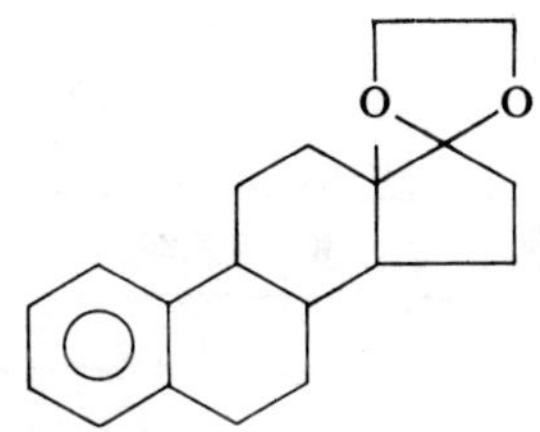

(Continued)

TABLE 1.1 (*continued*)

DIENE	ARENE UNIT IN π-(ARENE)$Cr(CO)_3$
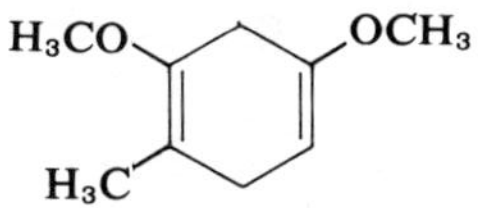	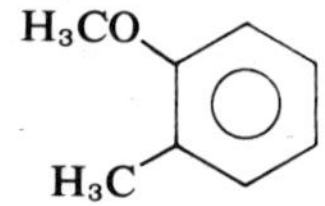 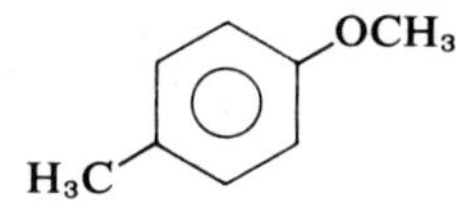
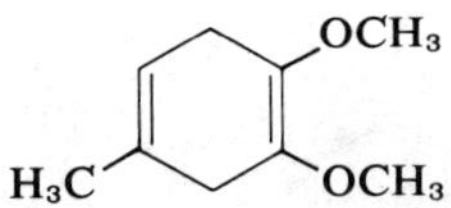	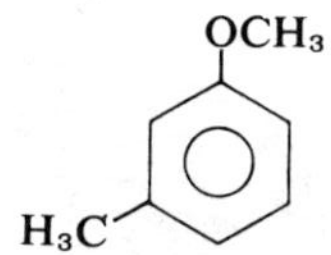 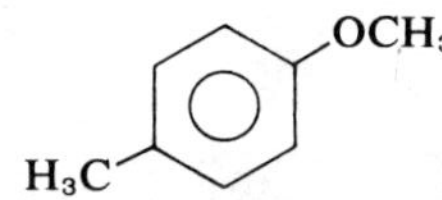
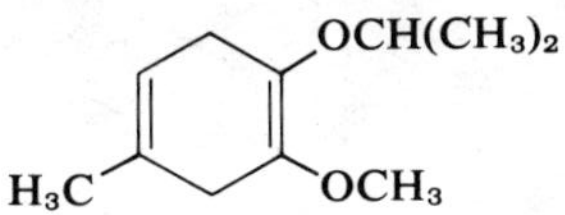	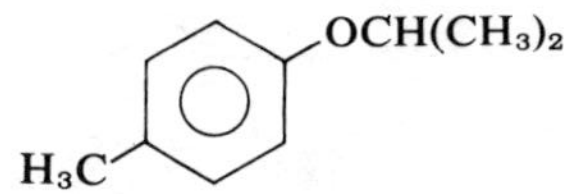
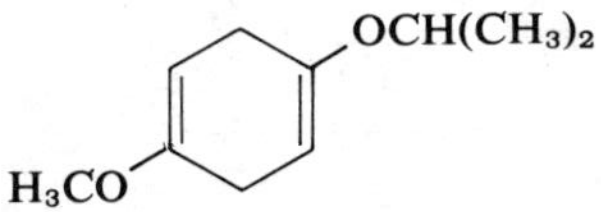 or 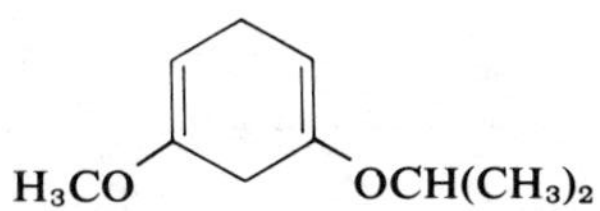	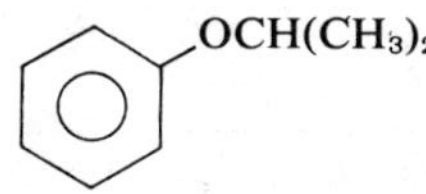

π-(Butadiene)chromium tetracarbonyl (**42**) has been prepared in 4.4% yield (based on Cr) by the interaction of chromium vapor, butadiene, and CO. The compound is not stable above 0°C; however, it has been obtained in a crystalline form and identified by mass, infrared, and NMR spectroscopies [113].

E. Carbenoid-Chromium Compounds

The reactions of chromium hexacarbonyl have been extensively studied [e.g., 114], in particular, in connection with the preparation of carbonyl-chromium π-complexes (see Section D of this chapter) and of polycarbonyl metallates {e.g., $Na_2[Cr(CO)_5]$} [115–117]. Fischer and Maasböl found, however, that one of the carbonyl groups of chromium hexacarbonyl undergoes nucleophilic attack by organolithium compounds to give anionic complexes, e.g., **43**, R = CH_3 and C_6H_5, Eq. (1-63). The complexes may be

$$Cr(CO)_6 \xrightarrow[\text{ether}]{\text{RLi}} Li\left[(CO)_5CrC\begin{smallmatrix}\diagup O\\ \diagdown R\end{smallmatrix}\right] \xrightarrow{(CH_3)_4NBr} (CH_3)_4N\left[(CO)_5CrC\begin{smallmatrix}\diagup O\\ \diagdown R\end{smallmatrix}\right] \quad (1\text{-}63)$$

43 **44**

isolated as their crystalline tetramethylammonium salts (**44**) [17]. Reaction of the latter (**44**) with methyl iodide gives the methoxycarbenoid-chromium compounds **45**, R = CH_3 and C_6H_5, in low yield [(a) in Eq. (1-64)]. Better

$$\mathbf{44} \xrightarrow[\text{(b) (1) } H^{\oplus}\text{, (2) } CH_2N_2]{\text{(a) } CH_3I} (CO)_5CrC\begin{smallmatrix}\diagup OCH_3\\ \diagdown R\end{smallmatrix} \quad (1\text{-}64)$$

45

yields of these compounds may be obtained by acidification of the lithium or tetramethylammonium salts (**43** or **44**) and subsequent treatment with diazomethane, e.g., (b) in Eq. (1-64) [17].

The alkoxy- and aryloxycarbenoid-chromium compounds may be obtained more conveniently by the reaction of a filtered solution of the lithium salt **43** with trialkyloxonium fluoroborates or aryldiazonium fluoroborates, Eq. (1-65), R′ = CH_3 or CH_2CH_3, R″ = C_6H_5 [118,118a,118b,119]. Also it has

$$\mathbf{43} \xrightarrow[\text{or } [R''N_2]BF_4]{[R'_3O]BF_4} (CO)_5CrC\begin{smallmatrix}\diagup OR' \text{ (or } R'')\\ \diagdown R\end{smallmatrix} \quad (1\text{-}65)$$

been found that the more reactive halides [e.g., $(CH_3)_3SiCl$ and CH_3COCl], in contrast to CH_3I, react with the lithium salt (**43**, R = CH_3) and tetramethylammonium salt (**44**) to give good yields of the corresponding carbenoid-chromium compounds, Eqs. (1-66) and (1-67), [120,121].

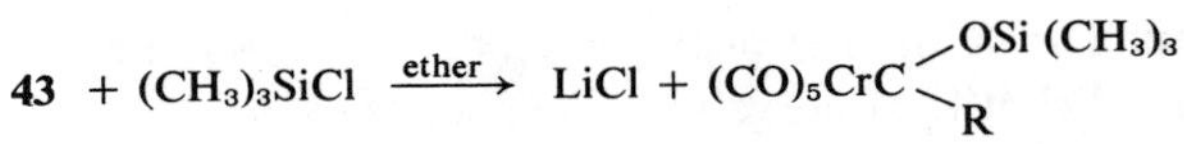 (1-66)

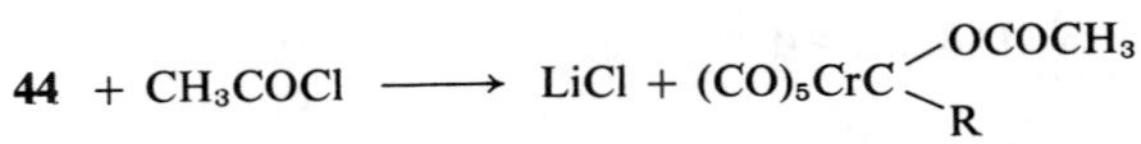 (1-67)

R = 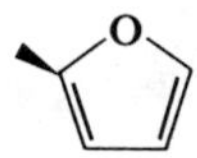

The hydroxycarbenoid compounds [$(CO)_5CrC(OH)CH_3$ and $(CO)_5CrC$-$(OH)C_6H_5$] may be obtained by treating the corresponding lithium salts (**43**, R = CH_3 or C_6H_5) with HBr [121a] and the hydroxycyclopropyl complex

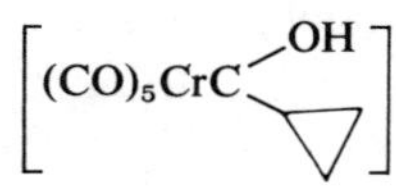

is obtained when the lithium salt **43**, R = 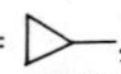, is treated with HBF_4 [121b].

Heteroorganolithium compounds [e.g., $LiN(CH_3)_2$ and $LiP(CH_3)_2$] also react with $Cr(CO)_6$ to give anionic complexes, which, when treated with triethyloxonium fluoroborate, give the carbenoid compounds. Thus, reaction with lithium dimethylamide gives the monocarbenoid compound, Eq. (1-68a), whereas reaction with lithium dimethyl phosphide gives the *cis*-bis-(carbenoid) compound, Eq. (1-68b) [118a].

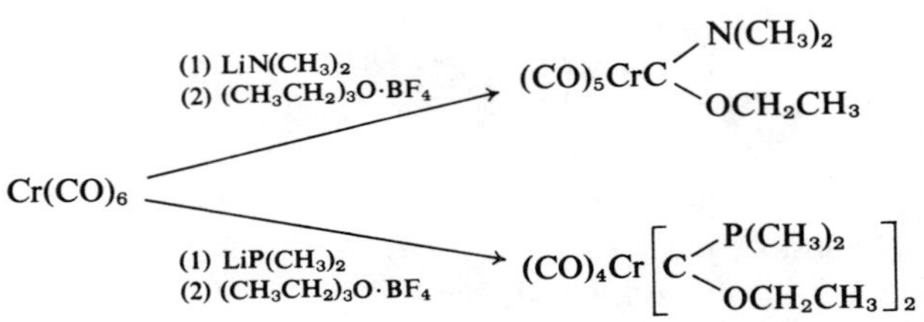 (1-68a) (1-68b)

Organomagnesium halides are reported to react with $Cr(CO)_6$, albeit more slowly than do the organolithium compounds to give anionic intermediates. These, upon alkylation with triethyloxonium fluoroborate, give the corresponding carbenoid compounds [122,123].

The aforementioned syntheses have been used to prepare a large variety of carbenoid-chromium compounds [4,124]. Furthermore, other carbenoid-chromium compounds may be prepared by the replacement of one of the five carbonyl groups bonded to chromium or by the replacement of the alkoxy unit bonded to the carbenoid carbon in **45** by PR_3, NHR, SR, etc. [Eqs. (1-69) and (1-70)]. These particular replacement reactions are discussed

$$(CO)_5CrC\begin{matrix}\diagup OCH_3 \\ \diagdown R'\end{matrix} \xrightarrow{PR_3} (CO)_4(R_3P)CrC\begin{matrix}\diagup OCH_3 \\ \diagdown R'\end{matrix} \quad (1\text{-}69)$$

$$(CO)_5CrC\begin{matrix}\diagup OCH_3 \\ \diagdown CH_3\end{matrix} \xrightarrow{HSC_6H_5} (CO)_5CrC\begin{matrix}\diagup SC_6H_5 \\ \diagdown CH_3\end{matrix} \quad (1\text{-}70)$$

in more detail in Chapter 3.

One of the carbonyl groups of π-(arene)chromium tricarbonyls likewise undergoes nucleophilic attack by organolithium compounds to give anionic complexes. Alkylation of these with trialkyloxonium fluoroborates gives the corresponding π-(arene)carbenoid-chromium dicarbonyls **46** [arene = C_6H_6, $CH_3C_6H_5$, 1,4-$(CH_3)_2C_6H_4$, 1,3,5-$(CH_3)_3C_6H_3$], Eq. (1-71) [125].

$$\pi\text{-(arene)}Cr(CO)_3 \xrightarrow[(2)\ [(CH_3)_3O]BF_4]{(1)\ LiC_6H_5} \pi\text{-(arene)}\,(CO)_2CrC\begin{matrix}\diagup OCH_3 \\ \diagdown C_6H_5\end{matrix} \quad (1\text{-}71)$$

46

Other carbenoid-chromium compounds have been prepared by some rather specific methods. Thus, the imidazolinylidene complex **47** has been prepared, in good yield, by the elimination of molecular hydrogen from the carbonylmetallate salts of the appropriate heteroaromatic cations **48** or **49**, Eq. (1-72) [126,127]. Irradiation of **47** furnishes the *cis*-bis(carbenoid)-chromium compound **50**, Eq. (1-73) in low yield [126]. The cis configuration

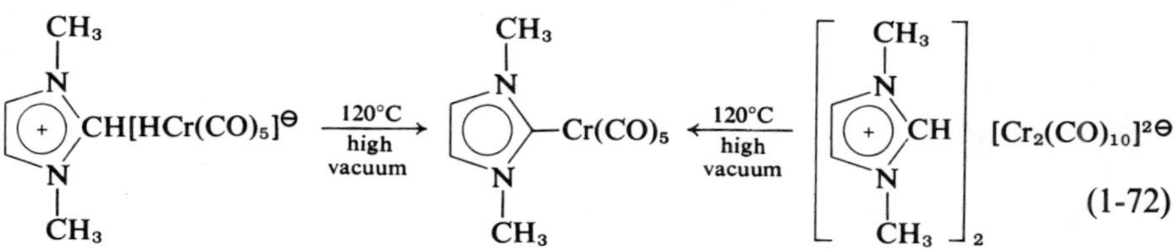

(1-72)

48 **47** **49**

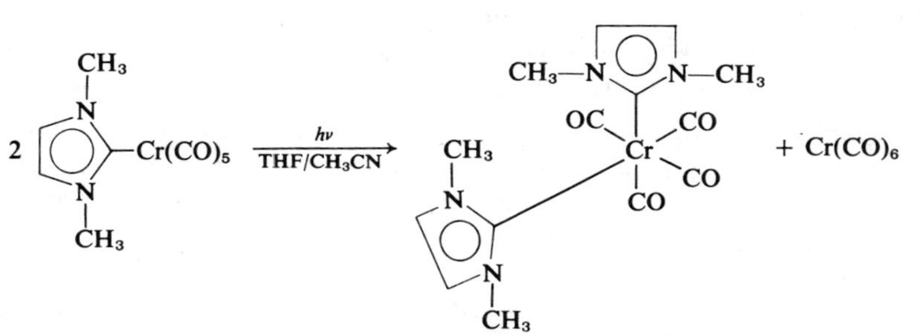

(1-73)

47 **50**

has been assigned to this compound on the basis of its infrared spectrum (four C≡O stretching frequencies) and its dipole moment [126].

The cyclopropenylidene carbenoid-chromium compound (**51**) may be prepared by the interaction of $Na_2Cr(CO)_5$ with either 1,1-dichloro-2,3-diphenylcyclopropene-2 (**52**) or 1-ethoxy-2,3-diphenylcyclopropene-2-enylium tetrafluoroborate (**53**), Eq. (1-74a) [128,129]. The aminocarbenoid compound (**54**), in which a hydrogen is bonded to $C_{carbene}$ is formed by the interaction of $Na_2Cr(CO)_5$ and chloroformiminium chloride, Eq. (1-74b) [129a].

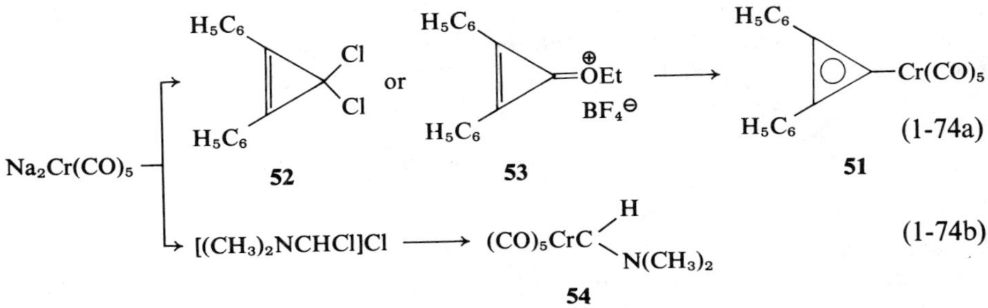

(1-74a)

(1-74b)

The cyclopropenylidenechromium complex (**51**) may be transformed into a new class of carbenoid-chromium compound, namely, the air-stable pyranylidene compound **55** (R = OCH_3, OCH_2CH_3, C_6H_5), Eq. (1-75) [129].

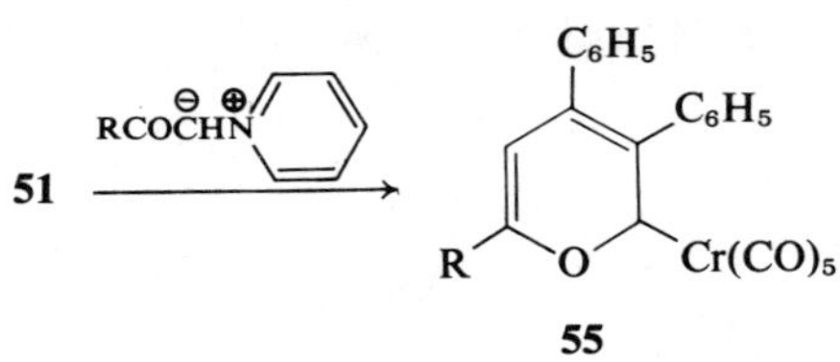

(1-75)

F. σ-Bonded Organochromium(III) and (II) Compounds

Within the last decade a large variety of σ-bonded organochromium compounds and complexes have been prepared in which the valency state of the chromium is formally II, III, or IV. Thus, the interaction of certain organomagnesium halides and chromic or chromous halides leads to the formation of the corresponding solvated σ-bonded organochromium(III) and (II) compounds of the general type $R_nCrCl_{3-n}(S)_x$, $R_nCr(II)Cl_{2-n}(S)_x$, where S represents a solvate molecule, e.g., Eqs. (1-76) and (1-77) (R =

$$3\ RMgBr + CrCl_3(THF)_3 \xrightarrow{THF} R_3Cr(III)(THF)_3 + \text{“3 MgBrCl”} \quad (1\text{-}76)$$

$$2\ RMgBr + CrBr_2(THF)_3 \xrightarrow{THF} [R_2Cr(II)]_n + \text{“2 MgBrCl”} \quad (1\text{-}77)$$

2-$CH_3OC_6H_4$) [130,131]. The interaction of organolithium compounds and $CrCl_3$ or $CrCl_2$ leads, on the other hand, to the formation of lithium poly-(organo)chromate complexes of the general type $Li_nCr(III)R_{3+n}$ and $Li_nCr(II)R_{2+n}$, e.g., Eq. (1-78), R = C_6H_5 [15] and finally the interaction of

$$6\ LiR + CrCl_3 \xrightarrow{\text{ether}} Li_3CrR_6 + LiCl \qquad (1\text{-}78)$$

alkyllithium compounds and chromium(IV) *tert*-butoxide furnishes the corresponding tetrakis(alkyl)chromium(IV) compounds, e.g., Eq. (1-79) [22].

$$(t\text{-BuO})_4Cr + 4\ Li(t\text{-Bu}) \longrightarrow (t\text{-Bu})_4Cr + 4(t\text{-BuO})Li \qquad (1\text{-}79)$$

The various methods available for the preparation of these different types of σ-bonded organochromium compounds are discussed separately. The present section treats of the preparation of simple solvated σ-bonded organochromium(III) and (II) compounds. The preparation of the lithium and sodium organochromium(II) and (III) complexes is discussed in Section G of this chapter and finally the preparation of σ-bonded organochromium(IV) compounds is reviewed in Section H of this chapter.

1. Mono(organo)chromium(III) Compounds

The known σ-bonded mono(organo)chromium(III) compounds fall into two categories. Those in the first, of general formula $RCr(III)Cl_2L_n$ (R = aryl, benzyl, and alkyl, L = tetrahydrofuran or pyridine, and n = 3 or 2), are sensitive to both air and moisture. Those in the second, of general formula $[RCr(III)(H_2O)_5]^{2\oplus}$ (R = benzyl, pseudobenzyl, methyl, and hydroxy- and halogenomethyl), are stable in aqueous solution but are, in the main, sensitive to oxygen. Several representatives of the former class of compounds have been isolated in the crystalline state, whereas the latter compounds have only been prepared in solution.

The general methods available for the preparation of both classes of σ-bonded mono(organo)chromium(III) compounds are discussed in the ensuing pages.

a. *Methods I and II: Grignard Synthesis and Ligand Exchange*

The interaction of an organomagnesium halide and $CrCl_3(THF)_3$, in equimolar ratios, in tetrahydrofuran (THF) has been shown to be a two-stage process [132,133]. The initial metathetical reaction, at low temperatures, leads to the solvated tris(organo)chromium compound, Eq. (1-80). When

$$3\ RMgCl + 3\ CrCl_3(THF)_n \xrightarrow[-70^\circ \text{ to } -40^\circ C]{\text{THF}} R_3Cr(THF)_n + 2\ CrCl_3(THF)_3 \qquad (1\text{-}80)$$

the temperature of the reaction is raised, there is a slow ligand redistribution between the latter and unreacted $CrCl_3(THF)_3$, leading finally to the σ-bonded mono(organo)chromium(III) dichloride, Eq. (1-81). From an organo-

$$R_3Cr(THF)_n + 2\ CrCl_3(THF)_3 \xrightarrow[-40° \text{ to } ?°C]{THF} RCrCl_2(THF)_n \qquad (1\text{-}81)$$

magnesium halide two syntheses of mono(organo)chromium(III) compounds are possible; the one, the Grignard synthesis (Method I), involves the interaction of an organomagnesium halide and $CrCl_3(THF)_3$ (in equimolar ratios) at the appropriate temperature, Eqs. (1-80) and (1-81) [e.g., 132–134]; the other, ligand exchange (Method II), involves the interaction of a preformed tris(organo)chromium(III) compound and $CrCl_3(THF)_3$, in the ratio 1:2, again at the appropriate temperature, Eq. (1-81) [133,134]. The successful application of either of these methods requires that the intermediate tris(organo)chromium(III) compound and the final product [i.e., the mono-(organo)chromium(III) dihalide] be stable at such temperatures that the reaction outlined in Eq. (1-81) can take place. Thus in the preparation of mono(benzyl)chromium(III) dichloride the thermal instability of the product is such that the reaction must be carried out below −10°C [132,133,135]. In contrast, in the preparation of solvated mono(decyl)Cr(III)Cl_2, the thermal instability of both the intermediate tris(decyl)chromium(III) and the final product are such that the reaction must be carried out between −70° and −40°C [136].

The choice of solvent is also of importance in these syntheses and is restricted to those in which the tris(organo)chromium(III) compound and the mono(organo)chromium dihalides are stable (e.g., tetrahydrofuran and 1,2-dimethoxyethane).

Both these methods should find a wide application in the preparation of σ-bonded mono(organo)chromium dihalides provided that the solvent and the temperature for the respective reactions be judiciously chosen.

A modification of Method II involves the interaction of preformed sodium or lithium poly(phenyl)chromium(III) complexes and $CrCl_3(THF)_3$ in tetrahydrofuran or 1,2-dimethoxyethane [137], Eqs. (1-82) and (1-83).

$$Na_2(C_6H_5)_5Cr + 4\ CrCl_3(THF)_3 \longrightarrow 5\ C_6H_5CrCl_2(THF)_3 + 2\ NaCl \qquad (1\text{-}82)$$

$$Li_3(C_6H_5)_6Cr + 5\ CrCl_3(THF)_3 \longrightarrow 6\ C_6H_5CrCl_2(THF)_3 + 3\ LiCl \qquad (1\text{-}83)$$

The latter interesting modification is of somewhat restricted applicability owing to the relative inaccessibility of the starting complexes and to the difficulties in separating the final mono(organo)chromium compound from the accompanying inorganic halide.

The tetrahydrofuran of solvation in the mono(organo)dichlorochromium-(III) tetrahydrofuranates may be displaced by pyridine (py) or 2,2′-bipyridyl. These nitrogenous ligands are known to stabilize carbon–metal σ-bonds and this has permitted the isolation of otherwise unstable mono(organo)chromium species, e.g., tris(pyridine)dichlorobenzylchromium(III) [**56**, Eq. (1-84)] [133].

$$C_6H_5CH_2MgCl + CrCl_3(THF)_3 \xrightarrow[-70^\circ \text{ to } -10^\circ C]{THF}$$

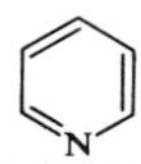

$$C_6H_5CH_2CrCl_2(THF)_n + MgCl_2(THF)_n \longrightarrow \underset{\mathbf{56}}{C_6H_5CH_2CrCl_2(C_5H_5N)_3} \quad (1\text{-}84)$$

b. Method III: Syntheses with Other Organometallic Compounds

There are reports of the use of other organometallic compounds (e.g., organoaluminum and organolithium compounds) in the metathetical synthesis of solvated monoorganochromium dihalides. Kurras [134,138] investigated the reaction of triphenylaluminum with chromic chloride. He found that only one of the phenyl groups of triphenylaluminum is transferred to chromium to give, as sole product, the mono(aryl) compound **57**, Eq. (1-85). Yamamoto and his colleagues extended this reaction to alkyl-

$$(C_6H_5)_3Al + CrCl_3 \xrightarrow{THF} \underset{\mathbf{57}}{C_6H_5CrCl_2(THF)_3} + (C_6H_5)_2AlCl \quad (1\text{-}85)$$

aluminum compounds and found that the products were again the mono-(organo)chromium dihalides **58**, R = CH_3, C_2H_5, n-C_3H_7, and i-C_4H_9, Eq. (1-86) [139]. However, in this case the thermal instability of the higher

$$CrCl_3(THF)_3 + R_3Al \text{ (or } R_2AlOCH_2CH_3) \xrightarrow{THF}$$

$$\underset{\mathbf{58}}{RCrCl_2(THF)_3} + R_2AlCl \text{ [or } RAl(OCH_2CH_3)Cl] \quad (1\text{-}86)$$

alkylchromium compounds is such that, even at low temperatures, it is extremely difficult to obtain pure crystalline alkylchromium dihalides free from chromous salts [139].

The reaction of dialkylaluminum halides with chromic chloride solvates is reported to lead to the formation of complexes, e.g., **59**, Eq. (1-87) [140].

$$(CH_3CH_2)_2AlCl + CrCl_3S_n \longrightarrow \underset{\mathbf{59}}{(CH_3CH_2CrCl_2)(CH_3CH_2AlCl_2)S_n} \quad (1\text{-}87)$$

Organolithium and sodium compounds usually react with chromic salts to give lithium poly(organo)chromium complexes (Section G of this chapter). Seidel and Fischer have found, however, that the interaction of lithium or sodium diphenylmethane and $CrCl_3$ (in 1:1 stoichiometric ratios) gives the mono(organo)chromium compound **60**, Eq. (1-88) (M = Li or Na) [141].

$$(C_6H_5)_2CHM + CrCl_3 \xrightarrow{THF} \underset{\mathbf{60}}{(C_6H_5)_2CHCrCl_2(THF)_2} + MCl \qquad (1\text{-}88)$$

The analogous triphenylmethylchromium dichloride could not be prepared. The reaction of sodium triphenylmethane with $CrCl_3$, even at very low temperatures, leads to the formation of solvated chromous chloride [141].

The foregoing syntheses with organoaluminum and organolithium (or sodium) compounds should be capable of being extended to the preparation of other mono(organo)chromium dihalides. Furthermore, in contrast to the Grignard synthesis (Method I), the above reagents also offer the possibility of using solvents other than diethyl ether and tetrahydrofuran.

The σ-bonded mono(organo)chromium compounds, which have so far been prepared by Methods I to III (or by modifications thereof), are listed in Table 1.2 [132–145].

c. *Method IV: Syntheses with Chromous Salts*

The reaction of organic halides with solutions of chromous salts was successfully modified by Anet and Leblanc for the preparation of the water-stable benzyl- [and, by inference, α-, 2-, and 4-methylbenzyl- and dichloromethylchromium(III) cations] [3, 146], Eq. (1-89) [R = $C_6H_5CH_2$, 2-$CH_3C_6H_4CH_2$, 4-$CH_3C_6H_4CH_2$, $C_6H_5C(CH_3)H$ and Cl_2CH].

$$RCl + 2\,Cr(II)(ClO_4)_2 \xrightarrow{HClO_4/N_2} \underset{\mathbf{61}}{[RCr(H_2O)_5]^{2\oplus}} + [Cr(III)(H_2O)_5Cl]^{2\oplus} \qquad (1\text{-}89)$$

The essential feature of this reaction is the use of chromous perchlorate in perchloric acid; this reacts with the benzyl halide in a two-step process to give the substitution-inert benzylchromium pentaaquate dication [**61**, Eq. (1-89), R = $C_6H_5CH_2$] [3,146]. The analogous reaction with $Cr(II)Cl_2$ in dilute acid gives only toluene and no organometallic compound [3]. Presumably the presence of complexed chloride ion in the intermediate organometallic compound $[C_6H_5CH_2CrCl(H_2O)_4]^{\oplus}$ facilitates the reduction of the latter via a chlorine-bridged intermediate [3].

The same benzylchromium pentaaquate dication can also be prepared by the following three routes:

(i) The interaction of phenyl-*tert*-butylhydroperoxide and chromous sulfate, Eq. (1-90) [147]. The related methylchromium pentaaquate dication

TABLE 1.2

σ-Bonded Mono(organo)chromium(III) Compounds of Type $RCrCl_2\,(S)_n$

COMPOUND[a]	METHOD OF PREPARATION[b]	STATE[c]	MAGNETIC MOMENT (μ_B)	REACTIONS, ETC.
$CH_3CrCl_2(THF)_3$	I [135] and III [139]	Green crystals *[135,139]	3.86 [135] 3.84 [139]	With $HgCl_2 \rightarrow CH_3HgCl$; with $I_2 \rightarrow CH_3I$ [135]; IR, UV, and thermal decomposition [139]
$CH_3CrCl_2(py)_3$	III, followed by pyridine	Green crystals *[139]	—	IR [139]
$CH_3CH_2CrCl_2(THF)_3$	III	Green crystals	3.77 [139]	IR, UV, thermal decomposition [139]
$CH_3CH_2CrCl_2(py)_3$	III, followed by pyridine	Green crystals *[139]	—	IR [139]
$CH_3CH_2CH_2CrCl_2(THF)_3$	III	Green crystals *[139]	—	IR, UV, thermal decomposition [139]
$(CH_3)_2CHCH_2CrCl_2(THF)_3$	III	Green crystals *[139]	—	IR, UV, thermal decomposition [139]
$CH_3(CH_2)_8CH_2CrCl_2(THF)_n$	I	Green solid [136]	—	With $HgCl_2 \rightarrow RHgCl$; deuterolysis; thermal decomposition [136]
$C_6H_5(CH_2)_2CD_2CH_2CrCl_2(THF)_n$	I	Green solution [142]	—	Thermal decomposition [142]

(Continued)

TABLE 1.2 (*continued*)

COMPOUND[a]	METHOD OF PREPARATION[b]	STATE[c]	MAGNETIC MOMENT (μ_B)	REACTIONS, ETC.
$C_6H_5(CH_2)_2CH(CH_3)CrCl_2(THF)_n$	I	Green solution [142]	—	Thermal decomposition [142]
$CH_3CH_2CrCl_2 \cdot CH_3CH_2AlCl_2(S)$	III	Yellow-brown solid *[140]	—	IR [140]
$C_6H_5CrCl_2(THF)_3$	II [137] III [134,138]	Green solid [134, 138] *[134]	3.76 [134] 3.73 [137]	With $HgCl_2 \rightarrow RHgCl$; thermal decomposition [134]; with $I_2 \rightarrow C_6H_5I$ [137]
$4\text{-}CH_3C_6H_4CrCl_2(THF)_3$	I	Green solid [143, 144] *[143]	3.87 [143]	With $HgCl_2 \rightarrow RHgCl$; UV and IR spectra [143]; x-ray structure [143,144]
$(C_6H_5)_2CHCrCl_2(THF)_2$	II and III (Li)	Blue solid *[141]	3.92 [141]	Thermal decomposition
$(CH_2{=}CH)CrCl_2(THF)_n$	I	In solution, brown [145]	—	Thermal decomposition [145]
$C_6H_5CH_2CrCl_2(THF)_n$	I	In solution, brown [132,133,135]	—	Thermal decomposition [132, 133,135]
$C_6H_5CH_2CrCl_2(py)_3$	I, followed by pyridine IV, with $CrCl_2(py)_2$	Brown solid *[133]	3.93 [133]	UV and IR spectra [133]; with $HgCl_2 \rightarrow RHgCl$ [133]
$2\text{-}ClC_6H_4CH_2CrCl_2(py)_3$	IV, with $CrCl_2(py)_2$	Yellow-orange solid *[133]	—	UV [133]
$4\text{-}ClC_6H_4CH_2CrCl_2(py)_3$	IV, with $CrCl_2(py)_2$	Red-brown solid *[133]	—	UV [133]

[a] THF = Tetrahydrofuran; S = solvating species; py = pyridine.
[b] Methods I–IV as described in text.
[c] Asterisk indicates that satisfactory analyses have been reported for the crystalline compounds.

$$C_6H_5CH_2\underset{\underset{CH_3}{|}}{\overset{\overset{CH_3}{|}}{C}}OOH \xrightarrow{CrSO_4} C_6H_5CH_2\underset{\underset{CH_3}{|}}{\overset{\overset{CH_3}{|}}{C}}O\cdot + Cr(III)$$

$$C_6H_5CH_2\underset{\underset{CH_3}{|}}{\overset{\overset{CH_3}{|}}{C}}O\cdot \longrightarrow C_6H_5\dot{C}H_2 + (CH_3)_2C{=}O \xrightarrow{CrSO_4}$$

$$[C_6H_5CH_2Cr(H_2O)_5]^{2\oplus} + (CH_3)_2C{=}O \qquad (1\text{-}90)$$

has also been obtained from the analogous reaction of *tert*-butylhydroperoxide with chromous perchlorate, Eq. (1-91) [148].

$$CH_3\underset{\underset{CH_3}{|}}{\overset{\overset{CH_3}{|}}{C}}OOH \xrightarrow{2\,Cr(ClO_4)_2} [CH_3Cr(H_2O)_5]^{2\oplus} + (CH_3)_2CO + Cr(III) \qquad (1\text{-}91)$$

(ii) The interaction of bis(phenylacetyl)peroxide and chromous perchlorate, Eq. (1-92) [149]. It is believed that in this reaction the phenylacetyl radical

$$(C_6H_5CH_2CO_2)_2 \xrightarrow{2\,Cr(ClO_4)_2}$$

$$[C_6H_5CH_2Cr(H_2O)_5]^{2\oplus} + CO_2 + [C_6H_5CH_2CO_2Cr(III)(H_2O)_n]^{2\oplus} \qquad (1\text{-}92)$$

generated by cleavage of the original peroxide by Cr(II) fragments into CO_2 and a benzyl radical, and this in turn reacts with a second molecule of chromous salt to give the benzylchromium pentaaquate dication, Eq. (1-93) [149].

$$C_6H_5CH_2\dot{C}O_2 \longrightarrow C_6H_5\dot{C}H_2 + CO_2 \xrightarrow{Cr(II)} [C_6H_5CH_2Cr(H_2O)_5]^{2\oplus} \qquad (1\text{-}93)$$

(iii) The controlled interaction of benzylbis(chloro)chromium(III) tetrahydrofuranate, prepared by the Grignard syntheses, and water at temperatures below $-5°C$, Eq. (1-94) [133,150].

$$C_6H_5CH_2CrCl_2(THF)_n \xrightarrow[<5°C]{H_2O} [C_6H_5CH_2Cr(H_2O)_5]^{2\oplus} \qquad (1\text{-}94)$$

The mechanism of the interaction of chromous salts and benzyl halides and a variety of substituted benzyl halides has been studied in detail [149]. Such studies are, however, complicated by the thermal instability of the benzylchromium compounds. Thus, though $[C_6H_5CH_2Cr(H_2O)_5]^{2\oplus}$ is stable for very long periods of time in an inert atmosphere at $-10°C$, at higher temperatures $(+20°C)$ it undergoes complete fragmentation to toluene and bibenzyl. Therefore, at 20°C, the overall reaction of chromous perchlorate and benzyl chloride is a superposition of the two processes corresponding to the formation of the solvated benzylchromium dication and its subsequent decomposition by hydrolysis and homolysis, Eq. (1-95).

$$C_6H_5CH_2Cl \xrightarrow[HClO_4]{Cr(ClO_4)_2} \begin{matrix} [C_6H_5CH_2Cr(H_2O)_5]^{2\oplus} \\ + \\ [Cr(III)Cl(H_2O)_5]^{2\oplus} \end{matrix} \longrightarrow \begin{matrix} C_6H_5CH_3 + Cr(II) \\ C_6H_5CH_3 + Cr(III) \\ (C_6H_5CH_2)_2 + Cr(II) \end{matrix} \tag{1-95}$$

However, by studying the rate of appearance of the benzylchromium cation (second-order rate constant 4.0×10^{-3} liter/mole sec) spectrometrically and the rate of disappearance of benzyl chloride (second-order rate constant 3.9×10^{-3} liter/mole sec) gas-chromatographically, it was found [149] that the formation of the benzylchromium compound is first order in benzyl halide and in chromous ion. All substituents in the aryl ring (e.g., 3-OCH_3, Cl, Br, CH_3, and 4-Br, CH_3) increase the rate, relative to the unsubstituted benzyl halide, and the relative rates of the different halides are 555:125:1 for iodide:bromide:chloride. The data were taken as evidence that the rate-controlling step in the formation of the benzylchromium cation is the homolysis of the carbon–halogen bond of the benzyl halide [149].

The course of the overall reaction can be formulated as the two-step process, outlined in Eqs. (1-96) and (1-97). The reaction is initiated by the

$$C_6H_5CH_2Cl + Cr(II) \longrightarrow \underset{\mathbf{62}}{C_6H_5\dot{C}H_2\cdots Cl\cdots Cr(II)} \xrightarrow{\text{slow}} C_6H_5\dot{C}H_2 + [Cr(III)Cl(H_2O)_5]^{2\oplus} \tag{1-96}$$

$$C_6H_5\dot{C}H_2 + Cr(II) \xrightarrow{\text{fast}} [C_6H_5CH_2Cr(H_2O)_5]^{2\oplus} \tag{1-97}$$

oxidative transfer of the halogen to Cr(II) in the complex **62** followed by the rapid reaction of the benzyl radical with a second molecule of chromous salt. The rate constant for the reaction of benzyl chloride and Cr(II) is $k_2 = 3.2 \times 10^{-3}$ liter/mole sec, at 27°C; and the activation parameters are ΔH^* 14.0 kcal/mole and ΔS^* − 14.3 e.u. [149].

This reaction of chromous salts with benzyl halides has been extended to the preparation of the pseudobenzylchromium compounds derived from the isomeric bromomethylpyridines. Thus the isomeric bromomethylpyridinium bromides and the related *N*-methylpyridinium bromides react with chromous sulfate solution to give the σ-bonded (pyridiomethyl)chromium(III) penta-aquates, **63–66** [151–154].

The interaction of other organic halides and chromous salts in a variety of media has been extensively investigated. The final products depend upon the

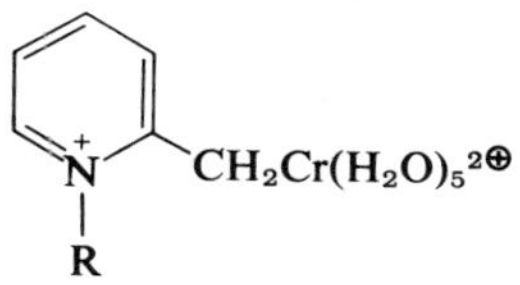

(R = H or CH_3)

63

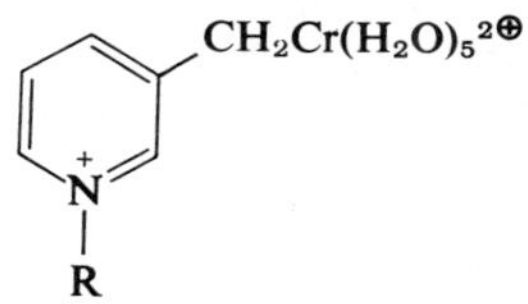

(R = H or CH_3)

64

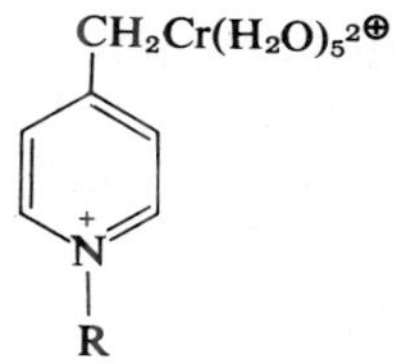

(R = H or CH_3)

65

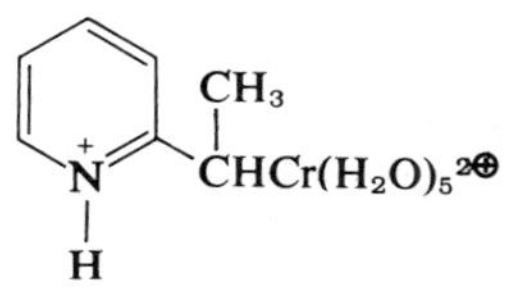

66

nature of the alkyl halide. Thus, with simple alkyl halides, either the alkane or the alkyl dimer is obtained [e.g., 155] while vicinal dihalides give the olefin as the major product [e.g., 156–159]. Organic halides that have a vicinal substituent that may be eliminated as an anion, also react with chromous salts to give the olefin, e.g., Eq. (1-98) (X = Cl, Y = OH, NH_2, OR, OCOR, etc.) [e.g., 160, 161]. Certain steroids give cyclopropyl derivatives,

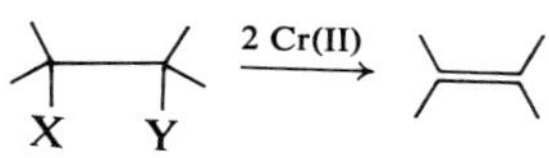

(1-98)

formed by an internal cyclization, together with other products formed, respectively, by reductive and vicinal elimination, e.g., Eq. (1-99) [162].

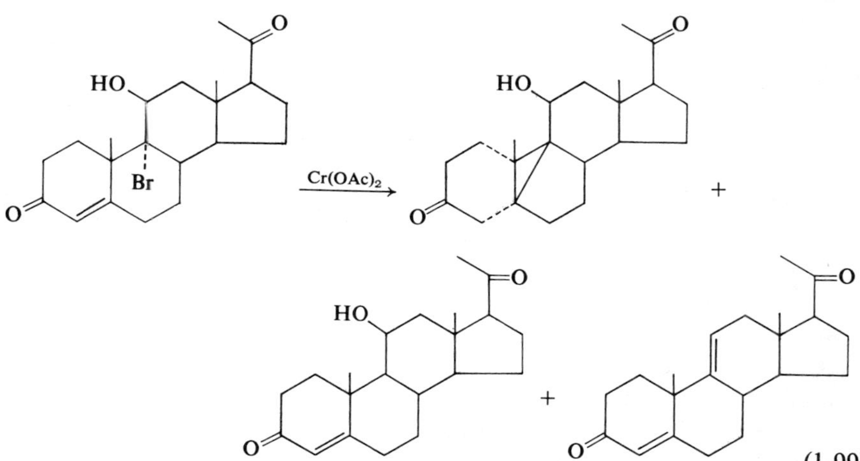

(1-99)

Cyclopropane and substituted cyclopropanes are produced by the reaction of the corresponding 1,3-dihalides with chromous salts, e.g., Eq. (1-100) ($X = Y =$ Br or Cl, R = H or CH_3) [163]. There has been considerable

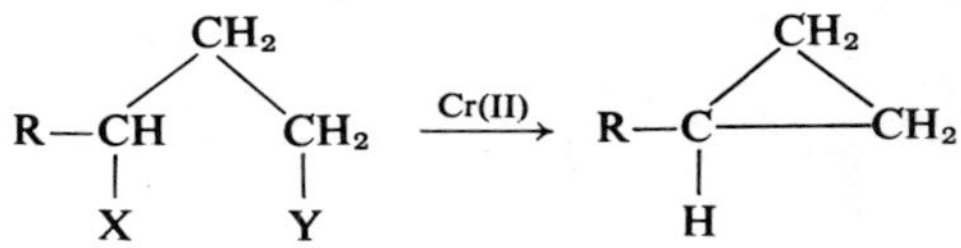

(1-100)

discussion concerning the mechanisms of these reactions and it has been suggested that σ-bonded alkylchromium(III) compounds are in effect key intermediates [164,165]. Kochi and his associates have found that whereas chromous salts in a variety of media react very sluggishly with alkyl halides, the Cr(II)–ethylenediamine (en) complex reacts rapidly with alkyl halides in aqueous dimethylformamide to give highly colored solutions containing, probably, a σ-bonded alkylchromium(III) compound [164,165]. The evidence for the presence of a σ-bonded alkylchromium(III) compound in these solutions is based upon the visible spectra of the solutions [e.g., λ_{max} 397 nm ($\varepsilon = 3000$) for the *tert*-butyl compound] and upon the fact that 6-bromohexene reacts with the $Cr(II)en_2$ complex to give finally a mixture of hexene-1 and methylcyclopentane, Eq. (1-101) [165]. Were the conversion of the

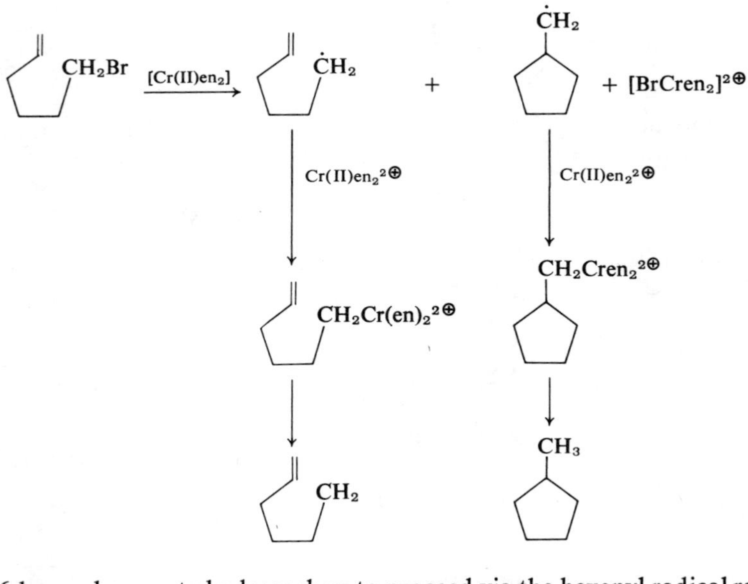

(1-101)

6-bromohexene to hydrocarbon to proceed via the hexenyl radical rather than via the σ-bonded organochromium complex, then the sole product from the reaction would be methylcyclopentane [165].

Relatively stable σ-bonded alkylchromium compounds can be obtained by the interaction of haloforms and chromous salts [146,166–168]. The precise products formed in these reactions are critically dependent upon the stoichiometry of the reagents. Thus, interaction of the haloforms and one equivalent of chromous perchlorate gives the *σ-bonded dihalogenomethylchromium(III) compounds* [**67**, Eq. (1-102) (X = Cl, Br, or I)] [146,167].

$$X_3CH + 2\,Cr(ClO_4)_2 \xrightarrow{HClO_4} \underset{\mathbf{67}}{[X_2CHCr(H_2O)_n]^{2\oplus}} + [Cr(III)X(H_2O)_5]^{2\oplus} \qquad (1\text{-}102)$$

With four to six equivalents of chromous ion, on the other hand, the products are the *σ-bonded monohalogenomethylchromium(III) compounds* [**68**, Eq. (1-103) (X = Cl, Br, or I)] [167]. The latter may also be prepared by the

$$X_3CH + \text{excess } Cr(ClO_4)_2 \xrightarrow{HClO_4} \underset{\mathbf{68}}{[XCH_2Cr(H_2O)_n]^{2\oplus}} + [Cr(III)X(H_2O)_5]^{2\oplus} \qquad (1\text{-}103)$$

direct interaction of methylene dihalides (CH_2X_2, X = Br or I) and one equivalent of chromous perchlorate. The presence of the carbon–chromium σ-bond in the monohalogenomethylchromium compounds **68** (X = Cl, Br, or I) was established by reaction with mercuric nitrate to give the corresponding halogenomethylmercuric chlorides [167].

The σ-bonded dihalogenomethylchromium(III) compounds **67** (X = Cl, Br, or I) are probably formed by the two-step oxidative addition already suggested for other organic halides, i.e., Eq. (1-104) (X = Cl, Br, or I). The

$$X_2CH\cdots X\cdots Cr(II)(H_2O)_n \longrightarrow X_2\dot{C}H + [XCr(III)(H_2O)_n]^{2\oplus} \xrightarrow{Cr(II)aq} \underset{\mathbf{67}}{[X_2CHCr(III)(H_2O)_n]^{2\oplus}} + [XCr(III)(H_2O)_n]^{2\oplus} \qquad (1\text{-}104)$$

further transformation of the dihalogenomethyl compounds, **67**, to the monohalogenochromium(III) compounds, **68**, is a novel reaction, Eq. (1-105), but may be interpreted in terms of the mechanisms suggested for

$$\mathbf{67} \xrightarrow{Cr(II)/H_2O} \mathbf{68} + XCr(III)(H_2O)_n{}^{2\oplus} \qquad (1\text{-}105)$$

the reduction of geminal dihalides by chromous ion. Thus, intramolecular halogen abstraction from **67** would give a transient carbenoid species **69**, which may be trapped subsequently by a second chromous ion to give the carbenoid-chromium(II) complex **70**, Eq. (1-106). Alternatively, intermolecular abstraction of halogen, by a second chromous ion, would give directly the organochromium(III) radical **71**, Eq. (1-107) [167,168]. The compounds **70** and **71** are valence tautomers.

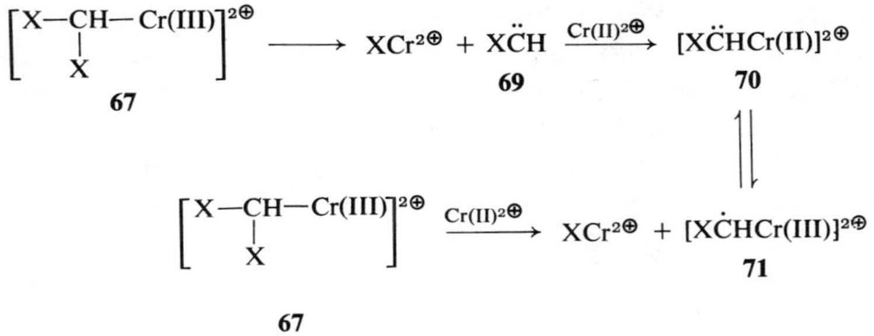

(1-106)

(1-107)

The subsequent transformation of **70** and **71** into the final σ-bonded monohalogenomethylchromium species **68** may proceed by a variety of routes. Thus internal hydrogen transfer in the aquated radical species **71** and subsequent fragmentation would liberate XCH_2, which may now be trapped by a second chromous ion to give **68**, Eq. (1-108). Alternatively,

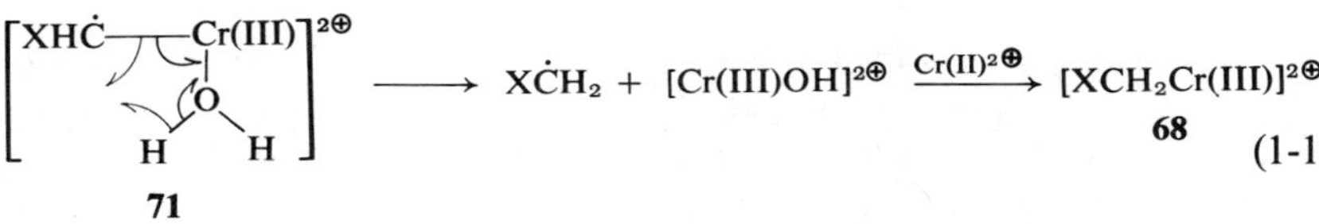

(1-108)

attack of a second chromous ion on **71** may give the bridged bischromium complex that could be protonated to **68**, Eq. (1-109) [167]. The mono-

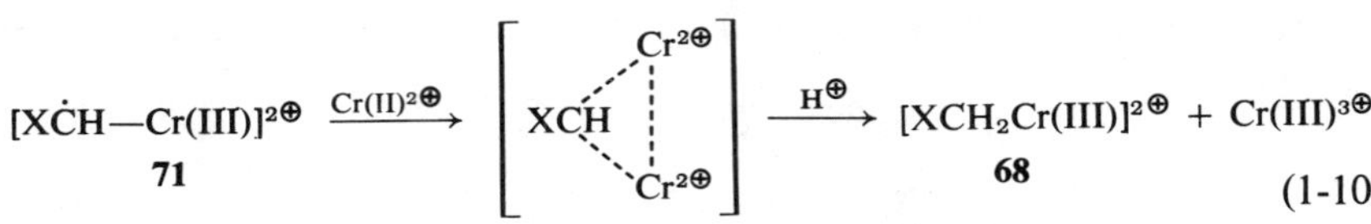

(1-109)

halogenomethyl compound **68** could also arise by the reductive protonation of the carbenoid complex **70**, Eq. (1-110) [168]. There are precedents for all

$$\underset{\mathbf{70}}{[X\ddot{C}H\cdots Cr(II)]^{2\oplus}} \xrightarrow{Cr(II)/H^{\oplus}} \underset{\mathbf{68}}{[XCH_2Cr(III)]^{2\oplus}} + Cr(III)^{3\oplus} \qquad (1\text{-}110)$$

the reactions cited and there is direct chemical evidence for the formation of both radical and carbenoid species in the reaction of isopropylidene bromide and chromous sulfate [168].

The present synthesis of σ-bonded mono(organo)chromium(III) compounds have the disadvantage that the reaction conditions and the subsequent purification of the products (chromatography on ion exchange resins) result in rather dilute solutions of the pure σ-bonded organochromium(III) compounds. Such solutions are ideally suited to the study of certain reactions of the carbon-chromium bond; however, they do not permit the isolation of crystalline organochromium compounds. Thus, all attempts to date to isolate a σ-bonded benzyl- or pyridinomethylchromium species either as the pentaaquate perchlorate or as a salt with a more complex anion have been unsuccessful [3,146,147,149,151–154]. A modification of the method, however, involving the use of $CrCl_2(py)_2$ in pyridine (py) finally led to the isolation of three crystalline σ-bonded tris(pyridine)mono(organo)dichlorochromium(III) compounds, Eq. (1-111) (R = H, *o*-Cl, and *p*-Cl) [133]. There

$$RC_6H_4CH_2Cl + 2\,CrCl_2(py)_2 \xrightarrow{py} RC_6H_4CH_2CrCl_2(py)_3 + CrCl_3(py)_3 \qquad (1\text{-}111)$$

are also reports of the isolation of somewhat impure solid σ-bonded mono(organo)chromium(III) compounds from the interaction of 2- and 3-(chloromethyl)pyridinium chlorides and the chromous ethylenediamine complex in perchloric acid [169].

These modifications should be capable of being extended to the preparation of other crystalline σ-bonded mono(benzylic) and mono(pseudobenzylic)-chromium compounds, and the mono(alkyl)chromium compounds. The foregoing results indicate that σ-bonded mono(organo)chromium compounds are formed by the rapid interaction of an organic radical, generated from substrate and one chromous ion, with a second chromous ion. It should therefore be possible to synthesize analogous organochromium(III) compounds by the direct interaction of organic radicals with chromous salts. Indeed, there are two reports of such syntheses.

Taube, Swinehart, and Schmidt [170] have prepared a series of organochromium pentaaquate cations, by the direct interaction of the appropriate organic radicals and Cr(II) in perchloric acid. These radicals were generated by the action of H_2O_2 on the organic substrate or by the decomposition of the *tert*-butoxy radical [e.g., Eqs. (1-112) and (1-113)]. The presence of the

$$CH_3OH + 2\,Cr(II) + H_2O_2 \longrightarrow [HOCH_2Cr(H_2O)_5]^{2\oplus} + [Cr(H_2O)_n]^{3\oplus} + H_2O \qquad (1\text{-}112)$$

$$[(CH_3)_3CO\cdot \longrightarrow \dot{C}H_3 + CH_3COCH_3] \xrightarrow{Cr(II)} [CH_3Cr(H_2O)_5]^{2\oplus} \qquad (1\text{-}113)$$

carbon–metal bond was deduced from the UV spectrum (λ_{max} 390 nm, ε_{max} 360), and, in the case of the methyl compound, from deuteriolysis (deuteriomethane).

Cohen and Meyerstein [171] have used pulsed radiolysis [172] to generate the radicals derived from a variety of organic substrates and determined the rate of formation of the alkylchromium compound. The order of reactivity of the radicals toward $Cr(H_2O)_6{}^{2\oplus}$ was

$$\cdot CH_2COOH > \cdot CH_2OH > CH_3\dot{C}HOH > (CH_3)_2\dot{C}OH > CH_3\dot{C}HOCH_2CH_3$$

Espenson and Shveima [173] have prepared a variety of alkylchromium(III) pentaaquate cations by the interaction of alkylbis(dimethylglyoximato)-cobalt complexes and $Cr^{2\oplus}$ in perchloric acid, Eq. (1-114) [R = CH_3, CH_2CH_3, $CH_2CH_2CH_3$, $CH_2C_6H_5$, $CH_2C(CH_3)_3$, $CH(CH_3)_2$, and $CH_3CH(CH_2)_5CH_3$; DMG = dimethylglyoximato). These reactions involve essen-

$$RCo(HDMG)_2H_2O + Cr(II)aq \xrightarrow{H^{\oplus}} [RCr(H_2O)_5]^{2\oplus} + Co^{2\oplus} + 2H_2DMG \quad (1\text{-}114)$$

tially the transfer of an alkyl group from cobalt to chromium, and the chemical evidence argues against a stepwise mechanism in which the Co–C bond undergoes homolysis and the organic radical is subsequently trapped by Cr(II); cf. Eqs. (1-96) and (1-97). The alkyl transfer, which follows second-order kinetics, is believed to involve either the transfer of a carbanion from Co to Cr (accompanied by an electron transfer from Cr to Co) or a bimolecular homolytic substitution at carbon [173].

2. Bis(organo)chromium(III) Compounds

Very few authentic σ-bonded bis(organo)chromium(III) compounds are known to date (see Table 1.3). The known compounds may, however, be divided into two classes. The first, of general formula $R_2Cr(III)ClL_n$ (R = allyl, aryl, and alkyl; L = tetrahydrofuran) are sensitive both to air and moisture. The second, of general formula $[R_2Cr(III)(LL)_2]^{\oplus}X^{\ominus}$ (R = aryl or alkyl; LL = 2,2′-bipyridyl or 1,10-phenanthroline; X = I, PBh_4), are stable both to oxygen and moisture. The methods available for the preparation of both classes of compounds are discussed in the following pages.

a. *Method I: Grignard Synthesis*

There are claims that bis(organo)chromium(III) monohalides may be prepared by the interaction of stoichiometric ratios (2:1) of a Grignard reagent and $CrCl_3(THF)_3$ in tetrahydrofuran [132,174,175], Eq. (1-115) (R = $C_6H_5CH_2$ and alkyl). It is, however, difficult to distinguish by the

$$2\,RMgX + CrCl_3(THF)_3 \xrightarrow{<0°C} R_2CrCl(THF)_n \quad (1\text{-}115)$$

usual analytical methods, between a distinct bis(organo)chromium(III) halide (R_2CrCl) and an equimolar mixture of tris- and mono(organo)chromium(III) compounds (i.e., $R_3Cr + RCrCl_2$). No crystalline organochromium compound was actually isolated from the reaction with benzylmagnesium chloride [Eq. (1-115), R = $C_6H_5CH_2$], and in the case of the reactions with alkyl Grignards, brown-green solids often contaminated with magnesium salts were isolated, Eq. (1-115) (R = alkyl) [132,174,175]. The evidence available suggests that the compounds prepared may be the bis(organo)-

chromium(III) halides; however, more definite confirmation (e.g., x-ray structure analysis) remains desirable.

b. Method II: Ligand Exchange

There is one reported instance of the synthesis of a solvated bis(organo)-chromium(III) chloride by the ligand exchange reaction. Thus, both tris-(lithium) hexa(phenyl)chromate and bis(sodium) penta(phenyl)chromate react with the appropriate quantity of $CrCl_3(THF)_3$, in 1,2-dimethoxyethane (DME) to give the solvated bis(phenyl)chromium(III) chloride **72**, Eqs. (1-116) and (1-117) [137]. The brown crystalline solid was shown by infrared

$$Li_3Cr(C_6H_5)_6 + 2\,CrCl_3(THF)_3 \longrightarrow 3\,(C_6H_5)_2CrCl(DME)_{1.5} + 3\,LiCl \quad (1\text{-}116)$$

72

$$2\,Na_2Cr(C_6H_5)_5 + 3\,CrCl_3(THF)_3 \longrightarrow 5\,(C_6H_5)_2CrCl(DME)_{1.5} + 4\,NaCl \quad (1\text{-}117)$$

72

spectroscopy to be the solvated bis(organo)chromium(III) chloride and not a mixture of tris(phenyl)chromium(III) and mono(phenyl)Cr(III)Cl_2. No x-ray structure data have been reported for this compound.

The extension of this synthetic method to the preparation of other σ-bonded bis(organo)chromium(III) halides has not been explored to date.

c. Method V: Reaction with HX *and* X_2

The selective hydrohalogenolysis or halogenolysis of one carbon–metal bond in a tris(organo)chromium compound has been effected in the case of π-tris(allyl)chromium [176]. In the case of gaseous hydrochloric acid, the reaction proceeds smoothly at −80°C in diethyl ether to give a good yield of π-bis(allyl)chromium chloride dimer, Eq. (1-118) [176]. The reaction of tris-

$$2\,(CH_2{=}CH{-}CH_2)_3Cr(III) \xrightarrow{HCl/-80^\circ C} [(CH_2{=}CHCH_2)_2CrCl]_2 \quad (1\text{-}118)$$

(allyl)chromium(III) with halogens (Cl_2 and I_2) is not so clear-cut. Thus, with chlorine a "complex mixture of products" is obtained; whereas with iodine, in diethyl ether solution at 0°C, π-bis(allyl)chromium(III) iodide dimer is formed, Eq. (1-119) [176]. Reaction with iodine in toluene or

$$2\,(CH_2{=}CHCH_2)_3Cr(III) \xrightarrow[ether]{I_2/0^\circ C} [(CH_2{=}CHCH_2)_2CrI]_2 \quad (1\text{-}119)$$

benzene solution leads, however, to chromous iodide and hexadiene, Eq. (1-120) [176]. The formation of these different products could be associated

$$2\,(CH_2{=}CHCH_2)_3Cr(III) \xrightarrow{I_2} 2\,CrI_2 + 3$$ 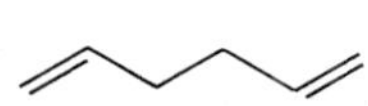 (1-120)

TABLE 1.3

σ-Bonded Bis(organo)chromium(III) Compounds

COMPOUND[a]	METHOD OF PREPARATION[b]	STATE[c]	MAGNETIC MOMENT (μ_B)	REACTIONS, ETC.
$(C_6H_5CH_2)_2CrCl(THF)_n$	I [132,174]	Red-orange solution	—	With D_2O [174]
$(CH_3)_2CrCl(THF)_n$	I [175]	Brown-green solid [175]	—	With I_2 [175]; with $AlCl_3$ and $(CH_3CH_2)_nAlCl_{3-n}$ polymerization catalyst for olefins [175]
$(C_2H_5)_2CrCl(THF)_n$	I [175]	Brown-green solid [175]	—	With I_2 [175], with $AlCl_3$ and $(CH_3CH_2)_nAlCl_{3-n}$ polymerization catalyst for olefins [175]
$(C_3H_7)_2CrCl(THF)_n$	I [175]	Brown-green solid [175]	—	With I_2 [175], with $AlCl_3$ and $(CH_3CH_2)_nAlCl_{3-n}$ polymerization catalyst for olefins [175]
$(i\text{-}C_3H_7)_2CrCl(THF)_n$	I [175]	Brown-green solid [175]	—	With I_2 [175], with $AlCl_3$ and $(CH_3CH_2)_nAlCl_{3-n}$ polymerization catalyst for olefins [175]
$(C_4H_9)_2CrCl(THF)_n$	I [175]	Brown-green solid [175]	—	With I_2 [175], with $AlCl_3$ and $(CH_3CH_2)_nAlCl_{3-n}$ polymerization catalyst for olefins [175]
$(i\text{-}C_4H_9)_2CrCl(THF)_n$	I [175]	Brown-green solid [175]	—	With I_2 [175], with $AlCl_3$ and $(CH_3CH_2)_nAlCl_{3-n}$ polymerization catalyst for olefins [175]
$(C_6H_{11})_2CrCl(THF)_n$	I [175]	Brown-green solid [175]	—	With I_2 [175], with $AlCl_3$ and $(CH_3CH_2)_nAlCl_{3-n}$ polymerization catalyst for olefins [175]
$(n\text{-}C_8H_{17})_2CrCl(THF)_n$	I [175]	Brown-green solid [175]	—	With I_2 [175], with $AlCl_3$ and $(CH_2CH_3)_nAlCl_{3-n}$ polymerization catalyst for olefins [175]
$(C_6H_5)_2CrCl(DME)_{1.5}$	II [137]	Brown solid* [137]	3.78 [137]	With $HgCl_2 \rightarrow C_6H_5HgCl$ [137]; thermal decomposition [137]; with $I_2 \rightarrow C_6H_5I$ [137]
$[(CH_2{=}CH{-}CH_2)_2CrCl]_2$	V [176]	Brown solid* [176]	—	With CH_3OH [176]
$[(CH_2{=}CH{-}CH_2)_2CrI]_2$	V [176]	Brown solid* [176]	—	With CH_3OH [176]

cis-$[(C_6H_5)_2Cr(bipy)_2]I$	VI [179,181, 184]	Red solid* [179,181]	3.77 [179] 3.82 [181]	With $HgCl_2 \rightarrow C_6H_5HgCl$ [179,181]; with I_2 [179]; IR [179]; UV [181]; polarography [179a]; x-ray [185]; other salts [179]
$[(C_6H_5)_2Cr(phen)_2]I$	VI [179]	Orange-yellow solid [179]	—	
cis-$[(2\text{-}CH_3OC_6H_4)_2Cr(bipy)_2]I$	VI [180,181]	Red crystal* [181]	3.80 [180, 181]	With $HgCl_2 \rightarrow 2\text{-}CH_3OC_6H_4HgCl$ [180,181]; UV [180,181]; x-ray [180,182]
cis-$[(3\text{-}CH_3OC_6H_4)_2Cr(bipy)_2]I$	VI [181]	Red crystal* [181]	3.69 [181]	With $HgCl_2 \rightarrow 3\text{-}CH_3OC_6H_4HgCl$ [181]; UV [181]
cis-$[(4\text{-}CH_3OC_6H_4)_2Cr(bipy)_2]I$	VI [181]	Red crystal* [181]	3.78 [181]	With $HgCl \rightarrow 4\text{-}CH_3OC_6H_4HgCl$ [181]; UV [181]
cis-$[(4\text{-}CH_3C_6H_4)_2Cr(bipy)_2]I$	VI [181]	Red crystal* [181]	3.72 [181]	With $HgCl_2 \rightarrow 4\text{-}CH_3C_6H_4HgCl$ [181]; UV [181]
cis-$[\{(CH_3)_3SiCH_2\}_2Cr(bipy)_2]I$	VI [183]	Dark red crystal* [183]	3.64 [183]	With $HgCl_2 \rightarrow (CH_3)_3SiCH_2HgCl$ [183]; UV [183]; x-ray [183]

[a] THF = tetrahydrofuran; bipy = 2,2′-bipyridyl; phen = 1,10-phenanthroline.

[b] Methods I–VI as described in text.

[c] Asterisk indicates that satisfactory analyses have been reported for crystalline compounds.

with the thermal instability of π-tris(allyl)chromium(III), since it is known that this compound is moderately stable in donor solvents, but undergoes thermal decomposition in a variety of other solvents to give bis(allyl)chromium(II) dimer and hexadiene, Eq. (1-121) [177,178]. The formation of

$(CH_2{=}CH{-}CH_2)_3Cr(III) \xrightarrow{0°C} [(CH_2{=}CHCH_2)_2Cr(II)]_2$ + 

(1-121)

chromous iodide from the reaction of tris(allyl)chromium(III) and iodine in toluene could therefore originate from the reaction of the bis(allyl)chromium(II) dimer, rather than from that of tris(allyl)chromium(III) compound, with iodine.

It would be of interest to ascertain whether the foregoing reactions with HCl and halogens are specific to allylchromium compounds or whether they may be used to prepare other bis(organo)chromium(III) halides from the corresponding sigma-bonded tris(organo)chromium(III) compounds.

d. Method VI: Oxidation of σ-Bonded Bis(organo)chromium(II) Compounds

Müller isolated and characterized an air- and water-stable ionic σ-bonded bis(organo)chromium(III) complex [179]. He found that crystalline tris(phenyl)chromium(III)(2,2′-bipyridyl) tetrahydrofuranate [**73**, Eq. (1-122)] underwent thermal rearrangement to a black solid. Hydrolysis of the latter *in air* gave the bis(phenyl)chromium(III)-bis(2,2′-bipyridyl) cation **74**, (isolated as its crystalline iodide), Eq. (1-122), along with the expected π-bis(arene) chromium(I) π-complexes, Eq. (1-122). Thermal rearrangement of

$(C_6H_5)_3Cr(2,2'\text{-bipyridyl})(THF)$
73

$\downarrow \Delta$

black intermediate $\xrightarrow[KI]{H_2O/O_2}$ $[(C_6H_5)_2Cr(2,2'\text{-bipyridyl})_2]^{\oplus}I^{\ominus}$
74

$+\ \pi\text{-}(C_6H_6)_2Cr(I)I^{\ominus} + \pi\text{-}(C_6H_6)\text{-}\pi\text{-}(C_6H_5C_6H_5)Cr(I)I^{\ominus}$

(1-122)

the analogous 1,10-phenanthroline (phen) complex $[(C_6H_5)_3Cr(phen)(THF)]$ and subsequent hydrolysis in air gave the corresponding air- and water-stable 1,10-phenanthroline complex, again isolated as its crystalline iodide $[(C_6H_5)_2Cr(III)(phen)_2]^{\oplus}I^{\ominus}$ [179].

The formation of these compounds could proceed via the fragmentation of the tris(organo)chromium(III) compound to the bis(organo)chromium(II) compound. Ligand redistribution in the latter could then give the coordinatively saturated bis(2,2′-bipyridyl), or bis(1,10-phenanthroline)chromium(II) complexes [**75**, Eq. (1-123)] (LL = 2,2′-bipyridyl or 1,10-phenanthroline).

$$2\,(C_6H_5)_3Cr(LL)(THF) \xrightarrow{\Delta} 2\,(C_6H_5)_2Cr(II)(LL)_2 + C_6H_5C_6H_5 \xrightarrow[KI]{O_2/H_2O}$$

75

2 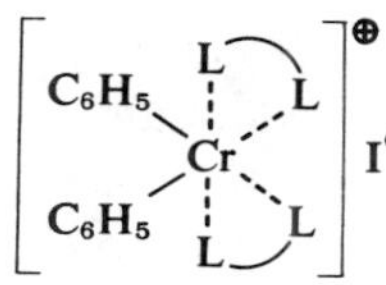 $I^{\ominus}$ (1-123)

74

These upon air oxidation would give the chromium(III) cations **74**, Eq. (1-123) (for steric reasons the cis configuration is the more probable). It follows therefore that other bis(organo)chromium(II) complexes analogous to **75** should result from the interaction of σ-bonded bis(organo)chromium(II) compounds and a bidentate nitrogenous ligand and that these should undergo air oxidation to give the ionic bis(organo)chromium(III) complexes analogous to **74**. Indeed, a number of such air- and water-stable complexes have now been synthesized directly. In the original synthesis pure bis-(2-methoxyphenyl)chromium(II) was converted to the bis(2,2′-bipyridyl) complex **76**, Eq. (1-124) (R = 2-$CH_3OC_6H_4$), which in turn was oxidized in air to the chromium(III) complex **77**, Eq. (1-124) (R = 2-$CH_3OC_6H_4$, LL = 2,2′-bipyridyl) [180,181]. The cis configuration was established by a single

$$2\,RMgX + CrBr_2(THF)_2 \longrightarrow [R_2Cr(II)]_n \xrightarrow{LL}$$

76

$$R_2Cr(LL)_2 \xrightarrow[KI]{O_2/H_2O}$$ 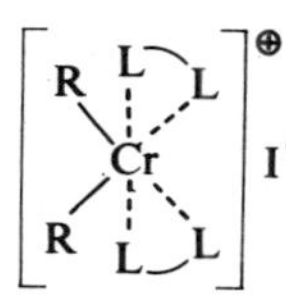 $I^{\ominus}$ (1-124)

77

crystal x-ray structure analysis [180,182]. One disadvantage to this synthesis is that the majority of bis(organo)chromium(II) compounds are rather unstable at room temperature. It was subsequently found that the isolation of the bis(organo)chromium(II) compound was not necessary. Thus certain organomagnesium halides react with $CrBr_2(THF)_2$ in the presence of 2,2′-bipyridyl to give directly the bis(organo)chromium(II) bipyridyl complexes. On air oxidation, these give the ionic *cis*-bis(organo)chromium(III)-bis-(2,2′-bipyridyl) compounds **77**, Eq. (1-125) {LL = 2,2′-bipyridyl, R = C_6H_5, 4-$CH_3C_6H_4$, 2-, 3-, and 4-$CH_3OC_6H_4$ [181], and $(CH_3)_3SiCH_2$ [183]}.

$$2\,RMgX + 2\,LL + CrBr_2(THF)_2 \longrightarrow$$

$$R_2Cr(II)(LL)_2 \xrightarrow[KI]{O_2/H_2O} \underset{\mathbf{77}}{[R_2Cr(III)(LL)_2]^{\oplus}I^{\ominus}} \qquad (1\text{-}125)$$

Two other syntheses of this new class of air- and water-stable σ-bonded bis(organo)chromium(III) compounds have been described [184]. The first uses phenyllithium in place of the organomagnesium halide. The second involves the selective homolysis of one carbon–chromium bond in a mixed tris(organo)chromium(III) compound, and reaction of the resulting bis(organo)chromium(II) compound with 2,2′-bipyridyl. Thus, crystalline solvated mono(diphenylmethyl)$CrCl_2$ reacts with phenyllithium to give the mixed tris(organo)chromium(III) compound **78**, Eq. (1-126) ($R = C_6H_5$) [184.] This compound, which was not isolated, undergoes thermal decomposition in the presence of 2,2′-bipyridyl to give the bis(phenyl)chromium(II)-bis(2,2′-bipyridyl) complex **79**, Eq. (1-127) (i.e.,

$$(C_6H_5)_2CHCrCl_2(THF)_2 \xrightarrow{LiR} \underset{\mathbf{78}}{(C_6H_5)_2CHCr(R)_2(THF)_n} \qquad (1\text{-}126)$$

$$(C_6H_5)_2CHCrR_2(THF)_n \xrightarrow{\Delta/LL} \underset{\mathbf{79}}{R_2Cr(II)(LL)_2} + (C_6H_5)_2\dot{C}H \xrightarrow[KI]{O_2/H_2O}$$

$$\underset{\mathbf{77}}{[R_2Cr(III)(LL)_2]^{\oplus}I^{\ominus}} \qquad (1\text{-}127)$$

the diphenylmethyl–chromium bond is selectively ruptured; the fate of the diphenylmethyl radical is not discussed). Aerial oxidation of **79** gives the ionic *cis*-bis(organo)chromium(III)-bis(2,2′-bipyridyl) complex **77**, Eq. (1-127) ($R = C_6H_5$; LL = 2,2′-bipyridyl) [184], isolated as its crystalline iodide.

It will be of interest to follow developments in this field to ascertain whether the syntheses described above can be modified to prepare the bis-(hydrido)chromium(III) compound **77** (R = H).

3. Tris(organo)chromium(III) Compounds

a. *Method I: Grignard Synthesis*

The successful preparation of σ-bonded tris(organo) chromium compounds was made possible by the vital observation that the products formed by the interaction of phenylmagnesium bromide and chromic chloride depend critically upon the temperature, stoichiometry, and solvent of the reaction [186,187]. Thus, in diethyl ether, with a 3:1 stoichiometry, a black pyrophoric organochromium species is obtained, which on hydrolysis gives a mixture of bis(arene)chromium π-complexes, Eq. (1-128). On the other hand, the

$$3\ C_6H_5MgBr + CrCl_3\ [\text{or}\ CrCl_3(THF)_3] \xrightarrow{Et_2O/0°C}$$

$$[\text{Black product}] \xrightarrow[KX]{H_2O/O_2} \pi\text{-}(C_6H_6)_2Cr(I)X^{\ominus} +$$

$$\pi\text{-}(C_6H_6)\text{-}\pi\text{-}(C_6H_5C_6H_5)Cr(I)X^{\ominus} + \pi\text{-}(C_6H_5\text{-}C_6H_5)_2Cr(I)X^{\ominus} \quad (1\text{-}128)$$

reagents, in the same ratio of 3:1, interact in tetrahydrofuran at −20° to +20°C to give the solvated σ-bonded tris(organo)chromium(III) compound **80**, Eq. (1-129) (R = C_6H_5 and n = 3) [186,187]. Following the discovery

$$3\ RMgBr + CrCl_3\ [\text{or}\ CrCl_3(THF)_3] \xrightarrow[\text{to} + 20°C]{THF/-20°} \underset{\mathbf{80}}{R_3Cr(THF)_n} + 3\ MgXCl(THF)_n \quad (1\text{-}129)$$

of this simple and direct synthesis, a large number and variety of solvated σ-bonded compounds of the type $R_3Cr(THF)_n$ have been prepared. Some of these have been obtained in the crystalline state (e.g., R = aryl and allyl) while others (e.g., R = vinyl, benzyl, and alkyl) have only been prepared in solution in tetrahydrofuran (see Table 1.4). The successful isolation of crystalline tris(organo)chromium(III) compounds depends upon their relative solubilities and thermal stabilities.

Many tris(organo)chromium(III) compounds are more soluble in tetrahydrofuran than the accompanying solvated magnesium halides. Their separation can be effected, in some instances, either by precipitating the magnesium salts as the insoluble complex with dioxane, e.g., in the preparation of tris(mesityl)chromium(III) [188] and tris(2-methoxyphenyl)chromium(III) [131], or by the selective extraction of the tris(organo)chromium compound with hexane {e.g., in the preparation of tris(allyl)chromium(III) [177,189]}.

The thermal stability of the tris(organo)chromium(III) may present a more serious difficulty. Broadly speaking, the majority of the tris(aryl)chromium(III) compounds are stable in the presence of tetrahydrofuran at temperatures of 20°C and above, though the crystalline organochromium compounds may lose the coordinated tetrahydrofuran on prolonged drying at room temperature, e.g., $(C_6H_5)_3Cr(THF)_3$ [187]. The other tris(organo)chromium(III) compounds, on the other hand, are less stable and many undergo irreversible fragmentation at temperature between −70° and +20°C, e.g., tris(allyl)chromium(III) undergoes fragmentation to biallyl and bis(allyl)chromium(II) dimer [177,178] and the tris(alkyl)chromium(III) systems undergo fragmentation to a hydridochromium species and alkane and isomeric alkenes [136].

The choice of solvent and the stoichiometry of the reagents are also of great importance in the Grignard synthesis, e.g., Eqs. (1-128) and (1-129), and the reaction is usually carried out in tetrahydrofuran as solvent, with a 3:1

TABLE 1.4

σ-Bonded Tris(organo)chromium(III) Compounds

COMPOUND[a]	METHOD OF PREPARATION[b]	STATE[c]	MAGNETIC MOMENT (μ_B)	REACTIONS, ETC.
$(C_6H_5)_3Cr3[MgBrCl][THF]_6$	I [186,187]	Red crystals* [186,187]	3.89 [186]	With $HgCl_2 \rightarrow C_6H_5HgCl$ [186,187]
$(C_6H_5)_3Cr(THF)_3$	I [186,187]	Red crystals* [186,187]	3.85 [187]	With $HgCl_2 \rightarrow C_6H_5HgCl$ [186,187,211]; with $CH_3C{\equiv}CCH_3$ [57,58,193]; with $CD_3C{\equiv}CCH_3$ [193,212,213]; with $CF_3C{\equiv}CCF_3$ [214]; with $CH{\equiv}CH$ [215]; with $C_6H_5C{\equiv}CC_6H_5$ [57,216]; with $CH_3OOC{\cdot}C{\equiv}CCOOCH_3$ [217]; with CO [13,218–220]; with ketones [218]; with $CH_3COCH_2COCH_3$ [218]; with $C_6H_5COOCH_3$ [221]; with $(CH_3)_2CHCHO$ [222]; with $CH_2{=}CH{-}CH{=}CH_2$ [223]; with 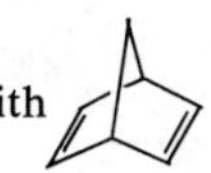 [224]; with 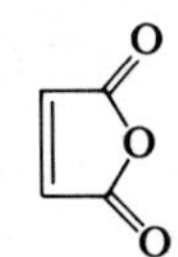 [224]; with H_2 [225,226]; with $CuCl_2$ and $CoCl_2$ [211]; rearrangement to π-complex [10,13,23–27,29,186,187,227–230]
$(C_6H_5)_3Cr(THF)_2$	I/*in vacuo* [26]	Green crystals* [26]	3.98 [26]	With $HgCl_2 \rightarrow C_6H_5HgCl$ [26]
$(C_6H_5)_3Cr(bipy)(THF)_3$	I, ligand exchange [194]	Red crystals* [194]	—	With $HgCl_2 \rightarrow C_6H_5HgCl$ [194]
$(C_6H_5)_3Cr(bipy)(THF)$	I, ligand exchange [194]	Red crystals* [194]	3.80 [194]	With $HgCl_2 \rightarrow C_6H_5HgCl$ [194]

$(C_6H_5)_3Cr(phen)(THF)$	I, ligand exchange [194]	Red crystals* [194]	3.78 [194]	With $HgCl_2 \rightarrow C_6H_5HgCl$ [194]
$(C_6H_5)_3Cr[P(C_6H_5)(n\text{-}C_4H_9)_2]_2$	I, ligand exchange [195]	Violet crystals* [195]	3.89 [195]	UV, visible and IR spectra [195]; with H_2O [195]; with THF [195]
$(C_6H_5)_3Cr[P(C_6H_5)(C_2H_5)_2]_2$	I, ligand exchange [195]	Violet crystals* [195]	—	UV, visible and IR spectra [195]; with H_2O [195]; with THF [195]
$(C_6H_5)_3Cr[P(C_2H_5)_3]_2$	I, ligand exchange [195]	—	—	—
$(C_6H_5)_3Cr[DME]_{1.75}$	II [231]	Red crystals* [231]	3.92 [231]	IR spectrum: hydrolysis [231]
$(C_6H_5)_2CHCr(C_6H_5)_2(THF)_n$	III [184]	Solution	—	Decomposition to $(C_6H_5)_2Cr(II)$; trapped as $(C_6H_5)_2Cr(bipy)_2$ [184];
$(4\text{-}CH_3C_6H_4)_3Cr(THF)_4$	I, [27]	Red crystals* [27]	3.8 [27]	With $HgCl_2 \rightarrow 4\text{-}CH_3C_6H_4HgCl$ [27]; rearrangement to π-complex [27,29]
$(4\text{-}CH_3C_6H_4)_3Cr(THF)_n$	I [28,193,232]	Red solid [28, 193,232]	—	Rearrangement to π-complex [28]; with $CH_3C{\equiv}CCH_3$ [193,232]; with olefins [203]
$(3\text{-}CH_3C_6H_4)_3Cr(THF)_3$	I, [27]	Red crystals* [27]	—	With $HgCl_2 \rightarrow 3\text{-}CH_3C_6H_4HgCl$ [27]; rearrangement to π-complex [27,29]; with $CH_3C{\equiv}CCH_3$ [193]
$(2\text{-}CH_3C_6H_4)_3Cr(THF)_3$	I [193,216]	Red solution [193,216]	—	With $CH_3C{\equiv}CCH_3$ [193]; with $C_6H_5C{\equiv}CC_6H_5$ [216]
$(4\text{-}CH_3OC_6H_4)_3Cr(THF)_n$	I [131]	Red solid** [131]	—	With $CH_3C{\equiv}CCH_3$ [131]; with $HgCl_2 \rightarrow 4\text{-}CH_3OC_6H_4HgCl$ [131]
$(3\text{-}CH_3OC_6H_4)_3Cr(THF)_n$	I [131]	Red solid** [131]	—	With $CH_3C{\equiv}CCH_3$ [131]; with $HgCl_2 \rightarrow 3\text{-}CH_3OC_6H_4HgCl$ [131]
$(2\text{-}CH_3OC_6H_4)_3Cr(THF)_n$	I [131]	Garnet crystals** [131]	3.9	With $CH_3C{\equiv}CCH_3$ and D_2O [131]; with $HgCl_2 \rightarrow 2\text{-}CH_3OC_6H_4HgCl$ [131] Thermal decomposition to $(2\text{-}CH_3OC_6H_4)_2Cr(II)$ [233]
$(4\text{-}BrC_6H_4)_3Cr(THF)_3$	I [27,234]	Red crystals* [27,234]	3.83 [27]	With $HgCl_2 \rightarrow (4\text{-}BrC_6H_4)HgCl$ [27,234]

(continued)

TABLE 1.4 (*continued*)

COMPOUND[a]	METHOD OF PREPARATION[b]	STATE[c]	MAGNETIC MOMENT (μ_B)	REACTIONS, ETC.
$(4\text{-}ClC_6H_4)_3Cr(THF)_{3.5}$	I [27]	Red crystals* [27]	3.98 [27]	With $HgCl_2 \rightarrow (4\text{-}ClC_6H_4)HgCl$ [27]
$(4\text{-}FC_6H_4)_3Cr(THF)_3$	I [27]	Red crystals* [27]	3.79 [27]	With $HgCl_2 \rightarrow (4\text{-}FC_6H_4)HgCl$ [27]
$(3\text{-}ClC_6H_4)_3Cr(THF)_3$	I [234]	Red crystals* [234]	3.97 [234]	With $HgCl_2 \rightarrow (3\text{-}ClC_6H_4)HgCl$ [27,234]
$(3\text{-}ClC_6H_4)_3Cr(py)_3$	I [234]	Yellow-brown crystal* [234]	3.95 [234]	—
$(2\text{-}BrC_6H_4)_3Cr(THF)_n$	I [235]	Solution	—	With $HgCl_2 \rightarrow (2\text{-}BrC_6H_4)HgCl$ [235]
$(2\text{-}ClC_6H_4)_3Cr(THF)_n$	I [235]	Solution	—	With $HgCl_2 \rightarrow (2\text{-}ClC_6H_4)HgCl$ [235]; with $C_6H_5C{\equiv}CC_6H_5$, $C_6H_5C{\equiv}CCH_3$, and $C_6H_5C{\equiv}CH$ [235]
$(4\text{-}C_6H_5C_6H_4)_3Cr(THF)_3$	I [27]	Red crystals* [27]	3.79 [27]	With $HgCl_2 \rightarrow (4\text{-}C_6H_5C_6H_4)HgCl$ [27]; rearrangement to π-complex [27]
$(2,4,6\text{-}(CH_3)_3C_6H_2)_3Cr(THF)$	I [188,217]	Blue crystals* [188,217]	3.74 [188]	With $HgCl_2 \rightarrow (2,4,6\text{-}(CH_3)_3C_6H_2)HgCl$ [188]; with $CH_3C{\equiv}CCH_3$ [193,217]; with D_2 [193]; with $CrCl_3(THF)_3$ [188]; attempted rearrangement [188,228,236]
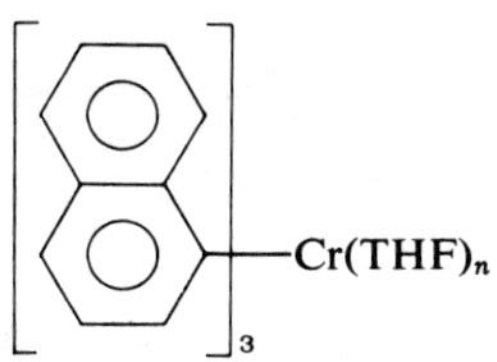	I [57]	Red solution [57]	—	With $CH_3C{\equiv}CCH_3$ [57]
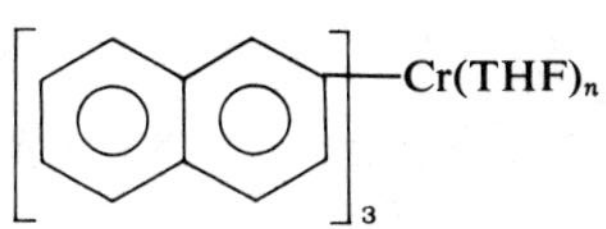	I [57]	Red solution [57]	—	With $CH_3C{\equiv}CCH_3$ [57]

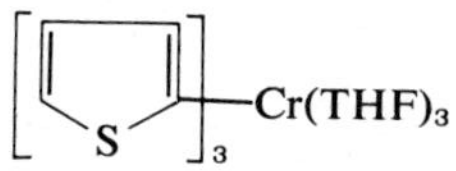	I [27]	Red crystals* [27]	3.84 [27]	With $HgCl_2 \rightarrow (2\text{-}C_4H_3S)HgCl$ [27]
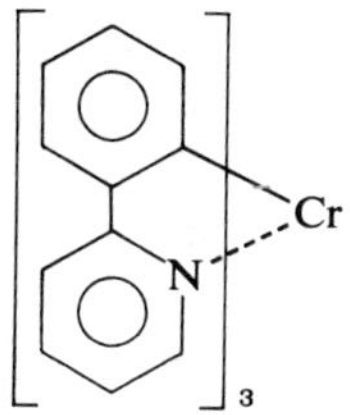	III [196]	Orange crystals* [196]	3.64 [196]	Stable to H_2O and O_2 [196]; hydrolysis to 2-phenylpyridine [196]
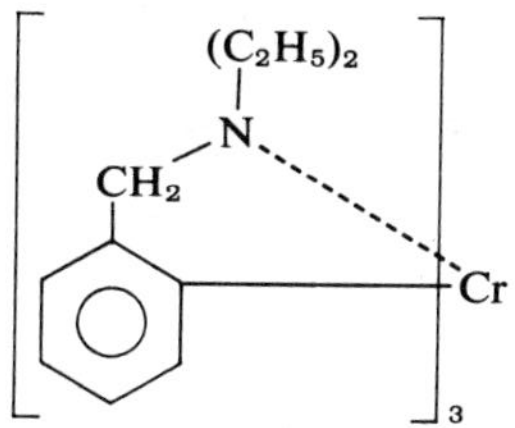	III [197]	Garnet crystals* [197]	—	Sensitive to H_2O and O_2 [197]; molecular weight [197]; with $H_2O \rightarrow C_6H_5CH_2N(Et)_2$ [197]
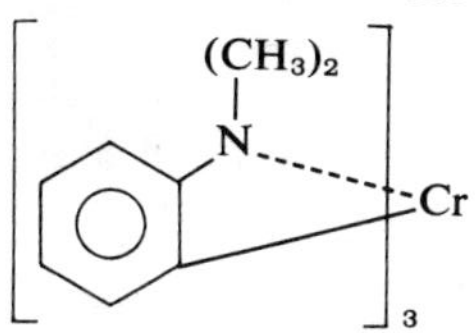	III [198,198a]	Red crystals* [198]	3.88 [198]	With $HgCl_2 \rightarrow [2\text{-}(CH_3)_2NC_6H_4]HgCl$ [198]
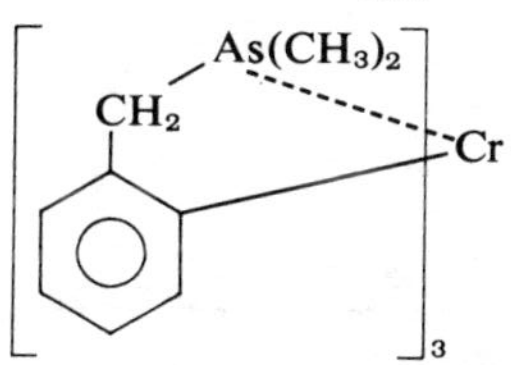	III [199]	Garnet crystals* [199]	3.80 [199]	UV spectrum [199]; stable to air and H_2O [199]; ligand field parameters [199]

(continued)

TABLE 1.4 (*continued*)

COMPOUND[a]	METHOD OF PREPARATION[b]	STATE[c]	MAGNETIC MOMENT (μ_B)	REACTIONS, ETC.
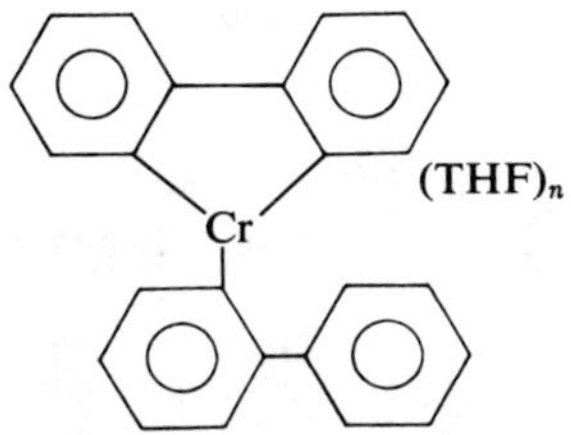	I [193]	Solution [193]	—	With $HgCl_2$ followed by $Br_2 \rightarrow$ 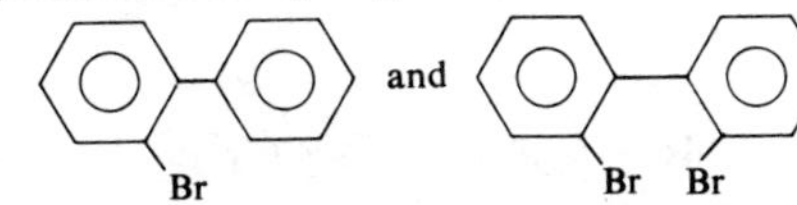 [193]; with $CH_3C{\equiv}CCH_3$ [193]; with CO ⟶ 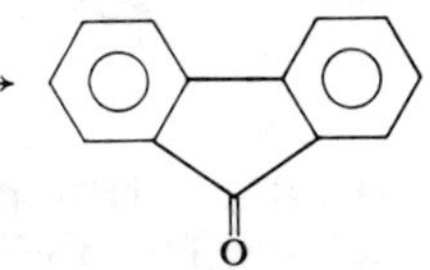 [193]
$(CH_2{=}CH)_3Cr(THF)_n$	I [145]	Red solution [145]	—	With $CH_3C{\equiv}CCH_3$; $C_6H_5C{\equiv}CC_6H_5$; and D_2O [145]
$(C_6H_5CH{=}CH)_3Cr(THF)_n$	I [211,237]	Solution [237]	—	With $C_6H_5C{\equiv}CH$, $C_6H_5C{\equiv}CCH_3$, and $C_6H_5C{\equiv}CC_6H_5$ [237]; with $HgCl_2$, $CoCl_2$, and $CuCl_2$ [211]
$[C_6H_5C({=}CH_2)]_3Cr(THF)_n$	I [211]	Solution [211]	—	With $HgCl_2$, $CoCl_2$, and $CuCl_2$ [211]
$[(C_6H_5)_2C{=}CH]_3Cr(THF)_n$	I [193]	Solution [193]	—	With $CH_3C{\equiv}CCH_3$ [193]
$[CH_2{=}CHCH_2]_3Cr$	I [176,177,189]	Red crystals* [176,189]	3.78 [177]	With iodine [176,177]; with H_2O [176,177]; with H_2 [176]; with $HgCl_2$ [211]; with $CO \rightarrow Cr(CO)_6$; with $P(C_6H_5)_3$ [176], with pyridine [177]; with $CH_3C{\equiv}CCH_3$ [176,238]; x-ray structure [239]; with $CH_2{=}CH_2$ and $CH_2{=}CHCH{=}CH_2$ [176]; mass spectrum [176,240]; polymerization of $CH_2{=}C(CH_3)COOEt$, $CH_2{=}CHCN$, $CH_2{=}CHCH{=}CH_2$, $C_6H_5CH{=}CH_2$ [241]; and

				mechanism of $CH_2{=}C(CH_3)COOEt$ polymerization [242]
$(CH_2{=}CHCH_2)_3Cr[P(C_6H_5)_3]_3$	Ligand exchange [176]	Black crystals* [176]	—	Alcoholysis [176]
$(CH_2{=}CHCH_2)_3Cr(py)_3$	Ligand exchange [177]	Brown crystal [177]	—	—
$(CH_2{=}C(CH_3)CH_2)_3Cr$	I [241,242]	Red solid [241,242]	—	Polymerization of $CH_2{=}C(CH_3)COOEt$, $CH_2{=}CHCN$, $CH_2{=}C(CH_3)CN$, [241,242]
$(CH_3CH{=}CHCH_2)_3Cr$	VII [205,206]	Red solutions [205,206]	—	Polymerization of $CH_2{=}CHCH{=}CH_2$ [205,206]
$(C_6H_5CH_2)_3Cr(THF)_n$	I [132,133]	Red solution [132,133]	—	With $CrCl_3(THF)_3$ [133]; with D_2O [174]
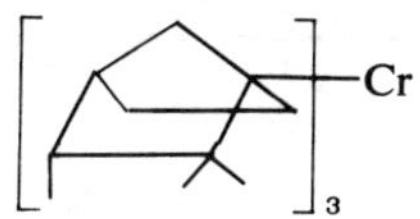	III in pentane [21]	Red brown crystal* [21]	3.53 [21]	With air [21]
$(CH_3)_3Cr(THF)_n$	I [59] III [243]	Red brown solution [59, 243]	—	With $C_6H_5C{\equiv}CC_6H_5$ [59,244–249]; with olefins [203]; with $CH_2{=}CHCH{=}CH_2$ [223]; in synthesis of π-bis(arene)chromium compounds [50–52]
$(CH_3CH_2)_3Cr(THF)_n$	I [59]	Red brown solution [59]	—	With $C_6H_5C{\equiv}CC_6H_5$ [59]; with $CH_3C{\equiv}CCH_3$ [57,58]; with $CH_2{=}CHCH{=}CH_2$ [223]; with olefins [203]; with ketones [218];
$(CH_3(CH_2)_2CH_2)_3Cr(THF)_n$	I [203]	Red brown solution [203]		With $CH_3C{\equiv}CCH_3$ and $CH_3(CH_2)_2C{\equiv}C(CH_2)_2CH_3$ [232]; with olefins [203]; with dienes [250]
$(CH_3CH_2(CH_3)CH)_3Cr(THF)_n$	I [250]	Blue solution [250]	—	With dienes [250]
$(CH_3(CH_2)_4CH_2)_3Cr(THF)_n$	I [211]	—	—	With $HgCl_2$,$CuCl_2$ and $CoCl_2$ [211]
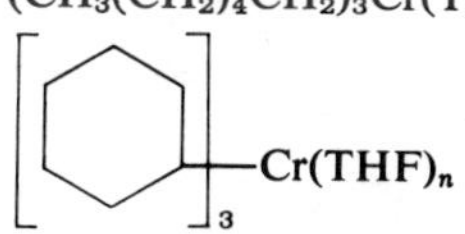	I [211]	—	—	With $HgCl_2$,$CuCl_2$ and $CoCl_2$ [211]

(continued)

TABLE 1.4 (*continued*)

COMPOUND[a]	METHOD OF PREPARATION[b]	STATE[c]	MAGNETIC MOMENT (μ_B)	REACTIONS, ETC.
$(C_6H_5CH_2CH_2)_3Cr(THF)_n$	I [211]	—	—	With $HgCl_2$, $CuCl_2$ and $CoCl_2$ [211]
$(C_6H_5(CH_2)_4)_3Cr(THF)_n$ and 1,1-dideuterio and 2,2-dideuterio compounds	I [142,251]	Red solution [142,251]	—	Hydrolysis [142,251]; thermal decomposition [136, 142,252]; with dienes [204]
$(C_6H_5CH_2CH_2CH(CH_3))_3Cr(THF)_n$	I [142]	Blue solution [142]	—	Thermal decomposition [142]
$(C_6H_5CH_2CH_2C(CH_3)_2)_3Cr(THF)_n$	I [253]	Red solution [253]	—	Thermal decomposition [253]
$(CH_3(CH_2)_8CH_2)_3Cr(THF)_n$	I [136,203]	Red solution [136,203]	—	Thermal decomposition [136]; with olefins [136,203];with $CH_2{=}CHCH{=}CH_2$ [250]

[a] THF = tetrahydrofuran; py = pyridine; phen = 1,10-phenanthroline; DME = 1,2-dimethoxyethane.

[b] Methods I–VII as described in text.

[c] One asterisk indicates that satisfactory combustion analyses have been obtained for crystalline specimens of the compound. Two asterisks indicate that satisfactory chemical analyses (e.g., reaction with D_2O or $HgCl_2$) have been obtained for crystalline specimens of the compounds. Many of the compounds listed (e.g., the vinyl-, some allyl-, benzyl-, and the alkylchromium compounds have not been isolated in a crystalline state.

stoichiometry of reagents. There are scattered reports of the successful preparation of tris(organo)chromium(III) compounds by the interaction of an organomagnesium halide and $CrCl_3(THF)_3$ in diethyl ether, e.g., tris-(allyl)chromium(III) [189] and tris(2-methoxyphenyl)chromium(III) [190]. However, recent results show that in diethyl ether solution certain alkyl-magnesium halides (of type RCH_2MgBr) and $CrCl_3(THF)_3$ in 4:1 stoichiometric ratios give tetra(organo)chromate(III) complexes [**81**, Eq. (1-130)]. The latter undergo subsequent oxidation (air) or "disproportionation" to give, in turn, σ-bonded organochromium(IV) compounds [**82**, Eq. (1-130), R = $(CH_3)_3C$, $C_6H_5(CH_3)_2C$, and $(CH_3)_3Si$] [20,191,192; see also Section H of this chapter]. The scant evidence available suggests that the role of

$$4\text{-}RCH_2MgX + CrCl_3(THF)_3 \xrightarrow{\text{ether}} \underset{\mathbf{81}}{[(RCH_2)_4Cr(III)]^{\ominus}} \xrightarrow{[O]} \underset{\mathbf{82}}{(RCH_2)_4Cr(IV)} \tag{1-130}$$

the solvent in the Grignard synthesis can be interpreted in terms of its donor properties. Thus, tetrahydrofuran is a sufficiently strong donor to stabilize the tris(organo)chromium(III) compounds and the latter are therefore the usual products of the Grignard synthesis in this solvent, Eq. (1-129). Diethyl ether, on the other hand, is a poorer donor solvent and will not stabilize the tris(organo)chromium(III) compounds, and they will therefore undergo secondary transformations. Thus, $(C_6H_5)_3Cr(III)$ undergoes irreversible fragmentation to a π-complex precursor, Eq. (1-128); also certain alkyl-chromium compounds react with excess organomagnesium halide to give the tetra(alkyl)chromate(III) complexes [precursors to tetra(organo)-chromium(IV) compounds], Eq. (1-130). Naturally, when no donor molecule is required to stabilize the tris(organo)chromium(III) compound, e.g., π-tris(allyl)chromium(III), their formation will be independent of the donor properties of the solvent used.

In theory, therefore the Grignard route, Method I, Eq. (1-129), offers a simple and effective synthesis of tris(organo)chromium(III) compounds. It is evident, however, that, in practice, many difficulties have to be overcome. Furthermore, particular attention has to be given to ascertain that the organochromium compound obtained (whether crystalline or in solution) is indeed the expected one. A striking example is the recent observation by Whitesides and Ehmann that the product from the interaction of 2-bi-phenylmagnesium bromide and $CrCl_3(THF)_3$ in tetrahydrofuran is not the expected tris(2-biphenyl)chromium(III), **83**, but the metallocyclic compound **84**, Eq. (1-131) [193]. The formation of **84** may proceed via abstraction of the *ortho*-hydrogen in one biphenyl unit by a second biphenyl unit originally bonded to chromium in the tris(2-biphenyl)chromium complex **83**, Eq.

(1-131) ($R = 2\text{-}C_6H_5C_6H_4$). The donor ligands in tris(organo)chromium(III)

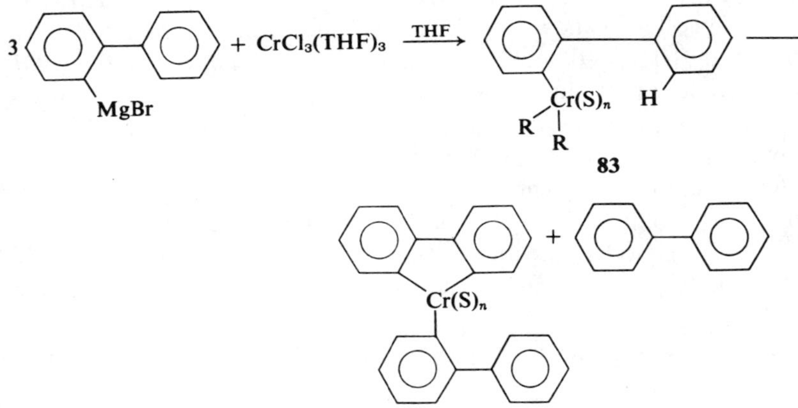

83

84

(1-131)

tetrahydrofuranates may be replaced totally or in part by certain nitrogenous or phosphine ligands (see Table 1.4). This exchange often confers stability to otherwise unstable organochromium(III) compounds. Thus, $(C_6H_5)_3Cr(THF)_3$, which is sensitive to air and moisture, can be converted to a 2,2′-bipyridyl (bipy) and a 1,10-phenanthroline (phen) compound, Eq. (1-132) (LL = bipy or phen) [194]. Both these compounds are moderately stable in

$$(C_6H_5)_3Cr(THF)_3 \xrightarrow{LL} (C_6H_5)_3Cr(LL)(THF) + 2\,THF \qquad (1\text{-}132)$$

air and are not readily hydrolyzed by acids [194].

The solvating tetrahydrofuran in $(C_6H_5)_3Cr(THF)_3$ may also be replaced by certain phosphines to give the pentacoordinate complexes $(C_6H_5)_3Cr(PR_3)_2$, $PR_3 = P(C_6H_5)(C_2H_5)_2$; $P(C_6H_5)(n\text{-}C_4H_9)_2$; $P(CH_2CH_3)_3$ [195]. π-Tris(allyl)chromium(III) with pyridine gives the σ-bonded tris(allyl)-tris(pyridine)chromium(III) [177].

The Grignard synthesis, therefore, with appropriate modifications (i.e., subsequent ligand exchange), is the best method available to date for the preparation of σ-bonded tris(organo)chromium(III) compounds.

b. Method III: Syntheses with Other Organometallic Compounds

There is one report concerning the attempted phenylation of $CrCl_3$ with dichlorophenylaluminum [44]. The final product, however, appears to be complex and to contain both σ- and π-bonded phenyl groups.

The reaction of lithium and sodium alkyl and aryls with $CrCl_3$ [or $CrCl_3(THF)_3$] usually results in the formation of the corresponding lithium or sodium poly(organo)chromate complexes (see Sections G and H of this chapter). However, the lithium derivatives of certain chelating compounds react with $CrCl_3(THF)_3$ in diethyl ether or tetrahydrofuran to give the lithium-free σ-bonded tris(organo)chromium(III) compounds [196–199]. These have been formulated as the hexacoordinated octahedral chromium complexes **85–88**, in which the chromium center is surrounded by three bidentate ligands.

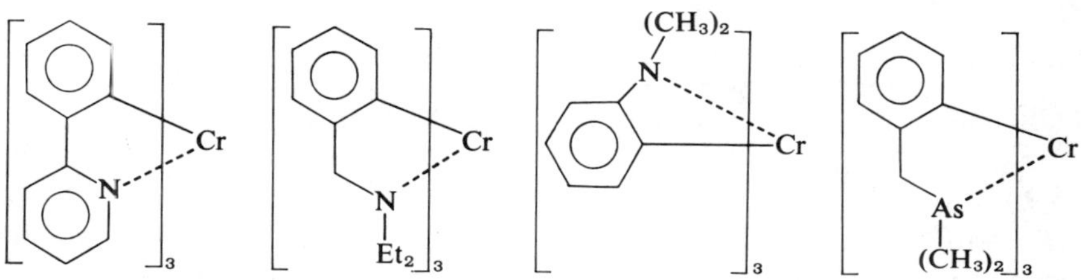

85 [196] **86** [197] **87** [198, 198a] **88** [199]

Hein and Schmiedeknecht have shown that the interaction of lithium hexa(phenyl)chromate(III) and $CrCl_3$ (in 1:1 stoichiometric ratios) in tetrahydrofuran, gives high yields of solvated σ-bonded tris(phenyl)chromium(III), Eq. (1-133) [200]. Therefore, it follows that σ-bonded tris(organo)-

$$Li_3[Cr(C_6H_5)_6] + CrCl_3 \xrightarrow{THF} 2\,(C_6H_5)_3Cr(THF)_3 + 3\,LiCl \qquad (1\text{-}133)$$

chromium(III) compounds should be the products of the interaction of organolithium compounds and $CrCl_3$ (in 3:1 stoichiometric ratios), in tetrahydrofuran solution, since, even if the lithium hexa(organo)chromate-(III) complex is the initial reaction product, it should react further with the remaining $CrCl_3$ to give the tris(organo)chromium(III) compound, Eq. (1-134).

$$6\,LiR + 2\,CrCl_3 \xrightarrow{THF} Li_3CrR_6 + CrCl_3 + 3\,LiCl \longrightarrow 2\,R_3Cr(THF)_n + 6\,LiCl \qquad (1\text{-}134)$$

c. *Method VII: Metal-Hydride Addition*

σ-Bonded organo transition metal compounds have been prepared by the addition of metal hydrides to olefins, e.g., Eq. (1-135) [$PR_3 = P(C_2H_5)_3$, X = Cl] [201,202]. The analogous reaction to give crystalline alkylchromium-(III) compounds has not been effected, though it has been suggested that

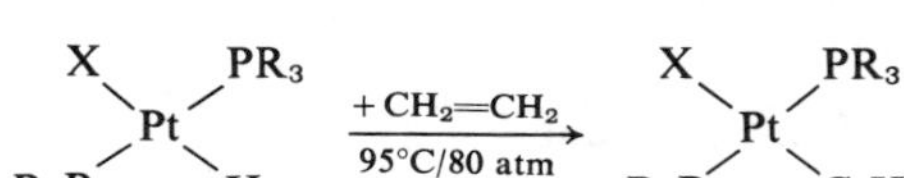

(1-135)

alkylchromium compounds are formed by the addition of a hydridochromium species to an olefin, and that they are intermediates in the isomerization of olefins by alkylchromium(III) systems [203,204]. No crystalline hydridochromium compound or alkylchromium compound was, however, isolated from these experiments.

Similarly, it has been reported that π-(crotyl)chromium compounds result from the interaction of a chromium hydride and butadiene, Eq. (1-136)

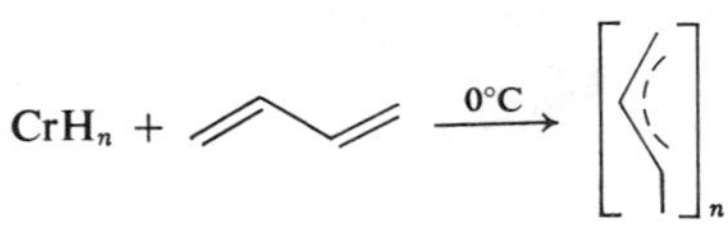

(1-136)

[205,206]. The chromium hydride (CrH_3?) was prepared by the interaction of an alkylmagnesium halide and $CrCl_3$ in diethyl ether in an atmosphere of hydrogen [205,206]. However, no crystalline hydridochromium compound or π-(crotyl)chromium compound was isolated. The presence of the π-(crotyl)chromium complex in the reaction mixture was supported by the fact that acid hydrolysis furnished a mixture of butenes [205,206].

The thermal instability of alkylchromium(III) compounds is such as to make their synthesis by the addition of hydridochromium compounds to olefins difficult. Thus, in the absence of stabilizing ligands, the equilibrium position at room temperature of the system outlined in Eq. (1-137) (cr = chromium and associated ligands) will be on the side of the olefin. The

$$RCH_2CH_2cr \rightleftharpoons RCH{=}CH_2 + Hcr \qquad (1\text{-}137)$$

use of stabilizing ligands (e.g., 2,2′-bipyridyl or a similar bidentate ligand) might, as in the case of cobalt compounds, e.g., Eq. (1-138) (bipy = 2,2′-bipyridyl, X = PF_6 or ClO_4) [207], be used to reverse this equilibrium.

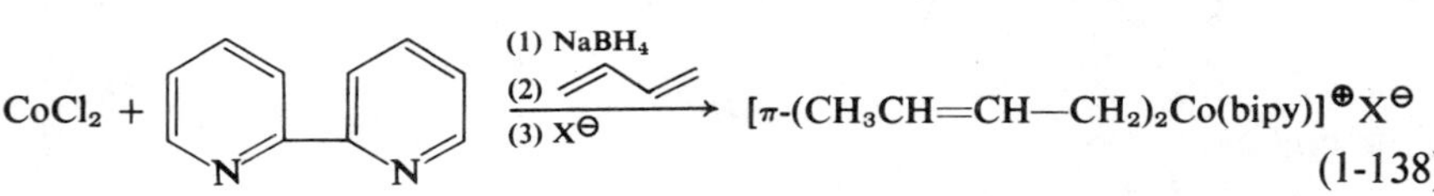

(1-138)

All the metathetical syntheses of σ-bonded mono-, bis-, and tris(organo)-chromium(III) compounds discussed in the foregoing pages (Methods I, II, and III) have used $CrCl_3$ [or $CrCl_3(THF)_3$] as a source of chromium. An alternative source of chromium(III) is tris(acetylacetonato)chromium(III) [$Cr(AcAc)_3$], a compound that is exceedingly soluble in hydrocarbon and ether solvents. There are many reports concerning the use of $Cr(AcAc)_3$ in combination with alkyl- or arylaluminum compounds in the preparation of catalysts for the polymerization of olefins and dienes, e.g., [208]. However, it would appear that these catalysts contain both chromium and aluminum and therefore cannot be considered as true σ-bonded organochromium(III) compounds [209]. Similarly, distinct alkylchromium(III) compounds could not be prepared by the interaction of $Cr(AcAc)_3$ and triethylaluminum or diethylaluminum ethoxide [139]. It would seem, therefore, that $Cr(AcAc)_3$ cannot replace $CrCl_3$ or $[CrCl_3(THF)_3]$ in the metathetical synthesis of σ-bonded organochromium(III) compounds. There is one report of the preparation of a chromium(III) π-complex **89** by the interaction of $Cr(AcAc)_3$ and cyclopentadienylmagnesium bromide, Eq. (1-139) [210].

$$C_5H_5MgBr + Cr(AcAc)_3 \xrightarrow{\text{benzene}} \quad (1\text{-}139)$$

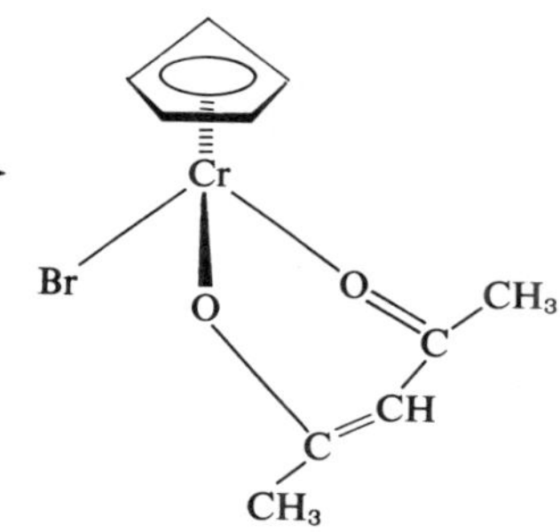

89

4. Organochromium(II) Compounds

In contrast to the large number of σ-bonded organochromium(III) compounds of the type $R_nCrCl_{3-n}(L)_x$ (Tables 1.2–1.4) there are relatively few σ-bonded organochromium(II) compounds of the type $R_nCr(II)Cl_{2-n}(L)_x$ known to date (see Table 1.5). It is not possible to deduce from the information available whether this is due to an instability inherent to the organochromium(II) compounds or whether it is due to excessive practical difficulties in their preparation. The methods available for the synthesis of organochromium(II) compounds that are now discussed include the direct synthesis from chromous halides, the direct thermal transformation of organochromium(III) compounds and some other less direct methods.

The chromous halides required for the direct synthesis of organochromium(II) compounds had to be prepared, in the past, by the high-temperature reactions outlined in Eqs. (1-140) and (1-141) [254]. Both these processes

$$2\,CrCl_3 + H_2 \xrightarrow{500°C} 2\,CrCl_2 + 2\,HCl \quad (1\text{-}140)$$

$$Cr + 2\,HCl \xrightarrow{1150°C} CrCl_2 + H_2 \quad (1\text{-}141)$$

are somewhat inconvenient for normal laboratory practice, and the resulting chromous chloride is often contaminated with HCl. A more convenient

TABLE 1.5

σ-Bonded Organochromium(II) Compounds

COMPOUND[a]	METHOD OF PREPARATION[b]	STATE[c]	MAGNETIC MOMENT (μ_B)	REACTIONS, ETC.
Compounds of type $RCrCl[L]_x$				
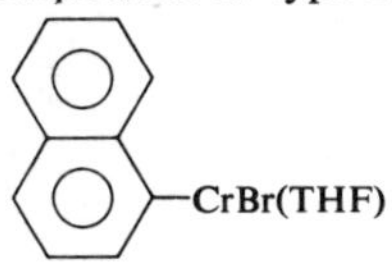	I [255]	Green crystals* [255]	4.04 [255]	With $HgCl_2 \rightarrow$ 1-naphthylHgCl + Hg_2Cl_2
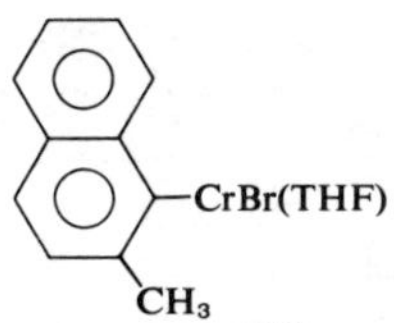	I [255]	Green crystals* [255]	4.24 [255]	With HgCl $\rightarrow$ 1-(2-methylnaphthyl) HgCl + Hg_2Cl_2 [255]
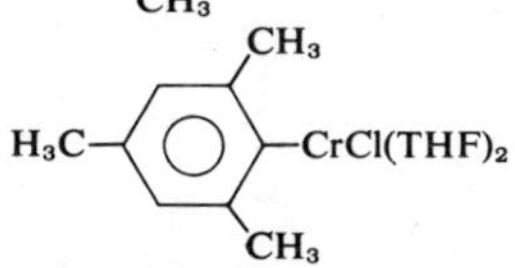	II [255]	Blue crystals* [255]	3.85 to 3.94 [255]	With $HgCl_2 \rightarrow$ mesitylHgCl + Hg_2Cl_2 [255]; with I_2, with H_2O [255]
Compounds of type $R_2Cr(L)_x$				
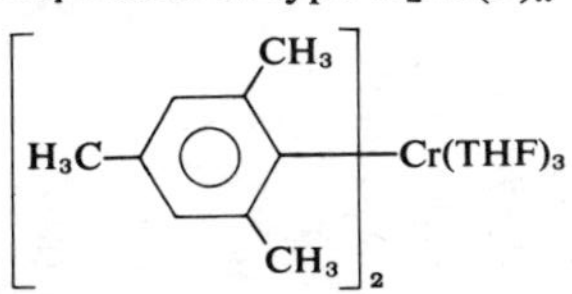	I [255,256]	Violet crystals* [255,256]	4.82 [255]	With $HgCl_2 \rightarrow$ mesitylHgCl + Hg_2Cl_2 [255,256]; with iodine $\rightarrow$ mesityl iodide [255]; with $H_2O \rightarrow$ mesitylene + H_2 [255]; with $CH_3C{\equiv}CCH_3$ and $C_6H_5C{\equiv}CC_6H_5$ [256]; attempted rearrangement to π-complex [228, 255,256]

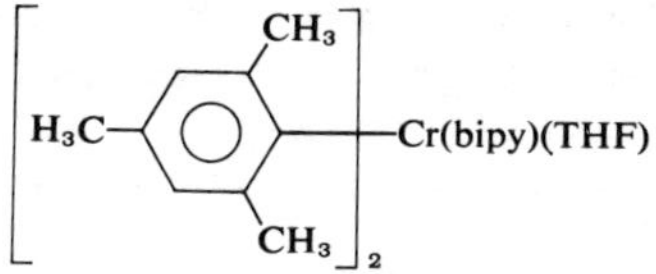	Ligand exchange [255]	Deep violet solid* [255]	4.93 [255]	With $HgCl_2 \rightarrow$ mesitylHgCl + Hg_2Cl_2 [255]; with $I_2 \rightarrow$ mesityl iodide [255]
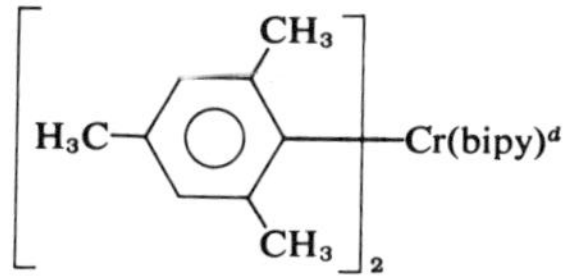	VIII high vac./ 130°C [255]	Deep violet* [255]	4.65 [255]	—
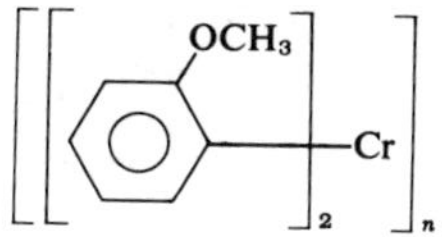	I [130] VII [198, 198a,233]	Yellow solid* [130, 233]	Diamagnetic [130, 198]	With $HgCl_2 \rightarrow (2\text{-}CH_3OC_6H_4)HgCl$ and Hg_2Cl_2 [130]; with $I_2 \rightarrow$ $[2CH_3OC_6H_4]_2$ [198]; with D_2O [130]; with 2,2′-bipyridyl [130]
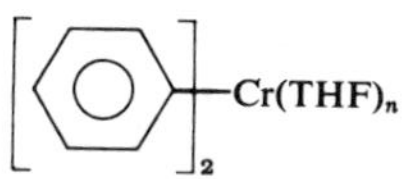	I [228]	Brown solution at −30°C [228]	—	Unstable at −20°C $\xrightarrow{H_2O}$; arene-π-complex mixture [228]
$[(CH_2{=}CHCH_2)_2Cr]_2$	I [178,189] VIII [177,178]	Brown crystals* [189]	Diamagnetic [178]	IR [178]; x-ray [264]; with $I_2 \rightarrow$ $CH_2{=}CHCH_2I$ [178]; with $CH_3C{\equiv}CCH_3$ [238]; polymerization of olefins [189,264]; with $LiCH_3$ [265]
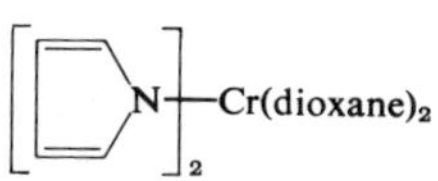	I [257,258]	Blue crystals* [257, 258]	4.86 [258]	IR [258]
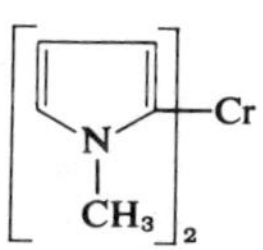	III [258]	Green crystals* [258]	—	IR [258]

(*continued*)

TABLE 1.5 *(continued)*

COMPOUND[a]	METHOD OF PREPARATION[b]	STATE[c]	MAGNETIC MOMENT (μ_B)	REACTIONS, ETC.
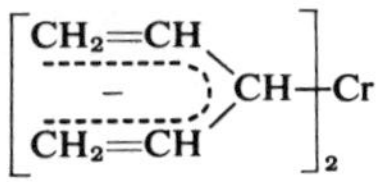	III [259]	Green crystals* [259]	2.74 [259]	With CO; with H_2O; with H_2; IR [259]
$(CH_3)_2Cr(S)_n$	III [243]	Dark solution [243]	—	Hydrolysis [243]
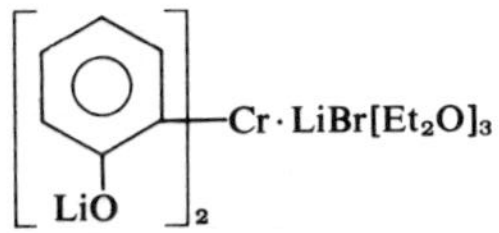	III (using $CrCl_3$) [198,266]	Orange-yellow solid* [198]	Diamagnetic [198, 266]	—
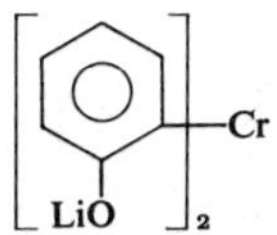	IX [198, 266]	Yellow solid [198, 266]	Diamagnetic [198]	
$(C_6H_5)_2Cr(2,2'\text{-bipyridyl})_2$	I [130] III [184] VIII [179,184]	Deep blue solid* [184]	3.22 [184]	With $HgCl_2 \rightarrow C_6H_5HgCl + Hg_2Cl_2$ [184]; oxidation to Cr(III) compound [130]; polarography [179a]
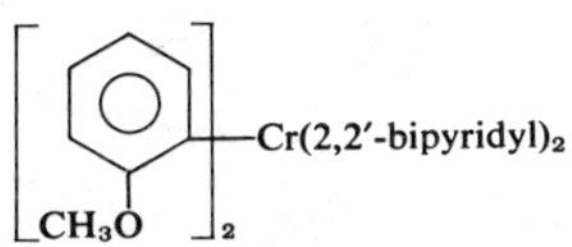	I [130] VIII [233]	Deep blue solid [130]	2.58 [130]	With $HgCl_2 \rightarrow (2\text{-}CH_3OC_6H_4)HgCl + Hg_2Cl_2$ [130]; oxidation to Cr(III) compound [130]
$CH_3Cr(CO)_3\pi\text{-}(C_5H_5)$	IXa [74]	Yellow solid [74]	Diamagnetic [267]	IR [74] [268]; NMR [80]
 $Cr(CO)_2\pi\text{-}(C_5H_5)$	IXb [260]	Orange yellow solid* [260]	Diamagnetic	NMR [261]
 $Cr(CO)_2\pi\text{-}(C_5H_5)$	IXc [262,263]	Yellow solid* [262, 263]	Diamagnetic [263]	IR and NMR [263]

[a] THF = tetrahydrofuran, bipy = 2,2′-bipyridyl.
[b] Methods I–IX as described in text.
[c] An asterisk indicates that satisfactory analyses have been obtained for crystalline specimens of the compounds.
[d] This compound is perhaps better formulated as bis(mesityl)bis(2,2′-bipyridyl)Cr(II) in analogy with other compounds.

preparation of $CrBr_2(THF)_2$ and $CrCl_2(THF)_2$, employing aqueous mineral acid, Eq. (1-142) (X = Cl or Br) [198], has been described.

$$2\,Cr + 2\,HX \text{ in } H_2O \longrightarrow CrX_2 + H_2 + H_2O \xrightarrow[\text{(2) THF}]{\text{(1) } -H_2O/\textit{in vacuo}} CrX_2(THF)_2 \quad (1\text{-}142)$$

a. σ-Bonded Mono(organo)chromium(II) Halides $RCr(II)Cl(L)_x$

i. Methods I and II: Grignard Synthesis and Ligand Exchange. There is very little information available to date concerning the interaction of various stoichiometric ratios of organomagnesium halides and chromous halides. However, it has been reported that both 1-naphthyl- and 1-(2-methyl-naphthyl)magnesium bromide interact with $CrCl_2$, in tetrahydrofuran (in 2:1 and 1:1 stoichiometric ratios) to give the mono(organo)chromium(II) compounds [**90**, Eq. (1-143), R = H or CH_3] [255]. The compounds were

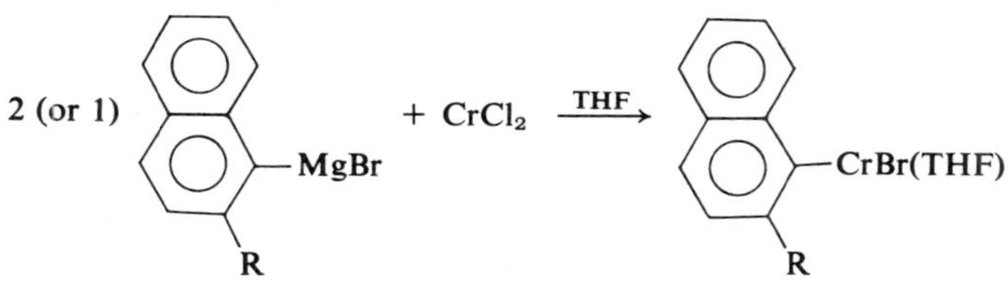

90

(1-143)

characterized by analysis, their reaction with $HgCl_2$ (to give the organomercuric halide and an equivalent quantity of calomel), and their magnetic moments [255].

It is reported that bis(mesityl)chromium(II) and $CrCl_2$ undergo ligand exchange in tetrahydrofuran to give a compound formulated as solvated mesitylchromium(II) chloride [**91**, Eq. (1-144)] [255]. The compound was

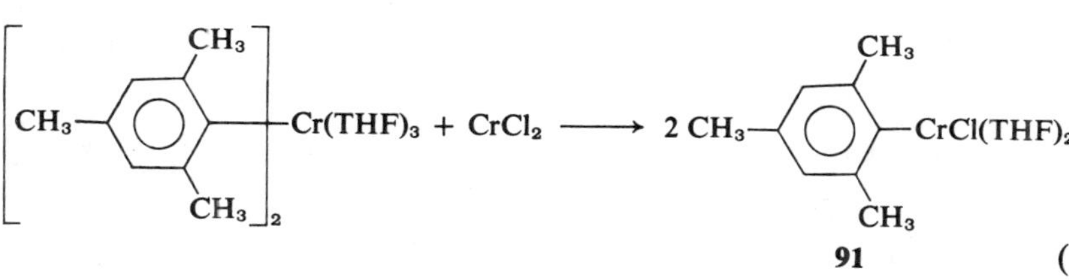

91 (1-144)

characterized by analysis, by its reaction with $HgCl_2$ (to give the organomercurial and an equivalent quantity of calomel), and by magnetic moment. These data, however, do not indicate whether the compound is indeed the monomer **91** or the bridged dimer, e.g., **92**, Eq. (1-145). The fact that the

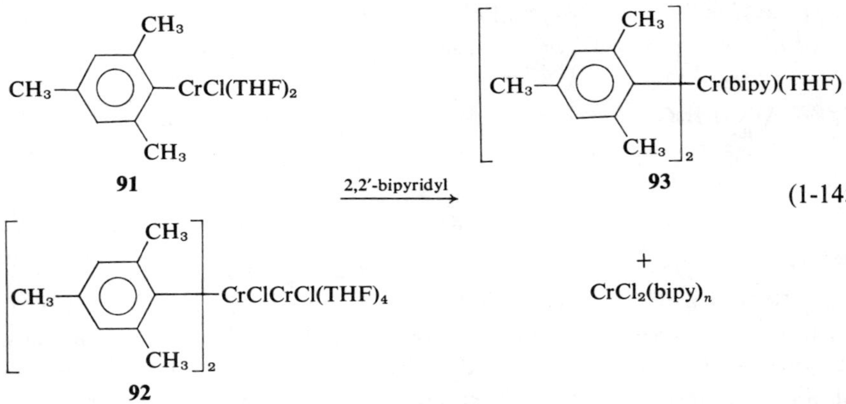

(1-145)

compound reacts with 2,2′-bipyridyl to give bis(mesityl)chromium(II)(bipy)THF [**93**, Eq. (1-145)] [255] could be taken as support for the dimeric structure.

No other σ-bonded mono(organo)chromium(II) halides have been prepared to date.

b. σ-Bonded *Bis(organo)chromium(II) Compounds* $R_2Cr(II)L_x$

i. Method I: Grignard Synthesis. The Grignard synthesis has been used successfully in the preparation of some bis(organo)chromium compounds. Thus:

(1) Mesitylmagnesium bromide and $CrCl_2$ interact in tetrahydrofuran to give the paramagnetic crystalline blue bis(mesityl) compound [**94**, Eq. (1-146)] [228,255,256].

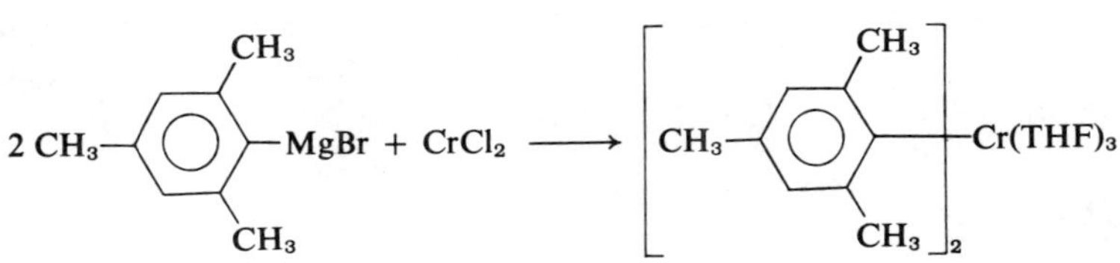

(1-146)

(2) 2-Methoxyphenylmagnesium bromide and $CrBr_2(THF)_2$ interact in tetrahydrofuran to give the diamagnetic bis(organo)chromium(II) compound **95**, Eq. (1-147) [130]. The diamagnetism observed for **95** has been taken as evidence that the compound possesses a polymeric structure with

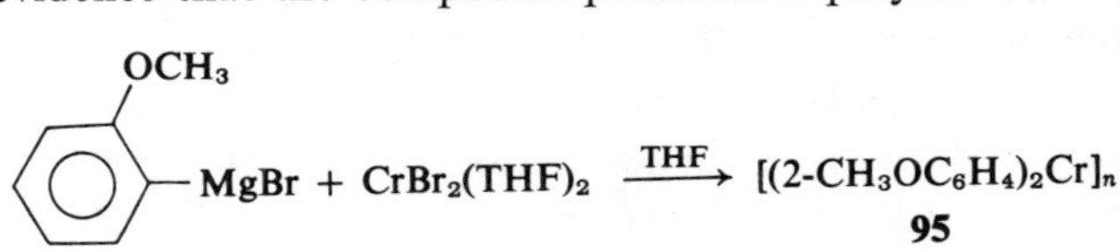

(1-147)

Cr–Cr bonds and bridging 2-methoxyphenyl groups (Chapter 3, Section F,3,c, [130]). This formulation, however, has not been confirmed by an x-ray structure determination.

(3) Allylmagnesium bromide and $CrCl_2$ interact in diethyl ether to give the diamagnetic bis(allyl)chromium(II) compound [**96**, Eq. (1-148)] [178].

$$(CH_2{=}CH{-}CH_2)MgBr + CrCl_2 \xrightarrow{Et_2O}$$ 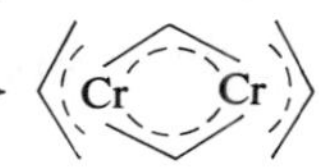 (1-148)

96

The diamagnetism of **96** was ascribed to the presence of Cr–Cr bonds in a dimeric species, and such a formulation was confirmed by a complete x-ray structure analysis (Chapter 2, Section F,3,a).

(4) Pyrrolylmagnesium bromide and $CrCl_2(THF)_2$ give, in ether solution, the blue bis(organo)chromium(II) compound [**97**, Eq. (1-149)] in which the pyrrole units are bonded to chromium via the nitrogen atom [257,258].

$$2\ C_4H_4NMgBr + CrCl_2(THF)_2 \xrightarrow[(2)\ dioxane]{(1)\ Et_2O}$$ 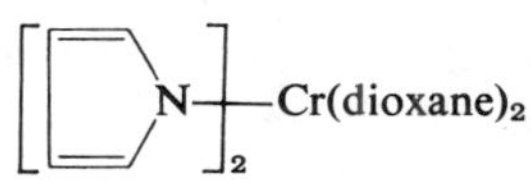 (1-149)

97

The *N*-methylpyrrole chromium compound, in which the pyrrole unit is bonded to chromium via a carbon atom, has been prepared by the interaction of *N*-methylpyrrolyl lithium and $CrCl_2$, Eq. (1-150) [258].

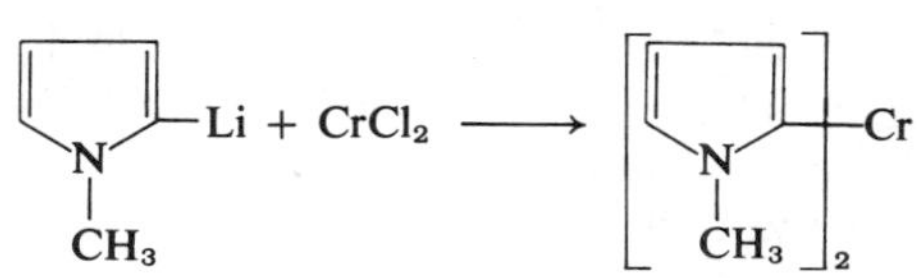 (1-150)

In the foregoing syntheses [i.e., (1) to (4)] either the organic ligands bonded to the chromium or the final products contain some feature that stabilizes the bis(organo)chromium(II) compounds. In the absence of such stabilizing features, it appears that the bis(organo)chromium(II) compounds undergo secondary transformations. Thus, it is reported that the product from the interaction of phenylmagnesium bromide and $CrCl_2$, in tetrahydrofuran, rapidly darkens at −30°C. Hydrolysis of the final reaction mixture does not give benzene [the expected hydrolysis product of bis(phenyl)chromium(II)] but instead a mixture of bis(arene)chromium π-complexes, Eq. (1-151) [228].

$$2\ C_6H_5MgBr + CrCl_2 \xrightarrow[(2)\ O_2/NaX]{(1)\ THF/-25°\ to\ -30°C}$$
$$\pi\text{-}(C_6H_6)_2Cr(I)X^{\ominus} + \pi\text{-}(C_6H_6)\text{-}\pi\text{-}(C_6H_5C_6H_5)Cr(I)X^{\ominus} \qquad (1\text{-}151)$$

One method of stabilizing σ-bonded organochromium(III) compounds of the type $R_3Cr(THF)_n$ consists in replacing the coordinating tetrahydrofuran by nitrogenous ligands, and it has been found that the same applies to some otherwise unstable bis(organo)chromium(II) compounds. Thus, the interaction of phenylmagnesium bromide (and other alkyl or aryl Grignards) and $CrBr_2(THF)_2$ in the presence of 2,2-bipyridyl leads to an isolable bis(organo)-chromium(II)-bis(bipyridyl) complex, e.g., **98**, Eq. (1-152) (bipy = 2,2′-bipyridyl) [130,183,185], Chapter 1, Section F,2,d.

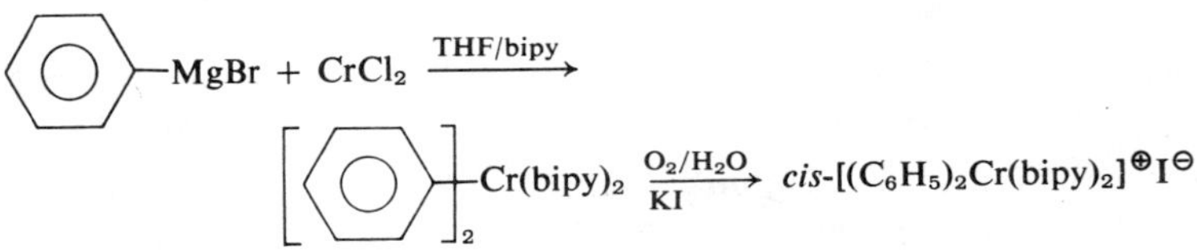

98 (1-152)

This is clearly a field in which further information is urgently required, expecially information that would enable a comparison to be made between the thermal stabilities, bond dissociation energies and molecular structure of organochromium(II) compounds and those of the corresponding organo-chromium(III) compounds.

ii. Method III: Synthesis with Other Organometallic Compounds. Lithium poly(organo)chromate(III) or (II) complexes, e.g., $Li[(2\text{-}CH_3OC_6H_4)_4Cr(III)]$ or $Li_2[(C_6H_5)_4Cr(II)]$, are the usual products formed by the reaction of organolithium compounds and chromous halides [184,198,198a; see also Section H of this chapter]. However, it has been reported that bis(phenyl)-chromium(II) may be synthesized by the interaction of lithium phenyl and $CrCl_2(THF)_2$ (in 2:1 stoichiometric ratio); the compound was not isolated but was converted *in situ* to the corresponding bis(2,2′-bipyridyl) complex [**99**, Eq. (1-153)] [184]. This synthesis, like the Grignard synthesis, should be

$$2\,LiC_6H_5 + CrCl_2(THF)_2 \xrightarrow{-70°C/THF} (C_6H_5)_2Cr(THF)_n \xrightarrow{bipy} \underset{\mathbf{99}}{(C_6H_5)_2Cr(bipy)_2} \qquad (1\text{-}153)$$

capable of being extended to the preparation of other bis(organo)chromium-(II) bis(2,2′-bipyridyl) complexes.

A compound $C_{10}H_{14}Cr$ has been prepared by the interaction of penta-dienylsodium and $CrCl_2$ in tetrahydrofuran. The compound which has a reduced magnetic moment for a Cr(II) complex, is monomeric, highly sensitive to air and moisture, and reacts with protic solvents to give 2 moles of pentadiene. It has been formulated as a σ-bonded bis(pentadienyl)-chromium(II) complex [**100**, Eq. (1-154)] [259].

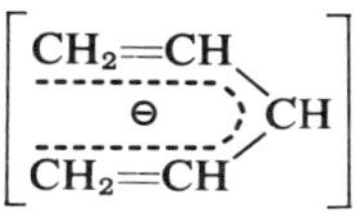 $\mathrm{Na + CrCl_2 \xrightarrow{THF/-30°C} (C_5H_7)_2Cr(II)}$ (1-154)

100

iii. Method VIII: Thermal Transformation of Organochromium(III) Compounds. σ-Bonded tris(organo)chromium(III) compounds and the related lithium or sodium poly(organo)chromates, when heated, undergo rather complex transformations [e.g., the σ- to π-rearrangement of some tris(aryl)chromium(III) compounds and the total fragmentation of others]. In some cases, however, a tris(organo)chromium(III) compound, or a lithium poly(organo) chromate complex may be transformed into the corresponding bis(organo)chromium(II) compound. Reference has already been made to the thermal transformation of π-tris(allyl)chromium(III) into bis(allyl)chromium(II) dimer, of $(C_6H_5)_3Cr(III)$ (bipy)(THF) into $(C_6H_5)_2Cr(II)(bipy)_2$, and of a mixed tris(organo)-chromium(III) compound into a bis(organo)chromium(II) compound, Eq. (1-155) [184]. Similarly, both tris(2-methoxyphenyl)chromium(III)

$$\mathrm{(C_6H_5)_2CrCH(C_6H_5)_2(THF)_n \xrightarrow[2,2'\text{-bipyridyl}]{\Delta/THF} (C_6H_5)_2Cr(II)(bipy)_2 + (C_6H_5)_2\dot{C}H} \quad (1\text{-}155)$$

and lithium tetra(2-methoxyphenyl)chromate(III) may be transformed thermally into bis(2-methoxyphenyl)chromium(II), Eq. (1-156) [233] and Eq. (1-157) [198] ($R = 2\text{-}CH_3OC_6H_4$). However, it is evident that the success

$$\mathrm{(2\text{-}CH_3OC_6H_4)_3Cr(THF)_n \xrightarrow{dioxane} [(2\text{-}CH_3OC_6H_4)_2Cr(II)]_n + R_2CHCHR_2 + RR} \quad (1\text{-}156)$$

$$\mathrm{Li[(2\text{-}CH_3OC_6H_4)_4Cr(III)] \longrightarrow [(2\text{-}CH_3OC_6H_4)_2Cr(II)]_n + Li(2\text{-}CH_3OC_6H_4) + RR} \quad (1\text{-}157)$$

of this method depends upon the thermal stability of the organochromium(II) compound, or upon the possibility of stabilizing it by subsequent reaction with a suitable ligand.

iv. Method IX: Miscellaneous Methods. Other organochromium(II) compounds have also been obtained by the following indirect methods.

(1) Sodium π-(cyclopentadienyl)chromium tricarbonyl with methyl iodide gives, in low yield, a volatile methylchromium compound, Eq. (1-158) [74].

$$\mathrm{\pi\text{-}(C_5H_5)Cr(CO)_3Na + CH_3I \xrightarrow{THF} CH_3Cr(CO)_3(\pi\text{-}C_5H_5)} \quad (1\text{-}158)$$

(2) The photo-catalyzed interaction of π-(cyclopentadienyl)chromium carbonyl dimer and butadiene leads to an organochromium(II) compound, identified (by NMR spectroscopy) as π-(cyclopentadienyl)-π-(crotyl)chromium(II) dicarbonyl [**101**, Eq. (1-159)] [260,261].

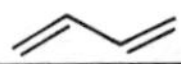

$$[\pi\text{-}(C_5H_5)Cr(CO)_3]_2 \xrightarrow[\text{heptane/UV/Hg}]{} \underset{\mathbf{101}}{\pi\text{-}(C_5H_5)\text{-}\pi\text{-}(C_4H_7)Cr(CO)_2} \qquad (1\text{-}159)$$

(3) The reaction of π-bis(cyclopentadienyl)chromium with carbon monoxide and hydrogen leads to $HCr(CO)_3$-π-C_5H_5 and, in low yield, to a bis(organo)-chromium(II) compound formulated (on the basis of NMR spectroscopy) as π-(cyclopentenyl)-π-(cyclopentadienyl)chromium dicarbonyl [**102**, Eq. (1-160)] [262,263].

$$(C_5H_5)_2Cr \xrightarrow[\text{14–15 hr/68°C}]{CO/H_2/\text{pressure}} \underset{\mathbf{102}}{\pi\text{-}(C_5H_5)\text{-}\pi\text{-}(C_5H_7)Cr(CO)_2} \qquad (1\text{-}160)$$

G. Lithium and Sodium Poly(organo)chromate(III) and (II) Complexes

Hein and his collaborators have made an extensive study of the reactions of organolithium and sodium compounds with chromic and chromous salts, and it is clear from their results that these reactions can lead to the formation of lithium or sodium poly(organo)chromate complexes of the type $M_3Cr(III)R_6$, $M_2Cr(III)R_5$, $MCr(III)R_4$, and $MCr(II)R_3$ (M = Li or Na) (see Table 1.6). However, the precise products formed in these reactions depend critically upon: the nature of the organic group in the organometallic reagent, the stoichiometry of the reagents, the ligands associated with the final product, and the solvent employed for the reaction.

The influence of the nature of the organometallic reagent is illustrated in the synthesis of σ-bonded organochromium(III) and (II) compounds by the interaction of the lithium derivatives of certain bidentate compounds and chromic salts, e.g., tris[(2-phenylyl)pyridine]chromium(III), and the related compounds **85–88** referred to on page 59.

The importance of the stoichiometry of the reagents and of the ligands associated with the final product has already been referred to in the syntheses of certain σ-bonded organochromium(III) and (II) compounds by the interaction of organolithium compounds and chromium salts (in precise stoichiometric ratios) both in the absence and presence of stabilizing ligands, e.g., $(C_6H_5)_2$ $CHCr(III)Cl_2(THF)_2$, **60** (page 32), and $(C_6H_5)_2Cr(II)(2,2'$-bipyridyl$)_2$, **99** (page 68).

The solvent may also influence the course of the reaction, as in the case of some organolithium compounds and $CrCl_3$ in diethyl ether or hydrocarbon solution giving finally σ-bonded tetra(organo)chromium(IV) compounds (see Section H of this chapter). Furthermore, in some instances *poly(organo)-chromate(III)* complexes are obtained when *chromium(II)* salts are used as starting material and vice versa, e.g., Eq. (1-161) [266] and Eq. (1-162) [200].

$$\underset{\text{(excess)}}{C_6H_5Li} + Cr(II)Cl_2 \xrightarrow{Et_2O} Li_3[Cr(III)(C_6H_5)_6][Et_2O]_n \qquad (1\text{-}161)$$

$$8\,C_6H_5Li + 2\,CrCl_3(THF)_3 \xrightarrow{Et_2O/-15°C} \{Li[Cr(II)(C_6H_5)_3]\}_2 + 6\,LiCl + C_6H_5C_6H_5 \qquad (1\text{-}162)$$

In other words, certain lithium (organo)chromate complexes undergo either oxidative or reductive processes rather readily to give more stable end-products. These oxidative or reductive processes (or disproportionations), as will be seen in the ensuing pages, can probably be ascribed to the instability of the intermediate lithium (organo)chromate compounds, occasioned by an unusual coordinative state within these compounds.

1. Poly(organo)chromate Complexes of the Type $M_3Cr(III)R_6$

The interaction of organolithium compounds and $CrCl_3$ in stoichiometric ratios exceeding 6:1 usually leads to the formation of tris(lithium) hexa-(organo)chromate(III) complexes, e.g., **103**, Eq. (1-163) ($R = C_6H_5$) [15,16, 198,269–272]. The compounds that have been prepared to date (Table 1.6)

$$6\,LiR + CrCl_3 \xrightarrow{Et_2O} \underset{\mathbf{103}}{Li_3[Cr(III)R_6][Et_2O]_n} + 3\,LiCl \qquad (1\text{-}163)$$

may be divided into two classes: the hexa(aryl)chromates(III) and the hexa-(alkyl)chromates. The known hexa(aryl)chromates(III) include various 3- and 4-substituted-aryl compounds but no 2-substituted-aryl ones, e.g., hexa(2-methylphenyl)chromate(III); they also include the interesting spiro-cyclic compound **104**, which contains three chromocycles [271]. The known

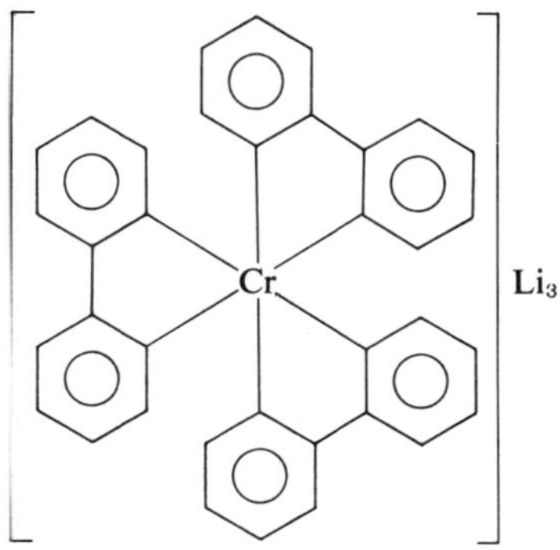

104

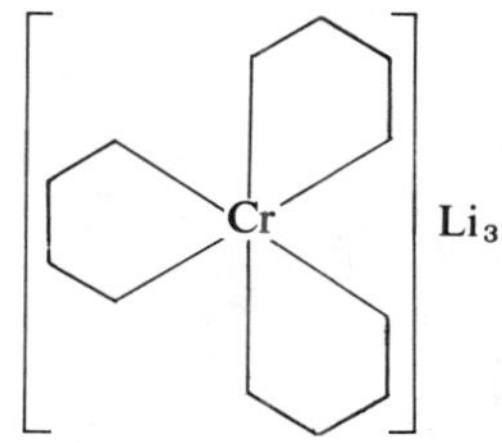

105

TABLE 1.6

σ-Bonded Lithium and Sodium Poly(organo)chromate(III) and (II) Complexes

Compound[a]	Method of preparation	State[b]	Magnetic moment (μ_B)	Reactions, etc.
Type $M_3Cr(III)R_6$				
1. $Li_3[Cr(C_6H_5)_6]2.5(C_2H_5)_2O$	$LiR + CrX_3$ [15, 266,270] $LiR + CrX_2$ [266]	Orange-yellow crystals* [15]	3.61 [15]	With $I_2 \rightarrow 3\ C_6H_5I + C_6H_5C_6H_5$ [15]; with H_2O [15]; with H_2 [15, 225,281,266,279]; with $LiAlH_4$ [282]; with 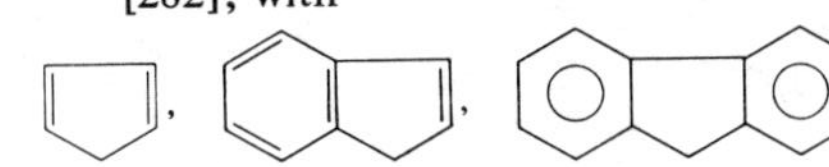 [278]; with DME $\rightarrow$ $Li_2[Cr(C_6H_5)_5]2.5\ [Et_2O]$ [275]; with $CrCl_3$ [137,200,231]
2. $Na_3[Cr(C_6H_5)_6][C_2H_5O]_n$	$NaR + Na_2[CrR_5]$ [275]	Yellow solid not isolated [275]	—	Only stable in presence of excess NaC_6H_5 [275]
3. $Li_3[Cr(3\text{-}ClC_6H_4)_6]LiBr[dioxane]_7$	$LiR + CrCl_3$ [16]	Yellow crystals* [16]	3.89 [16]	With $HgCl_2 \rightarrow (3\text{-}ClC_6H_4)HgCl$ [16]
4. $Li_3[Cr(4\text{-}BrC_6H_4)_6]LiBr[Et_2O]_{7.5}$	$LiR + CrCl_3$ [16]	Yellow crystals* [16]	3.21 [16]	With $HgCl_2 \rightarrow (4\text{-}BrC_6H_4)HgCl$ [16]
5. $Li_3[Cr(C_6H_5)_3(3\text{-}ClC_6H_4)_3][Et_2O]_{2.5}$	$LiR + (C_6H_5)_3Cr\text{-}(THF)_3$ [16]	Yellow crystals* [16]	3.82	With $HgCl_2 \rightarrow 3\text{-}ClC_6H_4HgCl$ [16] $+ C_6H_5HgCl$
6. $Li_3[Cr(4\text{-}CH_3C_6H_4)_6][Et_2O]_3$	$LiR + CrCl_3$ [266]	Yellow powder [266]	—	With $HgCl_2 \rightarrow 4\text{-}CH_3C_6H_4HgCl$ [266]; with $H_2O \rightarrow CH_3C_6H_5$ [266]
7. $Li_3[Cr(4\text{-}CH_3OC_6H_4)_6][THF]_6$ also the tri(etherate)	$LiR + CrCl_3$ [266]	Yellow crystals [266]	—	—
8. $Li_3[Cr(4\text{-}(CH_3)_2NC_6H_4)_6][THF]_3$	$RLi + CrCl_3$ [198a]	Yellow solid [198a]	—	—

9. $Li_3[Cr(4\text{-}(CH_3)_2NC_6H_4)_6][Et_2O]_3$	$RLi + CrCl_3$ [198]	Yellow solid* [198]	3.82	With $HgCl_2 \rightarrow (CH_3)_2NC_6H_4HgCl$ [198]
10. $Li_3[Cr(4\text{-}C_6H_5C_6H_4)_6][Et_2O]_3$	$RLi + CrCl_3$ [198, 269]	Yellow solid* [198]	—	With $H_2O \rightarrow$ biphenyl [198]
11. 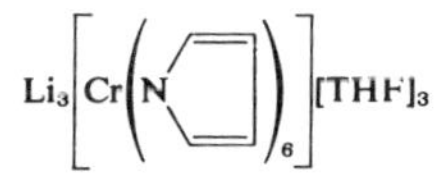	$RLi + CrCl_3$ [257, 258]	Red crystals* [257,258]	3.49 [257]	IR [258]
12. 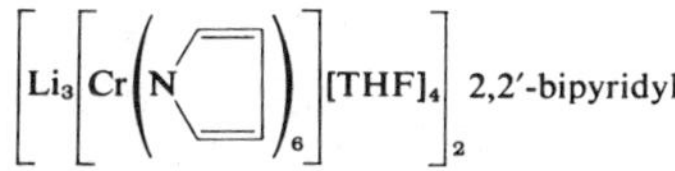	11 + 2,2′-bipyridyl [257,258]	Golden yellow* [257,258]	—	—
13. 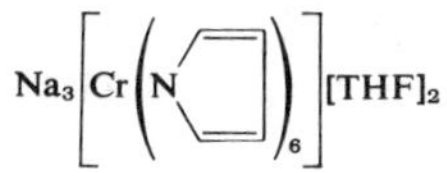	$NaR + CrCl_2$ [258]	Red crystals* [258]	—	—
14. 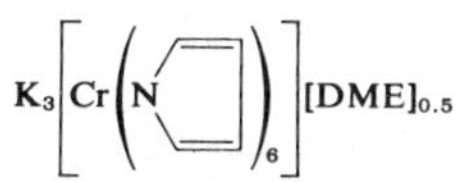	$KR + CrCl_3$ [258]	Red crystals* [258]	—	—
15. 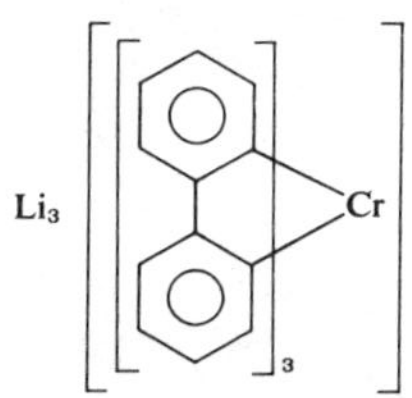	$LiR + CrCl_3$ [271]	Yellow-orange crystals** [271]	—	With H_2O [271]; with $HgCl_2 \rightarrow$ 2,2′-bis(chloromercury) biphenyl [271]
16. $Li_3[CrH(C_6H_5)_5][Et_2O]_3$	$1 + H_2$ [225,281]	Red crystals* [281]	—	With $HgCl_2 \rightarrow C_6H_5HgCl + (C_6H_5)_2Hg$ [281]
17. 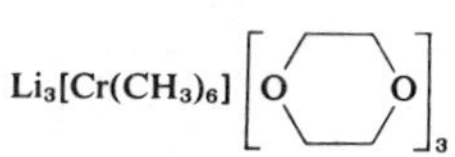	$LiR + CrCl_3$ [272, 273]	Red crystals* [272,273]	3.7 [273]	With H_2O, I_2 [273]; thermal decomposition [273]; x-ray [276, 276a]; with $C_6H_5C{\equiv}CC_6H_5$ [245]; ligand field parameters [273a]

(continued)

TABLE 1.6 *(continued)*

COMPOUND[a]	METHOD OF PREPARATION	STATE[b]	MAGNETIC MOMENT (μ_B)	REACTIONS, ETC.
18. $Li_3[Cr(1,4\text{-}C_4H_8)_3][Et_2O]_{2.5}$	$RLi + CrCl_3$ [274]	Yellow-orange solid*	3.9 [274]	With H_2O [274]
Type $M_2Cr(III)R_5$				
1. $Li_2[Cr(C_6H_5)_5][Et_2O]$	$Li_3[Cr(C_6H_5)_6]$ + THF or DME [275,277]	Blue-green crystals [277–279]	Cr(III) [277]	With $H_2 \rightarrow$ $Li_3Cr_2H_3(C_6H_5)_6$ 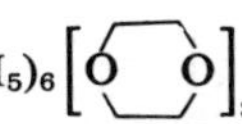 [279]; with $I_2 \rightarrow C_6H_5I + C_6H_5C_6H_5$ [277]; with $H_2O \rightarrow C_6H_6$ [277]
	$Li_3[Cr(C_6H_5)_6]$ + cyclopentadiene or fluorene [278,279]			
2. $Na_2[Cr(C_6H_5)_5][Et_2O]_3$	$NaR + CrCl_3$ [275]	Blue-green crystals* [275]	3.66 [275]	With H_2O, with I_2 [275]; with $HgCl_2 \rightarrow C_6H_5HgCl$ [275]; x-ray [283]
3. $Na_2[Cr(C_6H_5)_5][bipy]_2$	2 + 2,2′-bipyridyl	Olive-green solid [275]	3.90 [275]	With $HgCl_2 \rightarrow C_6H_5HgCl$ [275]
Type $MCr(III)R_4$				
1. $Li[Cr(C_6H_5)_4][DME]_4$	$Li[Cr(C_6H_5)_6]$ + $CrCl_3$ [231]	Cherry-red crystals* [231]	3.65 [231]	With $HgCl_2 \rightarrow C_6H_5HgCl$ [231]; with I_2 [231]; thermal decomposition [231]
2. $Na[Cr(C_6H_5)_4][DME]_4$	$Na_3[Cr(C_6H_5)_5]$ + $CrCl_3$ [231]	Cherry-red crystals* [231]	3.85 [231]	With $HgCl_2 \rightarrow C_6H_5HgCl$ [231]
3. $Li[Cr(2\text{-}CH_3OC_6H_4)_4][Et_2O]$	$RLi + CrCl_3$ [198]; $RLi + CrBr_2[THF]_2$ [198a]; $R_2Cr + LiR$ [198]	Red-brown crystals* [198]	3.89 [198]	Thermal decomposition to $[(2\text{-}CH_3OC_6H_4)_2Cr(II)]_n$ [198, 198a]; with $HgCl_2 \rightarrow 2\text{-}CH_3OC_6H_4HgCl$ [198]; with $I_2 \rightarrow 2\text{-}CH_3OC_6H_4I$ [198]

Compound	Preparation	Appearance	Magnetic moment	Reactions
Type $M_2Cr(II)R_4$				
1. 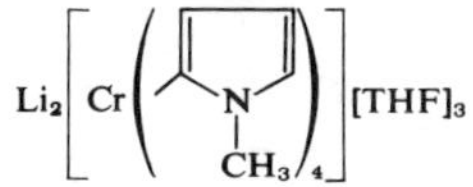	$RLi + CrBr_2(THF)_2$ [258]	Yellow crystals* [258]	4 unpaired electrons [258]	With H_2O [258]
2. $Li_2[Cr(C_6H_5)_4][THF]_4$	$RLi + CrCl_2(THF)_2$ [184]	Yellow crystals* [184]	4.73 [184]	With $HgCl_2 \rightarrow 4\text{-}C_6H_5HgCl$ [184]; with 2,2′-bipyridyl $\rightarrow Li_2[Cr(C_6H_5)_4][bipy]_2[THF]_2$ [184]
3. $Li_4[Cr_2(CH_3)_8][THF]_4$	$\Delta/Li_3Cr(CH_3)_6(S)_x$ [273]; $RLi + CrCl_2$ [273]; $(CH_2{=}CH{-}CH_2)_2Cr + LiR$ [265]	Yellow crystals* [273]	Diamagnetic 0.5 [273]	With H_2O, with I_2 [273,284]; x-ray structure [285]
4. $Li_4[Cr_2(1,4\text{-}C_4H_8)_4][THF]_4$	$LiR + CrCl_2$ [280]	Yellow crystals [280]	Diamagnetic [280]	—
5. $Li_4[Cr_2(1,4\text{-}C_4H_8)_4][Et_2O]_4$	$LiR + CrCl_2$ [280]	Yellow crystals [280]	Diamagnetic [280]	X-ray [286]
6. $Li_4[Cr_2(1,5\text{-}C_5H_{10})_4][LiBr \cdot 3Et_2O]_2$	$LiR + CrCl_3$ [274]	Yellow solid* [274]	Diamagnetic [274]	With H_2O [274]
Type $[MCr(II)R_3]_2$				
1. $Li_2[Cr_2(C_6H_5)_6][Et_2O]_3$	$Li_3[Cr(C_6H_5)_6] + CrCl_3$ [200]; $LiR + CrCl_3$ [4:1] [200]	Dark brown crystal* [200]	0.62 [200]	With $HgCl_2 \rightarrow C_6H_5HgCl + Hg_2Cl_2$ [200]; with I_2 [200]
2. $Li_2[Cr_2(C_6H_5)_6](LiCl)_3(Et_2O)_2$	$Li_3[Cr(C_6H_5)_6] + CrCl_3$ [200]	Brown crystal* [200]	0.83 [200]	With $HgCl_2 \rightarrow C_6H_5HgCl + Hg_2Cl_2$ [200]
3. $Na_2[Cr_2(C_6H_5)_6][Et_2O]_3$	$Na_2[Cr(C_6H_5)_5] + CrCl_3$ [200]	Dark brown crystal [200]	0.70 [200]	With $HgCl_2 \rightarrow C_6H_5HgCl + Hg_2Cl_2$ [200], with I_2 [200]
4. $Li_2[Cr_2(4\text{-}(CH_3)_2NC_6H_4)_6][Et_2O]_2$	$RLi + CrCl_3$ [4:1] [198,269]	Dark brown crystal [198]	0.35 [198]	—

[a] THF = tetrahydrofuran; bipy = 2,2′-bipyridyl; DME = 1,2-dimethoxyethane; C_4H_8 = $-(CH_2)_4-$; C_5H_{10} = $-(CH_2)_5-$.

[b] One asterisk indicates that satisfactory combustion analyses have been obtained for crystalline specimens of the compounds. Two asterisks indicate that satisfactory chemical analyses (i.e., hydrolysis, reaction with $HgCl_2$) have been obtained for crystalline specimens of the compounds.

hexa(alkyl)chromates(III) include hexa(methyl)chromate(III) [272,273], but no homologous ones [e.g., hexa(ethyl)chromate]; they also include the interesting spirocyclic compound **105** [274], which contains three chromocycles.

The mixed lithium hexa(organo)chromate complex (**106**) has been synthesized from tris(3-chlorophenyl)chromium(III), as outlined in Eq. (1-164):

$$3\ LiC_6H_5 + (3\text{-}ClC_6H_4)_3Cr(THF)_3 \xrightarrow{Et_2O} \underset{\mathbf{106}}{Li_3[(3\text{-}ClC_6H_4)_3Cr(C_6H_5)_3][Et_2O]_{2.5}} \quad (1\text{-}164)$$

thereby establishing a link between the σ-bonded octahedral tris(organo)-chromium(III) compounds and the octahedral σ-bonded hexa(organo)-chromates(III) [16].

Tris(lithium) hexa(phenyl)chromate(III) has also been prepared by the interaction of phenyllithium and $CrBr_3$, $CrCl_3(pyridine)_3$ or tris(acetyl-acetonato)chromium(III) [266], and, curiously enough, by the interaction of a large excess of phenyllithium and *chromous salts* [Eqs. (1-165) and (1-166) [266]. The mechanism whereby the chromium(II) is oxidized to chromium-(III) in the latter reaction is not known. It is possible that the coordinatively unsaturated tetra(phenyl)chromate(II) **107** is formed initially, Eq. (1-165). This compound reacts further with the phenyllithium, which is present in a large excess in the reaction mixture, to give the octahedral hexa(phenyl)-chromate(II) (**108**), which subsequently loses an electron to give the hexa-(phenyl)chromate(III), Eq. (1-166). Sodium hexa(phenyl)chromate(III) is

$$4\ LiC_6H_5 + CrCl_2 \longrightarrow \underset{\mathbf{107}}{Li_2[Cr(II)(C_6H_5)_4]} + 2\ LiCl \quad (1\text{-}165)$$

$$\mathbf{107} + \underset{(excess)}{LiC_6H_5} \longrightarrow \underset{\mathbf{108}}{Li_4[Cr(II)(C_6H_5)_6]} \xrightarrow{-e} Li_3[Cr(III)(C_6H_5)_6] \quad (1\text{-}166)$$

presumed to be the product formed by the interaction of sodium penta-(phenyl)chromate(III) and phenylsodium, Eq. (1-167) [275]. The product,

$$Na_2[Cr(C_6H_5)_5][Et_2O]_3 \underset{-NaC_6H_5}{\overset{+NaC_6H_5}{\rightleftharpoons}} Na_3[Cr(C_6H_5)_6][Et_2O]_x \quad (1\text{-}167)$$

which crystallizes from the reaction mixture in the form of fine yellow needles, has, however, not been obtained in a pure form since it appears that, in the absence of excess phenylsodium, the yellow solid loses phenylsodium to give the penta(phenyl)chromate, i.e., the reverse of Eq. (1-167) [275].

All the known σ-bonded hexa(organo)chromates(III) are exceedingly reactive toward air, water, and other protic solvents. They have been characterized and identified by analysis, chemical reactions (e.g., with $HgCl_2$, I_2, etc.), and magnetic moments. However, their precise formulation has presented certain problems since it was not possible from the chemical information available to say how many of the organic groups were directly bonded to the chromium center. The complexes could be thus either mixtures of the organolithium compound and the tris(organo)chromium(III) compound [i.e., (3 LiR)(R_3Cr(III)S_x] or distinct hexa(organo)chromate(III) complexes (i.e., $Li_3[CrR_6]S_x$). Results of a single crystal x-ray structure analysis of $Li_3Cr(CH_3)_6$(dioxane)$_3$ [276,276a] support the view that the complexes are best formulated as octahedral σ-bonded hexa(organo)chromate(III) compounds, i.e., **109**. The stability of such a molecule will depend

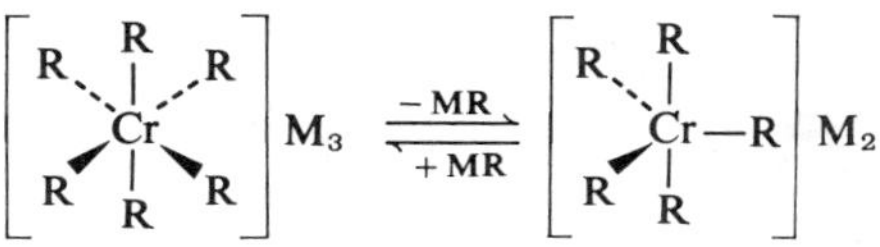
(1-168)

109 **110**

upon the steric requirements of the organic group R, the metal associated with R (i.e., Li or Na) and the solvating species.

In aryl compounds there will be steric interactions between substituents in the ortho positions of the aryl rings, which may explain why hexa(3- and 4-substituted-aryl)chromate(III) compounds may be prepared but not apparently the hexa(2-substituted-aryl)chromate(III) compounds.

In the case of the simple hexa(phenyl)chromate(III) complexes, the steric interaction may be relieved by the extrusion of metal–aryl (MR) to give the trigonal bipyramidal penta(phenyl)chromate(III) compound **110**. Reference has already been made to the fact that sodium hexa(phenyl)chromate(III) is only stable in the presence of excess phenylsodium, and that in the absence of the latter it loses phenylsodium to give the penta(phenyl)chromate(III) [**110**, Eq. (1-168), M = Na, R = C_6H_5]. Lithium hexa(phenyl)chromate(III), which is stable in diethyl ether, also loses one molecule of phenyllithium in a variety of other solvents (e.g., tetrahydrofuran, dioxane and 1,2-dimethoxyethane [275,277] to give the penta(phenyl)chromate(III) [**110**, Eq. (1-168), M = Li, R = C_6H_5].

The foregoing fragmentations of hexa(aryl)chromates(III) do not involve any change in the oxidation state of the chromium center; they involve merely a change in the coordination state of the chromium. The fragmentation of the hexa(alkyl)chromates(III), on the other hand, does involve a change in the valency state of the chromium [273,274]; for example, tris(lithium) hexa(methyl)chromate(III) undergoes fragmentation to the tetra(methyl)chromium(II) dimer **111**, Eq. (1-169) [273].

$$2\ Li_3[Cr(CH_3)_6][S]_x \xrightarrow{\text{ether}} \{Li_2[Cr(II)(CH_3)_4][Et_2O]_2\}_2 + 2\ LiCH_3 + 2\ \dot{C}H_3 \qquad (1\text{-}169)$$

2. Poly(organo)chromate Complexes of the Type $M_2Cr(III)R_5$

There are two penta(organo)chromates known to date: the bis(lithium) and bis(sodium) penta(phenyl)chromates. The lithium compound was first prepared by the interaction of cyclopentadiene, or fluorene on hexa(phenyl)-chromate(III), Eq. (1-170) [278,279]. The same compound was also prepared

$Li_3[Cr(C_6H_5)_6][Et_2O]_x$ +

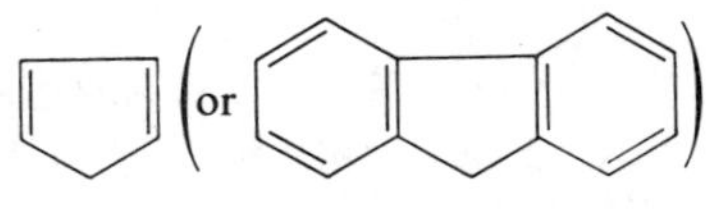

$\xrightarrow{\text{ether}}$

$$Li_2[Cr(C_6H_5)_5][Et_2O]_3 + LiC_5H_5 \text{ (or } LiC_{13}H_9) + C_6H_6 \qquad (1\text{-}170)$$

by the fragmentation of lithium hexa(phenyl)chromate(III) induced by the action of tetrahydrofuran or 1,2-dimethoxyethane [275,277], Eq. (1-168).

Sodium penta(phenyl)chromate has been prepared by the interaction of phenylsodium (in the presence of a small amount of phenyllithium) and chromic chloride in diethyl ether, Eq. (1-171) [275].

$$5\ NaC_6H_5 + CrCl_3(THF)_3 \longrightarrow Na_2[Cr(C_6H_5)_5][Et_2O]_3 + 3\ NaCl \qquad (1\text{-}171)$$

3. Poly(organo)chromate Complexes of the Type $MCr(III)R_4$

The scant information available concerning the few σ-bonded tetra(organo)-chromates(III) known suggests that, in the absence of stabilizing features, these compounds are in general unstable.

Tetra(2-methoxyphenyl)chromium(III) can be prepared either by the interaction of (2-methoxyphenyl)lithium and chromic (and chromous) halides in diethyl ether, Eq. (1-172) [198,198a], or by the action of (2-methoxyphenyl)lithium on bis(2-methoxyphenyl)chromium(II), Eq. (1-173)

$$CrCl_3 \text{ (or } CrCl_2) + (2\text{-}CH_3OC_6H_4)Li \xrightarrow{Et_2O} Li[Cr(2\text{-}CH_3OC_6H_4)_4][Et_2O]_3 \qquad (1\text{-}172)$$

$$[(2\text{-}CH_3OC_6H_4)_2Cr(II)]_n + (2\text{-}CH_3OC_6H_4)Li \longrightarrow Li[Cr(2\text{-}CH_3OC_6H_4)_4][Et_2O]_3 \qquad (1\text{-}173)$$

[198]. The mechanism of these reactions is unknown, particularly that of "oxidation" of the chromium(II) compounds to the poly(organo)chromium-(III) lithium complexes.

The interaction of phenyllithium or 4-dimethylaminophenyllithium and $CrCl_3$, in stoichiometric ratios of 4:1, in diethyl ether did not lead to the formation of the tetra(organo)chromate(III) complexes (**112**) but gave instead the dimeric tris(organo)chromate(II) complexes [**113**, Eq. (1-174), R = C_6H_5 or 4-$(CH_3)_2NC_6H_4$,] [198,200,269]. Similarly, the interaction of

$$\underset{}{8\,RLi + 2\,CrCl_3(THF)_3} \xrightarrow{Et_2O} \underset{\mathbf{112}}{2\,Li[CrR_4]} + 6\,LiCl \longrightarrow \underset{\mathbf{113}}{\{Li[Cr(II)R_3]\}_2} + RR \qquad (1\text{-}174)$$

sodium penta(phenyl)chromate or lithium hexa(phenyl)chromate and $CrCl_3$, in stoichiometric ratios of 4:1 or 2:1, respectively, did not give the tetra(phenyl)chromates(III) (**114** or **112**) but gave instead the dimeric tris-(organo)chromates(II) [**115** and **113**, Eqs. (1-175) and (1-176)] [200].

$$8\,Na_2[Cr(C_6H_5)_5]S_x + 2\,CrCl_3 \longrightarrow \underset{\mathbf{114}}{10\,Na[Cr(C_6H_5)_4]} + 6\,NaCl \longrightarrow \underset{\mathbf{115}}{5\{Na[Cr(C_6H_5)_3]\}_2} + 5\,C_6H_5C_6H_5 \qquad (1\text{-}175)$$

$$4\,Li_3[Cr(C_6H_5)_6]S_x + 2\,CrCl_3 \longrightarrow \underset{\mathbf{112}}{6\,Li[Cr(C_6H_5)_4]} + 6\,LiCl \longrightarrow \underset{\mathbf{113}}{3\{Li[Cr(C_6H_5)_3]\}_2} + 3\,C_6H_5C_6H_5 \qquad (1\text{-}176)$$

The formation of the dimeric tris(organo)chromates(II) in the foregoing reactions has been ascribed to the instability of the intermediate tetra(organo)-chromate(III) complexes **112** and **114** [198]. The coordination state of such intermediates is not known; however, in the absence of stabilizing ligands at the chromium center, two molecules could undergo reductive dimerization to give the tris(organo)chromate(II) dimer as outlined schematically in

$$[LiCrR_4]_2 \longrightarrow Li_2[R_3Cr{-}CrR_3] + RR \qquad (1\text{-}177)$$

Eq. (1-177). Were the complexes to be tetrahedral (*sd*^{3} hydribization), then loss of an electron would give directly the tetrahedral tetra(organo)chromium(IV) complex, Eq. (1-178). There are reports of the preparation of

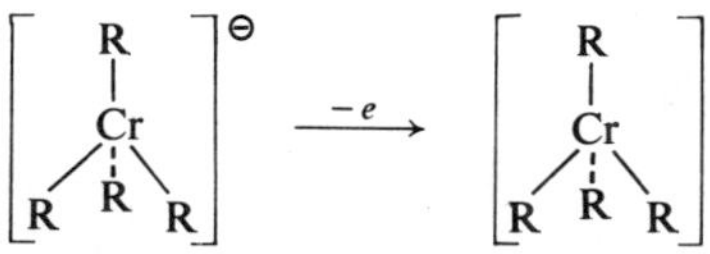

(1-178)

d^3 Cr(III) complex $\qquad$ d^2 Cr(IV) complex

tetra(organo)chromium(IV) compounds by the oxidation (or disproportionation) of tetra(organo)chromates(III) [20] (see Section H of this chapter).

The fate of tetra(organo)chromate(III) complexes will therefore depend upon the nature of the organic groups bonded to chromium (in particular their steric requirements), the nature of the solvent (in particular its donor capacity), and the nature of the reaction medium (in particular the presence of an oxidizing agent).

Hein and Schmiedeknecht [231] found that the donor properties of dimethylglycol ether (DME) were such as to stabilize sodium and lithium tetra(phenyl)chromates(III). Thus, whereas the interaction of bis(sodium) penta(phenyl)chromate or tris(lithium) hexa(phenyl)chromate and $CrCl_3$ in diethyl ether or tetrahydrofuran led consistently to the dimeric tris(organo)-chromium(II) complexes [Eqs. (1-175) and (1-176)] the same reactions in dimethylglycol ether (DME) led to the formation of crystalline $Na[Cr(C_6H_5)_4][DME]_4$ and $Li[Cr(C_6H_5)_4][DME]_4$ [231].

4. Poly(organo)chromate Complexes of the Type $M_2Cr(II)R_4$

There are two types of compounds of the general formula $M_2Cr(II)R_4$ known to date: the dimeric compounds $[M_2Cr(II)R_4]_2$ with magnetic moments corresponding to diamagnetic Cr(II) compounds and the monomeric compounds $M_2Cr(II)R_4$ with magnetic moments corresponding to high-spin Cr(II) compounds.

a. Dimeric Compounds

The interaction of methyllithium and $CrCl_3$ in diethyl ether leads to the formation of crystalline tris(lithium) hexa(methyl)chromate(III) [272]. This compound, though soluble in diethyl ether, undergoes a slow transformation in ether/dioxane solution to give lithium tetra(methyl)chromate(II) dimer, Eq. (1-179) [273]. The corresponding tetrakistetrahydrofuranate (**116**) can

$$12\ CH_3Li + 2\ CrCl_3 \xrightarrow{\text{rapid}} 2\ Li_3[Cr(CH_3)_6][Et_2O]_3 \xrightarrow{\text{slow}} Li_4[(CH_3)_4Cr{-}Cr(CH_3)_4][Et_2O]_4 + 2\ LiCH_3 + 2\ \dot{C}H_3 \quad (1\text{-}179)$$

be prepared either by (1) the direct interaction of methyllithium and chromous chloride in diethyl ether solution followed by crystallization of the product from tetrahydrofuran, Eq. (1-180) [273] or (2) by the interaction of π-bis-(allyl) chromium (II) dimer and methyllithium, Eq. (1-181) [265].

$$8\ CH_3Li + 2\ CrCl_2 \xrightarrow{Et_2O} [Li_2Cr(CH_3)_4]_2[Et_2O]_n \xrightarrow{THF} \underset{\mathbf{116}}{Li_4[(CH_3)_4Cr{-}Cr(CH_3)_4][THF]_4} \quad (1\text{-}180)$$

$$[(CH_2{=}CH{-}CH_2)_2Cr]_2 + 8\ CH_3Li \xrightarrow[(2)\ THF]{(1)\ Et_2O} Li_4[(CH_3)_4Cr{-}Cr(CH_3)_4][THF]_4 + 4\ LiC_3H_5 \quad (1\text{-}181)$$

The C_5 spirocyclic poly(organo)chromate(II) complex (**117**) is formed by the interaction of 1,5-dilithium pentane and $CrCl_3$ in ether, Eq. (1-182) [274].

$$LiCH_2(CH_2)_3CH_2Li + CrCl_3 \longrightarrow \underset{\mathbf{117}}{Li_4[(C_5H_{10})_2Cr—Cr(C_5H_{10})_2][LiBr\ 3\ Et_2O]_2} \qquad (1\text{-}182)$$

1,4-Dilithium butane, in contrast, reacts with $CrCl_3$ in diethyl ether to give the spirocyclic poly(organo)chromate(III) compound $Li_3Cr(C_4H_8)_3[Et_2O]_{2.5}$ (**105**) [274]. The C_4 spirocyclic chromate(II) compound (**118**) is formed by the interaction of 1,4-dilithium butane and chromous chloride, (1-183) [280].

$$2\ LiCH_2(CH_2)_2CH_2Li + CrCl_2 \xrightarrow[(2)\ THF]{(1)\ Et_2O} \underset{\mathbf{118}}{Li_4[(C_4H_8)_2Cr—Cr(C_4H_8)_2][THF]_4} \qquad (1\text{-}183)$$

Complete single crystal x-ray structure analyses have confirmed that the tetra(organo)chromates(II) (**116** and **118**) are dimeric; these structures are discussed more fully in Chapter 2, Section F,3,a.

b. Monomeric Compounds

Two monomeric tetra(organo)chromate(II) complexes have been prepared by the interaction of *N*-methylpyrrolyllithium or phenyllithium and $CrCl_2$:

$$4\ LiR + CrCl_2 \xrightarrow{THF} Li_2[CrR_4](THF)_n + 2\ LiCl \qquad (1\text{-}184)$$

R = 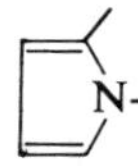 N—CH_3, $n = 3$ [258]; or R = C_6H_5, $n = 4$ [184]

Bis(lithium) tetra(phenyl)chromate(II) can also be prepared by the interaction of phenyllithium and a mixed tris(organo)chromium(III) compound [**119**], Eq. (1-185). The initial product in this reaction is assumed to be the mixed

$$\underset{\mathbf{119}}{(C_6H_5)_2CHCr(C_6H_5)_2(THF)_n} + 2\ LiC_6H_5 \longrightarrow$$

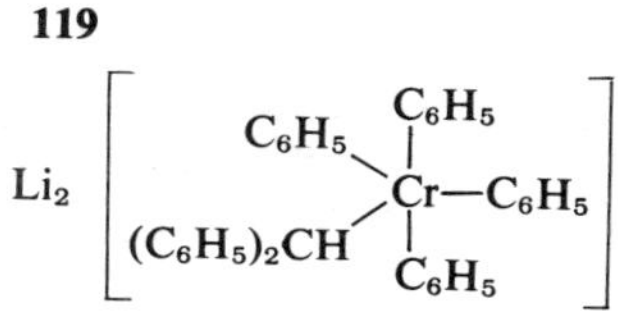

$$\longrightarrow Li_2[Cr(C_6H_5)_4] + \tfrac{1}{2}[(C_6H_5)_2CH]_2 \qquad (1\text{-}185)$$

120

penta(organo)chromium(III) complex **120**, which then undergoes fragmentation (by the selective homolysis of the diphenylmethyl–chromium bond) to give the stable bis(lithium) tetra(phenyl)chromate(II) [184].

5. Poly(organo)chromate Complexes of the Type $[MCr(II)R_3]_2$

The known tris(organo)chromates(II) have reduced magnetic moments and are therefore formulated as dimeric complexes $[MCr(II)R_3]_2$ containing

Cr–Cr bonds. These compounds have all been obtained indirectly from poly(organo)chromates(III) [presumably via the tetra(organo)chromate(III) complexes]. The preparation of compounds of the type $MCr(II)R_3$, by the interaction of organolithium or lithium or sodium poly(organo)chromates (III) with $CrCl_3$, in diethyl ether, has already been discussed in this chapter, Section G, 3. No direct synthesis of the present type of compound, by the interaction of chromous salts and organolithium compounds, has been described to date.

The tetra(potassium) hexa(pyrrolyl)chromate(II) complex (**121**) has been prepared as outlined in Eq. (1-186) [258]. This compound does not contain

(1-186)

121

a C–Cr σ-bond but a N–Cr bond; however, the fact that it has been prepared suggests that, with a suitable choice of organic group and of reaction conditions, it might be possible to prepare poly(organo)chromate(II) complexes of the type $M_4Cr(II)R_6$ and $M_3Cr(II)R_5$.

There is much fundamental work still to be done in the field of poly-(organo)chromate complexes; thus, it would be of interest to determine the factors that influence their stabilities and, in particular, that of the tetra-(organo)chromium(III) and chromate(II) complexes. It would also be interesting to determine the factors that influence the transformation of $MCr(III)R_4$ compounds into dimeric chromium(II) complexes, on the one hand, and into tetra(organo)chromium(IV) complexes, on the other. More information concerning the structures of the various poly(organo)chromate compounds, as determined by single crystal x-ray structure analyses, would also be of considerable help in understanding the various oxidation or reduction processes (or disproportionations) associated with many of these compounds.

H. Organochromium(IV) Compounds

Hagihara and Yamazaki isolated blue, crystalline chromium(IV) tetra-(*tert*-butoxide) from the reaction of π-bis(benzene)chromium(0) with di-(*tert*-butyl)peroxide, Eq. (1-187) [18]. Bradley and Basi prepared the chro-

$$\pi\text{-}(C_6H_6)_2Cr(0) \xrightarrow[90^\circ C/20\ hr]{(t\text{-}BuO)_2/\text{benzene}} \underset{(80\%)}{(t\text{-}BuO)_4Cr(IV)} + C_6H_5C_6H_5 \qquad (1\text{-}187)$$

mium(IV) amide **122** by the route outlined in Eq. (1-188), and found that it

$$LiN(CH_2CH_3)_2 + CrCl_3 \longrightarrow \underset{\textbf{122}\ (12\%)}{[(CH_3CH_2)_2N]_4Cr(IV)} \qquad (1\text{-}188)$$

reacted with tertiary alcohols and trialkylsilanols to give the corresponding alkoxides or siloxides [**123**, Eq. (1-189), R = R_3C or R_3Si] [19]. Bradley and

$$[(CH_3CH_2)_2N]_4Cr(IV) \xrightarrow{ROH} \underset{\textbf{123}}{(RO)_4Cr(IV)} + (CH_3CH_2)_2NH \qquad (1\text{-}189)$$

his colleagues also found that the chromium(IV) amide **122** could be prepared in high yield by the thermal disproportionation of the chromium(III) amide, Eq. (1-190). In this reaction, an electron is transferred from one chromium atom to another in a dimeric bridged complex [**124**, Eq. (1-190)] [287].

$2\,(R_2N)_3Cr(III) \longrightarrow$ 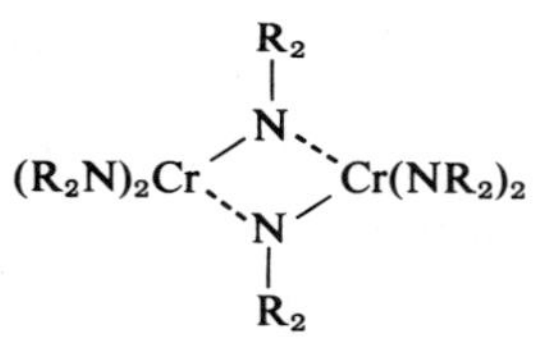 $\longrightarrow$

124

$$\underset{\substack{\textbf{122}\\(70\%)}}{(R_2N)_4Cr(IV)} + [(R_2N)_2Cr(II)]_n \qquad (1\text{-}190)$$

The driving forces in this disproportionation are the volatility and stability of the organochromium(IV) complexes. Thus, not only are the Cr(IV) compounds much more volatile than any of the other compounds present in the reaction mixture, but also they are much more stable since they possess a tetrahedral d^2 configuration [287].

The foregoing "organochromium(IV)" compounds all contain a hetero-atom bonded to chromium; but Wilkinson and his colleagues have been successful in preparing a tetra(organo)chromium(IV) compound containing a carbon atom bonded to chromium, i.e., tetrakis(trimethylsilylmethyl) chromium(IV) [**125**, Eq. (1-191)] [20,288]. The stability of this tetrakis-

$$4.5(CH_3)_3SiCH_2Li + CrCl_3(THF)_3 \xrightarrow{THF} Li\{Cr(III)[CH_2Si(CH_3)_3]_4\} \xrightarrow{(O)} \underset{\textbf{125}}{[(CH_3)_3SiCH_2]_4Cr(IV)} \qquad (1\text{-}191)$$

(alkyl)chromium(IV) compound can be ascribed to the absence of hydrogen atoms on the carbon beta to the metal center [20,288,289]. Thus the presence of a trimethylsilyl group on the carbon bonded to the chromium blocks two of the low-energy fragmentation routes characteristic of σ-bonded alkyl transition metal compounds, i.e., concerted β-metal-hydride elimination and β-hydrogen abstraction, Eqs. (1-192) and (1-193).

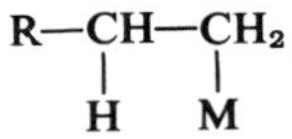 $\longrightarrow$ $RCH{=}CH_2 + MH$ (1-192)

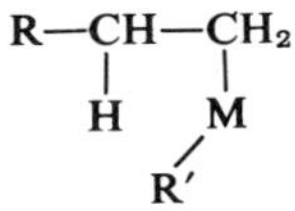 $\longrightarrow$ $RCH{=}CH_2 + R'H + M$ (1-193)

σ-Bonded alkyl transition metal compounds are unstable thermally because, apart from the simple homolytic bond-cleavage process, there are a variety of low-energy fragmentation routes whereby the carbon to metal bond may be destroyed, e.g., Eqs. (1-192) and (1-193). There are three ways in which these low-energy routes may be blocked, thereby stabilizing the σ-bonded alkyl transition metal (chromium) compound.

The first way is, as mentioned above, the use of "alkyl groups" lacking hydrogen (or other groups readily transferable to the metal) on the carbon atom β to the metal.

The second way is to block the coordination sites on the metal center, which are required for the metal-hydride elimination process, i.e., Eq. (1-192), with some suitable ligand, e.g., 2,2′-bipyridyl. In this case, however, the juxtaposition of the organic groups may be such as to permit fragmentation by β-hydrogen abstraction, i.e., Eq. (1-193).

The third way is to alter the geometry of the molecule so that not only are there no coordination sites on the metal center available for participation in the metal-hydride elimination process but also that the bulk of the organic groups and their juxtaposition are such as to prevent both concerted β-hydrogen abstraction and β-metal-hydride elimination. In this connection it is of interest to note that $(t\text{-Bu})_4Cr(IV)$ and $(n\text{-Bu})_4Cr(IV)$, which are both tetrahedral complexes containing hydrogen atoms on the carbon β to the chromium center, are stable at 30° and 0°C, respectively [22].

Since the synthesis of tetrakis(trimethylsilymethyl)chromium(IV) by Wilkinson and his collaborators, several tetra(alkyl)chromium(IV) compounds have been prepared (see Table 1.7). The compounds are all tetrahedral and have been prepared by the interaction of $CrCl_3(THF)_3$ or $Cr(t\text{-BuO})_4$ with organolithium or organomagnesium compounds.

1. From $CrCl_3(THF)_3$

Certain organolithium compounds and $CrCl_3(THF)_3$ interact (in stoichiometric ratios of 4:1) in hydrocarbon solvents or diethyl ether to give, by a disproportionation reaction, the tetra(organo)chromium(IV) compounds

[**126**, Eq. (1-194), R = $(CH_3)_3C$; $(CH_3)_3CCH_2$; $(C_2H_5)_3CCH_2$; 1-camphyl; 1-norbornyl] [21,22]. It would seem that this disproportionation reaction

$$4\,RLi + CrCl_3(THF)_3 \xrightarrow[\text{or ether}]{\text{pentane}} \underset{\mathbf{126}}{R_4Cr(IV)} \qquad (1\text{-}194)$$

does not occur in tetrahydrofuran [22], although it has been reported that the same stoichiometric ratios of reagents do interact in tetrahydrofuran to give, in certain cases, the lithium tetra(organo)chromate(III) complex, which may be subsequently oxidized to the tetra(organo)chromium(IV) compound, Eq. (1-195) {R = $CH_2Si(CH_3)_3$ [20]; $(C_6H_5)_3CCH_2$; $C_6H_5CH_2$ [192]}.

$$4\,RLi + CrCl_3(THF)_3 \xrightarrow{\text{THF}} Li[Cr(III)R_4] \xrightarrow{-e} Cr(IV)R_4 \qquad (1\text{-}195)$$

Similarly, certain organomagnesium halides and $CrCl_3(THF)_3$ (in stoichiometric ratios of 4:1) in diethyl ether give, possibly via the tetra(organo)-chromate(III) complexes, the tetra(organo)chromium(IV) compounds.

There are thus two processes involved in the synthesis of tetra(organo)-chromium(IV) compounds by the interaction of organolithium, or organomagnesium compounds with $CrCl_3(THF)_3$, the one a disproportionation reaction, the other an oxidation reaction.

The disproportionation reaction may proceed either by the mechanism outlined previously for the disproportionation of Cr(III) amides to Cr(IV) amides [see Eq. (1-190)] or via an octahedral dimeric complex as represented schematically in Eq. (1-196). The "oxidation reaction" may proceed via an

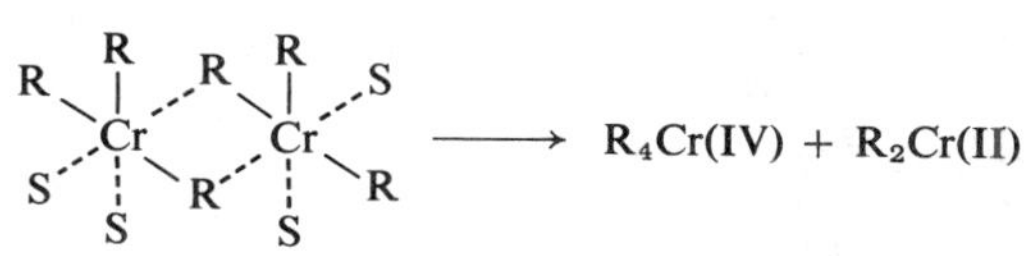

(1-196)

Cr(III)

intermediate tetrahedral tetra(organo)chromate(III) complex (**127**) [20], which by loss of an electron would give the tetra(organo)chromium(IV) compound directly, Eq. (1-197). It is pertinent to recall at this point that

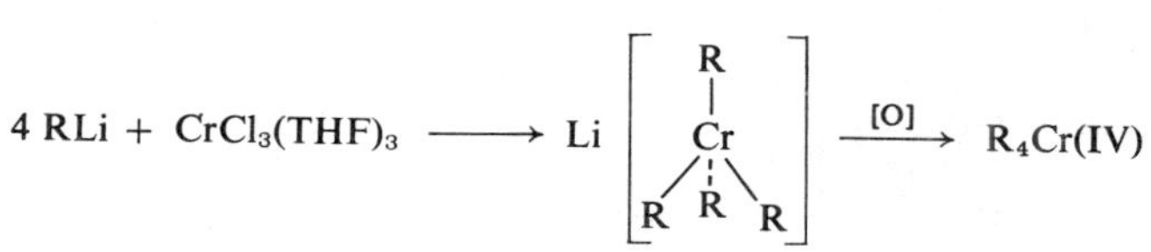

(1-197)

127

lithium chromate(III) complexes of the type $LiCrR_4$ may also undergo reductive dimerization to tris(organo)chromate(II) complexes (**128**); these,

TABLE 1.7

σ-Bonded Tetra(organo)chromium(IV) Compounds

COMPOUND[a]	METHOD OF PREPARATION	STATE[b]	MAGNETIC MOMENT (μ_B)	REACTIONS, ETC.
$[(CH_3)_3SiCH_2]_4Cr(IV)$	RLi + $CrCl_3$ in THF [20,288]; RMgBr + $CrCl_3$ in Et_2O [20, 288]	Purple crystals* [20]	2.9 [20]	With O_2 (reactive) [20]; with amines, phosphines, CS_2, ethylene $C_6H_5C{\equiv}CH$, $(C_6H_5)_3CBF_4$, CO (unreactive) [20]; with $(CN)_2C{=}C(CN)_2$, $CF_3C{\equiv}CCF_3$[20], NO, Cl_2, H_2O, HCl, H_2SO_4 [20]; electronic spectrum (19,000 cm^{-1}); IR and ESR [20]; polarography (−1.28 V) [20]
$[(CH_3)_3CCH_2]_4Cr(IV)$	RMgBr + $CrCl_3$ in Et_2O [192] RLi + $CrCl_3$ in pentane [22]	Maroon crystals* [192]	2.7 [192] 2.9 [22]	IR and ESR [192,291]; electronic spectrum [22,192]; mass spectrum [22]; with NO and [O] [192]; with amines, phosphines, $CH_2{=}CH_2$, CO (unreactive) [192]; with HCl and Cl_2 [192]
$[C_6H_5(CH_3)_2CCH_2]_4Cr(IV)$	RMgX + $CrCl_3$ in Et_2O [192] RLi + $CrCl_3$ in pentane [22]	Purple crystals* [192]	2.8 [192]	IR and ESR electronic spectra [192,291]; other reactions [192]; polarography, half-wave potential = −1.97 V [192], x-ray structure [292]
$[(C_6H_5)_3CCH_2]_4Cr(IV)$	RLi + $CrCl_3$ in THF [192]	Purple crystals* [192]	2.6 [192]	IR, ESR, and electronic spectra [192]
$[C_6H_5CH_2]_4Cr(IV)$	RLi + $CrCl_3$ in THF [192]	Red solution [192]		ESR [192]

$[CH_3]_4Cr(IV)$	RLi + Cr(t-BuO)$_4$ in pentane [22,192]	Red liquid		Thermally unstable [192]; electronic spectrum [192]
[1-Camphyl]$_4$Cr(IV)	RLi + $CrCl_3$ in pentane [21]	Red-brown crystals* [21]	2.96 [21]	Mass spectrum [21]; thermolysis and photolysis [21]; with air [21]
[1-Norbornyl]$_4$Cr(IV)	RLi + $CrCl_3$ in pentane [21]	Red-brown crystals* [21]	2.84 [21]	Mass spectrum; with air; thermolysis and photolysis [21]; ESR [291]
$[(C_2H_5)_3CCH_2]_4Cr(IV)$	RLi + $CrCl_3$ in pentane [22]			ESR [291]
$[(CH_3)_3C]_4Cr(IV)$	RLi + $CrCl_3$ in pentane or RLi + Cr (t-BuO)$_4$ [22]	Dark-red crystals* [22]	2.6 [22]	Mass, electronic, and ESR spectra [22, 291]; thermal decomposition [22]
$[(CH_3)_2CH]_4Cr(IV)$	RLi + Cr(t-BuO)$_4$ [22]	—		—
$[CH_3\text{-}CH_2CH(CH_3)]_4Cr(IV)$	As above [22]	—		—
$[CH_3(CH_2)_2CH_2]_4Cr(IV)$	As above [22]	—		ESR [291]
$[(CH_3CH_2)_2CH]_4Cr(IV)$	As above [22]	—		—
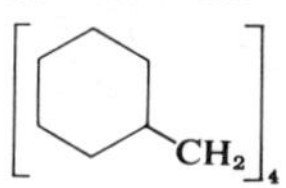 Cr(IV)	As above [22]	—		—

[a] THF = tetrahydrofuran; 1-camphyl = 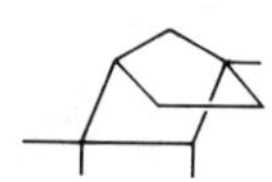 ; 1-norbornyl = 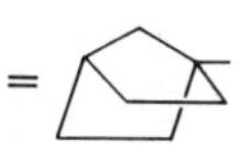 .

[b] An asterisk indicates that satisfactory combustion analyses have been obtained for crystalline specimens of the compounds.

in turn, can react with more of the organolithium reagent to give the dimeric tetra(organo)chromate(II) complex [**129**, Eq. (1-198)]. Kurras and Otto

$$\underset{}{2\,\mathrm{LiCrR_4}} \longrightarrow \underset{\mathbf{128}}{\mathrm{Li_2[R_3Cr(II)\text{-}Cr(II)R_3]}} + \mathrm{RR} \xrightarrow{2\,\mathrm{LiR}} \underset{\mathbf{129}}{\mathrm{Li_4[R_4Cr(II)\text{-}Cr(II)R_4]}} \qquad (1\text{-}198)$$

have described the preparation of such a tetra(alkyl)chromate(II) dimer [**129**, R = CH_3] by the interaction of methyllithium and $CrCl_3$ in diethyl ether [273].

It will be of interest to follow future developments in this area, in particular to ascertain whether distinct crystalline lithium tetra(organo)chromate(III) complexes can be oxidized directly to the tetra(organo)chromium(IV) compounds.

2. From Cr(*t*-BuO)$_4$

The interaction of chromium(IV) tetra(*tert*-butoxide) and various alkyl lithium compounds (ratio 4:1) *in pentane solution* furnishes by a methathetical reaction, the corresponding tetra(alkyl)chromium(IV) compounds in reasonable yields, e.g., Eq. (1-199) [R = $(CH_3)_3C$; $(CH_3)_2CH$; $CH_3(CH_2)_2CH_2$; $CH_3CH_2CH(CH_3)$; $(CH_3)_3CCH_2$; 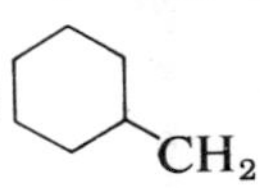 ; CH_3] [22,192]. These alkyl-

$$(t\text{-BuO})_4\mathrm{Cr(IV)} + 4\,\mathrm{RLi} \xrightarrow{\text{pentane}} \underset{60\%}{\mathrm{R_4Cr(IV)}} + 4\,(t\text{-BuO})\mathrm{Li} \qquad (1\text{-}199)$$

chromium(IV) compounds could not be prepared by the disproportionation reaction discussed under Section H,1 above [22]. Furthermore, in the reactions with $(t\text{-BuO})_4Cr$ the solvent, once again, plays a determining role. Thus, in the case of methyllithium, when donor solvents such as diethyl ether or tetrahydrofuran are used in place of pentane, the tetra(methyl)-chromium(IV) compound is not obtained. Also, Kochi and Tsuda found

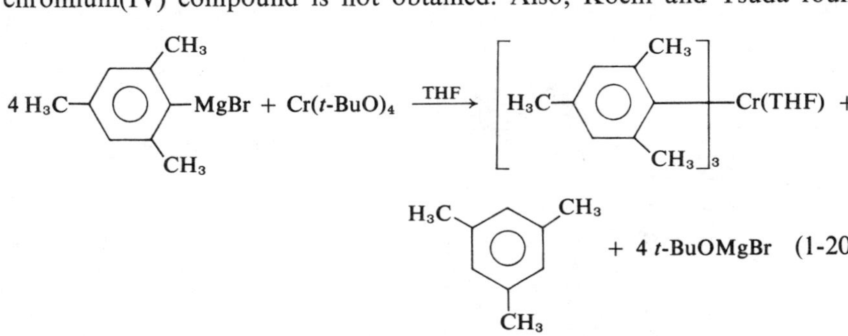

(1-200)

that certain aryl- and alkylmagnesium halides react with $Cr(IV)(t\text{-}BuO)_4$, in tetrahydrofuran solution, to give the corresponding σ-bonded tris(organo)-chromium(III) compounds, e.g., Eq. (1-200)–(1-203) [290].

$$8\ C_6H_5MgBr + 2\ Cr(t\text{-}BuO)_4 \xrightarrow{THF} 2\ (C_6H_5)_3Cr(THF)_3 + C_6H_5C_6H_5 + 8\ t\text{-}BuOMgBr \quad (1\text{-}201)$$

$$8\ CH_3CH_2MgBr + 2\ Cr(t\text{-}BuO)_4 \longrightarrow 2\ (CH_3CH_2)_3Cr(III)(THF)_n + 8\ t\text{-}BuOMgBr + CH_3CH_3 + CH_2{=}CH_2 \quad (1\text{-}202)$$

$$4\ CH_3MgBr + Cr(t\text{-}BuO)_4 \longrightarrow (CH_3)_3Cr(III)(THF)_n + CH_4 + 4\ t\text{-}BuOMgBr \quad (1\text{-}203)$$

The tetra(organo)chromium(IV) compounds that have been prepared by the two syntheses discussed above, are all tetra(alkyl) compounds and to date there are no reports of the synthesis of tetra(aryl)chromium(IV) compounds.

References

1. M. Berthelot, *Ann. Chim. Phys.* [4] **9**, 401 (1866).
2. G. M. Bennett and E. E. Turner, *J. Chem. Soc., London* **105**, 1057 (1914).
3. F. A. L. Anet and E. Leblanc, *J. Amer. Chem. Soc.* **79**, 2649 (1957).
4. "Gmelin's Handbook of Inorganic Chemistry," 8th ed., Suppl. Vol. 3. Verlag Chemie, Weinheim, 1971.
5. M. S. Kharasch and O. Reinmuth, "Grignard Reactions of Nonmetallic Substances." Prentice-Hall, Englewood Cliffs, New Jersey, 1954.
6. N. V. Kondyrev and D. A. Fomin, *Russ. J. Phys. Chem.* **47**, 190 (1915).
7. A. Job and R. Reich, *C. R. Acad. Sci.* **177**, 1439 (1923).
8. F. Hein, *Chem. Ber.* **52**, 195 (1919).
9. H. H. Zeiss, *Abstr., 126th Meet., Amer. Chem. Soc.* p. 29–O (1954).
10. H. H. Zeiss and M. Tsutsui, *J. Amer. Chem. Soc.* **79**, 3062 (1957).
11. W. Herwig and H. H. Zeiss, *J. Amer. Chem. Soc.* **79**, 6561 (1957).
12. W. Herwig and H. H. Zeiss, *J. Amer. Chem. Soc.* **81**, 4798 (1959).
13. H. H. Zeiss, "Organometallic Chemistry," Amer. Chem. Soc. Monogr. No. 147. Van Nostrand-Reinhold, Princeton, New Jersey, 1960.
14. E. O. Fischer and W. Hafner, *Z. Naturforsch. B* **10**, 665 (1955).
15. F. Hein and R. Weiss, *Z. Anorg. Allg., Chem.* **295**, 145 (1958).
16. F. Hein, J. Haehle, and G. Stolze, *Monatsber. Deut. Akad. Wiss. Berlin* **5**, 530 (1963).
17. E. O. Fischer and A. Maasböl, *Chem. Ber.* **100**, 2445 (1967).
18. N. Hagihara and H. Yamazaki, *J. Amer. Chem. Soc.* **81**, 3160 (1959).
19. J. S. Basi and D. C. Bradley, *Proc. Chem. Soc., London* p. 305 (1963).
20. W. Mowat, A. Shortland, G. Yagupsky, N. J. Hill, M. Yagupsky, and G. Wilkinson, *J. Chem. Soc., Dalton Trans.* p. 533 (1972).
21. B. K. Bower and H. G. Tennent, *J. Amer. Chem. Soc.* **94**, 2512 (1972).
22. W. Kruse, *J. Organometal. Chem.* **42**, C39 (1972).

23. H. H. Zeiss, P. J. Wheatley, and H. J. S. Winkler, "Benzenoid-Metal Complexes." Ronald Press, New York, 1966.
24. G. E. Coates, M. L. H. Green, and K. Wade, "Organometallic Compounds," Vol. 2. Methuen, London, 1968.
25. M. Tsutsui, M. Levy, A. Nakamura, M. Ichikawa, and K. Mori, *in* "Introduction to Metal-π-Complex Chemistry" (E. G. Rochow, ed.). Plenum, New York, 1970.
26. J. Haehle and G. Stolze, *Z. Naturforsch. B* **19**, 1081 (1964).
27. G. Stolze and J. Haehle, *J. Organometal. Chem.* **5**, 545 (1966).
28. R. P. A. Sneeden and H. H. Zeiss, *J. Organometal. Chem.* **4**, 355 (1965).
29. J. Haehle and G. Stolze, *J. Organometal. Chem.* **8**, 311 (1967).
30. F. Hein, *Chem. Ber.* **54**, 1905 (1921).
31. F. Hein, *Chem. Ber.* **54**, 2708 (1921).
32. F. Hein, *Chem. Ber.* **54**, 2727 (1921).
33. F. Hein, P. Kleinert, and E. Kurras, *Z. Anorg. Allg. Chem.* **289**, 229 (1957).
34. F. Hein and H. Mueller, *Chem. Ber.* **89**, 2722 (1956).
35. M. Tsutsui and H. H. Zeiss, *J. Amer. Chem. Soc.* **81**, 1367 (1959).
36. R. P. A. Sneeden, F. Glockling, and H. H. Zeiss, *J. Organometal. Chem.* **6**, 149 (1966).
37. R. P. A. Sneeden and H. H. Zeiss, unpublished results.
38. E. O. Fischer and W. Hafner, *Z. Anorg. Allg. Chem.* **286**, 146 (1956).
39. F. Hein and K. Kartte, *Z. Anorg. Allg. Chem.* **307**, 22 (1960).
40. W. Kleinwächter, *Monatsber. Deut. Akad. Wiss. Berlin* **3**, 640 (1961).
41. E. O. Fischer and J. Seeholzer, *Z. Anorg. Allg. Chem.* **312**, 244 (1961).
42. M. Tsutsui and M. N. Levy, *Proc. Chem. Soc., London* p. 117 (1963).
43. M. Tsutsui and G. Chang, *Can. J. Chem.* **41**, 1255 (1963).
44. G. Costa, A. Puxeddu, and G. Sartori, *Makromol. Chem.* **161**, 254 (1963).
45. R. W. Bush and H. R. Snyder, *J. Org. Chem.* **25**, 1240 (1960).
46. F. Hein and W. Kleinwächter, *Monatsber. Deut. Akad. Wiss. Berlin* **2**, 610 (1960).
47. G. A. Huttner, E. O. Fischer, and C. Elschenbroich, *J. Organometal. Chem.* **3**, 330 (1965).
48. Yu. A. Sorokin and G. G. Petukhov, *J. Gen. Chem. USSR* **35**, 2123 (1966).
49. G. G. Devyatykh, V. A. Umilin, and Yu. B. Zverev. *Bull. Acad. Sci. USSR, Div. Chem. Sci.* p. 1568 (1969).
49a. A. N. Nesmeyanov, L. P. Yur'eva, N. N. Zaitseva, E. I. Mysov, and A. M. Vainberg, *Bull. Acad. Sci. USSR, Div. Chem. Sci.* **22**, 169 (1973).
50. E. O. Fischer and J. Mueller, *Z. Naturforsch. B* **17**, 776 (1962).
51. E. O. Fischer and J. Mueller, *Z. Naturforsch. B* **18**, 413 (1963).
52. E. O. Fischer, J. Mueller, and P. Kuzel, *Rev. Chim., Acad. Rep. Pop. Roum.* **7**, 827 (1962).
53. E. O. Fischer and J. Mueller, *J. Organometal. Chem.* **1**, 464 (1964).
54. R. B. King and M. B. Bisnette, *Inorg. Chem.* **3**, 785 (1964).
55. E. O. Fischer, A. Reckziegel, J. Mueller, and P. Goeser, *J. Organometal. Chem.* **11**, P13 (1968).
55a. J. Müller and B. Mertschenk, *Chem. Ber.* **105**, 3346 (1972).
56. E. O. Fischer and J. Mueller, *Z. Naturforsch. B* **18**, 1137 (1963).
57. W. Herwig, W. Metlesics, and H. H. Zeiss, *J. Amer. Chem. Soc.* **81**, 6203 (1959).
58. H. H. Zeiss and W. Herwig, *J. Amer. Chem. Soc.* **80**, 2913 (1958).
59. M. Tsutsui and H. H. Zeiss, *J. Amer. Chem. Soc.* **81**, 6090 (1959).
60. H. H. Zeiss, *Bull. Soc. Chim. Fr.* p. 1500 (1963).

61. M. Tsutsui and H. H. Zeiss, *J. Amer. Chem. Soc.* **82**, 6255 (1960).
62. P. L. Timms, *Chem. Commun.* p. 1033 (1969).
62a. R. Middleton, J. R. Hull, S. R. Simpson, C. H. Tomlinson, and P. L. Timms, *J. Chem. Soc., Dalton Trans.* p. 120 (1973).
63. P. S. Skell, D. L. Williams-Smith, and M. J. McGlinchey, *J. Amer. Chem. Soc.* **95**, 3337 (1973).
64. H. P. Fritz, and L. Schaefer, *Z. Naturforsch. B* **19**, 169 (1964).
65. E. O. Fischer, K. Ulm, and P. Kuzel, *Z. Anorg. Allg. Chem.* **319**, 253 (1963).
66. E. O. Fischer and W. Hafner, *Z. Naturforsch. B* **10**, 140 (1955).
67. E. O. Fischer, W. Hafner, and H. O. Stahl, *Z. Anorg. Allg. Chem.* **282**, 47 (1955).
68. S. A. Keppie and M. F. Lappert, *J. Organometal. Chem.* **19**, P5 (1969).
69. R. B. King and A. Fronzaglia, *Inorg. Chem.* **5**, 1837 (1966).
70. R. B. King and A. Efraty, *J. Amer. Chem. Soc.* **94**, 3773 (1972).
71. J. Potenza, P. Giordano, D. Mastropaolo, A. Efraty, and R. B. King, *Chem. Commun.* p. 1333 (1972).
72. F. G. A. Stone and R. B. King, *Inorg. Syn.* **7**, 99 (1963).
73. E. O. Fischer, K. Ulm, and H. P. Fritz, *Chem. Ber.* **93**, 2167 (1960).
74. T. S. Piper and G. Wilkinson, *J. Inorg. Nucl. Chem.* **3**, 104 (1956).
75. T. S. Piper and G. Wilkinson, *J. Inorg. Nucl. Chem.* **2**, 38 (1956).
76. R. B. King, "Organometallic Syntheses," Vol. 1, p. 161. Academic Press, New York, 1965.
77. A. N. Nesmeyanov, R. B. Materikova, and N. S. Kochetkova, *Bull Acad. Sci. USSR, Div. Chem. Sci.* p. 1211 (1963).
78. A. N. Nesmeyanov, R. B. Materikova, N. S. Kochetkova, and R. V. Luk'yanova, *Bull. Acad. Sci. USSR, Div. Chem. Sci.* p. 896 (1969).
79. T. S. Piper and G. Wilkinson, *Chem. Ind.* (*London*) p. 1296 (1955).
80. A. Davison, J. A. McCleverty, and G. Wilkinson, *J. Chem. Soc., London* p. 1133 (1963).
81. J. M. Birmingham, *Advan. Organometal. Chem.* **2**, 365 (1964).
82. E. O. Fischer and J. Jira, *Z. Naturforsch. B* **8**, 327 (1953).
83. E. O. Fischer and W. Hafner, *Z. Naturforsch. B* **8**, 444 (1953).
84. V. F. Hnizda, Ger. Pat. 1,159, 949 (1963); *Chem. Abstr.* **60**, 12056 (1964).
85. G. Wilkinson, *J. Amer. Chem. Soc.* **76**, 209 (1954).
86. F. A. Cotton and G. Wilkinson, *Z. Naturforsch. B* **9**, 417 (1954).
87. A. F. Reid and P. C. Wailes, *Inorg. Chem.* **5**, 1213 (1966).
88. E. O. Fischer and K. Ulm, *Chem. Ber.* **95**, 692 (1962).
89. British Patent 768,083 (1957); *Chem. Abstr.* **51**, 15584 (1957).
90. E. O. Fischer and H. P. Koegler, *Z. Naturforsch. B* **13**, 197 (1958).
91. E. O. Fischer and S. Breitschaft, *Chem. Ber.* **99**, 2213 (1966).
92. E. O. Fischer and S. Breitschaft, *Angew Chem.* **75**, 94 (1963).
93. B. Nicholls and M. Whiting, *J. Chem. Soc., London* p. 551 (1959).
94. E. W. Abel, M. A. Bennett, R. Burton, and G. Wilkinson, *J. Chem. Soc., London* p. 4559 (1958).
95. J. D. Munro and P. L. Pauson, *J. Chem. Soc., London* p. 3475 (1961).
96. W. Strohmeier, *Chem. Ber.* **94**, 2490 (1961).
97. W. R. Jackson, B. Nicholls, and M. C. Whiting, *J. Chem. Soc., London* p. 469 (1960).
98. H. Werner and R. Prinz, *Chem. Ber.* **100**, 265 (1967).
99. P. L. Pauson, G. H. Smith, and J. H. Valentine, *J. Chem. Soc., C* p. 1061 (1967).
100. D. P. Tate, W. R. Knipple, and J. M. Augl, *Inorg. Chem.* **1**, 433 (1962).

101. H. Werner, K. Deckelmann, and U. Schönenberger, *Helv. Chim. Acta* **53**, 2002 (1970).
102. M. D. Rausch, G. A. Moser, E. J. Zaiko, and A. L. Lipman, Jr., *J. Organometal Chem.* **23**, 185 (1970).
103. K. Oefele, *Chem. Ber.* **99**, 1732 (1966).
104. P. L. Pauson and K. H. Todd, *J. Chem. Soc., C* p. 2315 (1970).
105. G. R. Knox, D. G. Leppard, P. L. Pauson, and W. E. Watts, *J. Organometal. Chem.* **34**, 347 (1972).
106. E. O. Fischer and K. Oefele, *Chem. Ber.* **90**, 2532 (1957).
107. E. O. Fischer and W. Fröhlich, *Chem. Ber.* **92**, 2995 (1959).
108. G. J. Leigh and E. O. Fischer, *J. Organometal. Chem.* **4**, 461 (1965).
109. A. J. Birch, P. E. Cross, and H. Fitton, *Chem. Commun.* p. 366 (1965).
110. E. O. Fischer, W. Berngruber, and C. G. Kreiter, *Chem. Ber.* **101**, 824 (1968).
111. R. B. King, *J. Organometal. Chem.* **8**, 139 (1967).
112. J. S. Ward and R. Pettit, *Chem. Commun.* p. 1419 (1970).
113. E. Koerner von Gustorf, O. Jaenicke, and O. E. Polansky, *Angew Chem., Int. Ed. Engl.* **11**, 532 (1972).
114. E. W. Abel and F. G. A. Stone, *Quart. Rev. Chem. Soc.* **24**, 498 (1970).
115. H. Behrens and R. Weber, *Z. Anorg. Chem.* **291**, 123 (1957).
116. H. Behrens and J. Vogl, *Chem. Ber.* **96**, 2220 (1963).
117. H. Behrens and W. Haag, *Chem. Ber.* **94**, 312 (1961).
118. R. Aumann and E. O. Fischer, *Chem. Ber.* **101**, 954 (1968).
118a. E. O. Fischer, *Pure Appl. Chem.* **30**, 353 (1972).
118b. E. O. Fischer and W. Kalbfuss, *J. Organometal. Chem.* **46**, C15 (1972).
119. M. Y. Darensbourg and D. J. Darensbourg, *Inorg. Chem.* **9**, 32 (1970).
120. E. Moser and E. O. Fischer, *J. Organometal. Chem.* **12**, P1 (1968).
121. J. A. Connor and E. M. Jones, *Chem. Commun.* p. 570 (1971).
121a. E. O. Fischer, G. Kreis, and F. R. Kreissl, *J. Organometal. Chem.* **56**, C37 (1973).
121b. J. A. Connor and E. M. Jones, *J. Chem. Soc., Dalton Trans.* p. 2119 (1973).
122. D. J. Darensbourg and M. Y. Darensbourg, *Inorg. Chem.* **9**, 1691 (1970).
123. D. J. Darensbourg and M. Y. Darensbourg, *Inorg. Chim. Acta* **5**, 247 (1971).
124. E. O. Fischer, C. G. Kreiter, H. J. Kollmeier, M. Müller, and R. D. Fischer, *J. Organometal. Chem.* **28**, 237 (1971).
125. H. J. Beck, E. O. Fischer, and C. G. Kreiter, *J. Organometal. Chem.* **26**, C41 (1971).
126. K. Oefele and M. Heberhold, *Angew. Chem., Int. Ed. Engl.* **9**, 739 (1970).
127. K. Oefele, *J. Organometal. Chem.* **12**, P42 (1968).
128. K. Oefele, *Angew. Chem., Int. Ed. Engl.* **7**, 950 (1968).
129. C. W. Rees and E. von Angerer, *Chem. Commun.* p. 420 (1972).
129a. B. Cetinkaya, M. F. Lappert, and K. Turner, *Chem. Commun.* p. 851 (1972).
130. R. P. A. Sneeden and H. H. Zeiss, *J. Organometal. Chem.* **47**, 125 (1973).
131. R. P. A. Sneeden and H. H. Zeiss, *J. Organometal. Chem.* **40**, 163 (1972).
132. F. Glockling, R. P. A. Sneeden, and H. H. Zeiss, *J. Organometal. Chem.* **2**, 109 (1964).
133. R. P. A. Sneeden and H. P. Throndsen, *J. Organometal. Chem.* **6**, 542 (1966).
134. E. Kurras, Dissertation, University of Jena (1959).
135. E. Kurras, *Monatsber. Deut. Wiss. Berlin* **5**, 378 (1963).
136. R. P. A. Sneeden and H. H. Zeiss, *J. Organometal. Chem.* **13**, 369 (1968).
137. F. Hein and K. Schmiedeknecht, *Z. Anorg. Allg. Chem.* **352**, 138 (1967).
138. E. Kurras, *Naturwissenschaften* **46**, 171 (1959).

139. K. Nishimura, H. Kuribayashi, A. Yamamoto, and S. Ikeda, *J. Organometal. Chem.* **37**, 317 (1972).
140. T. Arakawa, *Kogyo Kagaku Zasshi* **72**, 1733 (1969).
141. W. Seidel and K. Fischer, *Z. Chem.* **12**, 147 (1972).
142. R. P. A. Sneeden and H. H. Zeiss, *J. Organometal. Chem.* **16**, 449 (1969).
143. J. J. Daly, R. P. A. Sneeden, and H. H. Zeiss, *J. Amer. Chem. Soc.* **88**, 4287 (1966).
144. J. J. Daly and R. P. A. Sneeden, *J. Chem. Soc., A* p. 736 (1967).
145. H. P. Throndsen, W. Metlesics, and H. H. Zeiss, *J. Organometal. Chem.* **5**, 176 (1966).
146. F. A. L. Anet, *Can. J. Chem.* **37**, 58 (1959).
147. J. K. Kochi and F. F. Rust, *J. Amer. Chem. Soc.* **83**, 2017 (1961).
148. M. Ardon, K. Woolmington, and A. Pernick, *Inorg. Chem.* **10**, 2812 (1971).
149. J. K. Kochi and D. D. Davis, *J. Amer. Chem. Soc.* **86**, 5264 (1964).
150. J. K. Kochi and D. Buchanan, *J. Amer. Chem. Soc.* **87**, 853 (1965).
151. R. G. Coombes, M. D. Johnson, M. L. Tobe, N. Winterton, and L. Y. Wong, *Chem. Commun.* p. 251 (1965).
152. R. G. Coombes, M. D. Johnson, and N. Winterton, *J. Chem. Soc., London* p. 7029 (1965).
153. R. G. Coombes and M. D. Johnson, *J. Chem. Soc., A* p. 1805 (1966).
154. R. G. Coombes and M. D. Johnson, *J. Chem. Soc., A* p. 177 (1966).
155. C. E. Castro and W. C. Kray, Jr., *J. Amer. Chem. Soc.* **85**, 2768 (1963).
156. P. L. Julian, W. Cole, A. Magnani, and E. W. Meyer, *J. Amer. Chem. Soc.* **67**, 1728 (1945).
157. W. C. Kray, Jr. and C. E. Castro, *J. Amer. Chem. Soc.* **86**, 4603 (1964).
158. J. K. Kochi and D. M. Singleton, *J. Amer. Chem. Soc.* **90**, 1582 (1968).
159. D. M. Singleton and J. K. Kochi, *J. Amer. Chem. Soc.* **89**, 6547 (1967).
160. F. A. L. Anet and E. Isabelle, *Can. J. Chem.* **36**, 589 (1958).
161. J. K. Kochi, D. M. Singleton, and L. J. Andrews, *Tetrahedron* **24**, 3503 (1968).
162. D. H. R. Barton, N. K. Basu, R. H. Hesse, F. S. Morehouse, and M. M. Pechet, *J. Amer. Chem. Soc.* **88**, 3016 (1966).
163. J. K. Kochi and D. M. Singleton, *J. Org. Chem.* **33**, 1027 (1968).
164. J. K. Kochi and P. E. Mocaldo, *J. Amer. Chem. Soc.* **88**, 4094 (1966).
165. J. K. Kochi and J. W. Powers, *J. Amer. Chem. Soc.* **92**, 137 (1970).
166. H. Lux, D. Sarre, and I. Schaffelhofer, *Chem. Ber.* **97**, 2301 (1964).
167. D. Dodd and M. D. Johnson, *J. Chem. Soc., A* p. 34 (1968).
168. C. E. Castro and W. C. Kray, Jr., *J. Amer. Chem. Soc.* **88**, 4447 (1966).
169. C. T. Loo, L. Y. Goh, and S. H. Goh, *J. Chem. Soc., Dalton Trans.* p. 585 (1972).
170. W. Schmidt, J. H. Swinehart, and H. Taube, *J. Amer. Chem. Soc.* **93**, 1117 (1971).
171. H. Cohen and D. Meyerstein, *Chem. Commun.* p. 320 (1972).
172. W. Brackman, F. van de Craats, and P. J. Smit, *Rec. Trav. Chim. Pays-Bas* **83**, 1253 (1964).
173. J. H. Espenson and J. S. Shveima, *J. Amer. Chem. Soc.* **95**, 4468 (1973).
174. H. H. Zeiss and R. P. A. Sneeden, *Angew. Chem. Int. Ed. Engl.* **6**, 435 (1967).
175. T. Arakawa, *Kogyo Kagaku Zasshi* **70**, 1742 (1967); *Chem. Abstr.* **69**, 87158d (1968).
176. W. Oberkirch, Dissertation, Technische Hochschule, Aachen (1963).
177. E. Kurras and P. Klimsch, *Monatsber. Deut. Akad. Wiss. Berlin* **6**, 736 (1964).
178. E. Kurras and P. Klimsch, *Monatsber. Deut. Akad. Wiss. Berlin* **6**, 735 (1964).
179. H. Mueller, *Z. Chem.* **9**, 311 (1969).
179a. A. Rusina, H. P. Schröer, and A. A. Vlček, *Inorg. Chim. Acta* **3**, 411 (1969).

180. J. J. Daly, F. Sanz, R. P. A. Sneeden, and H. H. Zeiss, *Chem. Commun.* p. 243 (1971).
181. R. P. A. Sneeden and H. H. Zeiss, *J. Organometal. Chem.* **47**, 125 (1973).
182. J. J. Daly and F. Sanz, *J. Chem. Soc., Dalton Trans.* p. 2584 (1972).
183. J. J. Daly, F. Sanz, R. P. A. Sneeden, and H. H. Zeiss, *Helv. Chim. Acta* **56**, 503 (1973).
184. W. Seidel, K. Fischer, and K. Schmiedeknecht, *Z. Anorg. Allg. Chem.* **390**, 273 (1972).
185. J. J. Daly, F. Sanz, R. P. A. Sneeden, and H. H. Zeiss, *J. Chem. Soc., Dalton Trans.* p. 73 (1973).
186. W. Herwig and H. H. Zeiss, *J. Amer. Chem. Soc.* **79**, 6561 (1957).
187. W. Herwig and H. H. Zeiss, *J. Amer. Chem. Soc.* **81**, 4798 (1959).
188. G. Stolze, *J. Organometal. Chem.* **6**, 383 (1966).
189. P. Klimsch and E. Kurras, Ger. (East) Pat. 45,708 (1966).
190. R. P. A. Sneeden and H. H. Zeiss, unpublished results.
191. W. Mowat and G. Wilkinson, *J. Organometal. Chem.* **38**, C35 (1972).
192. W. Mowat, A. J. Shortland, N. J. Hill, and G. Wilkinson, *J. Chem. Soc., Dalton Trans.* p. 770 (1973).
193. G. M. Whitesides and W. J. Ehmann, *J. Amer. Chem. Soc.* **92**, 5625 (1970).
194. H. Müller, *Z. Chem.* **8**, 67 (1968).
195. K. Schmiedeknecht, W. Reichardt, and W. Seidel, *Z. Chem.* **11**, 432 (1971).
196. K. Madeja, E. Huesing, and N. H. Ahrens, *Z. Chem.* **7**, 22 (1967).
197. G. Baehr and H. Zohm, *Angew. Chem.* **75**, 94 (1963).
198. F. Hein and D. Tille, *Z. Anorg. Allg. Chem.* **329**, 72 (1964).
198a. F. Hein and D. Tille, *Monatsber. Deut. Akad. Wiss. Berlin* **4**, 414 (1962).
199. A. Tzschach and H. Nindel, *J. Organometal. Chem.* **24**, 159 (1970).
200. F. Hein and K. Schmiedeknecht, *J. Organometal. Chem.* **6**, 45 (1966).
201. J. Chatt and B. L. Shaw, *J. Chem. Soc., London* p. 5075 (1962).
202. J. Chatt, R. S. Coffey, A. Gough, and D. T. Thompson, *J. Chem. Soc., A* p. 190 (1968).
203. R. P. A. Sneeden and H. H. Zeiss, *J. Organometal. Chem.* **13**, 377 (1968).
204. R. P. A. Sneeden and H. H. Zeiss, *J. Organometal. Chem.* **27**, 89 (1971).
205. B. A. Dolgoplosk and E. I. Tinyakova, *Bull. Acad. Sci. USSR, Div. Chem. Sci.* p. 291 (1970).
206. O. K. Sharayev, A. V. Alferov, Ye. I. Tinyakova, and B. A. Dolgoplosk, *Vysokomol. Soedin A Ser.* **9**, 633 (1967).
207. C. Mestroni, A. Camus, and E. Mestroni, *J. Organometal. Chem.* **24**, 775 (1970).
208. A. B. Deshpande, R. V. Subramaniam, and S. L. Kapur, *J. Polym. Sci., Part A-1* **4**, 1799 (1966).
209. G. Sartori and G. Costa, *Z. Elektrochem.* **63**, 105 (1959).
210. J. C. Thomas, *Chem. Ind.* (*London*) p. 1388 (1956).
211. M. Sato and Y. Ishida, *Nippon Kagaku Zasshi* **89**, 1244 (1968); *Chem. Abstr.* **70**, 96898t (1969).
212. G. M. Whitesides and W. J. Ehmann, *J. Amer. Chem. Soc.* **90**, 804 (1968).
213. G. M. Whitesides and W. J. Ehmann, *J. Amer. Chem. Soc.* **91**, 3800 (1969).
214. J. L. Boston, D. W. A. Sharp, and G. Wilkinson, *J. Chem. Soc., London* p. 3488 (1962).
215. I. Hashimoto, M. Ryang, and S. Tsutsumi, *J. Org. Chem.* **33**, 3955 (1968).
216. Y. T. Huang, H. I. Tai, S. L. Ch'en, K. Y. Hou, and T. N. Ni, *Hua Hsueh Hsueh Pao* **31**, 149 (1965); *Chem. Abstr.* **63**, 5543 (1965).

217. W. Metlesics and H. H. Zeiss, *J. Amer. Chem. Soc.* **81**, 4117 (1959).
218. R. P. A. Sneeden, T. F. Burger, and H. H. Zeiss, *J. Organometal. Chem.* **4**, 397 (1965).
219. M. Ryang, K. Yoshida, and S. Tsutsumi, *Technol. Rep. Osaka Univ.* **16**, 205 (1966).
220. M. Ryang, K. Yoshida, H. Yokoo, and S. Tsutsumi, *Bull. Chem. Soc. Jap.* **38**, 636 (1965).
221. R. P. A. Sneeden and H. H. Zeiss, unpublished results.
222. T. F. Burger and H. H. Zeiss, unpublished results.
223. M. Tsutsui and J. Ariyoshi, *J. Polym. Sci. Part A* **3**, 1729 (1965).
224. W. Metlesics, P. J. Wheatley, and H. H. Zeiss, *J. Amer. Chem. Soc.* **84**, 2327 (1962).
225. F. Hein, H. Müller, and R. Weiss, *Handb. Int. Congr. Pure Appl. Chem., 17th, 1959* A 131, p. 21 (1959).
226. K. Clauss and H. Bestian, *Justus Liebigs Ann. Chem.* **654**, 8 (1962).
227. H. H. Zeiss and W. Herwig, *J. Amer. Chem. Soc.* **78**, 5959 (1956).
228. M. Tsutsui and H. H. Zeiss, *J. Amer. Chem. Soc.* **81**, 1367 (1959).
229. M. Tsutsui and M. N. Levy, *Z. Naturforsch. B* **21** 823 (1966).
230. H. H. Zeiss and W. Herwig, *Justus Liebigs Ann. Chem.* **606**, 209 (1957).
231. F. Hein and K. Schmiedeknecht, *J. Organometal. Chem.* **8**, 503 (1967).
232. R. P. A. Sneeden and H. H. Zeiss, *J. Organometal. Chem.* **20**, 153 (1969).
233. J. J. Daly, F. Sanz, R. P. A. Sneeden, and H. H. Zeiss, *J. Chem. Soc., Perkin Trans. 2* p. 1614 (1972).
234. F. Hein, J. Hähle, and G. Stolze, *Monatsber. Deut. Akad. Wiss Berlin* **5**, 528 (1963).
235. M. Sato and Y. Ishida, *Nippon Kagaku Zasshi* **91**, 173 (1970); *Chem. Abstr.* **73**, 25624b (1970).
236. M. Tsutsui, *Z. Chem.* **3**, 215 (1963).
237. M. Sato and Y. Ishida, *Bull. Chem. Soc. Jap.* **41**, 730 (1968).
238. R. P. A. Sneeden and H. H. Zeiss, *J. Organometal. Chem.* **28**, 259 (1971).
239. H. Dietrich, *Acta Crystallogr., Sect. A* **25**, Suppl., 160 (1969).
240. J. K. Becconsall, B. E. Job, and S. O'Brien, *J. Chem. Soc., A* p. 423 (1967).
241. D. G. H. Ballard, W. H. Janes, and T. Medinger, *J. Chem. Soc., B* p. 1168 (1968)
242. D. G. H. Ballard and T. Medinger, *J. Chem. Soc., B* p. 1176 (1968).
243. K. Clauss and C. Beermann, *Angew. Chem.* **71**, 627 (1959).
244. M. Michman and H. H. Zeiss, *J. Organometal. Chem.* **13**, P23 (1968).
245. M. Michman and H. H. Zeiss, *J. Organometal. Chem.* **15**, 139 (1968).
246. J. R. C. Light and H. H. Zeiss, *J. Organometal. Chem.* **21**, 391 (1970).
247. J. R. C. Light and H. H. Zeiss, *J. Organometal. Chem.* **21**, 517 (1970).
248. M. Michman and H. H. Zeiss, *J. Organometal. Chem.* **25**, 161 (1970).
249. M. Michman and H. H. Zeiss, *J. Organometal. Chem.* **25**, 167 (1970).
250. R. P. A. Sneeden and H. H. Zeiss, *J. Organometal. Chem.* **19**, 93 (1969).
251. R. P. A. Sneeden and H. H. Zeiss, *J. Label. Compounds* **5**, 54 (1969).
252. R. P. A. Sneeden and H. H. Zeiss, *Angew. Chem., Int. Ed. Engl.* **7**, 951 (1968).
253. R. P. A. Sneeden and H. H. Zeiss, *J. Organometal. Chem.* **26**, 101 (1971).
254. G. Brauer, "Handbuch der präparativen anorganischen Chemie," Vol. II, p. 1169. Enke, Stuttgart, 1962.
255. G. Stolze and J. Hähle, *J. Organometal. Chem.* **7**, 301 (1967).
256. M. Tsutsui and H. H. Zeiss, *J. Amer. Chem. Soc.* **82**, 6255 (1960).
257. D. Tille, *Z. Naturforsch. B* **21**, 1239 (1966).
258. D. Tille, *Z. Anorg. Allg. Chem.* **384**, 136 (1971).
259. U. Giannini, E. Pellino, and M. P. Lachi, *J. Organometal. Chem.* **12**, 551 (1968).
260. E. O. Fischer, H. P. Koegler, and P. Kuzel, *Chem. Ber.* **93**, 3006 (1960).

261. H. P. Fritz, H. Keller, and E. O. Fischer, *Naturwissenschaften* **48**, 518 (1961).
262. E. O. Fischer and K. Ulm, *Z. Naturforsch. B* **15**, 59 (1960).
263. E. O. Fischer and K. Ulm, *Chem. Ber.* **94**, 2413 (1961).
264. T. Aoki, A. Furusaki, Y. Tomiie, K. Ono, and K. Tanaka, *Bull. Chem. Soc. Jap.* **42**, 545 (1969).
265. E. Kurras and K. Zimmermann, *J. Organometal. Chem.* **7**, 348 (1967).
266. F. Hein, R. Weiss, B. Heyn, K. H. Barth, and D. Tille, *Monatsber. Deut. Akad. Wiss. Berlin* **1**, 541 (1959).
267. E. O. Fischer and U. Piesberger, *Z. Naturforsch. B* **11**, 758 (1956).
268. R. D. Fischer, *Chem. Ber.* **93**, 165 (1960).
269. F. Hein and D. Tille, *Monatsber. Deut. Akad. Wiss. Berlin* **4**, 725 (1962).
270. G. Brauer, "Handbuch der präparativen anorganischen Chemie," Vol. II, p. 1203. Enke, Stuttgart, 1962.
271. G. Wittig and K. D. Rümpler, *Justus Liebigs Ann. Chem.* **751**, 1 (1971).
272. E. Kurras and J. Otto, *Monatsber. Deut. Akad. Wiss. Berlin* **5**, 705 (1963).
273. E. Kurras and J. Otto, *J. Organometal. Chem.* **4**, 114 (1965).
273a. H. Dunken and G. Marx, *Z. Chem.* **6**, 436 (1966).
274. E. Kurras and J. Otto, *J. Organometal. Chem.* **3**, 479 (1965).
275. F. Hein and K. Schmiedeknecht, *J. Organometal. Chem.* **5**, 454 (1966).
276. J. Krausse, *Proc. Int. Conf. Coord. Chem., 9th, 1966* p. 168 (1966).
276a. J. Krausse and G. Marx, *J. Organometal. Chem.* **65**, 215 (1974).
277. F. Hein, B. Heyn, and K. Schmiedeknecht, *Monatsber. Deut. Akad. Wiss. Berlin* **2**, 552 (1960).
278. F. Hein and B. Heyn, *Monatsber. Deut. Akad. Wiss. Berlin* **4**, 220 (1962).
279. F. Hein and B. Heyn, *Monatsber. Deut. Akad. Wiss. Berlin* **4**, 223 (1962).
280. E. Kurras and J. Otto, *Proc. Int. Conf. Coord. Chem., 9th, 1966* p. 166 (1966).
281. F. Hein and R. Weiss, *Naturwissenschaften* **46**, 321 (1959).
282. F. Hein and E. Brink, *Monatsber. Deut. Akad. Wiss. Berlin* **4**, 209 (1962).
283. E. Müller, J. Krausse, and K. Schmiedeknecht, *J. Organometal. Chem.* **44**, 127 (1972).
284. E. Kurras and J. Otto, *Monatsber. Deut. Akad. Wiss. Berlin* **6**, 267 (1964).
285. J. Krausse, G. Marx, and G. Schödl, *J. Organometal. Chem.* **21**, 159 (1970).
286. J. Krausse and G. Schödl, *J. Organometal. Chem.* **27**, 59 (1971).
287. J. S. Basi, D. C. Bradley, and M. H. Chisholm, *J. Chem. Soc., A* p. 1433 (1971).
288. G. Yagupsky, W. Mowat, A. Shortland, and G. Wilkinson, *Chem. Commun.* p. 1369 (1970).
289. M. R. Collier, M. F. Lappert, and M. M. Truelock, *J. Organometal. Chem.* **25**, C36 (1971).
290. T. Tsuda and J. K. Kochi, *Bull. Chem. Soc. Jap.* **45**, 648 (1972).
291. G. E. Ward, W. Kruse, B. K. Bower, and J. C. W. Chien, *J. Organometal. Chem.* **42**, C43 (1972).
292. V. Gramlich and K. Pfeffercorn, *J. Organometal. Chem.* **61**, 247 (1973).

Chapter 2

CHARACTERIZATION AND IDENTIFICATION

A. Introduction

The isolation, characterization, and identification of pure organochromium compounds is perhaps one of the most challenging aspects of their chemistry. In many instances the compounds may be isolated and purified by conventional techniques. In other cases, however, additional difficulties are occasioned by the thermal instabilities of the compounds and by their extreme sensitivity to air and moisture. The characterization and identification of a particular compound involves a knowledge not only of the empirical formula but also of the ligands associated with the chromium center and the valency state of the latter (i.e., 0, I, II, III, or IV). Only a restricted number of methods are available for the identification of the ligands associated with chromium and many of these have been developed for the solution of the specific problem in hand. The use of NMR spectroscopy is restricted to the diamagnetic Cr(0) and Cr(II) compounds, and ESR spectroscopy has only been of use in the identification of Cr(I) and Cr(IV) compounds. The valency state of the chromium center can usually be determined by a combination of chemical methods and magnetic measurements. The most unambiguous method for the characterization and identification of a particular compound is single crystal x-ray analysis. The structure of at least one member of each of the known classes of organochromium compounds has been established by a full x-ray analysis.

The problems associated with the combustion analysis of air- and moisture-sensitive organometallic compounds have been reviewed and discussed by Schwarzkopf and Schwarzkopf [1].

The present chapter treats of the general methods available for the separation, characterization, and identification of the various known classes of organochromium compounds together with the relevant x-ray structure data. Single crystal x-ray structure analyses provide information concerning the conformation of molecules in the solid state and this need not be the same as that of the molecule in solution or in the gaseous state. The existence of such fluxional isomers is well documented. Therefore, in the discussions concerning the isomeric species possible for a given molecule, it is assumed that the latter is free of the constraining forces associated with the crystalline state.

B. π-Bis(arene)chromium Compounds

Most syntheses of, and reactions involving, π-bis(arene)chromium compounds lead to mixtures of products (see Chapter 1, Section B,1 and 2) [e.g., 2,3]. The main problems are, therefore, the separation of the individual products and the identification of the arene components present therein.

1. π-Bis(arene)chromium(0) Compounds: Separation and Pyrolysis

π-Bis(arene)chromium(0) compounds are, in general, extremely sensitive to air (O_2) and are readily oxidized by a variety of compounds (e.g., alkyl halides and halogens) to the corresponding π-bis(arene)chromium(I) complexes. The π-bis(arene)chromium compounds may be distilled [4,5], sublimed, or recrystallized (from inert solvents); however, all these manipulations must be performed in an inert atmosphere. The lower homologs of π-bis(arene)chromium(0) (e.g., arene = ethylbenzene, toluene) have been separated by gas chromatography [6,7]. The less air-sensitive substituted π-bis(arene)-chromium(0) compounds (e.g., arene = $C_6H_5COOCH_3$) can be separated by column chromatography on deoxygenated alumina [3].

The pyrolysis of simple π-bis(arene)chromium(0) compounds leads to the arene (99%) and a chromium mirror, Eq. (2-1). The pyrolysis may be effected

$$\pi\text{-}(C_6H_6)_2Cr(0) \xrightarrow[\textit{in vacuo}]{300°\text{–}500°C} 2\,C_6H_6 + Cr\ (\text{mirror}) \qquad (2\text{-}1)$$

in a reaction tube [8–10], in the injection block of a gas chromatography apparatus [3] or in a glass capillary furnace connected to a G.L.C. apparatus [7]. The resulting organic products may be readily separated and identified by gas chromatographic and spectroscopic methods.

The thermal decomposition of alkyl-substituted π-bis(arene)chromium(0) compounds is accompanied by some thermal cracking of the arene moieties thereby leading to the formation of other alkyl aromatic derivatives and carbon [11]. Thus, the pyrolysis products of pure π-$(C_2H_5C_6H_5)_2Cr$ were shown to consist of benzene (4.8%), toluene (0.2%), ethylbenzene (70.7%), diethylbenzene (23.7%), and other compounds (0.6%) [12]. The thermal stability of alkyl-substituted π-(arene)chromium(0) compounds rises with increasing alkyl substitution; thus, $[C_6H_6]_2Cr < [CH_3C_6H_5]_2Cr < [C_2H_5C_6H_5]_2Cr < [(CH_3)_3C_6H_3]_2Cr$ [12,13].

The mechanism of this pyrolysis has been the subject of several theoretical studies, e.g., [14]; however, the precise details of the mechanism remain obscure, and it is still not known whether an unimolecular homogeneous gas-phase decomposition is involved. The heat of decomposition of π-bis-(benzene)chromium(0) (ΔH_r) has been calculated to be 79.1 (78) kcal mole^{-1}, based on a heat of formation (ΔH_f) of 55 (56) kcal mole^{-1} [15,16], the mean bond-dissociation energy ($\bar{D}_o$) is thus 39.5 (39) kcal mole^{-1}. However, this value has little significance, since synchronous multifragment decompositions [as in Eq. (2-1)] are highly improbable. Such decompositions are more likely to proceed stepwise and there is no a priori method by which the relative magnitudes of the bond-dissociation energies in each step may be estimated. Clearly a closer investigation of the gas-phase decomposition of the readily available π-bis(arene)Cr(0) compounds is called for.

2. π-Bis(arene)chromium(I) Compounds: Separation and Cleavage

In view of the practical difficulties involved in the handling of π-bis(arene) chromium(0) compounds, it is often more expedient to convert them to the corresponding cationic chromium(I) compounds by simple air oxidation, as these are air- and water-stable, and form salts with a variety of anions (e.g., $PF_6^{\ominus}$, $B(C_6H_5)_4^{\ominus}$, $I^{\ominus}$, etc.). Mixtures of π-bis(arene)chromium(I) salts may be readily separated by selective extraction [e.g., 17], by crystallization or by paper chromatography using aqueous butanol as solvent [18, 19].

The arene moieties may be liberated from π-bis(arene)chromium(I) compounds either by pyrolysis of the crystalline salts (e.g., iodide or tetra-phenylborate)or by acidic or oxidative disrupture of chromium(I) hydroxides.

The pyrolysis of π-bis(arene)chromium(I) iodides proceeds stepwise [20–22]; at low temperatures (ca. 200°C) disproportionation occurs, Eq. (2-2),

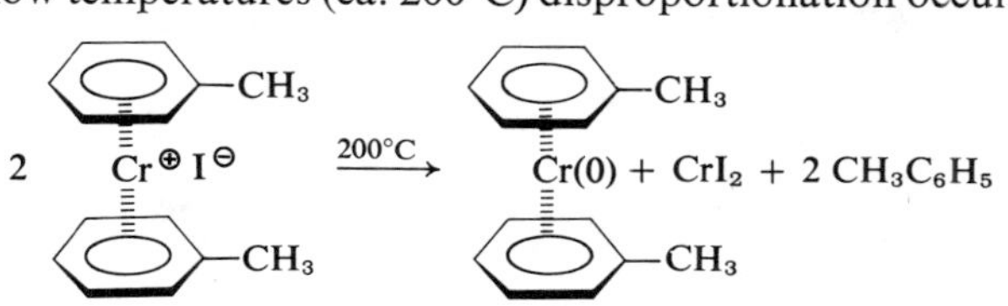

(2-2)

followed at higher temperatures (300°–500°C) by the decomposition of the π-bis(arene)chromium(0) compound to the free arene and chromium metal, Eq. (2-1).

The thermal decomposition of π-bis(arene)chromium(I) tetraphenylborates also proceeds stepwise; at low temperatures (ca. 200°C), the free arenes are

$$\pi\text{-}(C_6H_6)_2Cr^{\oplus}{}^{\ominus}B(C_6H_5)_4 \xrightarrow{200°C} 2$$ 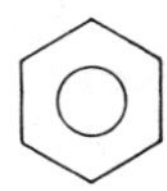 $$+\ CrB(C_6H_5)_4 \qquad (2\text{-}3)$$

formed [Eq. (2-3)]. At higher temperatures the $CrB(C_6H_5)_4$ decomposes to give biphenyl together with some benzene and $(CrB)_n$ [22,23]. Therefore, in the usual laboratory pyrolysis of π-bis(arene)chromium(I) tetraphenylborates, the arene products are always contaminated with benzene and biphenyl.

In both the above pyrolysis reactions, the final arene products are contaminated with "cracking products" from the arenes [23]. Thus, in the case of alkyl-substituted π-bis(arene)chromium(I) salts, these consist of homologous alkylarenes and higher-boiling colored aromatic condensation products. Therefore, it is better and more expedient to use acidic or oxidative decomposition of π-bis(arene)chromium(I) hydroxides. In the acidic decomposition, dilute HCl, H_2SO_4, and H_3PO_4 may be used, and the volatile arenes are trapped in a supernatant layer of hexane [2]. In the oxidative decomposition acidic H_2O_2 may be used [2] and when deuterated arenes are involved, any possible H/D scrambling may be avoided by the use of aqueous $Na_2Cr_2O_7$ [24,25].

3. NMR Spectroscopy

Bis-π-(arene)chromium(0) compounds are diamagnetic and thus NMR spectroscopic techniques can be used in the identification of the arene constituents [3,26].

The resonance signals of the protons of the aromatic ring are shifted upfield by 2 to 3 ppm. This is illustrated in the case of **1**. The aromatic protons of the complexed benzene ring have δ 4.16 (in contrast to 7.37 for the protons in benzene); in the triphenylmethylcarbinol part, the protons of the uncomplexed rings have δ 7.39 and 7.06, while those of the complexed

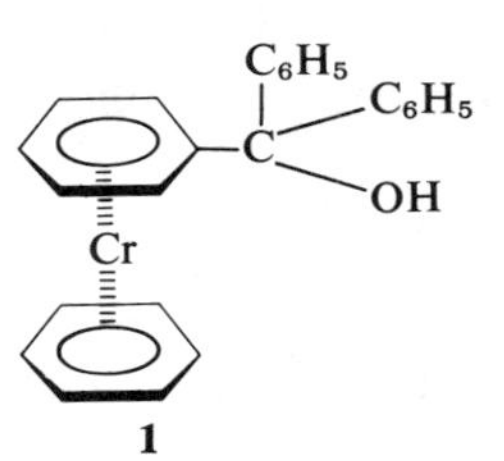

1

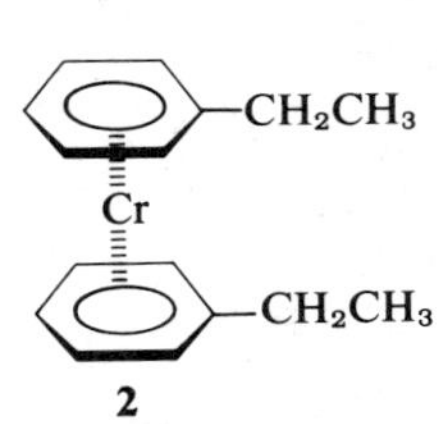

2

ring have δ 4.36 and 4.16 (the H of OH is at δ 3.42) [3]. This upfield shift of the arene protons can be interpreted in terms of a decrease in the deshielding effect of the π electrons occasioned by π-bonding of the arene ring to the chromium center.

In alkyl-substituted π-bis(arene)chromium(0) compounds, the resonance signals of the protons of the alkyl group are also shifted, slightly upfield (0.2 to 0.3 ppm). This is illustrated in the case of **2**; the aromatic protons are at δ 4.19 (in contrast to 7.23 for ethylbenzene); the CH_2 protons give a quadruplet (J = 7.5 Hz) centered at δ 2.30 (2.68 for ethylbenzene); the CH_3 protons give a triplet (J = 7.5 Hz) centered at δ 1.10 (1.25 for ethylbenzene) [26].

NMR spectroscopy is, however, of limited application owing to the low solubility of some π-bis(arene)chromium(0) compounds and to the extreme ease whereby they are oxidized by traces of air giving paramagnetic impurities.

4. ESR Spectroscopy

The ESR spectra of π-bis(arene)chromium(I) compounds have been studied and discussed [22,27–30]. π-Bis(benzene)chromium(I) iodide has a g value of 1.986 with a hyperfine splitting constant (a_H) of 3.5 gauss, the unpaired electron being mainly centered on chromium. The results of a study of the ESR spectra of alkyl-substituted π-bis(arene)chromium iodides {e.g., $[C_6H_6]_2CrI$, $[CH_3C_6H_5]_2CrI$, $[m\text{-}(CH_3)_2C_6H_4]_2CrI$, $[(CH_3)_4C_6H_2]_2CrI$} show that a_H rises progressively with the number of alkyl substituents (a_H = 3.5, 3.55, 3.60, 4.34, respectively). This has been ascribed to an increase in the electron density in the arene rings occasioned by the increasing number of alkyl substituents [29]. The usefulness of ESR spectroscopy in the actual identification of the arene constituents of an unknown π-bis(arene)-chromium(I) compound remains doubtful, though it is an excellent method for proving their presence in a given reaction mixture. In this way, it has been used to detect the formation of π-bis(arene)chromium(I) compounds in the reaction of tris(acetylacetonato)chromium(III) with various alkylaluminum compounds in hydrocarbon solvents (alkylbenzenes) [31,32]. Thus the ESR spectra for the system outlined in Eq. (2-4) indicate the formation of minor

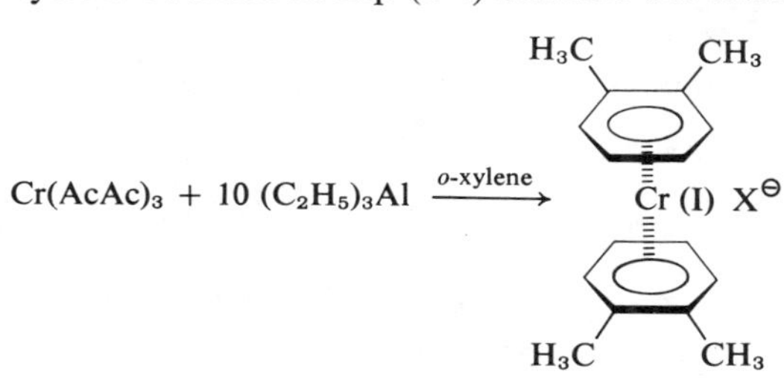

(2-4)

amounts of the π-bis(arene)chromium(I) compound [31]. Satisfactory yields of bis(arene)chromium(I) compounds can, however, be obtained by the interaction of $CrCl_3$ and $(C_2H_5)_3Al$ in aromatic hydrocarbon solvents (see Chapter 1, Section B,2).

5. IR and UV Spectra

The infrared, ultraviolet and visible spectra of π-bis(arene)chromium(0) and Cr(I) compounds have been extensively studied and discussed [e.g., 33–40]. These spectra, which are of fundamental importance to the interpretation of bonding within these molecules, are better suited to the characterization of π-complexes rather than to the unambiguous identification of the individual arene constituents. There are six main absorption regions in the IR spectra of π-bis(arene)chromium complexes:

(i)	3000–3060 cm^{-1}	C—H stretch
(ii)	1410–1430 cm^{-1}	C=C stretch
(iii)	1120–1140 cm^{-1}	C=C stretch
(iv)	995–1000 cm^{-1}	C—H deformation
(v)	740–790 cm^{-1}	C—H deformation
(vi)	330–490 cm^{-1}	metal–ring interaction

In general, the absorption frequencies of substituents in the π-complexed arene ring are lower than those of the same substituent in the free arene: e.g., in π-$(C_6H_5COC_6H_5)$-π-(C_6H_6)Cr(0), ν_{CO} is at 1610 cm^{-1} as compared with 1651 and 1641 cm^{-1} for ν_{CO} of benzophenone in solution and in the solid state.

6. Magnetic Measurement

The measurement of the magnetic moment of π-bis(arene)chromium complexes is essential in their characterization, insofar as it indicates the number of unpaired electrons per chromium center. The zerovalent complexes (no unpaired electrons) are diamagnetic, and the univalent complexes (one unpaired electron) all have an observed μ_{eff} of 1.70 to 1.80 μ_B [e.g., 41]. The unusual magnetic moment of 3.90 μ_B reported for the compound presumed to be π-$(C_6H_5C_6H_5)$(2,2′-bipyridyl)$_2$Cr(I) iodide [42] suggests that this substance is probably the air- and water-stable σ-bonded organochromium(III) compound $[(C_6H_5)_2Cr(2,2'\text{-bipyridyl})_2]^{\oplus}I^{\ominus}$ [43,44]

7. Mass Spectrometry

Mass spectrometry provides a convenient, rapid, and unambiguous method for identifying a variety of organochromium π-complexes [45,46]. In the case of π-bis(arene)chromium(0) compounds, a stepwise fragmentation is observed, Eq. (2-5) [47–49]. There are also peaks present in the mass spectrum

$$[\text{arene}]_2\text{Cr}^{\oplus} \longrightarrow [\text{arene}]\text{Cr}^{\oplus} + \text{arene} \longrightarrow \text{Cr}^{\oplus} + \text{arene} \qquad (2\text{-}5)$$

due to ions derived from the free ligands, formed in a thermal cracking process [47,49,50], and from the doubly charged molecular ion peak, e.g., Eq. (2-6) [47,48]. The ionization potentials (5.2–5.9 eV) of π-bis(arene)-chromium complexes lie below that of the free chromium atom (6.76 eV)

$$[\text{arene}]_2\text{Cr}^{2\oplus} \longrightarrow [\text{arene}]\text{Cr}^{\oplus} + [\text{arene}]^{\oplus} \qquad (2\text{-}6)$$

[49] and are susceptible to changes in the substituents present in the aromatic ring. The dissociation energies for substituted π-bis(arene)chromium compounds show that substitution in the arene ring, particularly by alkyl groups, increases the strength of the arene–metal interaction [49,51]. The stability sequence is $C_6H_6 < (CH_3)_3C_6H_3 \leqslant C_6H_5C_6H_5 \ll (CH_3)_6C_6$. The strengthening of the arene–metal interaction can be attributed to increased electron density in the arene rings occasioned by the donor capacity of the substituents. It follows, therefore, that in mixed π-complexes, the selective loss of one ligand will be favored. This is illustrated in Eq. (2-7) [49] and Eq. (2-8)

$$[C_6H_5\cdot C_6H_5]\text{Cr}[C_6H_6]^{\oplus} \longrightarrow [C_6H_5\cdot C_6H_5]\text{Cr}^{\oplus} + C_6H_6 \qquad (2\text{-}7)$$

$$[C_6H_6]\text{Cr}[C_5H_5]^{\oplus} \longrightarrow [C_5H_5]\text{Cr}^{\oplus} + C_6H_6 \qquad (2\text{-}8)$$

[48]. However, the stability of the charged organic fragment will also play a determining role in the overall fragmentation process.

8. X-RAY STRUCTURE ANALYSIS

π-Bis(benzene)chromium(0) has been the object of several independent single crystal x-ray structure analyses [27,52–56]. The interpretation of the results and the discussions attending them emphasize the many difficulties that can be incurred in carrying out a highly accurate full x-ray structure analysis. All the results confirm the sandwich structure **3** shown below. The point of controversy, that has been reviewed and discussed [27,57] concerns whether or not the benzene rings have full D_{6h} symmetry, i.e. (among other points), whether or not the C–C bond lengths in the arene ring are equal. The final results of a low-temperature x-ray structure analysis [56], together with electron diffraction data [58,59] and analysis of the vapor-phase infrared spectrum [40] favor sixfold symmetry and no alteration in the C–C bond lengths in the benzene rings, as in **3**. The adjacent C–C bond lengths are 1.417(3) and 1.416(3) Å [as against 1.395(1) Å in C_6H_6]. The rings are eclipsed

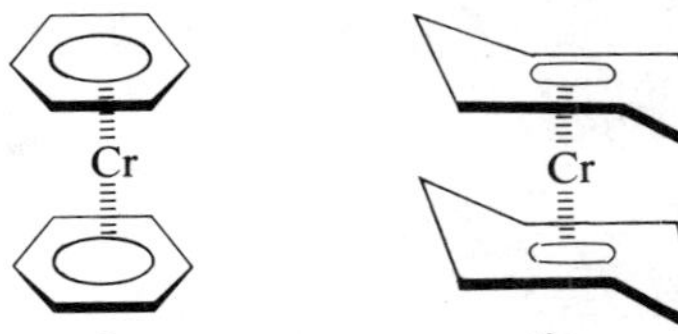

3 4

and the chromium is equidistant from all 12 carbon atoms, the average Cr–C bond length is 2.142 Å [range 2.140(2) to 2.143(2)] [56]. It has been pointed out, however, that the apparent D_{6h} symmetry in the crystal structure determination could be due to orientational disorder in the crystal [56].

In contrast to the above findings, the results of a neutron diffraction study have been interpreted in terms of D_{3d} symmetry. The structure that has been proposed has nonplanar benzene rings, as depicted in **4** (C_{3i} symmetry?); the two different Cr–C distances are 2.095 and 2.137 Å and the two different C–C distances are 1.424 and 1.406 Å. The standard deviations for the Cr–C and C–C distances are 0.032 and 0.055 Å, respectively [60].

The only other π-bis(arene)chromium compound to have been studied by x-ray crystallographic methods is the cation [π-bis(toluene)chromium(I)]$^{\oplus}$, as its iodide **5a** (X = $I^{\ominus}$) [61], as its 1:1 complex with 7,7,8,8-tetracyano-quinodimethane **6** (X = $C_{12}H_4N_4^{\ominus}$) [62,63] and as its 1:2 complex with 7,7,8,8-tetracyanoquinodimethane **5b** (X = $C_{12}H_4N_4^{\ominus}C_{12}H_4N_4$) [63,63a]. The sandwich structure is confirmed and the average C–C_{ring} distances and Cr–C_{ring} distances are given in Table 2.1.

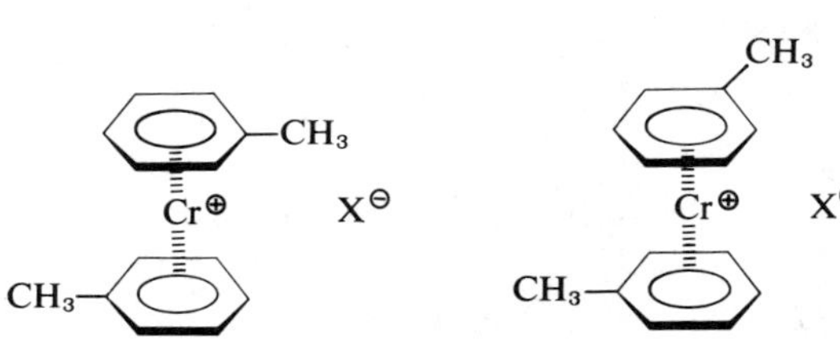

(**5a**) X = I
(**5b**) X = $C_{12}H_4N_4 \cdot C_{12}H_4N_4$
(**6**) X = $C_{12}H_4N_4$

One interesting feature of these structures is that the conformations of the π-bis(toluene)chromium ion are not the same. Thus, though in all three compounds the carbon atoms of the rings are eclipsed, in **5a** and **5b** the cations are centrosymmetric so that the toluene rings adopt a conformation with *anti*-methyl groups; in **6**, however, the methyl groups are skew with respect to each other. In the latter structure, however, some disorder in the location of the methyl groups is reported and, therefore, an interpretation of these conformational differences must await further data. It would be of

TABLE 2.1

Average Bond Distances in π-Bis(arene)chromium Compounds[a]

COMPOUND	RING C–C	$Cr–C_{ring}$	REF.
$(C_6H_6)_2Cr(0)$	1.417(3)	2.142(2)	[56]
$(CH_3C_6H_5)_2Cr(I)(I^{\ominus})$	1.42(3)	2.08(1)	[61]
$(CH_3C_6H_5)_2Cr(I)(C_{12}H_4N_4^{\ominus})$	1.38(4)	2.11(3)	[62]
$(CH_3C_6H_5)_2Cr(I)(C_{12}H_4N_4^{\ominus})(C_{12}H_4N_4)$	1.40(4)	2.18(3)	[63a]

[a] Bond distances are given in Å.

interest to discover what relationship exists between these and the electrical properties of the compounds [63].

In π-bis(arene)chromium compounds, the energy barrier to rotation of the arene portions about the axis perpendicular to their planes will be low. Therefore, in substituted π-bis(arene)chromium compounds, different conformers (rotamers) can exist, but the number of isolable stereoisomers will be restricted. Compounds possessing neither a center nor a plane of symmetry may exist in enantiomeric forms. For example (*ortho*-$ABC_6H_4)_2Cr$ could exist both in the meso form (**7**) and in the racemic forms (**8** and **9**).

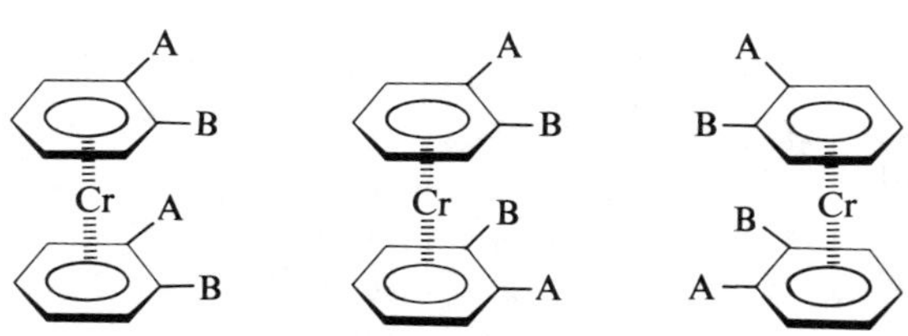

7 **8** **9**

This aspect of the chemistry of π-bis(arene)chromium compounds has not been actively studied.

C. π-(Cyclopentadienyl)chromium Compounds

1. Separation and Purification

The π-(cyclopentadienyl)chromium compounds whose preparation is discussed in the preceding chapter form a rather heteroclite family of compounds in which not only the valency state of the chromium center varies from 0 to III {e.g., π-(C_5H_5)-π-$(C_7H_7)Cr(0)$ [64,65], π-(C_5H_5)-π-$(C_6H_6)Cr(I)$ [66], π-$(C_5H_5)_2Cr(II)$ [67,68], π-$(C_5H_5)Cr(III)Cl_2(THF)$ [69]}, but also the ionic character of the compounds vary from the nonvolatile salts {e.g.,

[π-(C_5H_5)$CrCl_3$]NH_4 [69]} to the highly volatile π-complexes [e.g., π-(C_5H_5)$_2$Cr] [67,68]. The sensitivity of the compounds toward oxygen also varies from highly reactive [e.g., π-(C_5H_5)$_2$Cr] [70], through moderately stable [e.g., π-(C_5H_5)Cr(NO)$_2$$CH_3$] [71], to stable {e.g., [π-(C_5H_5)-π-(C_7H_6R)Cr]PF_6} [72]. However, the solutions of almost all the compounds (or their ionic precursors) are unstable in air. The separation and purification of π-(cyclopentadienyl)chromium compounds is, therefore, best effected under anaerobic conditions by the usual techniques of crystallization and, when applicable, by vacuum sublimation. There are no reports of the systematic investigation of the application of chromatographic techniques to the separation of the compounds in question. Some of the more stable species [e.g., π-(C_5H_5)Cr(NO)$_2$R] have, however, been purified by column chromatography on alumina [71]. The compounds have been characterized and identified by a combination of spectroscopic methods and magnetic measurements.

2. Spectroscopic Methods

The *infrared spectra* of π-(cyclopentadienyl)metal compounds have been reviewed and discussed [e.g., 33, 73, 74] and absorptions in the six main regions listed below have been used in their identification.

(i)	3027–3100 cm^{-1}	C—H stretch
(ii)	1400–1440 cm^{-1}	C=C stretch
(iii)	1090–1110 cm^{-1}	C=C stretch
(iv)	990–1009 cm^{-1}	C—H deformation
(v)	710–830 cm^{-1}	C—H deformation
(vi)	330–450 cm^{-1}	metal–ring interaction

In the case of nitrosyl compounds, the N–O stretching frequency lies in the range 1700–1850 cm^{-1} [75] and with carbonyl compounds the C–O stretching frequency lies within the range 1750–2030 cm^{-1} [76].

NMR spectroscopy is of great value in the identification of the constituent parts of the diamagnetic π-(cyclopentadienyl)chromium compounds. The resonance of the protons of the π-cyclopentadienyl unit is in the region δ 5.8 to 3.1 [77]. Thus, in the case of π-(C_5H_5)Cr(CO)$_3$H, the resonance signal for the C_5H_5-ring protons is at δ 4.78 and that of the hydrogen bonded to chromium is at δ 5.46 upfield from (CH_3)$_4$Si [78]. In the case of π-(cyclopentadienyl)-π-(cycloheptatrienyl)chromium compounds, the chemical shifts of the protons of the complexed ligands are shown in **10**, **11**, and **12** [64,72].

The *mass spectra* of a number of π-(cyclopentadienyl)chromium compounds have been reported, and the subject has been reviewed [45,46].

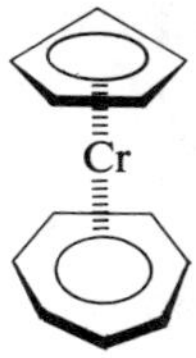

10

$\delta(C_5H_5)$ 3.62
$\delta(C_7H_7)$ 5.41

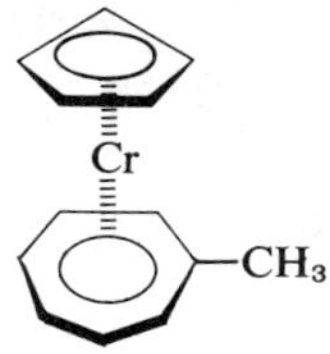

11

$\delta(C_5H_5)$ 3.9
$\delta(C_7H_6)$ 5.65
$\delta(CH_3)$ 2.9

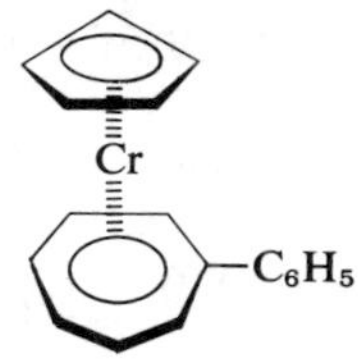

12

$\delta(C_5H_5)$ 3.9
$\delta(C_7H_6)$ 5.77, 6.06
$\delta(C_6H_5)$: *o* 7.92
m, p 7.47

π-Bis(cyclopentadienyl)chromium undergoes stepwise fragmentation, Eq. (2-9) [79]. However, in some mixed ligand species, transformations within

$$\pi\text{-}(C_5H_5)_2Cr^{\oplus} \longrightarrow \pi\text{-}(C_5H_5)Cr^{\oplus} + C_5H_5\cdot \longrightarrow Cr^{\oplus} + C_5H_5\cdot \qquad (2\text{-}9)$$

the parent molecular ion are observed. Thus, for π(cyclopentadienyl)-π-(cycloheptatrienyl)chromium, three fragmentation processes are observed [48,80]; the first is the successive loss of the C_7 and C_5 units, Eq. (2-10); the second is the extrusion of acetylene, Eq. (2-11); the third process involves an interesting rearrangement in which two C_6H_6 units, probably benzene [80], are generated, Eq. (2-12).

$$\pi\text{-}(C_5H_5)\text{-}\pi\text{-}(C_7H_7)Cr^{\oplus} \xrightarrow{-C_7H_7} \pi\text{-}(C_5H_5)Cr^{\oplus} \xrightarrow{-C_5H_5} Cr^{\oplus} \qquad (2\text{-}10)$$

$$\pi\text{-}(C_5H_5)\text{-}\pi\text{-}(C_7H_7)Cr^{\oplus} \xrightarrow{-C_2H_2} \pi\text{-}(C_5H_5)_2Cr^{\oplus} \longrightarrow \text{as in Eq. (2-9)} \qquad (2\text{-}11)$$

$$\pi\text{-}(C_5H_5)\text{-}\pi\text{-}(C_7H_7)Cr^{\oplus} \xrightarrow{-C_6H_6} \pi\text{-}(C_6H_6)Cr^{\oplus} \xrightarrow{-C_6H_6} Cr^{\oplus} \qquad (2\text{-}12)$$

The mass spectrum of the dimer $[\pi\text{-}(C_5H_5)Cr(CO)_3]_2$ shows no molecular ion peak, the ions of highest mass are $[\pi\text{-}(C_5H_5)Cr(CO)_3H]^{\oplus}$ (*m/e* 202) and $[\pi\text{-}(C_5H_5)Cr(CO)_3]^{\oplus}$ (*m/e* 201) [81].

The mass spectra of some bridged π-(cyclopentadienyl)chromium complexes (e.g., **13a**, **13b**, **13c**) are of particular interest. They show, besides the systematic loss of two NO groups and other fragments, a peak at *m/e* 182. This must be ascribed to $[\pi\text{-}(C_5H_5)_2Cr]^{\oplus}$ [82,83], since further fragmentation

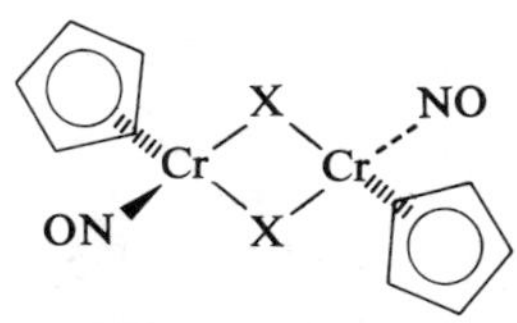

13a, X = SCH_3
13b, X = SC_6H_5
13c, X = OCH_3

yields ions with *m/e* 117 {[π-(C_5H_5)Cr]$^{\oplus}$} and 52 {[Cr]$^{\oplus}$}. The formation of the positively charged ion of chromocene **15** from the bridged species **13a** proceeds via extrusion of CrS_2 from **14**, Eq. (2-13). The presence of a metastable peak at *m/e* 111.1 is in accordance with such a fragmentation (i.e., *m/e* 298 → *m/e* 182).

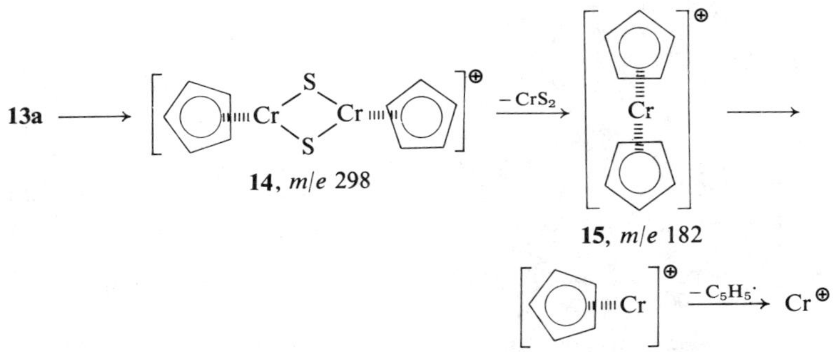

(2-13)

3. Magnetic Measurement

Magnetic measurements have been particularly useful in characterizing π-(cyclopentadienyl)chromium compounds, the relevant data are summarized in Table 2.2. The diamagnetism observed for π-(cyclopentadienyl)chromium tricarbonyl confirms that it is a dimeric species with a metal-to-metal bond. The magnetic moments found for the chromium(I) species (1.63 to 2.15 μ_B) agree well with the value expected for compounds with one unpaired electron (1.73 μ_B). The value found for π-$(C_5H_5)_2$Cr corresponds to that expected for a low-spin chromium(II) compound containing two unpaired electrons (2.83 μ_B). All the chromium(III) compounds have values (3.79 to 4.0 μ_B) in keeping with those of compounds containing three unpaired electrons (3.87 μ_B). The value of 3.97 μ_B found for [π-$(C_5H_5)_2$Cr][π-(C_5H_5)Cr$(CO)_3$] suggests that this is a salt derived from the [π-$(C_5H_5)_2$Cr(III)] cation and the diamagnetic anion [π-(C_5H_5)Cr$(CO)_3$].

4. X-ray Structure Analysis

π-Bis(cyclopentadienyl)chromium is isostructural with π-$(C_5H_5)_2$Fe and a Cr–C_{ring} distance of 2.22 Å has been quoted [87]. This is significantly longer than the average Cr–C_{ring} distances in π-bis(arene)chromium(0) and (I) compounds (2.08–2.18 Å) (see Table 2.1) [25].

Single crystal x-ray structure determinations of several other π-(cyclopentadienyl)chromium compounds have been effected. The compounds include:

(i) the monomeric species **16** (X = Cl and NCO);

(ii) the bridged solvated lithium-π-cyclopentadienylchromium(III) complex $\{Li[\pi\text{-}(C_5H_5)CrCl_3(THF)_2]\}_2$dioxane in which the two π-$(C_5H_5)CrCl_3$ units (**17**) are linked, through one chlorine, by a lithium–dioxane–lithium bridge [i.e., π-$(C_5H_5)CrCl_2Cl \cdots Li(THF)_2 \cdots$O O$\cdots Li(THF)_2 \cdots ClCl_2Cr$-($\pi$-$C_5H_5$)];

16

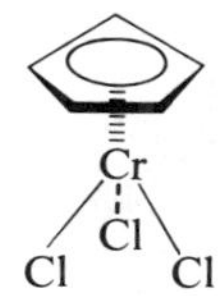

17

TABLE 2.2

Magnetic Measurement Data (in μ_B) for π-(Cyclopentadienyl)chromium Compounds

COMPOUND	MAGNETIC MOMENT	REF.
Diamagnetic compounds		
π-(C_5H_5)-π-$(C_7H_6R)Cr$		
R = H	—	[64,65]
R = CH_3, C_6H_5	—	[72]
$[\pi\text{-}(C_5H_5)Cr(CO)_3]_2$	—	[84]
π-$(C_5H_5)Cr(CO)_3H$	—	[84]
π-$(C_5H_5)Cr(NO)_2Cl$	—	[75]
π-$(C_5H_5)Cr(NO)_2R$		
R = CH_3, CH_2Cl, C_6H_5	—	[71]
Chromium(I) species		
π-(C_5H_5)-π-$(C_6H_6)Cr$	1.70	[66]
$[\pi\text{-}(C_5H_5)\text{-}\pi\text{-}(C_7H_6R)Cr]PF_6$		
R = H	2.0	[65]
R = CH_3	2.1	[72]
R = C_6H_5	2.15	[72]
π-(C_5H_5)-π-$(C_7H_8)Cr$	1.63	[85]
π-(C_5H_5)-π-$(C_8H_{10})Cr$	1.69	[85]
Chromium(II) species		
π-$(C_5H_5)_2Cr$	3.20	[86]
Chromium(III) species		
π-$(C_5H_5)CrBr_2L$		
L = py	3.79	[69]
L = THF	3.96	[69]
π-$(C_5H_5)CrCl_2THF$	4.0	[69]
π-$(C_5H_5)_2Cr(I)$	3.87	[70]
$[\pi\text{-}(C_5H_5)_2Cr][\pi\text{-}(C_5H_5)Cr(CO)_3]$	3.97	[84]

(iii) the bridged dimeric complexes **18a** to **18e**; in **18a**, **18b**, and **18c**, the NO groups are trans and in the amide-bridged compounds **18d** and **18e** they are cis and trans, respectively [88].

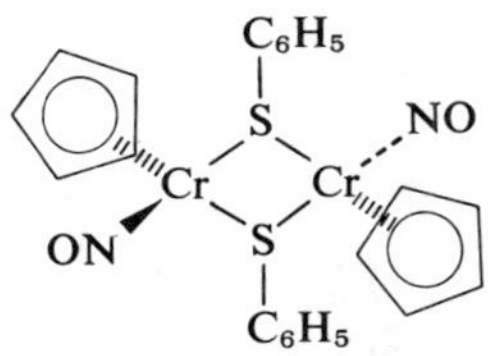

18a

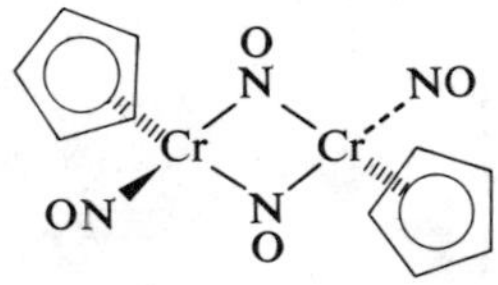

18b

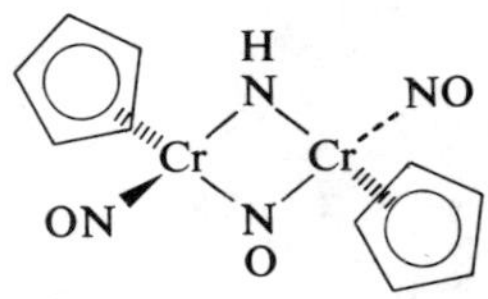

18c

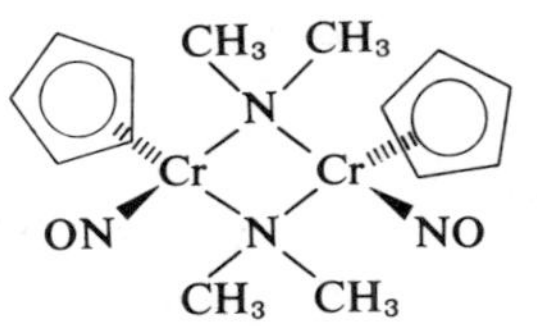

18d

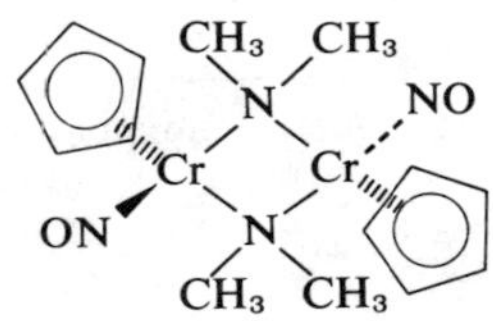

18e

The relevant bond lengths that are given in Table 2.3 indicate that the Cr–C_{ring} distances fall within the range 2.19–2.26 Å and are again longer than the Cr–C_{ring} distances found in π-bis(arene)chromium species, but of the same order as those found in π-(arene)chromium tricarbonyls (see Table 2.4).

The Cr–Cl distances in **16** (X = Cl) and **17** are in the range 2.290–2.346 Å, which is in good agreement with the values found in the octahedral *trans*-4-$CH_3C_6H_4CrCl_2(THF)_3$ [2.319(3) Å] [94] and *fac*-$[NH_2CH_2CH_2]_2NHCrCl_3$ [2.320(7) Å] [95]. All these values are somewhat larger than that found for the average Cr–Cl distance (2.232 Å) in the trigonal bipyramidal $[(CH_3)_3N]_2$-$CrCl_3$ [96,97], (Chapter 2, Section F,3,b). In the case of the lithium complex in which two units of **17** are linked, via chlorine, by a lithium–dioxane–lithium bridge, the distance (2.346 Å) from chromium to the chlorine involved in the bridging is significantly greater than that to the two nonbridging chlorines (average 2.302 Å) [90a].

The metallocycles, , in the bridged compounds **18a** to **18e** are

TABLE 2.3

***Bond Distances in π-(Cyclopentadienyl)chromium Compounds*[a]**

COMPOUND	Cr–C_{ring}	Cr–$N_{(NO)}$	Cr–X	REF.
π-$(C_5H_5)Cr(NO)_2Cl$	2.20[b] (2.12–2.25)	1.710 (1.704–1.717)	X = Cl 2.311(5)	[89]
π-$(C_5H_5)Cr(NO)_2NCO$	2.199(4) (2.173–2.227)	1.716(3)	X = N 1.982(4)	[90]
[Li-π-$(C_5H_5)CrCl_3(THF)_2]_2$ dioxane	2.26(2) (2.21–2.32)		X = Cl 2.316(5) (2.290–2.346)	[90a]
trans-[π-$(C_5H_5)Cr(NO)SC_6H_5]_2$	2.21(1) (2.167–2.253)	1.662(7)	Cr–Cr 2.950(2)	[91]
cis-[π-$(C_5H_5)Cr(NO)N(CH_3)_2]_2$	2.26[b] (2.21–2.32)	1.64(2)	2.72(1)	[92]
trans-[π-$(C_5H_5)Cr(NO)N(CH_3)_2]_2$	2.26 (2.23–2.28)	1.63(2)	2.67(1)	[92]
trans-[π-(C_5H_5)Cr(NO)(NH)(NO)(ON)Cr(π-C_5H_5)]	2.24(5)	1.672(23)[c]	2.650(4)	[92a]
trans-[π-(C_5H_5)Cr(NO)(NO)(NO)(ON)Cr(π-C_5H_5)]	2.209(4)	1.690(3)[c]	2.615(1)	[93]

[a] Bond distances are given in Å.

[b] Disorder associated with the cyclopentadienyl ring.

[c] These are the nonbridging Cr–NO distances; the Cr–NO (bridging) distances are 1.936(12) Å and 1.960(3) Å, respectively. The upper row of figures is the average bond distances, the lower row of figures in parentheses is the range of the respective bond distances.

almost planar, and the Cr–Cr distance varies from 2.95 Å in the sulfur bridged compound to 2.61 Å in **18b** (Table 2.3). These distances are significantly longer than the Cr–Cr distance found in chromous acetate (2.362 Å) [98], which is also a bridged dimer. The value of 2.97 Å is the same, however, as that found in the dinuclear chromium carbonyl anion $[(CO)_5Cr–Cr(CO_5]^{2\ominus}$ [Cr–Cr 2.97(1) Å] [99].

The bridged chromium compounds are diamagnetic, and this diamagnetism can be interpreted in terms of the coupling of the odd electron on each chromium center to give a chromium-to-chromium bond. The value of 2.97 Å would then represent the length of a Cr–Cr single bond. The variations in this length, observed in the case of the other bridged chromium compounds, could then be ascribed, at least in part, to differences in the covalent radii of the bridging atoms.

D. Chromium Carbonyl π-Complexes

1. Separation and Cleavage

The majority of chromium tricarbonyl π-complexes are stable in air and may be readily purified by column chromatography and crystallization [e.g. 100,101]. Certain π-(arene)chromium tricarbonyl complexes may be separated (and estimated) by gas chromatography [102,103]. The application of this method is, however, restricted, since some of the compounds undergo thermal decomposition [to arene and $Cr(CO)_6$] under the conditions of analysis [102,104,105]. The use of high-speed column liquid chromatography for the separation of isomeric π-(arene)chromium tricarbonyls [e.g., the isomers of π-(2,3-dimethylnaphthalene)$Cr(CO)_3$] has been described [104].

The π-bonded organic part of chromium tricarbonyl π-complexes may be liberated by oxidative or thermal decomposition or by acid treatment [100]. These rather drastic reaction conditions can occasion secondary transformations in the organic groups originally bonded to chromium. A milder displacement may be achieved either by photolysis [e.g., 106] or by ligand replacement [e.g., 100]. In the latter reaction, the chromium tricarbonyl π-complex is allowed to react with a more powerful electron donor [e.g., pyridine, $(C_6H_5)_3P$, etc.,].

2. NMR Spectroscopy

NMR spectroscopy has been extensively used in establishing both the identity and steric arrangement of the organic unit in tricarbonyl chromium π-complexes [e.g., 107]. As a general rule, the signals of protons in a ligand π-complexed to $Cr(CO)_3$ lie at higher fields than those of the same protons in the free ligand; in the case of π-(arene)$Cr(CO)_3$ compounds, the shift of the aromatic protons is of the order of 2 ppm and that of protons in alkyl substituents adjacent to the aromatic ring is of the order of 0.2 ppm [108, 109]; in π-(cycloheptatriene)chromium carbonyls the triene protons are approximately 0.5 to 2 ppm upfield [110]. This upfield shift of the aromatic protons can be ascribed to a reduction in the ring current and an overall withdrawal of π electrons from the arene ring by the $Cr(CO)_3$ unit, the overall effect being a reduction in the deshielding of the aromatic protons and, to a much lesser extent, of the protons in alkyl substituents, i.e., upfield shifts.

These differences in chemical shifts allow the distinction to be made between various isomeric compounds. Thus, in the reaction of 2,3-dimethylnaphthalene with $Cr(CO)_6$, two isomers are formed, one (**19**) with the unsubstituted ring complexed to chromium, the other (**20**) with the substituted ring complexed to chromium. In the case of **19**, the protons of the complexed

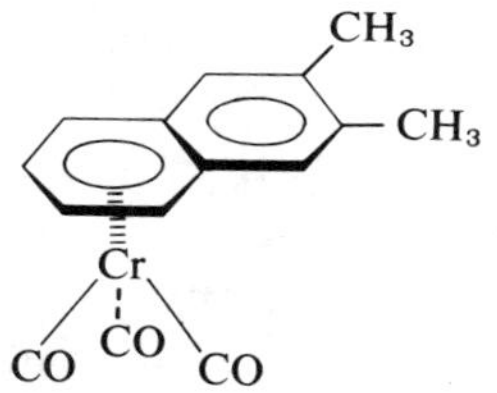

19

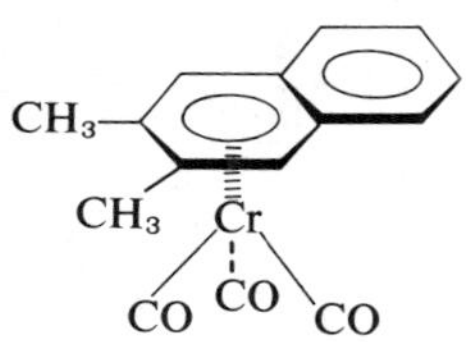

20

aromatic ring show signals at δ 6.07 and 5.46 (the same protons in **20** are at δ 7.54 and 7.38) and the signal of the protons of the uncomplexed ring in **19** are at δ 7.33 (the same protons in **20** are at δ 6.10). The signals of the methyl groups in the uncomplexed ring of **19** are at δ 2.30, while those of **20** are slightly upfield at 2.25 [109]. Similarly, in the exo and endo isomers of π-(7-propargylcycloheptatriene)$Cr(CO)_3$ (**21** and **22**, R = $CH_2C{\equiv}CH$) and related compounds, differences in the chemical shifts of the hydrogens at positions 1, 6, and 7 have been used to allocate configurations. Thus in endo compounds, the signals for the protons 1 and 6 (δ ~ 3) are at higher fields than

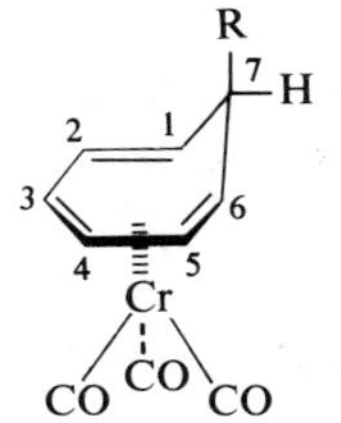

21

exo isomer

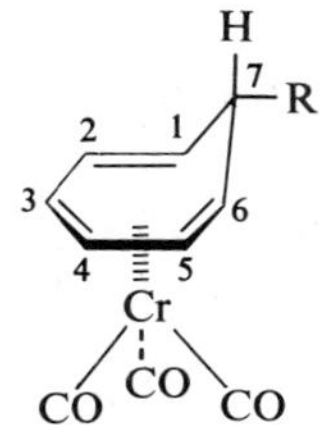

22

endo isomer

those for the exo isomer ($\delta \sim 4$); also in the endo isomer the axial H_7 is shielded by the triene system and its resonance occurs at higher fields ($\delta \sim 2$) than the resonance of the equatorial H_7 of the exo isomer ($\delta \sim 3$) [110].

3. Infrared Spectroscopy

The solution and solid-state spectra of π-(arene)chromium tricarbonyls have been studied and discussed both from a practical and a theoretical standpoint [e.g., 27,111–115a]. The two regions of the spectra that are of importance in the characterization and identification of this group of compounds are the absorptions due to the $Cr(CO)_3$ unit and those associated with the arene unit. In solution, the former unit is characterized by two intense C–O stretch absorptions, the one around 1900–1950 cm^{-1} and the other around 1970–2000 cm^{-1} [111,112]. In the solid state (KBr disc) there are three or more absorptions in the region 1820–1960 cm^{-1} [112–115]. The changes in the spectrum of an arene molecule when complexed to a $Cr(CO)_3$ unit have been discussed [111] and the main points may be summarized as follows:

(i) Region 3080–3100 cm^{-1}: the C–H stretching frequencies of the free arene usually merge into one weak band in the complex.

(ii) Region 1650–1450 cm^{-1}: the two C=C stretching frequencies of the free arene are shifted to lower frequencies (by 60–100 and 24–45 cm^{-1}, respectively) on complexing to $Cr(CO)_3$.

(iii) Region 850–670 cm^{-1}; the C–H deformation frequencies that are characteristic of unsubstituted and substituted arenes are not found in the spectra of the corresponding π-(arene)chromium tricarbonyls, but instead, a weak band in the region 785–800 cm^{-1} is observed.

(iv) Region 670–600 cm^{-1}; absorptions in this region have been ascribed to C–O modes.

(v) Region 530–475 cm^{-1}; absorptions in this region have been ascribed to $Cr–C_{ring}$ interactions.

4. Mass Spectrometry

Mass spectrometry is a very convenient method for the characterization and identification of π-(arene)- and π-(olefin)chromium tricarbonyls and the subject has been reviewed [45,46].

The overall fragmentation process involves the successive loss of CO groups followed by the loss of the organic ligand, e.g., Eq. (2-14) [116]. The ionization potentials of the π-(arene)chromium tricarbonyls lie from 1.5 to 2.0 eV higher than those of the corresponding π-bis(arene)chromium compounds [49,117,118]. The mass spectra of some substituted arene compounds also include peaks from species arising from the rearrangement of the parent molecular ion [46]. Similarly, with some π-(triene)$Cr(CO)_3$

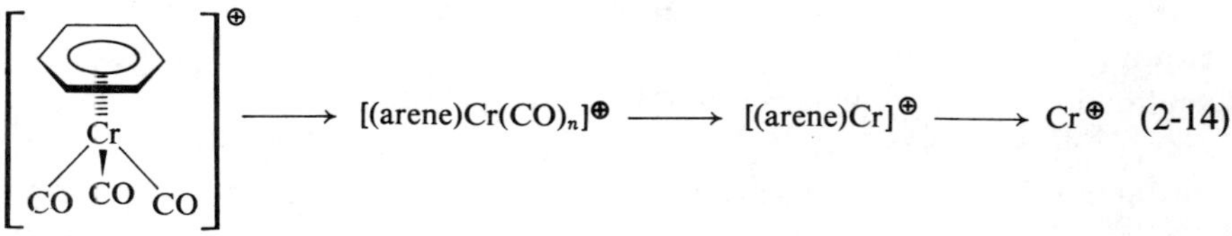

compounds and π-(olefin)$Cr(CO)_4$ compounds rearrangement of the parent molecular ion is observed along with the usual fragmentation via successive loss of carbonyl groups. Thus, the mass spectrum of π-(norbornadiene)Cr-$(CO)_4$ shows the presence of the ion $[(C_6H_6)Cr(CO)_4]^{\oplus}$ formed by CH_2 extrusion from the parent ion [45]. The extrusion of an acetylene unit from the parent ion is observed in the mass spectrum of π-(cyclooctatetraene)Cr-$(CO)_3$ [116], Eq. (2-15).

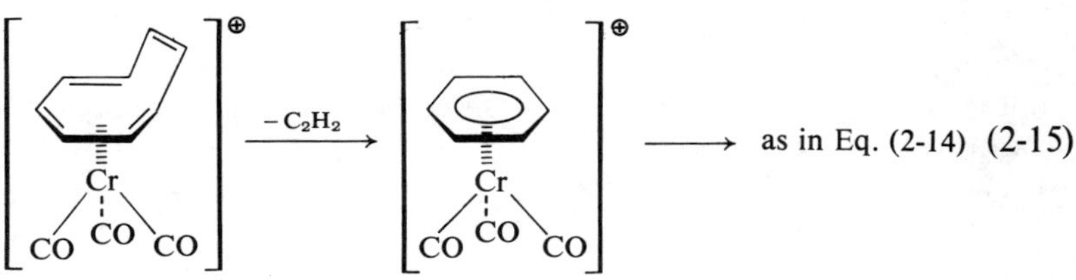

5. X-RAY STRUCTURE ANALYSIS

a. *π-(Arene)chromium Tricarbonyls*

There have been numerous single crystal x-ray analyses of π-(arene)-chromium tricarbonyl compounds [e.g., 119–134], and in all cases the "piano-stool" arrangement of the arene and $Cr(CO)_3$ units has been confirmed. Structure data for some representative compounds are listed in Table 2.4. The Cr–C_{CO} bond lengths are in the range 1.76–1.85 Å and are always shorter than those found in $Cr(CO)_6$ (1.92 Å). The distances from chromium to the aromatic carbons lie in the range 2.20–2.25 Å and would seem to be insensitive to the electronic nature of the substituents in the arene rings. Thus, with an electron-withdrawing group [e.g., π-(CH_3-$OOCC_6H_5$)$Cr(CO)_3$] the Cr–C_{ring} distance is 2.22 Å [122] and with an electron-donating group [e.g., π-($CH_3OC_6H_5$)$Cr(CO)_3$], it is 2.23 Å [133]. The Cr–C_{ring} distances in π-areneCr$(CO)_3$ compounds are consistently longer than the same distances in π-bis(arene)chromium compounds (i.e., 2.08 to 2.18 Å, Table 2.1). This difference can be ascribed to the greater electron-withdrawing power of the $Cr(CO)_3$ unit as compared to that of the $Cr(C_6H_6)$ unit. In the case of simple monocyclic arene compounds, the chromium is usually equidistant from the six carbons of the arene rings. However, with certain polycyclic π-(arene)chromium carbonyls (e.g., π-naphthalene- [123], π-anthracene- [127], π-phenanthrene- [124,125], and π-dihydrophenanthrene$Cr(CO)_3$ [126], compounds 3–6 Table 2.4), the chromium is not equidistant from all six carbons of the arene ring complexed to chromium and the distances from the chromium to the four peripheral carbons (2.206–2.220 Å) are shorter than those from Cr to the two carbons at the bridge head (2.289–2.240 Å). This can be due to differences in the electron densities on the bridge head carbons and the peripheral carbons occasioned by conjugation of the former with the remaining arene ring(s).

There are no unusual features in the C–C bond distances in the π-complexed arene rings; they are all equal within the limits of experimental error and lie in the range 1.40–1.43 Å (Table 2.4).

The x-ray structure analyses have revealed that the π-(arene)chromium carbonyls adopt a variety of conformations. There are two fundamental conformations, the staggered **23** and the eclipsed **24**, and in the case of monosubstituted arenes there are two possible eclipsed conformations, the one, **25**, with the CO groups eclipsing the substituent and the meta carbon atoms

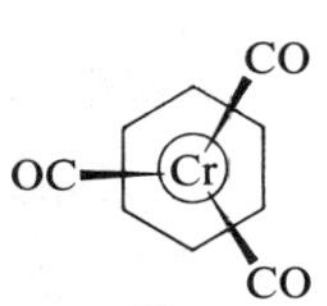

23

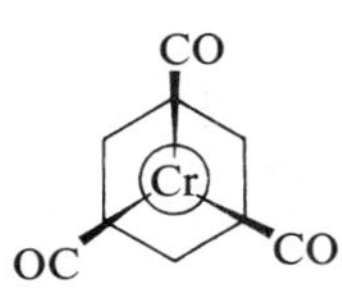

24

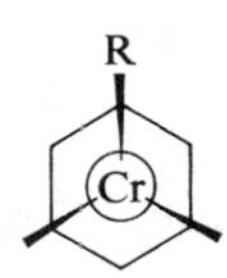

25

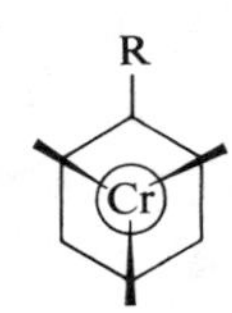

26

of the ring, the other, **26**, with the CO groups eclipsing the carbon atoms ortho and para to the substituent.

All the unsubstituted π-(arene)chromium tricarbonyls examined to date, and π-(hexamethylbenzene)$Cr(CO)_3$ have staggered conformations (Table 2.4, compounds 1 to 9). In contrast, substituted π-(arene)chromium tricarbonyls have been found to possess various conformations. It has been suggested that electron-releasing substituents favor the eclipsed conformation **25**, whereas electron-withdrawing substituents favor the eclipsed conformation **26** [122,132,133]. This is substantiated by the structures of π-(anisyl)- and π-(*o*-toluidine)$Cr(CO)_3$, on the one hand, and π-(methylbenzoate)Cr-$(CO)_3$, on the other, (Table 2.4, compounds 12, 11, and 10, respectively).

TABLE 2.4

Bond Distances in π-(Arene)chromium Tricarbonyls

COMPOUND	CONFORMATION	$Cr–C_{ring}$		$C_{ring}–C_{ring}$	$Cr–C_{CO}$	REF.
1. 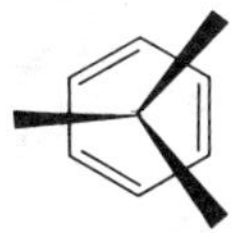	staggered	2.221 (2.216–2.230)		1.401 (1.370–1.421)	1.841 (1.841–1.842)	[120] [119]
2. 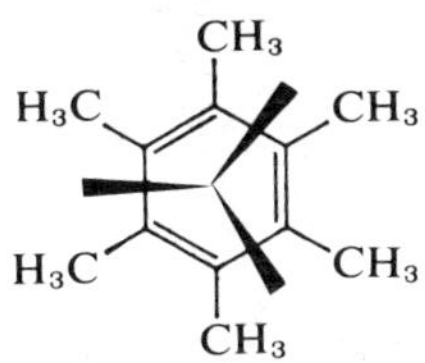	staggered	2.233 (2.211–2.253)		1.417 [1.370–1.461]	1.814 [1.789–1.831]	[121]
3. 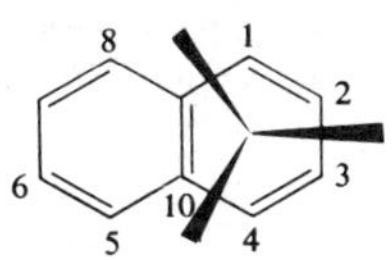	staggered	$Cr–C_{1-4}$ 2.201 (2.186–2.214)	$Cr–C_{9-10}$ 2.321 (2.306–2.337)	1.406 (1.375–1.447)	1.821 (1.815–1.830)	[123]
4. 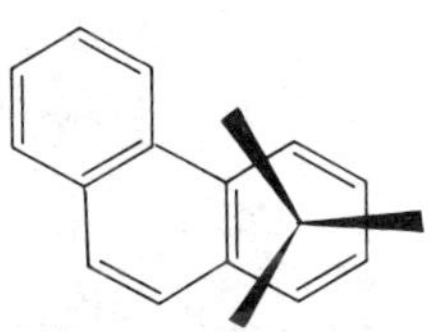	staggered	$Cr–C_{1-4}$ 2.209 (2.206–2.212)	$Cr–C_{11-12}$ 2.289	1.412 (1.385–1.440)	1.843 (1.828–1.857)	[124] [125]

(continued)

TABLE 2.4 (*continued*)

COMPOUND	CONFORMATION	$Cr–C_{ring}$		$C_{ring}–C_{ring}$	$Cr–C_{CO}$	REF.
5. 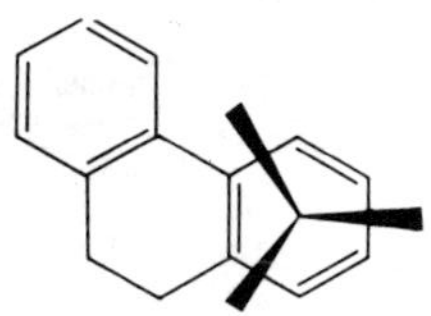	staggered	$Cr–C_{1-4}$ 2.206 (2.200–2.209)	$Cr–C_{11-12}$ 2.240 (2.232–2.248)	1.403 (1.383–1.422)	1.833 (1.821–1.843)	[126]
6. 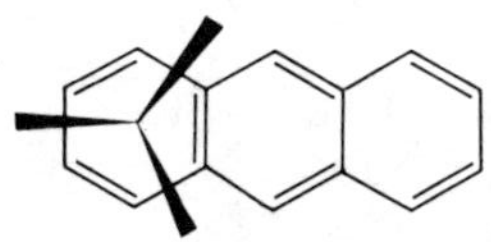	staggered	$Cr–C_{1-4}$ 2.219 (2.215–2.221)	$Cr–C_{11-12}$ 2.332 (2.324–2.340)	1.418 (1.372–1.449)	1.824 (1.812–1.834)	[127]
7. 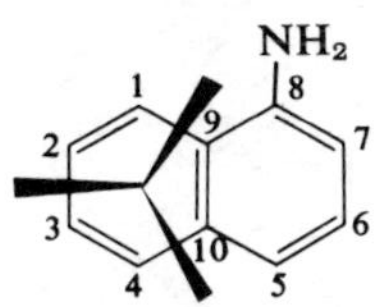	staggered	$Cr–C_{1-4}$ 2.22 (2.19–2.24)	$Cr–C_{9-10}$ 2.30	1.43 (1.41–1.45)	1.76 (1.68–1.83)	[128]
8. 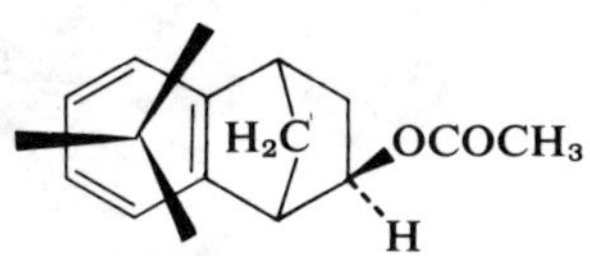	staggered	—		1.43 (1.41–1.47)	1.84 (1.83–1.85)	[129]
9. 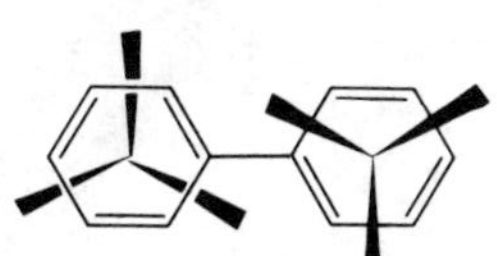	staggered	2.27 (2.20–2.34)		1.41 (1.37–1.44)	—	[130] [57]

10. 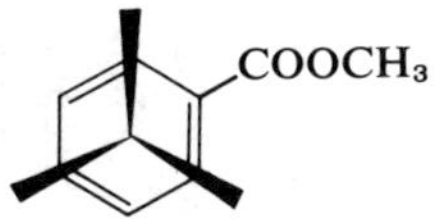	eclipsed	2.22 (2.20–2.25)	1.41 (1.39–1.45)	1.85 (1.81–1.88)	[122]
11. 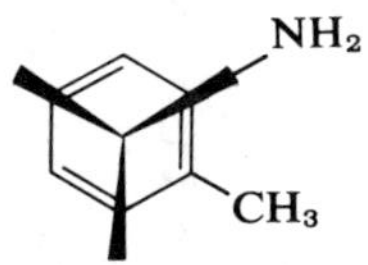	eclipsed	2.25 (2.20–2.37)	1.40 (1.37–1.44)	1.78 (1.76–1.80)	[132]
12. 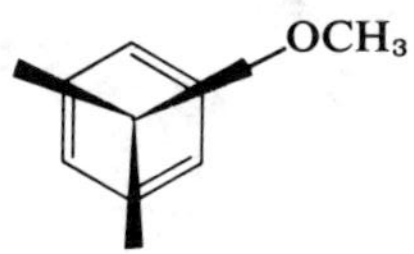	eclipsed	2.23 (2.20–2.27)	1.41 (1.38–1.44)	1.79 (1.77–1.82)	[133]
13. (+) 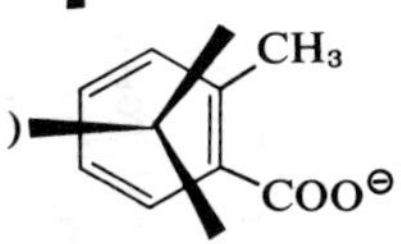	staggered	2.21	—	1.81	[134]
14. (−) 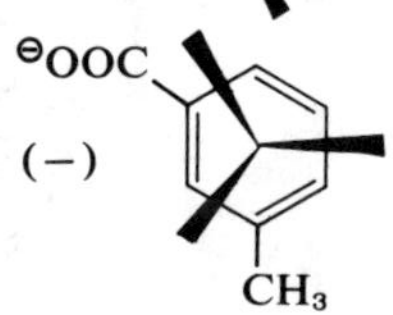	staggered	2.20	—	1.83	[134]

[a]  Represents the $Cr(CO)_3$ unit. The upper row of figures is the average bond distances, the lower row of figures in parentheses is the range of the respective bond distances. Bond distances are given in Å.

This is, however, an oversimplified picture, since π-(acetophenone)$Cr(CO)_3$ has a staggered conformation [131]:

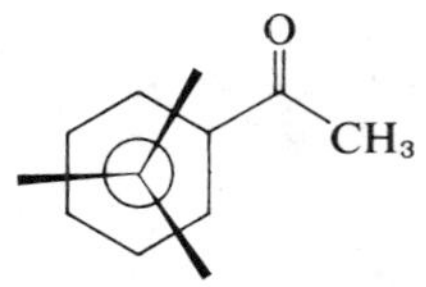

Similarly, the anionic compounds π-[2-$OOC(CH_3)C_6H_4$]$Cr(CO)_3$ and π-[3-$OOC(CH_3)C_6H_4$]$Cr(CO)_3$ (Table 2.4, compounds 13 and 14) have staggered conformations. It is probable that both steric and electronic factors influence the relative stabilities of the various conformers.

In simple π-(arene)chromium tricarbonyl compounds, the arene ring will be free to rotate about the axis perpendicular to the plane of the arene ring. The introduction of bulky substituents into the arene portion of the molecule should increase the energy barrier to this free rotation and could therefore lead to stable conformers. Evidence for restricted rotation in substituted π-(arene)chromium carbonyl compounds has been obtained from variable temperature NMR spectroscopy [e.g., 135–138]. Thus in simple methyl-substituted compounds, the arene ring rotates rapidly on the NMR time-scale [139], whereas π-(isopropylbenzene)$Cr(CO)_3$ adopts the preferred conformation **25** [R = $(CH_3)_2CH$] and π-(*tert*-butylbenzene)$Cr(CO)_3$ adopts the conformation **26** [R = $(CH_3)_3C$] [136,138].

The stabilities of various conformers of π-(arene)chromium carbonyl compounds have also been interpreted in terms of electronic and steric factors [e.g., 136].

Substituted π-(arene)chromium tricarbonyl compounds, which possess neither a plane nor a center of symmetry [e.g., (o-R,R$'C_6H_4$)$Cr(CO)_3$ and (m-R,R$'C_6H_4$)$Cr(CO)_3$], may exist in enantiomeric forms as shown in **27**

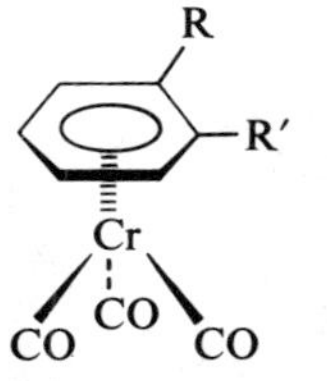

27

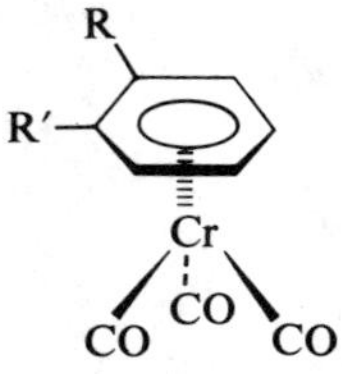

28

and **28**. The preparation and resolution of such chiral compounds has been discussed and reviewed [140]. Many substituted π-(arene)chromium tricarbonyls have been resolved [e.g., 141–143] and the absolute configurations of (2- and 3-methylbenzoic acid)$Cr(CO)_3$ (Table 2.4, compounds 13 and 14) have been established by single crystal x-ray structure analysis [134].

b. π-(Cycloheptatriene)Cr(CO)$_3$

Substituted π-(cycloheptatriene)chromium tricarbonyl compounds can exist either as the 7-exo or the 7-endo isomers **29a** and **29b**. A single crystal

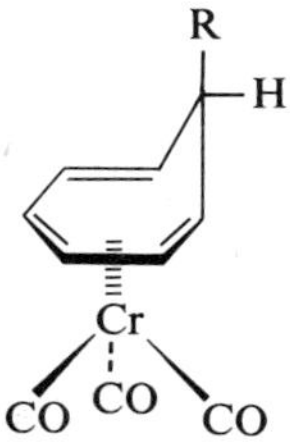

29a

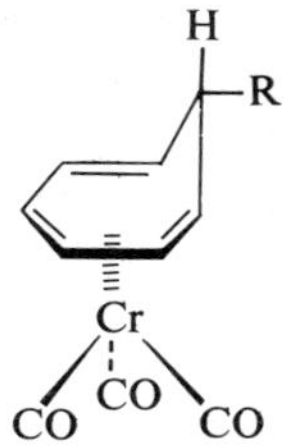

29b

x-ray structure analysis of π-(7-phenylcycloheptatriene)Cr(CO)$_3$ proved that it has the 7-exo configuration **29a** [144], thereby substantiating the steric correlations established in this series of compounds (Chapter 2, Section D,2 and Chapter 3, Section D,2,b). The structure analysis reveals that six carbon atoms of the conjugated triene system are planar with alternating C–C bond lengths (1.36–1.37 Å and 1.41–1.46 Å, respectively); the Cr–C_{ring} distances (2.20–2.31 Å) and the Cr–C_{CO} distances (1.79–1.82 Å) are similar to those found in π-(arene)chromium tricarbonyls. The Cr(CO)$_3$ unit adopts a staggered conformation with respect to the C_6-triene system.

c. π-(Hexamethylbicyclo[2.2.0]hexa-2,5-diene)Cr(CO)$_4$

The chromium tetracarbonyl complexes of various dienes have been prepared. A single crystal x-ray structure analysis of π-(hexamethyl-Dewar-benzene)Cr(CO)$_4$ [145] showed that the chromium is octahedrally coordinated to four carbonyl groups and to the two double bonds of the hexamethyl-Dewar-benzene (hexamethylbicyclo[2.2.0]hexa-2,5-diene). This is illustrated in **30** (the six methyl substituents have been omitted for clarity). The Cr–$C_{double\ bond}$ distances [average 2.33 (1) Å], which are shorter than the non-bonded Cr–$C_{bridgehead}$ distances [average 2.80(2) Å] are longer than the Cr–C_{ring} distances in π-(arene)chromium carbonyls (see Table 2.4). This could be caused by the orbital overlap between two unconjugated double bonds and the chromium center being less effective than the overlap between a conjugated cyclic triene system and chromium. However, in **30** there

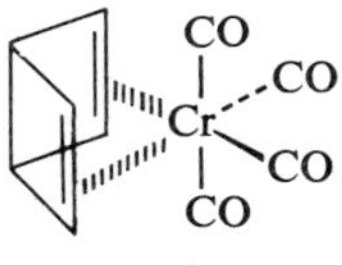

30

are also strong nonbonding interactions between the chromium and the bridgehead carbon atoms (i.e., repulsive forces). The lengthening of the Cr–$C_{\text{double bond}}$ may well be a result of these two opposing effects, namely the bonding interactions between the C=C and Cr and the repulsive forces between the chromium and the bridgehead carbons. The Cr–C_{CO} bond distances [average 1.85(2) Å] fall within the range found in π-(arene)- and π-(triene)chromium tricarbonyls [145].

E. Carbenoid-Chromium Pentacarbonyls

1. Separation and Cleavage

The majority of the title compounds are relatively stable in air; however, they are readily oxidized in solution. Their separation and purification therefore require anaerobic conditions. The compounds can be purified by the usual techniques of crystallization, sublimation, and chromatography using hexane or benzene as eluant [e.g., 146,147].

Carbenoid-chromium pentacarbonyls contain highly reactive species complexed to chromium and, therefore, their cleavage to give the "carbene" per se is not feasible. The carbenoid unit may be displaced, however, from the chromium center by a variety of reagents (e.g., pyridine, HI, and CO) and it may be trapped by interaction with certain activated olefins (e.g., methyl crotonate). These reactions will be discussed more fully in Chapter 3.

2. NMR Spectroscopy

NMR spectroscopy has been extensively used to characterize the organic fragment present in carbenoid-chromium pentacarbonyls [e.g., 146–151]. In the case of the methoxy- and ethoxycarbenoid compounds, **31** and **32**,

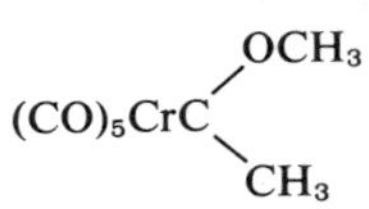

$\delta(OCH_3)$ 4.55
$\delta(CH_3)$ 2.85

31

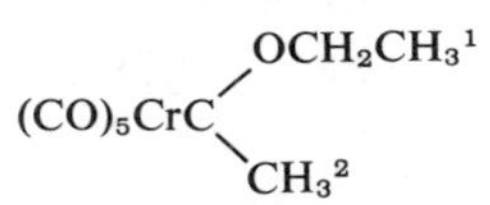

$\delta(OCH_2)$ 4.90 to 4.98
$\delta(CH_3{}^1)$ 1.64 to 1.61
$\delta(CH_3{}^2)$ 2.90 to 2.93

32

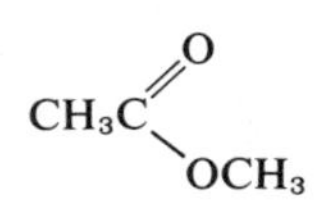

$\delta(OCH_3)$ 2.65
$\delta(CH_3)$ 2.0

33

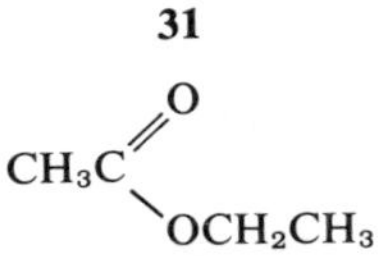

$\delta(OCH_2)$ 4.05
$\delta(CH_3)$ 2.0

34

CH_3OCH_3

$\delta(OCH_3)$ 3.24

35

$CH_3CH_2OCH_2CH_3$

$\delta(OCH_2)$ 3.37
$\delta(CH_3)$ 1.14

36

the chemical shifts of the protons on the carbon adjacent to the oxygen atom [149,150] lie closer to the values for the corresponding protons in esters (**33**, **34**) than they do to those in ethers (**35**, **36**). This is indicative of the electron-withdrawing character of the carbene carbon and supports the idea that the carbenoid unit bonded to chromium has certain analogies with an ester group (see Chapter 3, Section E).

Variable temperature NMR spectroscopy has been used to detect the presence of cis and trans isomers of $(CO)_5CrC(OCH_3)CH_3$, and the chemical shifts for the appropriate isomers in deuterioacetone at $-40°C$ are shown in **37** and **38**, respectively [151]. The corresponding *cis*- and *trans*-methylaminocarbenoid chromium pentacarbonyls have been isolated and identified

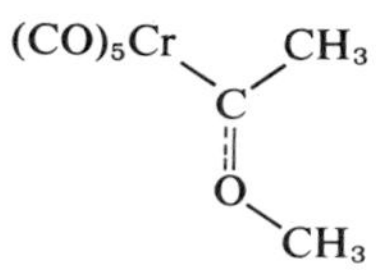

$\delta(OCH_3)$ 4.39
$\delta(CH_3)$ 3.17
cis isomer
37

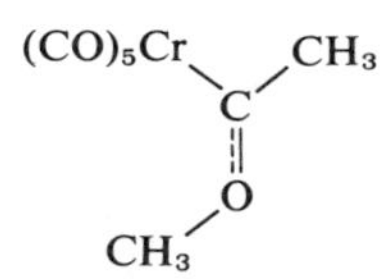

$\delta(OCH_3)$ 4.89
$\delta(CH_3)$ 3.00
trans isomer
38

by NMR spectroscopy [147], and the chemical shifts for the appropriate isomers are shown in **39** and **40**.

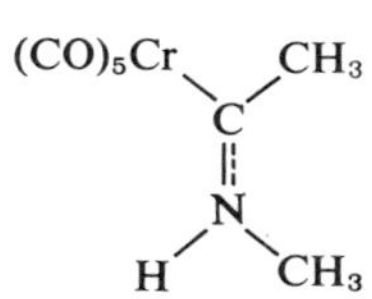

$\delta(CCH_3)$ 2.05
$\delta(NCH_3)$ 1.80
39

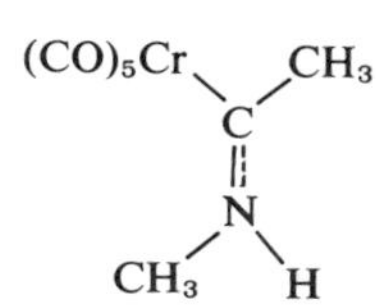

$\delta(CCH_3)$ 2.15
$\delta(NCH_3)$ 2.80
40

3. Infrared Spectroscopy

The infrared spectra of carbenoid-chromium pentacarbonyls have been discussed [e.g., 146,150,152–156]. They show a varying number of bands (2 to 4) in the CO region (1920–2070 cm^{-1}) depending upon the carbenoid unit bonded to chromium. This has been ascribed to a departure from true C_{4v} symmetry occasioned by the presence of the carbenoid ligand. The infrared spectra of certain compounds containing cyclic carbenoid units, e.g., **41**, have only two carbonyl frequencies at 2060 and 1930 cm^{-1} [157,158]. Analysis of the infrared spectrum (CO region) was used to ascribe the cis configuration to the bis(carbenoid)chromium tetracarbonyl **42** [159].

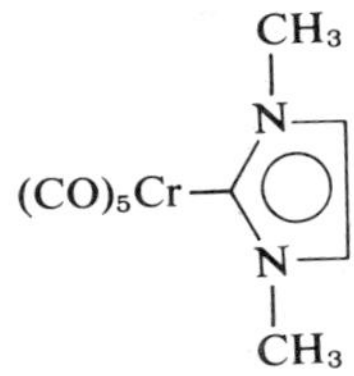

41

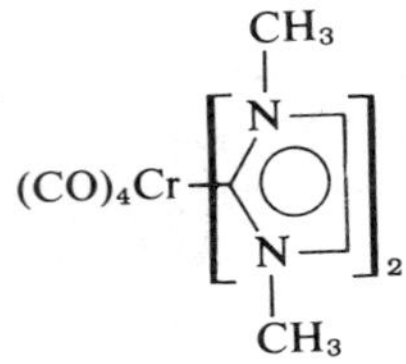

42

4. Mass Spectrometry

Mass spectrometry has been extensively used in the characterization and identification of carbenoid-chromium pentacarbonyls and bis(carbenoid)-chromium tetracarbonyls [e.g., 157–163]. The main fragmentation pattern involves the successive loss of CO and finally the fragmentation of the carbenoid-chromium or the bis(carbenoid)chromium ion. This is illustrated in Eqs. (2-16) and (2-17) [157,159].

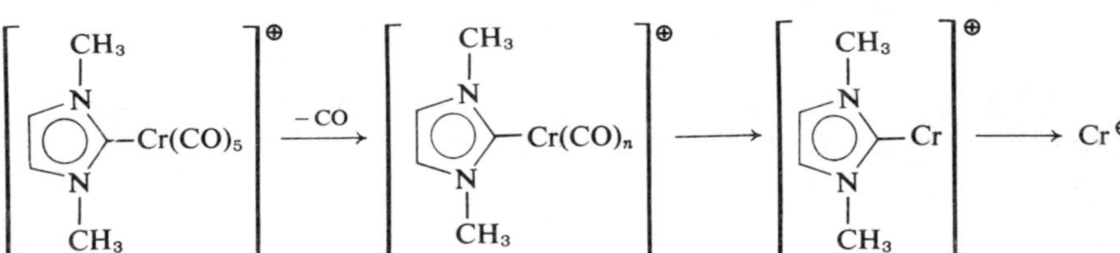

(2-16)

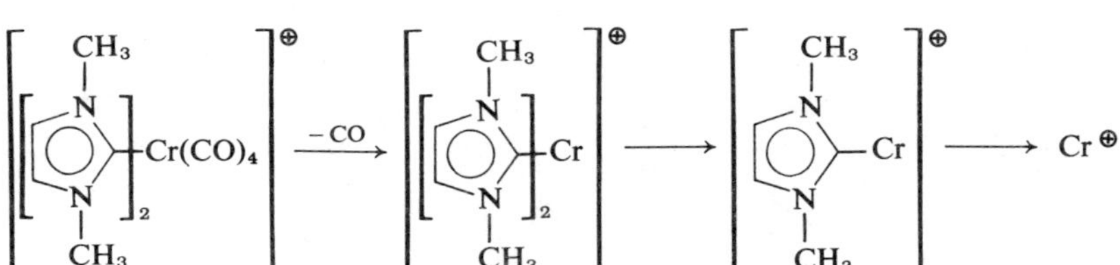

(2-17)

5. X-ray Structure Analysis

Single crystal x-ray analyses of several carbenoid-chromium compounds [164–171] (for a review, see Cotton and Lukehart [156]) have established that the chromium center and the carbenoid unit are coplanar and that the bonding to the metal is through the carbenoid carbon. This is illustrated in the structures **43** [164], **44** [167], and **45** [165]. In compounds containing hetero-

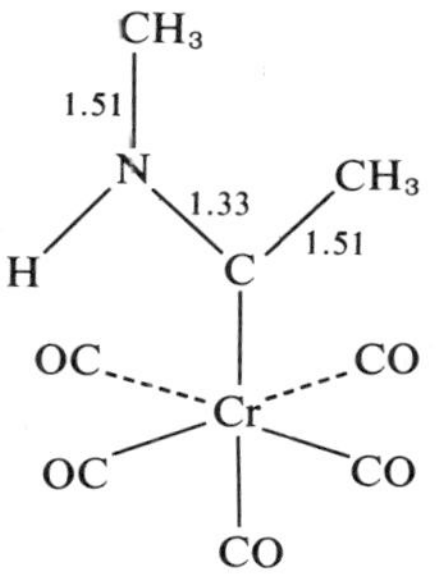

43

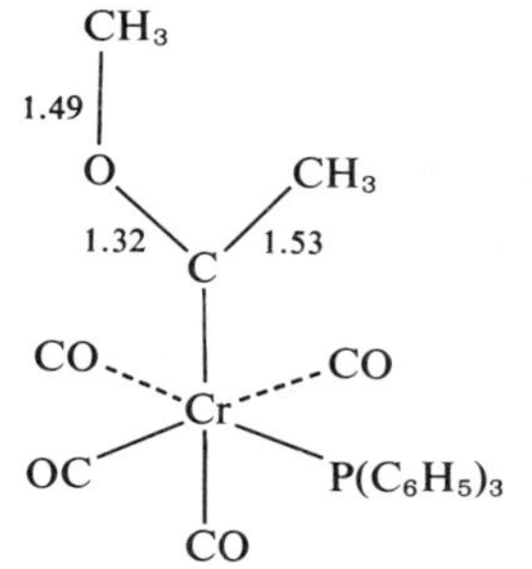

44

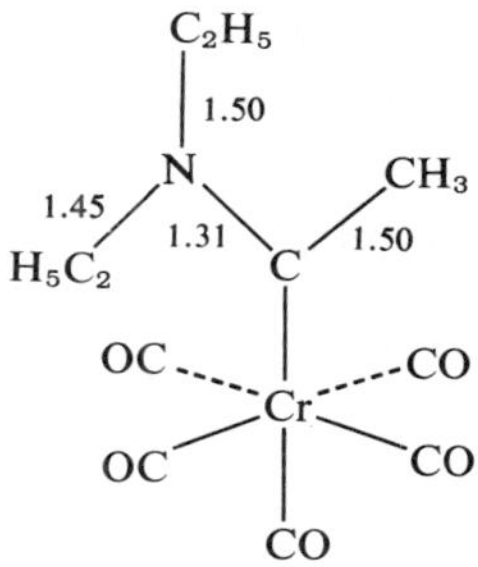

45

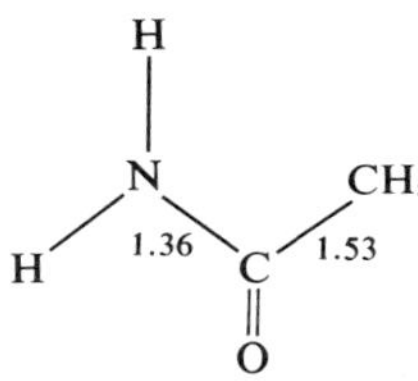

46

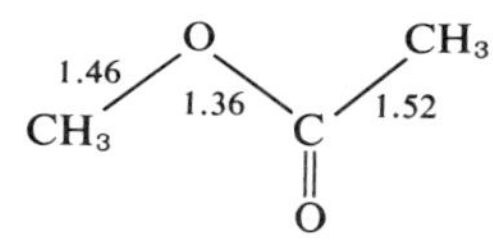

47

atoms [e.g., **43** to **45**], the $C_{carbene}$–heteroatom distance is consistently shorter than the heteroatom–C_{sp^3} distance [thus in **43** and **45**, it is 1.33 and 1.31 Å, as against 1.51 and 1.50 Å, and in **44** it is 1.32 Å as against 1.49 Å]. Furthermore, these bond distances are very close to those found in amides and esters (see **46** and **47**). In a compound containing oxygen and nitrogen (two heteroatoms) linked to $C_{carbene}$, i.e., **48**, the $C_{carbene}$–heteroatom distances are again short [168] and show a striking similarity to those found in ethyl carbamate **49** [172]. This consistent shortness of the $C_{carbene}$–heteroatom bond indicates that it possesses considerable double-bond character [156,164–171]. There will be, therefore, an appreciable barrier to rotation about the $C_{carbene}$–heteroatom bond, i.e., there should be cis and trans forms of such carbenoid-chromium compounds. The existence of such

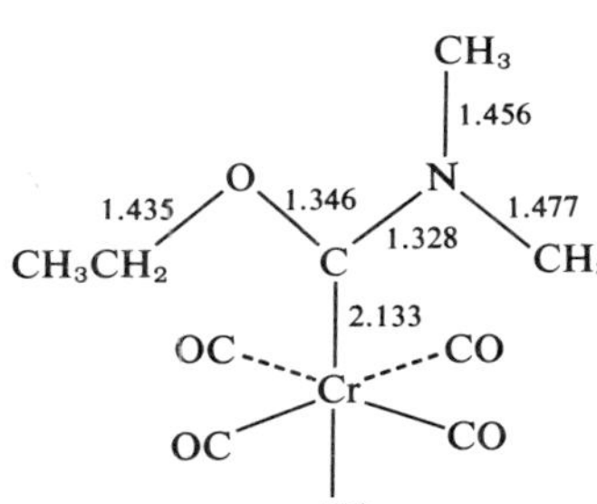

48

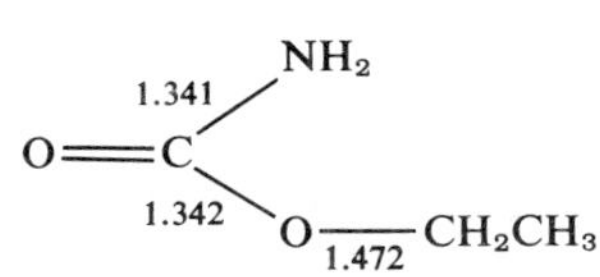

49

isomeric species has been established by x-ray crystallography, by NMR spectroscopy, and by chemical interconversion. Thus, a structure analysis of phenyl(methoxy)carbenoidCr(CO)$_5$ (**50**) shows that the CH_3 and C_6H_5 are disposed trans around the $C_{carbene}$–O bond [166] in contrast to methyl-(methoxy)Cr(CO)$_4$(P(C_6H_5)$_3$) (**44**), where the two CH_3 groups are cis disposed. It is of interest to note that the angle subtended by the organic substituents at $C_{carbene}$ seems to be larger in the cis-substituted compounds than in the trans-substituted ones.

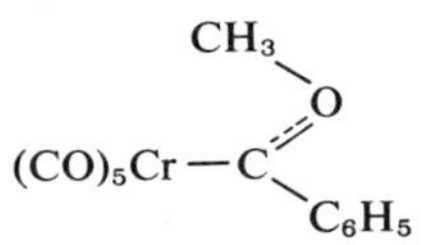
50

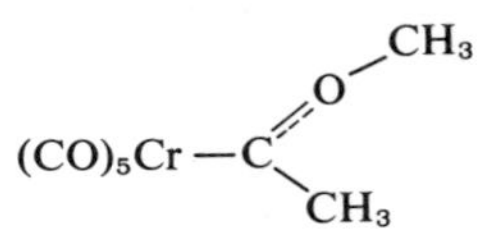
51

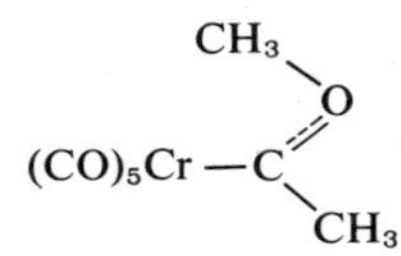
52

The occurrence of cis to trans interconversion in methyl(methoxy)-carbenoid-chromium pentacarbonyl (i.e., **51** to **52**) has been revealed by variable-temperature NMR spectroscopy [151] and, the isomerization of *cis*-methyl(methylamino)carbenoidCr(CO)$_5$ (**53**) [164], to the trans compound (**54**) has been effected by treatment with (*t*-BuO)K, Eq. (2-18) [147].

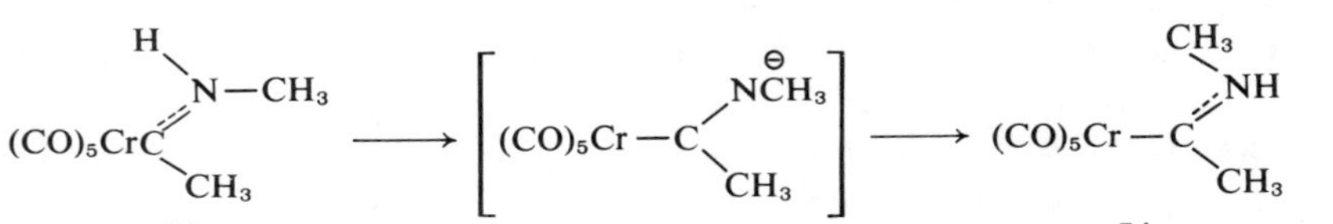
53 **54**

(2-18)

The bulk of a substituent bonded to the carbenoid-carbon may be such as to impede rotation around the substituent–$C_{carbenoid}$ bond thereby giving rise to a second type of isomers (atropisomers). Thus the reaction of 1-naphthyl(methoxy)carbenoidCr(CO)$_5$ with (+)-α-phenylethylamine gives mainly the cis isomer (**55**) together with some of the trans isomer [**56**, Eq.

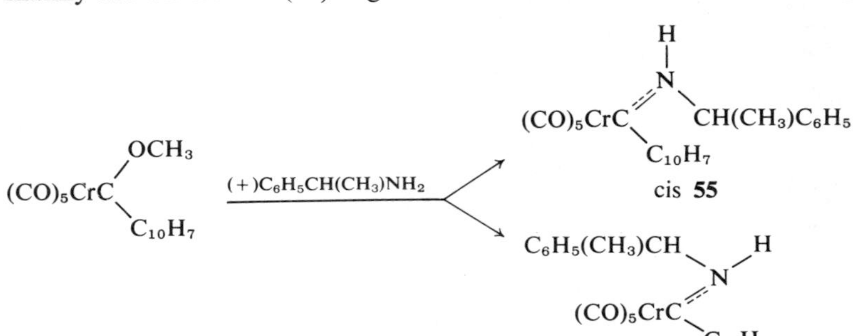
(2-19)

trans **56**

(2-19), $C_{10}H_7$ = 1-naphthyl] [173]. In both of these isomers the bulk of the 1-naphthyl group is such that it is not free to rotate about the $C_{carbene}$–naphthyl bond and, in the case of the cis isomer **55**, two atropisomers (**55a** and **55b**) (with $[\alpha]_{546}^{20} = -168°$ and $+46°$ respectively) have been isolated [173].

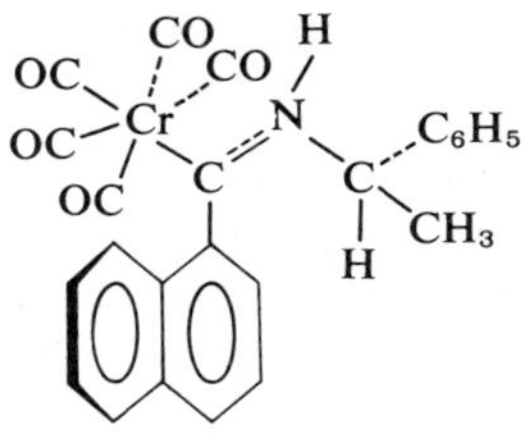

55a

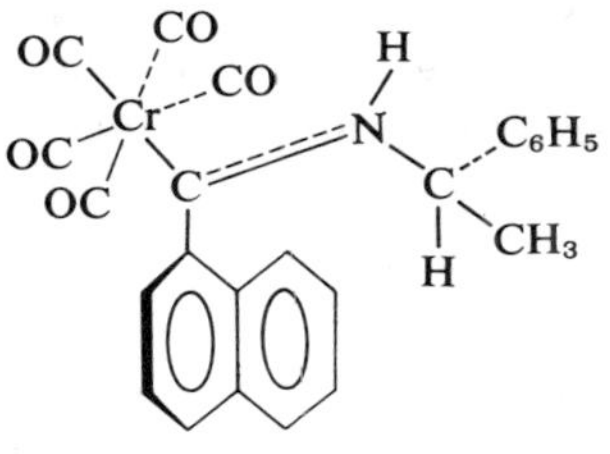

55b

The results accumulated so far from structural investigations of carbenoid-chromium compounds (Table 2.5) indicate significant variations in the lengths of the Cr–$C_{carbene}$ bond (2.00–2.16 Å). Thus, when oxygen or sulfur is bonded to $C_{carbene}$, the Cr–$C_{carbene}$ bond (2.00–2.04 Å) is shorter than when nitrogen is bonded to $C_{carbene}$ (Cr–$C_{carbene}$ 2.09–2.16 Å). These differences have been interpreted in terms of changes in the double-bond character of the Cr–$C_{carbene}$ bond [e.g., 164–171]. The bonding in carbenoid-chromium compounds (**57**) is believed to involve donation from the sp^2-hybridized carbenoid-carbon to the chromium atom (σ-contribution). Superimposed on this is a competitive back-donation (to $C_{carbene}$) from both the chromium center (d_π–p_π) and the heteroatom (p_π–p_π) (represented schematically in **57**).

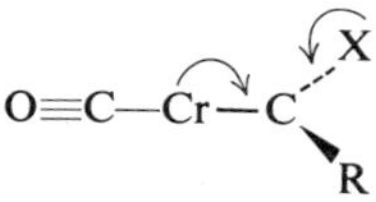

57

It follows that when the heteroatom is a strong donor (e.g., nitrogen) the back-donation Cr to $C_{carbene}$ will be weak and, conversely, when the heteroatom is a weak donor (e.g., oxygen), back-donation from chromium will be strong. That is, the double-bond character of Cr–$C_{carbene}$ will be at a maximum when weak donor atoms are attached to $C_{carbene}$ and will be progressively reduced with increasing donor capacity of the atoms attached to the carbenoid-carbon [146,156,164–171]. Thus, in carbenoid-chromium compounds with oxygen as heteroatom (Table 2.5, 4 and 5) the Cr–$C_{carbene}$ bond is shorter than that in compounds with the stronger donor, nitrogen as heteroatom (Table 2.5, 1 to 3). In the compound containing two heteroatoms bonded to $C_{carbene}$ (Table 2.5, 8), the Cr–$C_{carbene}$ bond length is 2.13 Å, which suggests a reduced back-donation from chromium. When there is no heteroatom bonded to $C_{carbene}$ (e.g., 7, Table 2.5), a short Cr–$C_{carbene}$ bond length is observed (2.05 Å), suggesting appreciable back-donation from chromium [171].

It might be expected that variations in the Cr–$C_{carbene}$ bond lengths in the various carbenoid-chromium compounds would be reflected in the lengths of the *trans*-Cr–C_{CO} bonds, in the sense that decreased back-donation from Cr to $C_{carbene}$ could be accompanied by increased back-donation in the *trans*-Cr–C_{CO} bond. Actually, a shortening in the length of the Cr–C_{CO} bond trans to the $C_{carbene}$ is observed in several cases (Table 2.5). Though this bond shortening is small, it is accompanied by a corresponding lengthening of the C_{CO}–O bond. In $CH_3(CH_3O)CCr(CO)_4P(C_6H_5)_3$, however, the length of the Cr–C_{CO} bond trans to $C_{carbene}$ is 1.86(2) Å and is the same as that of Cr–C_{CO} trans to $P(C_6H_5)_3$ [1.85(2)]; the lengths of the other two mutually *trans*-Cr–C_{CO} bonds are 1.89(2) and 1.87(2) Å, respectively [167].

In the case of carbenoid compounds containing only four carbonyl ligands, the fifth ligand and carbenoid unit may be either cis- or trans-disposed around the chromium center, as illustrated in **58** and **59**. Such cis and trans

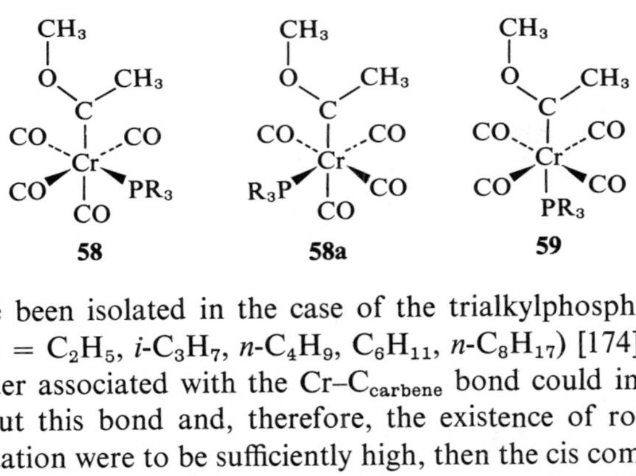

58 **58a** **59**

isomers have been isolated in the case of the trialkylphosphine complexes **58** and **59** (R = C_2H_5, *i*-C_3H_7, *n*-C_4H_9, C_6H_{11}, *n*-C_8H_{17}) [174]. The double-bond character associated with the Cr–$C_{carbene}$ bond could imply restricted rotation about this bond and, therefore, the existence of rotamers. If the barrier to rotation were to be sufficiently high, then the cis compound should exist in the two isomeric forms **58** and **58a**. To date no evidence has been adduced for the existence of such isomers, although in the case of a π-(arene)-carbenoid chromium compound **60** (R = C_6H_5), the infrared spectrum

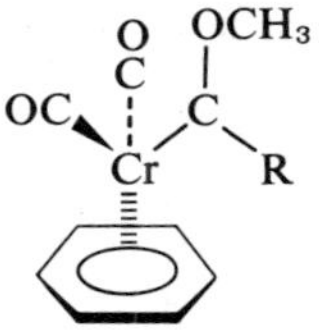

60

TABLE 2.5

Bond Distances and Angles in Carbenoid-Chromium Compounds

COMPOUND	Bond distances (Å)					Angle at $C_{carbene}$ (in degrees)	REF.
	$Cr{-}C_{carbene}$	$C_{carbene}{-}C_{carbene}$	$C_{carbene}{-}O(N,S)$	*trans*-Cr–CO[a]	equ-Cr–CO[b]		
1. $(CO)_5CrC(N(C_2H_5)_2)(CH_3)$	2.16(1)	1.50(1)	1.31(1)	1.85(1)	1.90 (1.89–1.91)	114.3(4)	[165]
2. $(CO)_5CrC(NHCH_3)(CH_3)$ *cis*	2.09(2)	1.51	1.33(3)	1.81(2)	1.87(2) (1.85–1.89)	—	[164,165]
3. $(CO)_5CrC(NHC_6H_{11})(COCH_3{=}CH_2)$ *trans*	2.15(1)	1.51(2)	1.32(2)	1.88(2)	1.88(2)	109.8(11)	[170]
4. $(CO)_5CrC(OCH_3)(C_6H_5)$ *trans*	2.04(3)	1.47(4)	1.33(2)	1.87(3)	1.89(3) (1.86–1.91)	104.0(2)	[166]
5. $(CO)_4[(C_6H_5)_3P]CrC(OCH_3)(CH_3)$ *cis*	2.00(2)	1.53(3)	1.32(2)	1.86(2)	1.87(2) (1.84–1.89)	114.0(1)	[167]
6. $(CO)_5CrC(SC_6H_5)(CH_3)$ *cis*	2.020(3)	1.490(4)	1.690(3)	1.899(3)	1.910(4) (1.901–1.914)	116.7(3)	[169]
7. $(CO)_5CrC$ (cyclopropenylidene, C_6H_5, C_6H_5)	2.05(1)	1.39,1.40	—	—	—	—	[171]
8. $(CO)_5CrC(OC_2H_5)(N(CH_3)_2)$	2.133(4)	$C_{carbene}{-}N$ 1.328(5)	$C_{carbene}{-}O$ 1.346(5)	1.847(5)	1.887(5) (1.873–1.897)	105.7(3)	[168]

[a] Carbonyl group trans to the carbenoid unit.
[b] Parentheses indicate range of the respective bond lengths.

(ν_{CO}) has been interpreted in terms of the presence of different rotamers arising from restricted rotation about the Cr–$C_{carbene}$ bond [175]. Theoretical considerations indicate that in molecules of the type $(CO)_5M$(carbene) the metal orbitals may form a symmetrical orbital such that the carbenoid unit can rotate freely while preserving the double-bond character in the M–$C_{carbene}$ bond. The orientation adopted by the carbenoid unit would then be determined solely by steric factors [176]. The concept of restricted rotation seems more plausible, however, and it will be of interest to see whether distinct isomeric species (due to restricted rotation about the Cr–$C_{carbene}$ bond) can be isolated.

F. σ-Bonded Organochromium(II) and (III) Compounds and Lithium or Sodium Complexes

1. Separation and Purification

When questions of separation and purification are considered, the known σ-bonded organochromium(III) and (II) compounds and lithium organochromate complexes may be divided into three main categories.

The first category is the air- and water-stable compounds of the general formula $[R_2Cr(bipyridyl)_2]^{\oplus}X^{\ominus}$ (R = aryl or trimethylsilylmethyl) [44,177–179]. These may be subjected to the usual crystallization techniques from a variety of polar solvents (e.g., CH_3OH, CH_3NO_2, $C_6H_5NO_2$) and to thin-layer chromatography.

The second category is the water-stable benzylic or pseudobenzylic species $[RCr(H_2O)_5]^{2\oplus}X^{\ominus}$. These compounds are sensitive to oxygen and have not been isolated as crystalline solids. Solutions of their pure perchlorates can, however, be obtained by chromatography on ion-exchange resins [180,181].

The third category includes the majority of the known σ-bonded organochromium(III) and (II) compounds, the lithium and sodium poly(organo)-chromate complexes, which are all extremely sensitive to oxygen and protic species (e.g., water, alcohols, β-diketones, etc.,). In view of the reactivity of these compounds, their purification is usually effected by recrystallization from diethyl ether or tetrahydrofuran in evacuated Schlenk tubes or in a rigorously inert atmosphere. The temperature range in which these crystallizations may be effected is determined by the thermal stabilities of the compound in question. Thus, $(C_6H_5)_3Cr(THF)_3$ and $CH_3CrCl_2(THF)_3$ may be handled between 0° and 40°C [182,183], whereas *n*-decyl$CrCl_2(THF)_n$ and analogous alkyldichlorochromium(III) compounds must be handled below −40°C [184,185].

Not all the known compounds in this category have been isolated in a crystalline state and many of the crystalline solvated compounds that have been isolated have a tendency to lose both their coordinated and their solvating species (tetrahydrofuran or diethyl ether) on drying in high vacuum.

2. Characterization and Identification

The main methods used for the characterization and identification of σ-bonded organochromium(III) and (II) compounds and lithium or sodium poly(organo)chromate complexes are the following: combustion, chemical and spectroscopic analyses, and magnetic moments.

a. Combustion Analyses

Combustion analyses and their associated problems have been reviewed and discussed by Schwarzkopf and Schwarzkopf [1].

b. Chemical Analyses

i. Reaction with Mercuric Chloride. This reaction is an excellent method for the characterization and identification of σ-bonded organochromium compounds. Thus, whereas π-bis(arene)chromium(I) compounds react with mercuric iodide to give complex salts without disrupture of the organochromium unit [186] [Eq. (2-20)], all the known σ-bonded organochromium(III) and (II) compounds react with mercuric chloride to give quantitatively (by cleavage of the C–Cr bond) the organomercuric halide and a Cr(III) species, Eqs. (2-21) and (2-22) [e.g., 44,180,182,187]. In the case of organochromium(II) compounds, the excess mercuric chloride is simultaneously reduced to mercurous chloride, e.g., Eq. (2-22) [e.g., 44, 187]. There are also

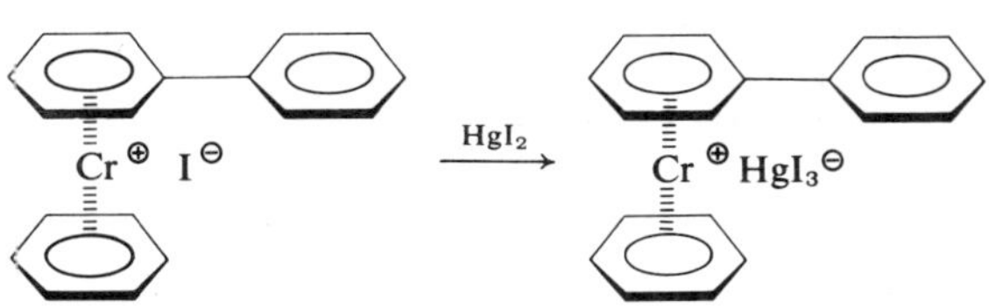

(2-20)

$$R_n(III)CrCl_{3-n}(S)_x \xrightarrow{HgCl_2/THF} n\,RHgCl + CrCl_3(THF)_3 \qquad (2\text{-}21)$$

$$R_2Cr(II)(S)_x \xrightarrow{HgCl_2/THF} 2\,RHgCl + CrCl_3(THF)_3 + Hg_2Cl_2 \qquad (2\text{-}22)$$

extensive reports that σ-bonded organochromium(III) and (II) lithium and sodiumpoly(organo)chromate complexes react with $HgCl_2$ to give quantitatively the corresponding organomercuric halide [e.g., 188–190]. Here again, the reaction of the organochromium(II) compounds with mercuric chloride is accompanied by the reduction of the excess $HgCl_2$ to mercurous chloride.

This reaction is not only an excellent general method for establishing quantitatively the presence of C–Cr bonds but it has the additional advantage of locating the position of the C–Cr link within the organic group, e.g., Chapter 1, Section F,3,a [191].

The mechanism of the reaction of pyridinomethylchromium pentaaquates with a variety of mercuric species has been studied in detail [192,193]. The results confirm direct electrophilic attack of the mercuric species on the carbon bonded to chromium, e.g., Eq. (2-23) (R = 2-, 3-, or 4-pyridino) without the formation of a halogen-bridged intermediate, i.e., Eq. (2-24).

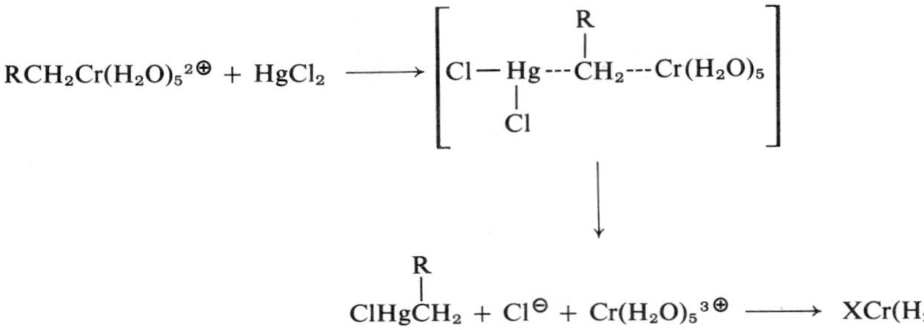

(2-23)

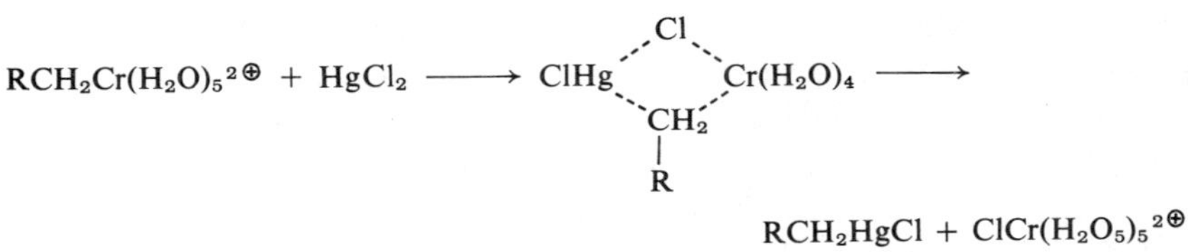

(2-24)

Such a mechanism explains why compounds that are stable to water and contain no readily replaceable ligands {e.g., $[R_2Cr(bipy)_2]I$} react instantaneously with solutions of mercuric halides [44].

An interesting modification of the above method involves the reaction of the organomercuric halide with bromine to give the corresponding organobromo compound. The latter may then be separated and estimated by gas-chromatographic analysis [191].

ii. Reaction with D_2O. Deuteriolysis forms a convenient method for providing the presence and the location of C–Cr bonds in σ-bonded organochromium(III) and (II) compounds and complexes [e.g., 181,184]. The resulting deuterated hydrocarbons may be isolated by gas chromatography, their isotopic compositions established by mass spectrometry, and the location of the deuterium atom, ascertained in many instances by NMR and IR spectroscopies. The reaction with deuterium oxide proceeds quantitatively to give monodeuterated compounds, and no hydrogen–deuterium exchange has been observed in the cases studied to date. This method has the advantage that the chromium content of the compound may also be estimated, either by conventional ways or by conversion to the volatile and crystalline tris-(acetylacetonato)chromium.

The carbon-to-chromium bond in organochromium(III) compounds may also be cleaved by D_2 gas (to give the monodeuterioorganic compound), though in some instances [e.g., tris(mesityl)chromium(III)] this reaction is accompanied by H/D exchange (to give dideuterio compounds) [191].

iii. Reaction with Acetylacetone. This reaction constitutes a simple, direct method for analyzing σ-bonded organochromium(III) compounds [194]. Acetylacetone is sufficiently acidic to cause cleavage of the carbon-to-chromium bond, Eq. (2.25). The products of the reaction [i.e., the hydro-

$$R_3Cr(S)_x + CH_3COCH_2COCH_3 \longrightarrow 3\,RH + Cr(AcAc)_3 \qquad (2\text{-}25)$$

carbon and $Cr(AcAc)_3$] can then be identified and analyzed by gas chromatography [194a]. This reaction has not been extensively used to date but it could provide a promising field for further investigation.

iv. Reaction with Iodine. The reaction of organomagnesium halides with iodine is known to proceed erratically. In some cases, relatively high yields of the corresponding organic iodide (ca. 90%) are obtained; while in other cases, the product is a mixture of the dimeric hydrocarbon and organic iodide [195]. Therefore, it is not surprising that the reaction of σ-bonded organochromium compounds should also give erratic results. This is illustrated by the following reactions:

Tris(allyl)chromium(III) reacts with iodine, in tetrahydrofuran, to give allyl iodide [196]. If there is insufficient iodine, the products (in toluene) are biallyl and CrI_2 [197]. Diphenylchlorochromium(III) with iodine gives quantitatively iodobenzene [198]. The two σ-bonded organochromium(II) compounds bis(mesityl)- and bis(2-methoxyphenyl)chromium(II) behave differently toward iodine, Eqs. (2-26) and (2-27) [187,199]. In lithium-

$$[2,4,6\text{-}(CH_3)_3C_6H_2]_2Cr(THF) \xrightarrow{I_2/THF} CrI_3(THF)_n + 2[2,4,6\text{-}(CH_3)_3C_6H_2I] \qquad (2\text{-}26)$$

$$[(2\text{-}CH_3OC_6H_4)_2Cr]_n \xrightarrow{I_2/THF} CrI_2 + (2\text{-}CH_3OC_6H_4)_2 \qquad (2\text{-}27)$$

(organo)chromate(III) complexes, the hexamethyl compound behaves differently from the hexaphenyl one, Eqs. (2-29) and (2-30) [200,201]. In all

$$Li_3Cr(CH_3)_6 \xrightarrow{I_2} 6\,CH_3I + CrI_3 + 3\,LiI \qquad (2\text{-}28)$$

$$Li_3Cr(C_6H_5)_6 \xrightarrow{I_2} 3\,C_6H_5I + \tfrac{3}{2}C_6H_5C_6H_5 + CrI_3 + 3\,LiI \qquad (2\text{-}29)$$

cases, however, the organochromium compounds are hydrolyzed to give quantitatively the appropriate arene [e.g., 201]. It is evident from information available that the results of iodometric titrations of σ-bonded organochromium compounds have to be interpreted with caution. In each case, the precise products formed, both organic and inorganic, and the overall mass balance must be determined.

c. Spectroscopic Methods

Neither NMR nor ESR spectroscopies have been of much use in the characterization of diamagnetic σ-bonded organochromium(II) or paramagnetic organochromium(II) and (III) compounds. In the case of the former method, there are very few diamagnetic Cr(II) compounds known, and of these, few are sufficiently soluble in inert solvents to give satisfactory NMR spectra. The reports available concerning the use of ESR spectroscopy suggest that no satisfactory signal has yet been obtained at room temperature for solutions of organochromium(III) or Cr(II) compounds [e.g., 187,200].

Infrared spectroscopy has been used, in a qualitative manner, for the characterization of many σ-bonded organochromium compounds (as mulls prepared in anaerobic conditions in fluorocarbons and nujol) [e.g., 177,179, 196,202–204]. In the case of the absorption of ligands bonded or coordinated to chromium, the expected shift to lower frequencies is observed. Thus, in π-allylchromium(III) and (II) compounds the bands at 1510 and 1522 cm^{-1} have been assigned to the C=C double bond [196,202]. With coordinated tetrahydrofuran, the absorptions at 1076 and 917 are shifted to 1020 and 860 cm^{-1}, respectively [203,204]. The Cr–C stretching frequencies in $[Li_2Cr(CH_3)_4(THF)_2]_2$ have been ascribed to absorptions at 430, 450, and 480 cm^{-1} [204].

The ultraviolet and visible spectra of several σ-bonded organochromium compounds have been reported [e.g., 44,181,205–207], and the ligand field parameters for $Li_3Cr(CH_3)_6(dioxane)_3$ have been calculated [205].

Ultraviolet and visible spectroscopy have been used extensively in the characterization of σ-bonded organochromium(III) pentaaquates and related compounds. Pyridinomethylchromium pentaaquates have characteristic absorptions at λ_{max} 220–280, 308–335, and 534–550 nm [181]. The benzyl and substituted benzylchromium pentaaquates have a characteristic absorption band in the region 322–370 nm, which has been used to study the kinetics of their formation and cleavage [208]. Alkylchromium compounds, formed by the reduction of alkylhalides with a Cr(II)-ethylenediamine complex are characterized by an absorption band in the region 360–420 nm [209]. The precise value for the true molecular extinction coefficient of these absorptions remains uncertain since no compound in the above series has been isolated in the crystalline state.

d. Magnetic Measurement

Magnetic susceptibility measurements are of great value in characterizing σ-bonded organochromium(III) and (II) compounds and lithium poly-(organo)chromates. The method has been particularly useful in detecting the presence of metal-to-metal bonds in some organochromium(II) compounds. Some representative magnetic moments are given in Table 2.6.

All the measured magnetic moments for σ-bonded organochromium(III) compounds and related lithium organochromate(III) complexes fall in the range 3.5–3.9 μ_B, e. g.,Table 2.6, compounds 1 to 8. These values are in good agreement with those expected for a d^3 system with three unpaired electrons (3.87 μ_B).

In the case of σ-bonded organochromium(II) compounds and complexes, magnetic measurements have revealed the existence of high-spin, low-spin, and diamagnetic compounds. Thus, the blue $[(2,4,6\text{-}CH_3)_3C_6H_2]_2Cr(THF)_3$ and related compounds, the blue bis(*N*-pyrrolyl)chromium(II) and the lithium organochromate(II) complex have moments in the range 4.65–4.82 μ_B. (Table 2.6, compounds 9–11), corresponding to high-spin Cr(II) species with four unpaired electrons (4.9 μ_B).

Other organochromium(II) compounds (Table 2.6, compounds 12–14), in particular the σ-bonded bis(organo)chromium-bis(bipyridyl) complexes have moments in the range 2.58–3.22 μ_B, corresponding to low-spin chromium(II) species with two unpaired electrons (2.8 μ_B).

Other σ-bonded organochromium(II) compounds and lithium (organo)-chromate(II) complexes have been found to be virtually diamagnetic. Thus, compounds 15–18 (Table 2.6) have magnetic moments in the range 0–0.6 μ_B. This lowering of the magnetic moment in such d^4 compounds was interpreted in terms of metal–metal bonds in polymeric or dimeric complexes and this has been confirmed for the three dimeric species by single crystal x-ray analysis (see Section F,3,a of this chapter).

3. X-RAY STRUCTURE ANALYSIS

a. σ-Bonded Organochromium(II) Compounds and Complexes

X-ray structure analyses of σ-bonded organochromium(II) compounds have revealed several interesting and novel features. Thus, they have confirmed that $[Li_2Cr(CH_3)_4(THF)_2]_2$, **61** [204], $[Li_2Cr(C_4H_8)_2(EtO)_2]_2$, **62** [214], and $[bis(allyl)Cr]_2$, **63** [215,216] are dimeric and contain a Cr–Cr bond. In all three compounds the carbon atoms bonded to chromium are eclipsed. The lithium chromocyclopentane complex **62** has D_{2d} symmetry so that the individual chromium units are rotated through 90° (about the Cr–Cr axis) with respect to each other. In the dimeric bis(allyl)chromium, the four

TABLE 2.6

Magnetic Moments of Some Representative σ-Bonded Organochromium Compounds

COMPOUND	OBSERVED MAGNETIC MOMENT (μ_B)	REF.
Chromium(III)		
1. 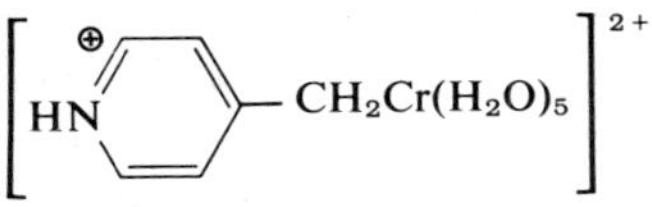	3.80	[181]
2. $C_6H_5CH_2CrCl_2(pyridine)_3$	3.93	[206]
3. 4-$CH_3C_6H_4CrCl_2(THF)_3$	3.87	[207]
4. $[(C_6H_5)_2Cr(bipyridyl)_2]I$	3.82	[44]
	3.77	[177]
5. $(C_6H_5)_2CrCl[CH_3OCH_2CH_2OCH_3]_{1.5}$	3.78	[210]
6. $(C_6H_5)_3Cr(THF)_3$	3.85	[182]
7. $Li_3Cr(CH_3)_6$ 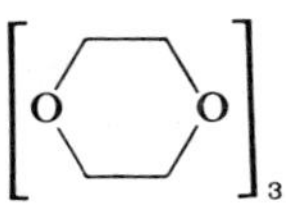	3.75	[200]
8. $LiCr(2\text{-}CH_3OC_6H_4)_4[(C_2H_5)_2O]$	3.89	[199]
Chromium(II)		
9. 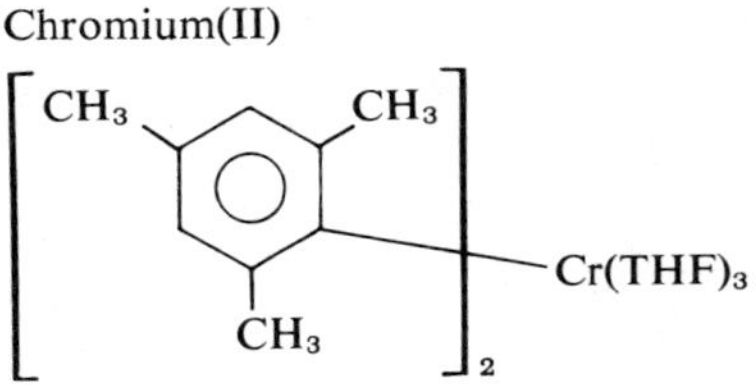	4.82	[187]
10. 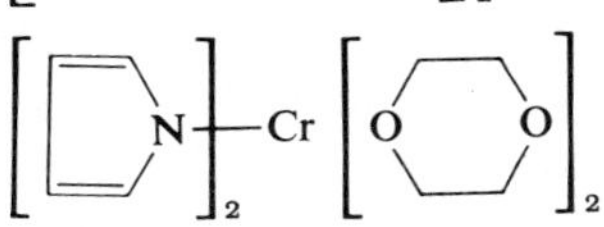	4.86	[211]
11. $Li_2Cr(C_6H_5)_4(THF)_4$	4.73	[212]
12. $(C_5H_7)_2Cr$	2.74	[213]
13. $(2\text{-}CH_3OC_6H_4)_2Cr(bipyridyl)_2$	2.58	[44]
14. $(C_6H_5)_2Cr(bipyridyl)_2$	3.22	[212]
15. $[(2\text{-}CH_3OC_6H_4)_2Cr]_n$	0.54	[199]
16. $[Li_2Cr(CH_3)_4(THF)_2]_2$	Diamagnetic	[200]
17. $[(C_3H_5)_2Cr]_2$	0.44	[202]
18. 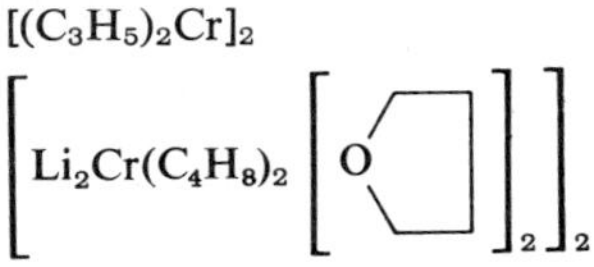	0.35	[214]
19. 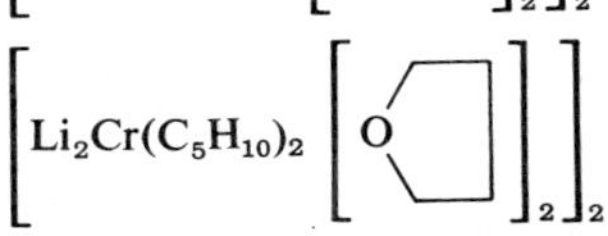	0.3	[214]

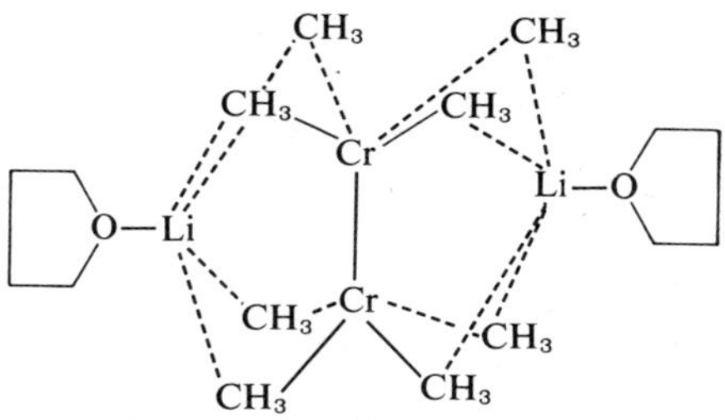

$Li_4Cr_2(CH_3)_8 \cdot 4$ THF

61

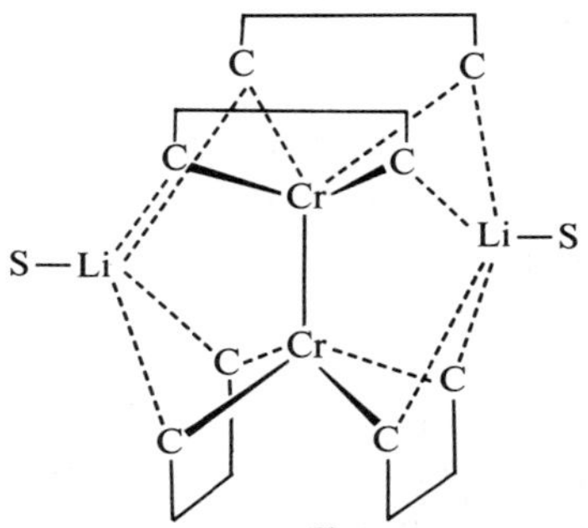

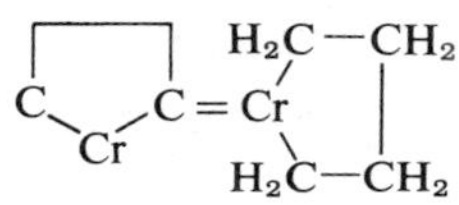

$S = O[C_2H_5]_2$

$Li_4Cr_2(C_4H_8)_4 \cdot 4[(C_2H_5)_2O]$

62

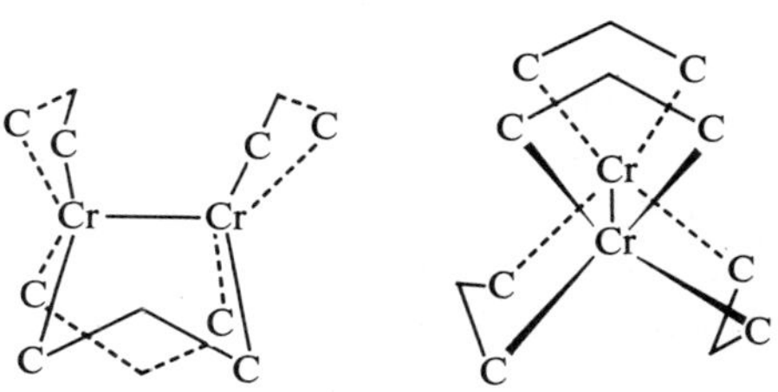

$[Cr(C_3H_5)_2]_2$

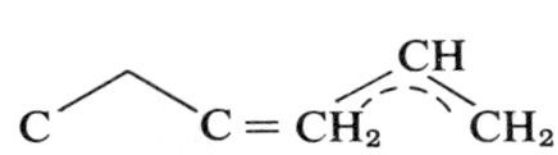

63

carbon atoms joined to chromium are not coplanar. The most striking feature in the three structures is the shortness of the Cr–Cr distances. These fall in the range 1.97–1.98 Å (see Table 2.7) and are all much shorter than those observed in Cr metal (2.50 Å) and in hydrated chromous acetate [2.362(1) Å] [98]. The shortness of the chromium–chromium distances can be interpreted in terms of a multiple bond, a quadruple bond [204,214], analogous to that postulated for $[ReX_8]^{2\ominus}$ ions [217]. The eclipsed conformation of the groups bonded to chromium is then a prerequisite for maximum orbital overlap between the metal centers [217]. Alternatively, the conformation of the carbon atoms bonded to chromium, in the lithium complexes **61** and **62** may be a result of the interaction of the lithium atoms with the carbons bonded to chromium. In effect, each of the four lithium atoms interacts with four of the carbons bonded to chromium, and more particularly with two on each of the chromium centers. This is depicted for two of the lithium atoms in **61** and **62**. The structure data for the lithium complexes **61** and **62** confirm that the alkyl groups are σ-bonded to chromium. The Cr–C bond lengths fall in the range 2.159–2.287(31) Å (see Table 2.7) suggesting that the Cr–C_{sp^3} distance in such square pyramidal Cr(II) species should be around 2.2 Å. However, since the carbon centers in these complexes interact with both chromium and lithium atoms, this value is not necessarily a general one for Cr–C_{sp^3} bonds.

The determination of the full structure of dimeric bis(allyl)chromium(II) **63** revealed that the four allyl groups are orientated tetrahedrally about the midpoint of the Cr–Cr bond. Two of the allyl groups are bonded to individual chromium centers (π-allyl groups) and the other two bridge the two chromium centers (μ-allyl groups). This is represented schematically in **63**. There appears to be a slight systematic difference in the distances from the chromium centers to the terminal carbon atoms of the π-allyl group (average Cr–C 2.24 Å), and from the chromium centers to the terminal carbon atoms of the bridging allyl groups (average Cr–C, 2.16 Å) [215]. A similar apparent systematic difference is also observed in π-bis(allyl)molybdenum dimer, the average Mo–C distance for the terminal carbon atoms of the π-allyl unit is 2.30 Å, while that for the terminal carbon atoms of the μ-allyl unit is 2.23 Å [218]. The average C–C bond length in the allyl groups of the chromium compound is 1.45 Å, which is close to the average C–C bond length (1.41 Å) found in the molybdenum compound [218].

b. σ-Bonded Mono(organo)chromium(III) Compounds

There are two main types of octahedral σ-bonded organochromium(III) compounds known to date, those of the type $RCrCl_2(L)_3$, and the pentaaquobenzylchromium cation (and related species) $[RCr(H_2O)_5]^{2\oplus}$. No crystalline specimen of the latter class of compounds has been isolated to date, so there is no structural information available in this field. The former class of compounds can possess one of three isomeric structures, i.e., the trans form, **64**, the cis form, in which R is trans to chlorine, **65**, and the chiral cis form, in which R is trans to L, **66**. The only crystallographic data

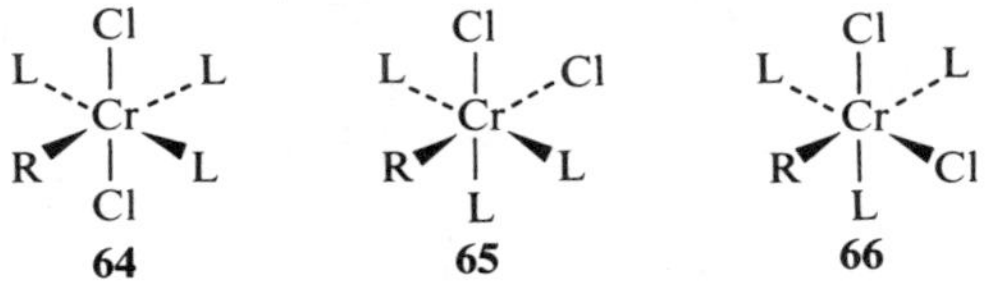

64 **65** **66**

available to date concerning these compounds of the type $RCrCl_2(L)_3$ are those for *p*-tolyldichlorochromium tris(tetrahydrofuranate). A complete single crystal x-ray structure analysis of $[(4\text{-}CH_3C_6H_4)CrCl_2(THF)_3]$ has shown that it is the *trans*-dichloro compound **64** ($R = 4\text{-}CH_3C_6H_4$ and L = THF), with the *p*-tolyl group σ-bonded to chromium [94,207]. The carbon-to-chromium bond length (2.014 Å) is of the same order of magnitude as that found in carbenoid-chromium compounds, in which oxygen is bonded to the carbenoid unit (2.00–2.04 Å, see Table 2.5) [48], but is significantly shorter than the Cr–C bond lengths in octahedral aryl complexes of the type $[R_2Cr(bipy)_2]I$ (2.087–2.106 Å, see Table 2.7). The difference in these $Cr\text{–}C_{sp^2}$ bond lengths in octahedral arylchromium complexes could be attributed to differences in back-donation between the arene ring and the chromium. However, as will be seen later in Section F,3,c, the lengths of carbon-to-chromium bonds depend upon the environment at the chromium center [219].

The salient feature that emerges from the present analysis is the lengthening of the chromium–oxygen bond trans to the σ-bonded *p*-tolyl group ($Cr\text{–}O_2$, 2.214 Å) compared to the two other mutually trans chromium–oxygen bonds ($Cr\text{–}O_1$ and $Cr\text{–}O_3$ average, 2.045 Å; see **67**; O_1, O_2, and O_3 represent the three solvating species 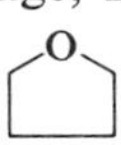). It follows that the tetrahydrofuran that is

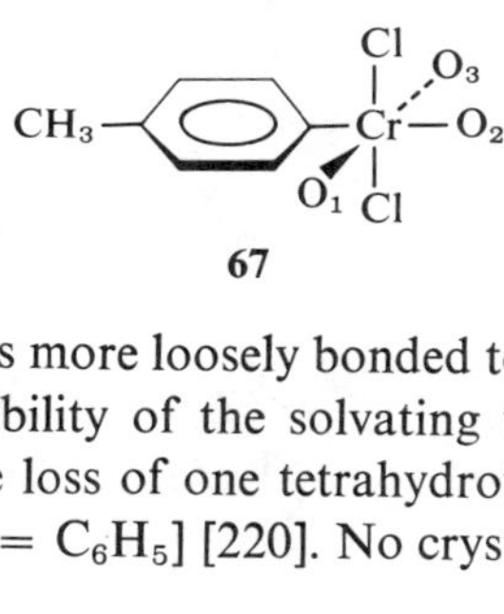

67

trans to the *p*-tolyl group is more loosely bonded to chromium than the other two. This difference in lability of the solvating tetrahydrofuran molecules may explain the reversible loss of one tetrahydrofuran from tris(phenyl)Cr-$(THF)_3$ [**68**, Eq. (2-30), $R = C_6H_5$] [220]. No crystal structure determination

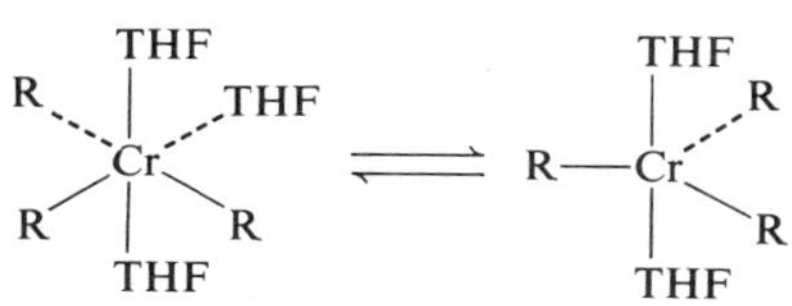

68 **69** (2-30)

of tris(phenyl)$Cr(THF)_3$ has been reported to date; however, assuming that it is the meridional isomer **68**, then the selective loss of the labile tetrahydrofuran (i.e., that trans to the phenyl group) would give the trigonal bipyramidal bistetrahydrofuranate **69**, Eq. (2-30). Complete x-ray structure analyses have

TABLE 2.7

Bond Distances in σ-Bonded Organochromium Compounds[a]

ORGANOCHROMIUM(II) COMPOUNDS	Cr–C	Cr–Cr			REF.
1. $[Li_2Cr(CH_3)_4 \cdot 2THF]_2$	2.199(13)	1.980(5)			[204]
2. $[Li_2Cr(C_4H_8)_2 \cdot 2\ Et_2O]_2$	2.238(22) (2.159–2.287)	1.975(5)			[214]
3. $[(C_3H_5)_2Cr]_2$	2.20(6) (2.15–2.27)	1.97(6)			[215]
4. $[(C_3H_5)_2Cr]_2$	—	1.98(6)			[216]
ORGANOCHROMIUM(III) and (IV) COMPOUNDS	**Cr–C**	**Cr–X (TRANS TO C)**	**Cr–X (TRANS TO X)**	**ANGLE AT RING C**	**REF.**
5. *trans*-4-$CH_3C_6H_4CrCl_2(THF)_3$	2.014(10)	2.214(7)	2.045(8) (2.044–2.046)	114.0(10)	[94,207]
6. *cis*-$[(2\text{-}CH_3OC_6H_4)_2Cr(bipy)_2]^{\oplus}I^{\ominus}$	2.101(12) (2.095–2.107)	2.156(10) (2.155–2.157)	2.071(10) (2.069–2.073)	114.4(11)	[178,225]
7. *cis*-$[(C_6H_5)_2Cr(bipy)_2]^{\oplus}I^{\ominus}$	2.087(4)	2.147(4)	2.087(4)	115.2(4)	[219]

8. *cis*-$\{[(CH_3)_3SiCH_2]_2Cr(bipy)_2]\}^{\oplus}I^{\ominus}$	2.106(9) (2.113–2.100)	2.156(7) (2.146–2.166)	2.103(7) (2.100–2.106)		[179]
9. *fac*-$\left[2\text{-}\begin{matrix}H_2C\text{—}O\\ \mid\\ H_2C\text{—}O\end{matrix}\!>\!CHC_6H_4\right]_3Cr$	2.037 (2.026–2.045)	2.157 (2.133–2.176)		116	(226a)
10. $Li_3Cr(CH_3)_6(dioxane)_3$	2.300(15)				[228,228a]
11. $Cr(CH_2C(CH_3)_2C_6H_5)_4$[b]	Cr–C 2.01–2.07				[239]
12. $Na_2[Cr(C_6H_5)_5][(C_2H_5)_2O]_3[C_4H_8O]$	Cr–C Equatorial 2.03(6) [2.01–2.06] Apical 2.17(7) [2.12–2.22]				[230]

[a] X = O in compounds 5 and 9 and X = *N* in compounds 6 to 8. The bond lengths are in Å; the upper row of figures is the average bond distances; the lower row of figures in parentheses is the range of the respective bond distances; angles are given in degrees.

[b] An organochromium(IV) compound.

shown that the pentacoordinate chromium(III) compounds bis(trimethylamine)$CrCl_3$ and $Na_2Cr(C_6H_5)_5 \cdot 3[(C_2H_5)_2O][THF]$ do have trigonal bipyramidal structures. Details for the former compound, including the relevant bond distances from the chromium center to the five surrounding atoms, are given in **70** [96,97]. The structure of the sodium organochromate-(III) complex is discussed in detail later in Section F,3,e of this chapter.

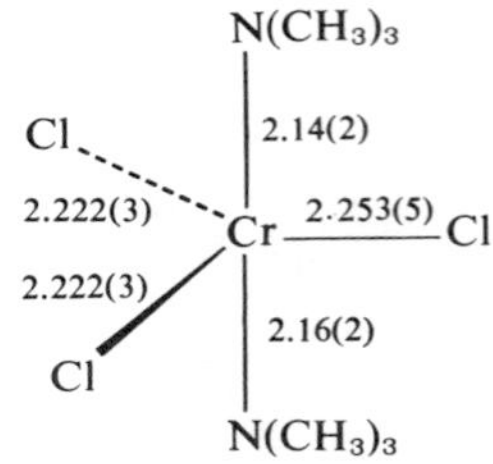

70

A final feature of interest in the structure of **67** is the angular distortion at the carbon atom bonded to chromium. This angle (114.0°) is smaller than that normally associated with the trigonal C_{sp^2} angle (120°). Similar distortions are observed in other σ-bonded octahedral arylchromium(III) compounds (see Table 2.7) and in other σ-bonded aryl transition metal compounds [e.g., 221,222].

c. σ-Bonded Bis(organo)chromium(III) Compounds

Two main classes of these compounds have been described to date; those of the type $R_2CrCl(L)_n$ and the air- and water-stable cationic complexes $[R_2Cr(bipy)_2]^{\oplus}X^{\ominus}$ (see Chapter 1, Section F,2). The available structural data concern the latter class.

Octahedral σ-bonded organochromium cations of the type $[R_2Cr(N–N)_2]^{\oplus}X^{\ominus}$, where N–N is a bidentate nitrogen-containing ligand can exist either as the trans isomer **71** or as the chiral cis isomer **72**. The precise configuration adopted will clearly depend upon the steric requirements of

71 **72**

the organic groups bonded to chromium and of the bidentate ligand. The trans configuration **71** (R = Cl; N–N = 2,2′-bipyridyl) was originally assigned to the cationic dichloro-bis(2,2′-bipyridyl)chromium(III) [223]. However, a reconsideration of the electronic spectral data and of the steric requirements of the two bipyridyl units led to the conclusion that it was indeed the cis isomer **72** (R = Cl; N–N = 2,2′-bipyridyl) [224].

A synthetic route to bis(organo)-bis(2,2′-bipyridyl)chromium(III) halides has recently been developed [44,178] and has permitted the systematic investigation of the structure of these compounds with special reference to the organic groups bonded to the chromium.

The results of single crystal x-ray structure analyses of three compounds $[R_2Cr(bipy)_2]^{\oplus},I^{\ominus}$ where R = C_6H_5 [219] (**73**) 2-$CH_3OC_6H_4$ [178,225] (**74**), and $(CH_3)_3SiCH_2$ [179] (**75**) have shown that they are all cis σ-bonded organochromium compounds (i.e., derivatives of **72**). The unit cells of the phenyl compound (monoclinic, $C\ 2/_c$), the 2-methoxyphenyl compound (triclinic, $P\ \bar{1}$) and the trimethylsilymethyl compound (monoclinic $P\ 2_{1/c}$), all contain equal numbers of *d* and *l* molecules. To date, however, no member

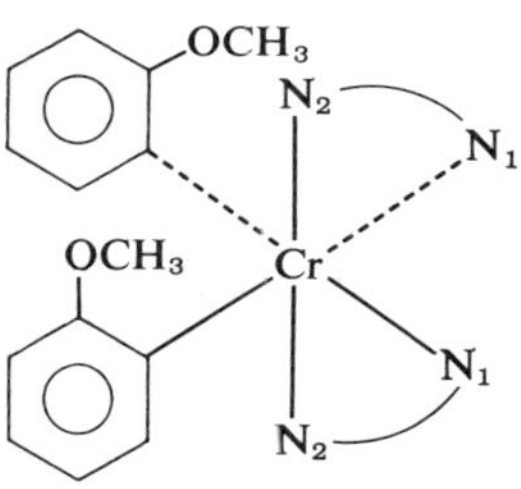

74

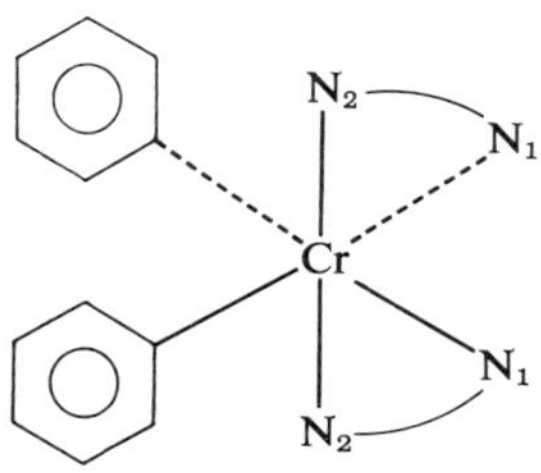

73

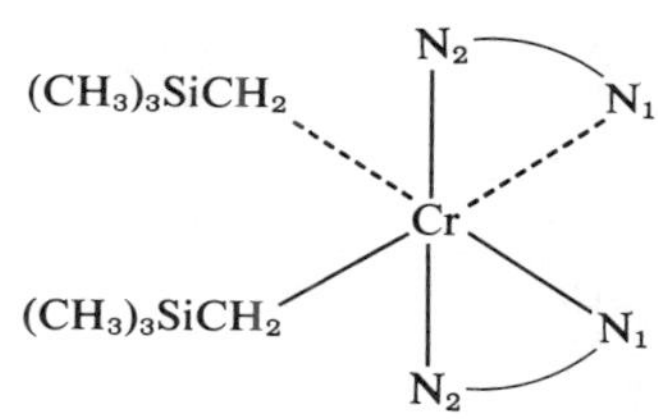

75

of this class of compound has been resolved into pure optical isomers. The bond distances from chromium to the six surrounding atoms are given in Table 2.7. The most striking feature is that the respective bond lengths are equal, within experimental limits, for all three compounds. The carbon–chromium distances fall within the range 2.087–2.106 Å and are much longer than the Cr–C_{sp^2} bond in 4-$CH_3C_6H_4CrCl_2(THF)_3$. The chromium–nitrogen bond lengths fall into two groups: the longer bonds (range 2.147–2.156 Å) involve the nitrogen atoms trans to carbon (i.e., Cr–N_1), and the shorter involve the nitrogen atoms trans to each other (i.e., Cr–N_2). The latter fall in the range 2.071–2.103 Å and are similar to the Cr–N bond distances (2.08(3) Å) in tris(bipyridyl)chromium(0) [226]. The ring angles subtended at the carbon atom, which is bonded to chromium in the two arylchromium compounds, are again appreciably less than the normal value for benzene (see Table 2.7).

The results show that the chromium–carbon bond length is not affected either by substituting an electron-releasing group in the aryl ring bonded to chromium or by going from an arylchromium compound to an alkylchromium compound. This suggests either that back-donation between carbon and chromium has only minor effects on the bond length or, less plausibly, that any such back-donation is the same for an aryl group as for a trimethylsilylmethyl group bonded to chromium. In contrast, however, the nature of the ligand trans to the carbon does have an appreciable effect on the carbon–chromium bond length. Thus, a comparison of the Cr–C_{sp^2} bond lengths in 4-$CH_3C_6H_4CrCl_2(THF)_3$ (**67**) with those in [(aryl)$_2$Cr(bipyridyl)$_2$]I (**73** and **74**) (see Table 2.7) show that there is an inverse relationship between the Cr–C_{sp^2} bond lengths and the electronegativity of the atom trans to the carbon, i.e., the Cr–C trans to nitrogen is longer than that trans to the more electronegative oxygen. This suggests that the influence of "back-donation" on carbon–metal bonds has perhaps been overemphasized and that the electronegativity and steric requirements of the trans ligand also play a determining role in carbon–metal bond lengths.

d. σ-Bonded Tris(organo)chromium(III) Compounds

Many of the known solvated σ-bonded tris(organo)chromium(III) compounds have been formulated as hexacoordinate octahedral d^3 complexes (cf. Chapter 1, Table 1.4), one exception being the tetracoordinate tris(mesityl)chromium(III)monotetrahydrofuranate. Octahedral σ-bonded tris(organo)chromium(III) compounds can have either the facial **76a** or the

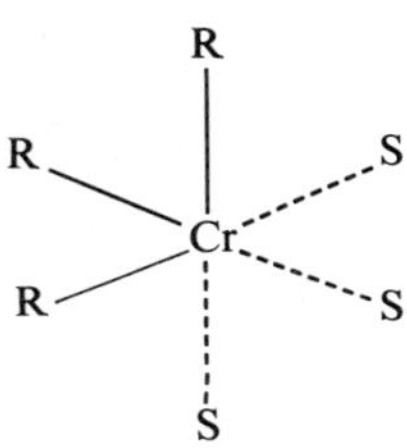

facial

76a

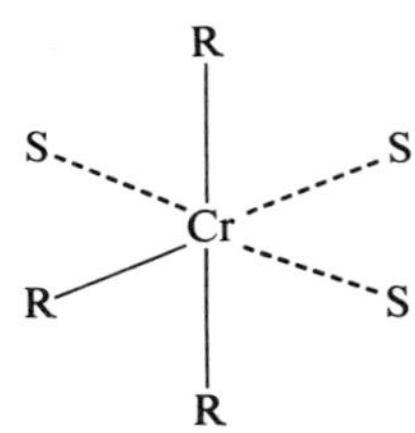

meridional

76b

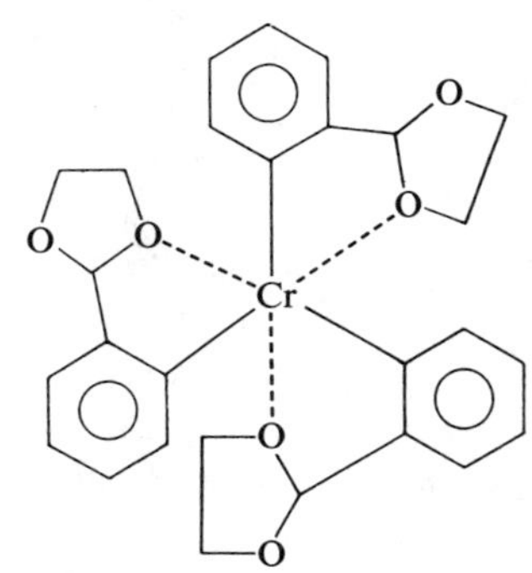

76c

meridional structure **76b**. A single crystal x-ray structure analysis of tris(2-dioxolanophenyl)chromium(III)

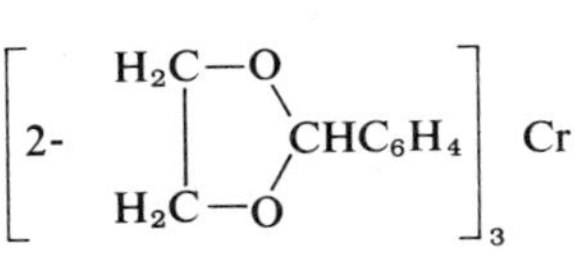

shows that it is an octahedral σ-bonded tris(organo)chromium(III) compound with the facial configuration **76c** [226a], i.e., all the carbon atoms that are bonded to chromium are trans to oxygen atoms and there are no carbon atoms trans to each other. The carbon-to-chromium bond lengths fall within the range 2.026–2.045(10) Å and are slightly longer than the Cr–C_{sp^2} bond length in 4-$CH_3C_6H_4CrCl_2(THF)_3$ (**67**) but are shorter than the Cr–C_{sp^2} and Cr–C_{sp^3} bond lengths in octahedral σ-bonded bis(organo)chromium(III) compounds (see Table 2.7). The chromium-to-oxygen bond lengths fall within the range 2.133–2.176(9) Å and are somewhat shorter than the Cr–O bond length trans to the aryl group in **67**, though longer than the two mutually trans Cr–O bond lengths in **67**. Compound **76c** is inert toward O_2, H_2O, CH_3OH, acetone, nitriles, and ethylene oxide, though it does act as a host toward the last three compounds in the crystalline state giving stable solvates. This inertness could be associated with the fact that the molecule has the "symmetrical trischelate" structure **76c** and that such a structure effectively inhibits the approach of the substrate to the chromium center. It is of considerable interest, therefore, to determine the structures of the more reactive and labile σ-bonded tris(organo)chromium compounds [e.g., $(C_6H_5)_3Cr(THF)_3$] in order to ascertain whether they are the meridional isomers **76b**.

e. Miscellaneous σ-Bonded Organochromium Compounds and Chromium Complexes

There is a preliminary report concerning the structure of π-tris(allyl)-chromium(III) in which it is stated that the three allyl groups are π bonded to the central chromium atom [227].

A complete single crystal x-ray structure analysis of $Li_3Cr(CH_3)_6(dioxane)_3$ confirms that the chromium center is surrounded octahedrally by six methyl groups, as in **77** [228,228a]. The average Cr–C bond length of 2.300(15) Å

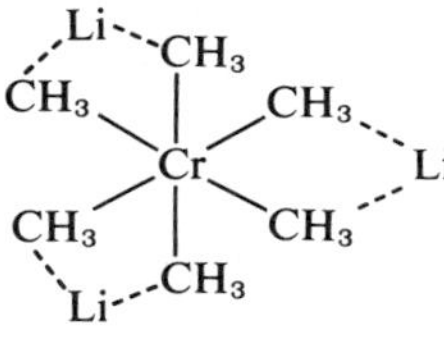

77

is considerably longer than any other chromium-to-carbon (either Cr–C_{sp^3} or Cr–C_{sp^2}) bond length known to date (see Table 2.7).

A complete single crystal x-ray structure analysis has been carried out on solvated sodium pentaphenylchromate [229,230]. This compound, which is prepared by the interaction of phenylsodium and $CrCl_3(THF)_3$ in diethyl ether [231] (see Chapter 1, Section G,2) crystallizes with varying amounts of diethyl ether and tetrahydrofuran depending upon the method of preparation. There always appears to be four molecules of ether per chromium. The x-ray analysis was carried out on a sample containing three diethyl ether molecules and one tetrahydrofuran, i.e., $Na_2Cr(C_6H_5)_5[(C_2H_5)_2O]_3[THF]$. The $[Cr(C_6H_5)_5]^{2-}$ unit possesses a distorted trigonal bipyramidal structure represented schematically in **78a** and **78b**. The determination of the present

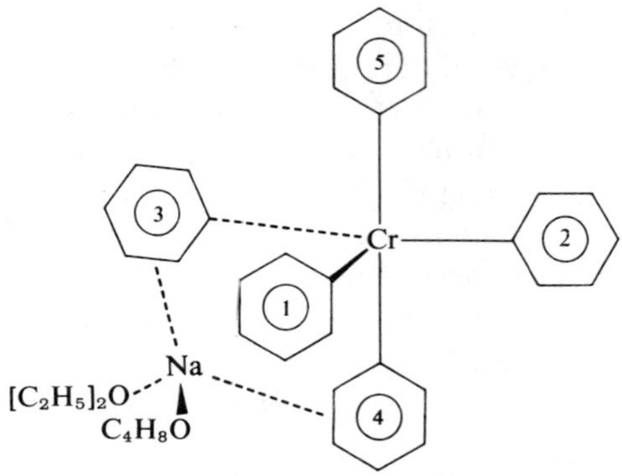

78a

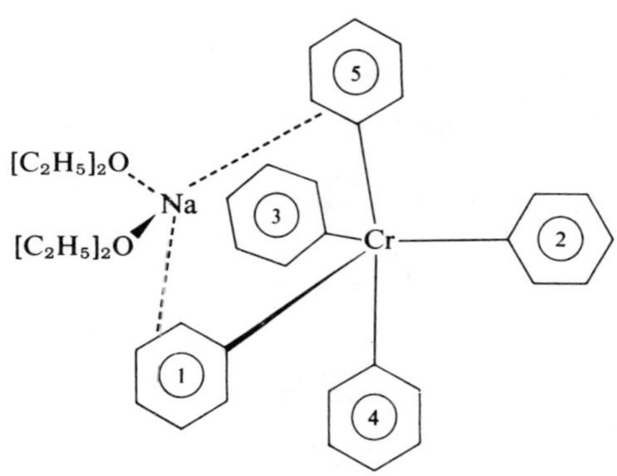

78b

structure was, however, rendered particularly difficult by the sensitivity of the compound, particularly to x-rays, and to the thermal motions within the crystal (the final R-factor was 17.2%) [230].

The five phenyl groups are σ-bonded to chromium(III) and each sodium atom is associated with one apical and one equatorial phenyl ring together with two solvating ether molecules (i.e., two diethyl ethers for one sodium and one diethyl ether and one tetrahydrofuran for the second sodium, **78a** and **78b**, respectively). The chromium–C_{phenyl} distances seem to fall into two groups: those to the equatorial aryl rings (i.e., 1,2 and 3) and those to the apical ones (i.e., 4 and 5). The chromium–C_{phenyl} distances of the former are of the order of 2.03(6) Å, [analogous to the Cr–C_{aryl} bond distance found in 4-$CH_3C_6H_4CrCl_2(THF)_3$] and are shorter than those to the apical aryl rings [Cr–C_{phenyl} 2.17(7) Å] [analogous to the Cr–C_{aryl} bond distance found in bis(aryl)-bis(2,2′-bipyridyl)chromium(III) compounds (see Table 2.7)].

Bradley and his colleagues have isolated the chromium(III) and chromium(II) compounds having the unusual coordination number of three and four, respectively. The details of their structure are included here because it might be possible with a suitable choice of organic ligand to prepare the corresponding σ-bonded organochromium compounds.

Tris[bis(isopropyl)aminato]chromium(III) is a highly air-sensitive compound possessing the planar trigonal structure **79** with a Cr–N (amide) distance of 1.87 Å [232].

Nitroso-tris[bis(trimethylsilyl)aminato]chromium(II) has the trigonal pyramidal structure **80** [NR = $N(Si(CH_3)_3)_2$] with a Cr–N (amide) bond distance of 1.790 Å [233].

Bis[bis(trimethylsilyl)aminato]chromium(II)-bis(tetrahydrofuranate) has the trans square planar structure **81** [NR = $N(Si(CH_3)_3)_2$], with a Cr–N (amide) bond distance of 2.089 Å and Cr–O (tetrahydrofuran) distances of 2.090 Å [233] similar to those found for the mutually trans Cr–O(THF) distances [2.045(8) Å] in (4-$CH_3C_6H_4$)$CrCl_2(THF)_3$ [94].

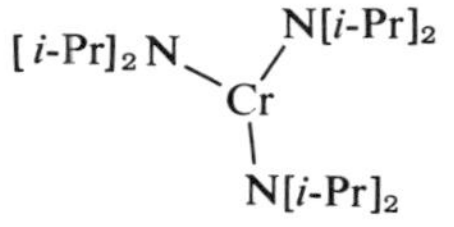

79

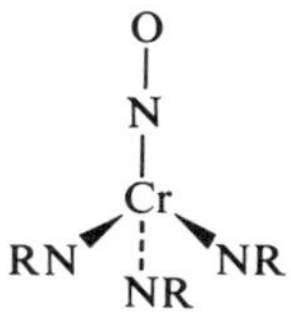

80

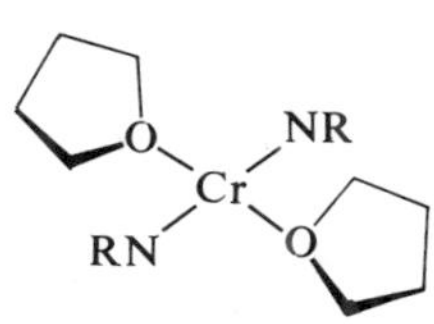

81

G. σ-Bonded Tetra(organo)chromium(IV) Compounds

1. Separation and Purification

The sensitivity of tetra(organo)chromium(IV) compounds toward air and moisture and their thermal stabilities depend upon the nature of the organic ligands. Thus, $[(CH_3)_3SiCH_2]_4Cr(IV)$, $(CH_3)_4Cr(IV)$, $[(CH_3)_3CCH_2]_4Cr(IV)$, and $[C_6H_5(CH_3)_2CCH_2]_4Cr(IV)$ are reported to be sensitive to oxygen [234,235], whereas tetra(1-norbornyl)chromium(IV) and tetra(1-camphyl)-chromium(IV) in octane solution appear to be stable to oxygen [236] and $[(C_6H_5)_3CCH_2]_4Cr(IV)$ may be handled in air [235]. Tetra(*tert*-butyl)-chromium(IV) and the corresponding neopentyl compound are stable in solution at room temperature, while a variety of other tetra(alkyl)chromium-(IV) compounds (e.g., alkyl = CH_3, n-C_4H_8, s-C_4H_8 and cyclohexylmethyl) are stable in pentane [235,237] only at lower temperatures.

The thermally stable compounds may be isolated by extraction with hydrocarbon solvents (e.g., pentane) and subjected to column chromatography under anaerobic conditions on basic alumina. They may be purified by crystallization from hydrocarbon solvents or by sublimation in high vacuum [e.g., 234,235,237].

2. Characterization and Identification

There are no reports of the reaction of tetra(organo)chromium(IV) compounds with $HgCl_2$. The trimethylsilylmethyl and other tetra(alkyl)-chromium(IV) compounds react with concentrated acids to give, respectively, $(CH_3)_4Si$ [234] and the alkanes [235] together with Cr(III) salts. Photolysis of tetra(1-norbornyl)chromium(IV) gives norbornane and 1,1′-bis(norbornane) [236].

The magnetic moments of the known tetra(organo)chromium(IV) compounds lie in the range 2.6–2.96 μ_B, a value that agrees well with that expected for a d^2-Cr(IV) complex with two unpaired electrons [234–237].

Mass spectrometry has been used to characterize σ-bonded tetra(alkyl)-chromium(IV) compounds [235–237] and of the compounds studied [e.g., R_4Cr(IV), R = 1-camphyl, 1-norbornyl, neopentyl and *tert*-butyl] the first three showed peaks at *m/e* corresponding to the tetraalkyl, whereas the last one [i.e., tetra(*tert*-butyl)chromium(IV)] showed *m/e* 223 corresponding to (*tert*-butyl)$_3$Cr$^{\oplus}$ [236,237].

Infrared, ESR and electronic spectroscopies have been used *inter alia* to characterize and identify σ-bonded tetra(alkyl)chromium(IV) compounds [234–238]. Thus, with tetra(trimethylsilylmethyl)chromium(IV), the characteristic absorptions of the $(CH_3)_3SiCH_2$ units have been used to establish their presence in the organometallic compound. Absorptions in the region 550–430 cm^{-1} have been ascribed to Cr–C vibrations [234] and absorptions in the region 370–390 cm^{-1} have been ascribed to Cr–C vibrations in the tetra(*tert*-butyl) and tetra(neopentyl) compounds [237]. The electronic spectra of σ-bonded tetra(alkyl)chromium(IV) compounds show absorptions in the 485–535 nm region, together with charge transfer bands. The variable-temperature ESR spectra of various tetra(alkyl)chromium(IV) compounds have been discussed [234,235,238]. The combined data from the electronic and ESR spectra establish that the tetra(alkyl)chromium(IV) compounds known to date all possess a slightly distorted tetrahedral configuration [234–238].

A preliminary report of a single crystal x-ray structure analysis confirms the tetrahedral array of the organic ligands in $Cr(CH_2C(CH_3)_2C_6H_5)_4$ [239]. The four chromium-to-carbon bond distances are 2.07, 2.06, 2.05, and 2.01(0.026) Å. They are of the same order of magnitude as the Cr–C bond distances found in octahedral σ-bonded mono-, and bis-, and tris-organochromium(III) compounds (Table 2.7), but they are considerably shorter than the Cr–C bond distances found in some lithium alkylchromate(II) complexes (Table 2.7).

References

1. O. Schwarzkopf and F. Schwarzkopf, *in* "Characterization of Organometallic Compounds" (M. Tsutsui, ed.) Vol. 1, p. 35. Wiley (Interscience), New York, 1969.
2. Yu. A. Sorokin and G. G. Petukhov, *J. Gen. Chem. USSR* **35**, 2123 (1966).
3. E. O. Fischer and H. Brunner, *Chem. Ber.* **98**, 175 (1965).
4. G. G. Devyatykh, V. A. Umilin, and Yu. B. Zverev, *Bull. Acad. Sci. USSR Div. Chem. Sci.*, p. 1568 (1969).
5. V. A. Umilin, Yu. B. Zverev, and V. K. Vanchagova, *Bull. Acad. Sci. USSR Div. Chem. Sci.*, **21**, 1503 (1972).
6. G. G. Devyatykh, V. A. Umilin, Yu. V. Zverev, and I. A. Frolov, *Bull. Acad. Sci. USSR Div. Chem. Sci.*, p. 207 (1969).
7. V. A. Umilin, Yu. B. Zverev, and I. I. Ermolaev, *Bull. Acad. Sci. USSR, Div. Chem. Sci.* **20**, 2585 (1971).
8. G. A. Razuvaev, G. G. Petukhov, and A. N. Artemov, *J. Gen. Chem. USSR* **39**, 2435 (1969).
9. H. H. Zeiss and W. Herwig, *Justus Liebigs Ann. Chem.* **606**, 209 (1957).
10. E. O. Fischer and W. Hafner, *Z. Anorg. Allg. Chem.* **286**, 146 (1956).
11. B. Gribov, V. P. Rumyantseva, N. N. Travkin, A. S. Pashinkin, B. I. Kozyrkin, and B. A. Salamantin, *Proc. Acad. Sci. USSR Chem. Sect.*, **194**, 677 (1970).
12. N. N. Travkin, V. P. Rumyantseva, and B. G. Gribov, *J. Gen. Chem. USSR*, **41**, 1929 (1971).
13. N. N. Travkin, B. G. Gribov, V. P. Rumyantseva, B. I. Kozyrkin, and B. A. Salamantin, *J. Gen. Chem. USSR* **40**, 2669 (1970).
14. G. A. Domrachev and N. N. Vyshinskii, *Proc. Acad. Sci. USSR Chem. Sect.*, **194**, 679 (1970), and references cited therein.
15. J. D. Cox and G. Pilcher, "Thermochemistry of Organic and Organometallic Compounds." Academic Press, New York 1970.
16. J. A. Connor, H. A. Skinner, and Y. Virmani, *J. Chem. Soc., Faraday Trans.* **69**, 1218 (1973).
17. H. H. Zeiss and M. Tsutsui, *J. Amer. Chem. Soc.* **79**, 3062 (1957).
18. G. Stolze and J. Hähle, *J. Organometal. Chem.* **5**, 545 (1966).
19. F. Hein and H. Müller, *Chem. Ber.* **89**, 2722 (1956).
20. B. G. Gribov, N. N. Travkin, G. M. Tabrina, V. P. Rumyantseva, B. S. Salamantin, B. I. Kozyrkin, and A. S. Pashinkin, *Proc. Acad. Sci. USSR* **187**, 535 (1969).
21. W. Ludwig and H. P. Schröer, *Monatsber. Deut. Akad. Wiss. Berlin* **9**, 197 (1967).

22. B. G. Gribov, B. I. Kozyrkin, A. D. Krivospitskii, and G. K. Chirkin, *Proc. Acad. Sci. USSR Chem. Sect.*, **204**, 386 (1972).
23. G. A. Razuvaev, A. A. Koksharova, G. G. Petukhov, and S. F. Zhil'tsov, *Proc. Acad. Sci. USSR Chem. Sect.*, **187**, 609 (1969).
24. E. O. Fischer and H. Brunner, *Chem. Ber.* **95**, 1999 (1962).
25. A. N. Nesmeyanov, L. P. Yur'eva, and S. N. Levchenko, *Proc. Acad. Sci. USSR Chem. Sect.*, **190**, 24 (1970).
26. A. N. Egorochkin S. Ya. Khorshev, A. I. Burov, V. A. Umilin, and Yu. B. Zverev, *Bull. Acad. Sci. USSR Div. Chem. Sci.*, p. 2671 (1968).
27. H. H. Zeiss, P. J. Wheatley, and H. J. S. Winkler, "Benzenoid-Metal Complexes." Ronald Press, New York, 1966.
28. R. D. Feltham, *J. Inorg. Nucl. Chem.* **16**, 197 (1961).
29. B. G. Gribov, B. I. Kozyrkin, A. D. Krivospitskii, and G. K. Chirkin, *Proc. Acad. Sci. USSR Chem. Sect.*, **193**, 457 (1970).
30. R. Prins, *J. Chem. Phys.* **50**, 4804 (1969).
31. E. Angelescu, C. Nicolau, and Z. Simon, *J. Amer. Chem. Soc.* **88**, 3910 (1966).
32. G. Olivé and S. Olivé, *Z. Phys. Chem.* **56**, 223 (1967).
33. K. Nakamoto, *in* "Characterization of Organometallic Compounds" (M. Tsutsui, ed.), Vol. 1, p. 73. Wiley (Interscience). New York, 1969.
34. "Gmelin's Handbuch der anorganischen Chemie," 8th ed., Suppl. Verlag Chemie, Weinheim, 1971.
35. H. P. Fritz, *Advan. Organometal. Chem.* **1**, 240 (1964).
36. E. O. Fischer and H. P. Fritz, *Angew. Chem.* **73**, 353 (1961).
37. S. Yamada, H. Yamazaki, H. Nishikawa, and R. Tsuchida, *Bull. Chem. Soc. Jap.* **33**, 481 (1960).
38. R. G. Snyder, *Spectrochim. Acta* **15**, 807 (1959).
39. H. P. Fritz and J. Manchot, *J. Organometal. Chem.* **2**, 8 (1964).
40. J. Brunvoll, S. J. Cyvin, and L. Schäfer, *J. Organometal. Chem.* **27**, 69 (1971).
41. E. O. Fischer and U. Piesbergen, *Z. Naturforsch. B* **11**, 758 (1956).
42. M. Tsutsui, *Z. Chem.* **2**, 214 (1962).
43. H. Müller, *Z. Chem.* **9**, 311 (1969).
44. R. P. A. Sneeden and H. H. Zeiss, *J. Organometal. Chem.* **47**, 125 (1973).
45. M. I. Bruce, *Advan. Organometal. Chem.* **6**, 273 (1968).
46. M. Cais and M. S. Lupin, *Advan. Organometal. Chem.* **8**, 211 (1970).
47. S. Pignataro and F. P. Lossing, *J. Organometal. Chem.* **10**, 531 (1967).
48. J. Müller and P. Göser, *J. Organometal. Chem.* **12**, 163 (1968).
49. G. E. Herberich and J. Müller, *J. Organometal. Chem.* **16**, 111 (1969).
50. S. Pignataro and F. P. Lossing, *J. Organometal. Chem.* **11**, 571 (1968).
51. G. G. Devyatykh, N. V. Larin, and P. E. Gaivoronskii, *Proc. Acad. Sci. USSR Chem. Sect.*, **198**, 424 (1971).
52. E. Weiss and E. O. Fischer, *Z. Anorg. Allg. Chem.* **286**, 142 (1956).
53. F. Jellinek, *J. Organometal. Chem.* **1**, 43 (1963).
54. F. A. Cotton, W. A. Dollase, and J. S. Wood, *J. Amer. Chem. Soc.*, **85**, 1543 (1963).
55. J. A. Ibers, *J. Chem. Phys.* **40**, 3129 (1964).
56. E. Keulen and F. Jellinek, *J. Organometal. Chem.* **5**, 490 (1966).
57. P. J. Wheatley, *Perspect. Struct. Chem.* **1**, 1 (1967).
58. A. Haaland, *Acta Chem. Scand.* **19**, 41 (1965).
59. L. Schäfer, J. F. Southern, S. J. Cyvin, and J. Brunvoll, *J. Organometal. Chem.* **24**, C13 (1970).
60. E. Förster, G. Albrecht, W. Dürselen, and E. Kurras, *J. Organometal. Chem.* **19**, 215 (1969).

61. O. V. Starovskii and Y. T. Struchkov, *Proc. Acad. Sci. USSR Chem. Sect.*, **135**, 1329 (1960).
62. R. P. Shibaeva, L. O. Atovmyan, and L. P. Rozenberg, *Chem. Commun.* p. 649 (1969).
63. R. P. Shibaeva and L. O. Atovmyan, *J. Struct. Chem (USSR)* **13**, 514 (1972).
63a. R. P. Shibaeva, L. O. Atovmyan, and M. N. Orfanova, *Chem. Commun.* p. 1494 (1969).
64. R. B. King and M. B. Bisnette, *Inorg. Chem.* **3**, 785 (1964).
65. E. O. Fischer and S. Breitschaft, *Angew. Chem.* **75**, 94 (1963).
66. E. O. Fischer and H. P. Kögler, *Z. Naturforsch. B* **13**, 197 (1958).
67. E. O. Fischer, W. Hafner, and H. O. Stahl, *Z. Anorg. Allg. Chem.* **282**, 47 (1955).
68. F. A. Cotton and G. Wilkinson, *Z. Naturforsch. B* **9**, 417 (1954).
69. E. O. Fischer, K. Ulm, and P. Kuzel, *Z. Anorg. Allg. Chem.* **319**, 253 (1963).
70. E. O. Fischer and K. Ulm, *Chem. Ber.* **95**, 692 (1962).
71. T. S. Piper and G. Wilkinson, *J. Inorg. Nucl. Chem.* **3**, 104 (1956).
72. E. O. Fischer and S. Breitschaft, *Chem. Ber.* **99**, 2213 (1966).
73. H. P. Fritz, *Chem. Ber.* **92**, 780 (1959).
74. H. P. Fritz and R. Schneider, *Chem. Ber.* **93**, 1171 (1960).
75. T. S. Piper and G. Wilkinson, *J. Inorg. Nucl. Chem.* **2**, 38 (1956).
76. R. D. Fischer, *Chem. Ber.* **93**, 165 (1960).
77. M. L. Maddox, S. L. Stafford, and H. D. Kaesz, *Advan. Organometal. Chem.* **3**, 1 (1965).
78. A. Davison, J. A. McCleverty, and G. Wilkinson, *J. Chem. Soc., London* p. 1133 (1963).
79. J. Müller and L. D'Or, *J. Organometal. Chem.* **10**, 313 (1967).
80. R. B. King, *Appl. Spectrosc.* **23**, 148 (1969).
81. R. B. King, *J. Amer. Chem. Soc.* **88**, 2075 (1966).
82. F. J. Preston and R. I. Reed, *Chem. Commun.* p. 51 (1966).
83. F. J. Preston and R. I. Reed, *Org. Mass. Spectrom.* **1**, 71 (1968).
84. E. O. Fischer and W. Hafner, *Z. Naturforsch. B* **10**, 140 (1955).
85. E. O. Fischer and J. Müller, *Z. Naturforsch. B* **18**, 1137 (1963).
86. F. Engelmann, *Z. Naturforsch. B* **8**, 775 (1953).
87. E. Weiss and E. O. Fischer, *Z. Anorg. Allg. Chem.* **284**, 69 (1956).
88. M. Ahmad, R. Bruce, and G. Knox, *Z. Naturforsch. B* **21**, 289 (1966).
89. O. L. Carter, A. T. McPhail, and G. A. Sim, *J. Chem. Soc., A* p. 1095 (1966).
90. M. A. Bush and G. A. Sim, *J. Chem. Soc., A* p. 605 (1970).
90a. B. Müller and J. Krausse, *J. Organometal. Chem.* **44**, 141 (1972).
91. A. T. McPhail and G. A. Sim, *J. Chem. Soc., A* p. 1858 (1968).
92. M. A. Bush and G. A. Sim, *J. Chem. Soc., A* p. 611 (1970).
92a. L. Y. Y. Chan and F. W. B. Einstein, *Acta Crystallogr., Sect. B* **26**, 289 (1966).
93. J. L. Calderon, S. Fontana, E. Frauendorfer, and V. W. Day, *J. Organometal. Chem.* **64**, C10 (1974).
94. J. J. Daly and R. P. A. Sneeden, *J. Chem. Soc., A* p. 736 (1967).
95. A. D. Fowlie, D. A. House, W. T. Robinson, and S. S. Rumball, *J. Chem. Soc., A* p. 803 (1970).
96. G. W. A. Fowles, P. T. Greene, and J. S. Wood, *Chem. Commun.* p. 971 (1967).
97. P. T. Greene, B. J. Russ, and J. S. Wood, *J. Chem. Soc., A* p. 3636 (1971).
98. F. A. Cotton, B. G. DeBoer, M. D. La Prade, J. R. Pipal, and D. A. Ucko, *J. Amer. Chem. Soc.*, **92**, 2926 (1970).
99. L. B. Handy, J. K. Ruff, and L. F. Dahl, *J. Amer. Chem. Soc.* **92**, 7312 (1970).
100. B. Nicholls and M. C. Whiting, *J. Chem. Soc., London* p. 551 (1959).

101. J. D. Munro and P. L. Pauson, *J. Chem. Soc., London* p. 3479 (1961).
102. H. Veening, N. J. Graver, D. B. Clark, and B. R. Willeford, *Anal. Chem.* **41**, 1655 (1969).
103. C. Segard, B. Roques, C. Pommier, and G. Guiochon, *Anal. Chem.* **43**, 1146 (1971).
104. J. M. Greenwood, H. Veening, and B. R. Willeford, *J. Organometal. Chem.* **38**, 345 (1972).
105. J. S. Keller, H. Veening, and B. R. Willeford, *Anal. Chem.* **43**, 1516 (1971).
106. G. Jaouen and R. Dabard, *Tetrahedron Lett.* p. 1015 (1971).
107. E. O. Fischer, H. A. Goodwin, C. G. Kreiter, H. D. Simmons, K. Sonogashira, and S. B. Wild, *J. Organometal. Chem.* **14**, 359 (1968).
108. R. V. Emanuel and E. W. Randall, *J. Chem. Soc., A* p. 3002 (1969).
109. B. Deubzer, H. P. Fritz, C. G. Kreiter, and K. Öfele, *J. Organometal. Chem.* **7**, 289 (1967).
110. P. L. Pauson, G. H. Smith, and J. H. Valentine, *J. Chem. Soc., C* p. 1061 (1967).
111. R. E. Humphrey, *Spectrochim. Acta* **17**, 93 (1961).
112. H. P. Fritz and J. Manchot, *Spectrochim. Acta* **18**, 171 (1962).
113. H. J. Buttery, G. Keeling, S. F. A. Kettle, I. Paul, and P. J. Stamper, *J. Chem. Soc., A* p. 2224 (1969).
114. H. J. Buttery, G. Keeling, S. F. A. Kettle, I. Paul, and P. J. Stamper, *J. Chem. Soc., A* p. 471 (1970).
115. H. J. Buttery, G. Keeling, S. F. A. Kettle, I. Paul, and P. J. Stamper, *J. Chem. Soc., A* p. 2077 (1969).
115a. S. F. A. Kettle and I. Paul, *Advan. Organometal. Chem.* **10**, 199 (1972).
116. R. B. King, *Appl. Spectrosc.* **23**, 536 (1969).
117. J. Müller, *J. Organometal. Chem.* **18**, 321 (1969).
118. G. Huttner and E. O. Fischer, *J. Organometal. Chem.* **8**, 299 (1967).
119. M. F. Bailey and L. F. Dahl, *Inorg. Chem.* **4**, 1314 (1965).
120. P. Corradini and G. Allegra, *J. Amer. Chem. Soc.* **81**, 2271 (1959).
121. M. F. Bailey and L. F. Dahl, *Inorg. Chem.* **4**, 1298 (1965).
122. O. L. Carter, A. T. McPail, and G. A. Sim, *J. Chem. Soc., A* p. 1619 (1967).
123. V. Kunz and W. Nowacki, *Helv. Chim. Acta* **50**, 1052 (1967).
124. K. W. Muir, G. Ferguson, and G. A. Sim, *J. Chem. Soc., B* p. 467 (1968).
125. H. Deuschl and W. Hoppe, *Acta Crystallogr.* **17**, 800 (1964).
126. K. W. Muir and G. Ferguson, *J. Chem. Soc., B* p. 476 (1968).
127. F. Hanic and O. S. Mills, *J. Organometal. Chem.* **11**, 151 (1968).
128. O. L. Carter, A. T. McPhail, and G. A. Sim, *J. Chem. Soc., A* p. 1866 (1968).
129. H. Lüth, I. F. Taylor, and E. L. Amma, *Chem. Commun.* p. 1712 (1970).
130. P. Corradini and G. Allegra, *J. Amer. Chem. Soc.* **82**, 2075 (1960).
131. Y. Dusausoy, J. Protas, J. Besançon, and J. Tirouflet, *C.R. Acad. Sci.* **270**, 1792 (1970).
132. O. L. Carter, A. T. McPhail, and G. A. Sim, *J. Chem. Soc., A* p. 288 (1967).
133. O. L. Carter, A. T. McPhail, and G. A. Sim, *J. Chem. Soc., A* p. 822 (1966).
134. M. A. Bush, T. A. Dullforce, and G. A. Sim, *Chem. Commun.* p. 1491 (1969).
135. D. E. F. Gracey, W. R. Jackson, W. B. Jennings, S. C. Rennison, and R. Spratt, *Chem. Commun.* p. 231 (1966).
136. W. R. Jackson, W. B. Jennings, S. C. Rennison, and R. Spratt, *J. Chem. Soc., B* p. 1214 (1969).
137. G. Barbieri and F. Taddei, *Chem. Commun.* p. 312 (1970).
138. W. R. Jackson, W. B. Jennings, and R. Spratt, *Chem. Commun.* p. 593 (1970).

139. J. T. Price and T. S. Sorensen, *Can. J. Chem.* **46**, 515 (1968).
140. K. Schlögl, *Top. Stereochem.* **1**, 39 (1965).
141. A. Mandelbaum, Z. Neuwirth, and M. Cais, *Inorg. Chem.* **2**, 902 (1963).
142. R. Dabard, A. Meyer, and G. Jaouen, *C.R. Acad. Sci.* **268**, 201 (1969).
143. H. Falk and K. Schlögl, *Monatsh. Chem.* **99**, 578 (1968).
144. P. E. Baikie and O. S. Mills, *J. Chem. Soc., A* p. 2704 (1968).
145. G. Huttner and O. S. Mills, *J. Organometal. Chem.* **29**, 275 (1971).
146. E. O. Fischer and A. Maasböl, *Chem. Ber.* **100**, 2445 (1967).
147. E. Moser and E. O. Fischer, *J. Organometal. Chem.* **15**, 147 (1968).
148. E. O. Fischer, *Pure Appl. Chem.* **24**, 407 (1970).
149. E. Moser and E. O. Fischer, *J. Organometal. Chem.* **13**, 209 (1968).
150. M. Y. Darensbourg and D. J. Darensbourg, *Inorg. Chem.* **9**, 32 (1970).
151. C. G. Kreiter and E. O. Fischer, *Angew. Chem., Int. Ed. Engl.* **8**, 761 (1969).
152. J. A. Connor and E. M. Jones, *J. Organometal. Chem.* **31**, 389 (1971).
153. D. J. Darensbourg and M. Y. Darensbourg, *Inorg. Chim. Acta* **5**, 247 (1971).
154. J. A. Connor and E. O. Fischer, *J. Chem. Soc., A* p. 578 (1969).
155. E. O. Fischer and L. Knauss, *Chem. Ber.* **103**, 1262 (1970).
156. F. A. Cotton and C. M. Lukehart, *Progr. Inorg. Chem.* **16**, 487 (1972).
157. K. Öfele, *J. Organometal. Chem.* **12**, P42 (1968).
158. K. Öfele, *Angew. Chem., Int. Ed. Engl.* **7**, 950 (1968).
159. K. Öfele and M. Herberhold, *Angew. Chem., Int. Ed. Engl.* **9**, 739 (1970).
160. E. Moser and E. O. Fischer, *J. Organometal. Chem.* **12**, P1 (1968).
161. U. Klabunde and E. O. Fischer, *J. Amer. Chem. Soc.* **89**, 7141 (1967).
162. E. O. Fischer and R. Aumann, *Chem. Ber.* **101**, 963 (1968).
163. E. O. Fischer and R. Aumann, *Angew. Chem., Int. Ed. Engl.* **6**, 181 (1967).
164. P. E. Baikie, E. O. Fischer, and O. S. Mills, *Chem. Commun.* p. 1199 (1967).
165. J. A. Connor and O. S. Mills, *J. Chem. Soc., A* p. 334 (1969).
166. O. S. Mills and A. D. Redhouse, *J. Chem. Soc., A* p. 642 (1968).
167. O. S. Mills and A. D. Redhouse, *J. Chem. Soc., A* p. 1274 (1969).
168. G. Huttner and B. Krieg, *Chem. Ber.* **105**, 67 (1972).
169. R. J. Hoare and O. S. Mills, *J. Chem. Soc., Dalton Trans.* p. 653 (1972).
170. G. Huttner and S. Lange, *Chem. Ber.* **103**, 3149 (1970).
171. G. Huttner, S. Schelle, and O. S. Mills, *Angew. Chem., Int. Ed. Engl.* **8**, 515 (1969).
172. B. H. Bracher and R. W. H. Small, *Acta Crystallogr.* **23**, 410 (1967).
173. H. Brunner, E. O. Fischer, and M. Lappus, *Angew. Chem., Int. Ed. Engl.* **10**, 924 (1971).
174. E. O. Fischer, H. Fischer, and H. Werner, *Angew. Chem., Int. Ed. Engl.* **11**, 644 (1972).
175. H. J. Beck, E. O. Fischer, and C. G. Kreiter, *J. Organometal. Chem.* **26**, C41 (1971).
176. G. E. Coates, M. L. H. Green, and K. Wade, "Organometallic Compounds," Vol. 2, p. 219. Methuen, London, 1968.
177. H. Müller, *Z. Chem.* **9**, 311 (1969).
178. J. J. Daly, F. Sanz, R. P. A. Sneeden, and H. H. Zeiss, *Chem. Commun.* p. 243 (1971).
179. J. J. Daly, F. Sanz, R. P. A. Sneeden, and H. H. Zeiss, *Helv. Chim. Acta* **56**, 503 (1973).
180. F. A. L. Anet and E. Leblanc, *J. Amer. Chem. Soc.* **79**, 2649 (1957).
181. R. G. Coombes, M. D. Johnson, and N. Winterton, *J. Chem. Soc., London* p. 7029 (1965).
182. W. Herwig and H. H. Zeiss, *J. Amer. Chem. Soc.* **81**, 4798 (1959).

183. E. Kurras, *Monatsber. Deut. Akad. Wiss. Berlin* **5**, 378 (1963).
184. R. P. A. Sneeden and H. H. Zeiss, *J. Organometal. Chem.* **13**, 369 (1968).
185. K. Nishimura, H. Kuribayashi, A. Yamamoto, and S. Ikeda, *J. Organometal. Chem.* **37**, 317 (1972).
186. F. Hein and E. Kurras, *Z. Anorg. Allg. Chem.* **290**, 179 (1957).
187. G. Stolze and J. Hähle, *J. Organometal. Chem.* **7**, 301 (1967).
188. F. Hein, J. Hähle, and G. Stolze, *Monatsber. Deut. Akad. Wiss. Berlin* **5**, 530 (1963).
189. F. Hein and K. Schmiedeknecht, *J. Organometal. Chem.* **6**, 45 (1966).
190. G. Wittig and K. D. Rümpler, *Justus Liebigs Ann. Chem.* **751**, 1 (1971).
191. G. Whitesides and W. J. Ehmann, *J. Amer. Chem. Soc.* **92**, 5625 (1970).
192. R. G. Coombes and M. D. Johnson, *J. Chem. Soc., A* p. 1805 (1966).
193. D. Dodd, M. D. Johnson, and D. Vamplew, *J. Chem. Soc., B* p. 1841 (1971).
194. R. P. A. Sneeden, T. F. Burger, and H. H. Zeiss, *J. Organometal. Chem.* **4**, 397 (1965).
194a. P. Jacquelot and G. Thomas, *Bull. Chem. Soc. Fr.* p. 1261 (1973).
195. M. S. Kharasch and O. Reinmuth, "Grignard Reactions of Nonmetallic Substances." Prentice-Hall, Englewood Cliffs, New Jersey, 1954.
196. E. Kurras and P. Klimsch, *Monatsber. Deut. Akad. Wiss. Berlin* **6**, 736 (1964).
197. W. Oberkirch, Dissertation Technische Hochschule, Aachen (1963).
198. F. Hein and K. Schmiedeknecht, *Z. Anorg. Allg. Chem.* **352**, 138 (1967).
199. F. Hein and D. Tille, *Z. Anorg. Allg. Chem.* **329**, 72 (1964).
200. E. Kurras and J. Otto, *J. Organometal. Chem.* **4**, 114 (1965).
201. F. Hein and R. Weiss, *Z. Anorg. Allg. Chem.* **295**, 145 (1958).
202. E. Kurras and P. Klimsch, *Monatsber. Deut. Akad. Wiss. Berlin* **6**, 735 (1964).
203. G. Stolze, *J. Organometal. Chem.* **6**, 383 (1966).
204. J. Krausse, G. Marz, and G. Schödl, *J. Organometal. Chem.* **21**, 159 (1970).
205. H. Dunken and G. Marz, *Z. Chem.* **6**, 436 (1966).
206. R. P. A. Sneeden and H. P. Throndsen, *Chem. Commun.* p. 509 (1965).
207. J. J. Daly, R. P. A. Sneeden, and H. H. Zeiss, *J. Amer. Chem. Soc.* **88**, 4287 (1966).
208. J. K. Kochi and D. D. Davis, *J. Amer. Chem. Soc.* **86**, 5264 (1964).
209. J. K. Kochi and J. W. Powers, *J. Amer. Chem. Soc.* **92**, 137 (1970).
210. F. Hein and K. Schmiedeknecht, *Z. Anorg. Allg. Chem.* **352**, 138 (1967).
211. D. Tille, *Z. Anorg. Allg. Chem.* **384**, 136 (1971).
212. W. Seidel, K. Fischer, and K. Schmiedeknecht, *Z. Anorg. Allg. Chem.* **390**, 273 (1972).
213. U. Giannini, E. Pellino, and M. P. Lachi, *J. Organometal. Chem.* **12**, 551 (1968).
214. J. Krausse and G. Schödl, *J. Organometal. Chem.* **27**, 59 (1971).
215. T. Aoki, A. Furusaki, Y. Tomiie, K. Ono, and K. Tanaka, *Bull. Chem. Soc. Jap.* **42**, 545 (1969).
216. G. Albrecht and D. Stock, *Z. Chem.* **7**, 321 (1967).
217. F. A. Cotton, *Inorg. Chem.* **4**, 334 (1965).
218. F. A. Cotton and J. R. Pipal, *J. Amer. Chem. Soc.* **93**, 5441 (1971).
219. J. J. Daly, F. Sanz, R. P. A. Sneeden, and H. H. Zeiss, *J. Chem. Soc., Dalton Trans.* p. 73 (1973).
220. J. Hähle and G. Stolze, *Z. Naturforsch. B* **19**, 1081 (1964).
221. M. R. Churchill, *Perspect. Struct. Chem.* **3**, p. 91 (1970).
222. M. R. Churchill and T. A. O'Brien, *J. Chem. Soc., A* p. 1110 (1969).
223. W. A. Baker and M. G. Phillips, *Inorg. Chem.* **5**, 1042 (1966).
224. J. G. Gibson and E. D. McKenzie, *J. Chem. Soc., A* p. 2637 (1969).

225. J. J. Daly and F. Sanz, *J. Chem. Soc., Dalton Trans.* p. 2584 (1972).
226. G. A. Albrecht, *Z. Chem.* **3**, 182 (1963).
226a. J. J. Daly, F. Sanz, R. P. A. Sneeden, and H. H. Zeiss, *Helv. Chim. Acta*, **57**, 1863 (1974).
227. H. Dietrich, *Acta Crystallogr., Sect. A* **25**, S160 (1969).
228. J. Krausse, *Proc. Int. Conf. Coord. Chem., 9th 1966* p. 168 (1966).
228a. J. Krausse and G. Marx, *J. Organometal. Chem.* **65**, 215 (1974).
229. J. Krausse, *Z. Chem.* **9**, 472 (1969).
230. E. Müller, J. Krausse, and K. Schmiedeknecht, *J. Organometal. Chem.* **44**, 127 (1972).
231. F. Hein and K. Schmiedeknecht, *J. Organometal. Chem.* **5**, 454 (1960).
232. D. C. Bradley, M. B. Hursthouse, and C. W. Newing, *Chem. Commun.* p. 411 (1971).
233. D. C. Bradley, M. B. Hursthouse, C. W. Newing, and A. J. Welch, *Chem. Commun.* p. 567 (1972).
234. W. Mowat, A. Shortland, G. Yagupsky, M. Yagupsky, and G. Wilkinson, *J. Chem. Soc., Dalton Trans.* p. 533 (1972).
235. W. Mowat, A. J. Shortland, N. J. Hill, and G. Wilkinson, *J. Chem. Soc., Dalton Trans.* p. 770 (1973).
236. B. K. Bower and H. G. Tennent, *J. Amer. Chem. Soc.* **94**, 2512 (1972).
237. W. Kruse, *J. Organometal. Chem.* **42**, C39 (1972).
238. G. A. Ward, W. Kruse, B. K. Bower, and J. C. W. Chien, *J. Organometal. Chem.* **42**, C43 (1972).
239. V. Gramlich and K. Pfefferkorn, *J. Organometal. Chem.* **61**, 247 (1973).

Chapter 3

REACTIONS

A. Introduction

The chemical reactions and the stabilities of the five main groups of organochromium compounds differ widely from one group to another. Therefore, before treating of the reactions of these individual groups of compounds in detail, some of the more general factors that influence their reactions will be discussed.

The bonding of organic ligands to chromium, and other transition metal compounds, has been discussed extensively in many reference texts [e.g., 1,2]. There are, broadly speaking, two main types of bonding; that involved in compounds with π-type ligands (e.g., arenes, trienes, olefins, CO, and carbenoid units) and that involved in compounds with simple donor ligands (e.g., aryl and alkyl).

The bonding of an arene, triene, cyclopentadienyl, or olefin ligand to chromium consists of two interrelated electron donations, the one from the ligand to the metal and the other from the metal to the ligand. An analogous situation pertains in the bonding of carbonyl, thiocarbonyl, and carbenoid ligands, though in these cases a pair of electrons is donated from the ligand to the metal (e.g., a σ-donation). These related bonding situations imply that when such ligands are bonded to chromium they should be susceptible to attack by nucleophiles, and this is indeed observed in many instances [e.g., nucleophilic attack on the arene and triene ligands in π-(arene)$Cr(CO)_3$ and π-(cycloheptatrienyl)$Cr(CO)_3$ and nucleophilic attack on the carbonyl ligands in $Cr(CO)_6$, π-$(C_6H_6)Cr(CO)_3$, and π-$(C_5H_5)Cr(CO)_2NO$].

The bonding of aryl, alkyl, and other simple organic groups to chromium is best regarded as a transfer of electrons from the ligand to the metal (a σ-bond) with no significant back-donation from the metal to the ligand. The electron density in this case is cylindrically symmetrical along the line connecting C_{ligand} to chromium. This bonding situation implies that σ-bonded organochromium compounds could be a source of anions, radicals, and perhaps cations, and indeed this is observed in many cases (e.g., reactions with $HgCl_2$, and in thermal decomposition reactions).

The electronic configuration and the coordination state of the chromium center, together with the fact that these may be changed by the respective gain or loss of an electron or ligand, are also determining factors in the reactions of organochromium compounds.

The reversible gain or loss of an electron implies that organochromium compounds should undergo redox reactions, and this is indeed observed for π-bis(arene) and π-bis(cyclopentadienyl)chromium compounds and tetra-(alkyl)chromate(III) complexes.

The reversible gain or loss of a ligand and the accompanying change engendered in the geometry of the compound are of fundamental importance to the reactions of most organochromium compounds. Thus, the acquisition of a ligand may lead to the formation of a bridged dimer wherein the juxtaposition of two of the organic groups is such as to facilitate their reductive elimination (e.g., the disproportionation of $[(CH_3CH_2)_2N]_3Cr(III)$ to $[(CH_3CH_2)_2N]_4Cr(IV)$ [3], of $Li_3Cr(III)(CH_3)_6$ to $Li_4Cr(II)_2(CH_3)_8$ (Section of G,1,a of this chapter), and of $(C_3H_5)_3Cr(III)$ to $[(C_3H_5)_2Cr(II)]_2$ (Section F,3,c,ii of this chapter). The loss of a ligand will liberate a site on the chromium center necessary for the coordination of the substrate, thereby enabling the reaction to proceed [e.g., hydrolysis of compounds of the type $R_nCrCl_{3-n}(THF)_3$]. This loss of the ligand may proceed either by simple thermal dissociation [e.g., $(C_6H_5)_3Cr(THF)_3 \rightleftharpoons (C_6H_5)_3Cr(THF)_2 + THF$] or by photocatalyzed dissociation [e.g., the exchange of the CO in π-(arene)-$Cr(CO)_3$ compounds (Section D,1,a of this chapter)].

The nature of the ligands around the chromium center will also have a profound influence on the stability and the chemical reactions of σ-bonded organochromium compounds. Certain strong donor ligands or bidentate ligands not only stabilize a σ-bonded organochromium compound but also reduce its reactivity toward certain reagents, since they effectively block the sites on the chromium center that are required for coordination of the substrate and therefore prevent the occurrence of the reaction. Thus, $(C_6H_5)_2$-$CrCl(1.5\ DME)$ is very reactive toward both O_2 and H_2O, whereas

$[(C_6H_5)_2Cr(bipy)_2]^{\oplus}I^{\ominus}$ may be recrystallized, in air, from aqueous ethanol; also $(C_6H_5)_3Cr(THF)_3$ is very reactive toward O_2, H_2O, acetylenes, etc., whereas the trischelate compound

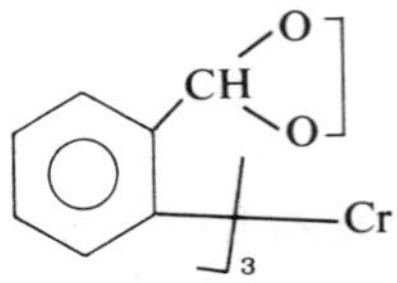

is inert toward these substrates.

The chemical reactions may occur either at the chromium center or at the ligand and, in the former case, free coordination sites at chromium are required, while in the latter case they are not. This explains, for instance, why certain σ-bonded organochromium(III) and (II) compounds are stable to water (a reaction that necessitates a free coordination site at chromium), whereas all the known σ-bonded organochromium compounds react with $HgCl_2$ (a reaction that occurs at the ligand).

The subtle interplay of reactions occurring at the chromium center and at a ligand is illustrated in the chemistry of the organochromium π-complexes: π-(arene)$_2$Cr, π-$(C_5H_5)_2Cr$, π-$(C_6H_6)Cr(CO)_3$, π-$(C_7H_8)Cr(CO)_3$, π-(C_7H_7)-$Cr(CO)_3$.

The reactions occurring at chromium (e.g., redox reactions) are dominant in the chemistry of the first two π-complexes; in the case of the π-C_7-compounds, the "redox reaction" occurs at the ligand (i.e., hydride abstraction or addition). The reactions occurring at the ligand (e.g., nucleophilic and electrophilic substitution) are dominant only in the chemistry of the three carbonyl π-complexes. An interesting feature of the latter reactions at the ligand is that the structure of the π-complexes is such as to impose a high degree of stereospecificity on these reactions.

Concerning the "mechanisms" of the reactions that will be discussed, it is important to bear in mind that in the majority of cases the complexity of the reacting systems and of the reaction products is such as to preclude the possibility of establishing a satisfactory material balance and of carrying out any meaningful kinetics. So, though the precise mechanism of many of these reactions remains obscure, many valid rationales (based on product analysis) have been proposed to account for the formation of the observed products. Such rationales are important to the development of organochromium chemistry and are therefore included in the ensuing text.

B. π-Bis(arene)chromium Compounds

The chemistry of bis(arene)chromium(0) and chromium(I) π-complexes is somewhat limited, since it appears that though the compounds are per se relatively stable, they undergo decomposition under the conditions usually associated with aromatic nucleophilic, electrophilic, and radical substitution [4].

Thus, attempts to effect the amidation of π-bis(toluene)chromium(0) with sodium amide failed [4].

Attempts to effect the acylation of the arene units of π-bis(toluene)-chromium(0) by the Friedel and Crafts reaction (with CH_3COCl and C_6H_5COCl) also failed. The organic products consisted of the corresponding ketone, but the π-complex had undergone complete disrupture, Eq. (3-1) [4].

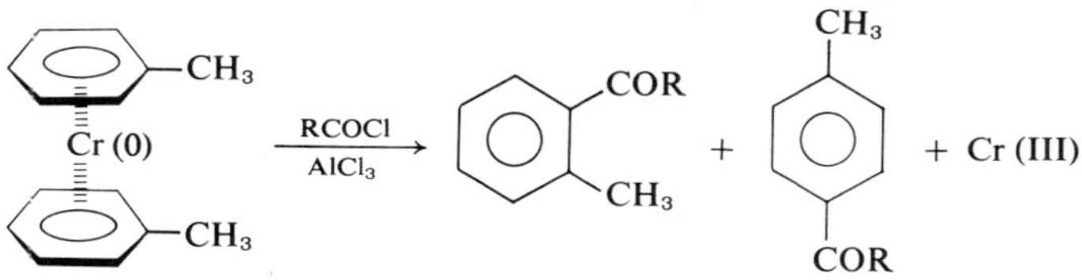

(3-1)

Similarly, attempts to effect the mercuration of π-bis(toluene)chromium(0) failed; the complex was oxidized by the mercuric salt to the Cr(I) cation [4].

Attempted radical substitution of π-(arene)chromium(0) compounds, under mild conditions, resulted in the oxidation of the Cr(0) complex to the Cr(I) cation [4]. The same reaction under more forcing conditions led to the disrupture of the π-complex and to the formation of the tetrakis(alkoxy)-chromium(IV) compound, Eq. (3-2) [5].

$$\pi\text{-}(C_6H_6)_2Cr(0) \xrightarrow{[(CH_3)_3CO]_2} Cr(IV)[OC(CH_3)_3]_4 + C_6H_6 \qquad (3\text{-}2)$$

π-Bis(arene)chromium compounds do undergo, however, several reactions, involving the arene ligands and/or the chromium center, in which the complex is not destroyed. These reactions together with the uses of π-bis(arene)-chromium compounds in catalytic reactions are discussed in the ensuing pages.

1. Reactions Involving the Arene Ligands

a. Ligand Replacement

The arene units of a bis(arene)chromium π-complex may be exchanged for other arene units. This ligand replacement reaction is catalyzed by $AlCl_3$ and is an equilibrium reaction, Eq. (3-3) (e.g. arene = biphenyl, mesitylene, tetralin, and 4,4′-bitolyl) [6–8]. The composition of the final bis(arene)chromium product will, therefore, depend upon the concentration of the respective arene components and upon their stabilities toward $AlCl_3$

(i.e., their resistance to isomerization, etc.,). The significance of this ligand replacement reaction in the "catalysis" of the Fischer–Hafner synthesis of π-bis(arene)chromium compounds has already been discussed (Chapter 1).

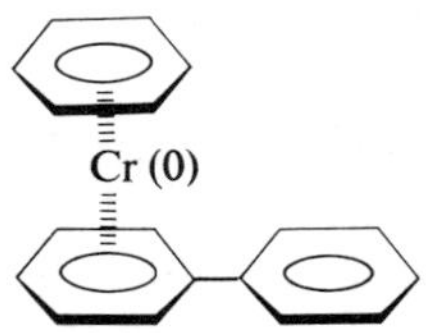

$$\xrightarrow[170°–200°C]{\text{arene/AlCl}_3} (\text{arene})_2\text{Cr(0)} + C_6H_6 + C_6H_5C_6H_5 \quad (3\text{-}3)$$

One of the arene units of a bis(arene)chromium π-complex may be replaced by three carbonyl groups, by reaction with $Cr(CO)_6$, to give π-(arene)-chromium tricarbonyl compounds, e.g., Eq. (3-4) [9].

$$\pi\text{-}(C_6H_6)_2\text{Cr(0)} \xrightarrow{\text{Cr(CO)}_6} \underset{(25\%)}{\pi\text{-}(C_6H_6)\text{Cr(CO)}_3} \quad (3\text{-}4)$$

b. Hydrogen–Deuterium Exchange

The hydrogen atoms of π-bis(arene)chromium(0) and π-bis(arene)-chromium(I) complexes readily undergo isotopic exchange in the presence of a base. Thus the aromatic and, in part, the aliphatic protons of the zero valent complexes [π-bis(benzene)chromium(0), π-bis(toluene)chromium(0) and π-bis(ethylbenzene)chromium(0)] are replaced by deuterium when the complexes are treated with sodium ethoxide in deuterioethanol (C_2H_5OD). The aromatic and, in part, the aliphatic protons of the cationic complexes [π-bis(benzene)chromium(I) iodide and π-bis(toluene)chromium(I) iodide] are exchanged for deuterium under even milder conditions, namely, upon treatment with KOD in D_2O [10]. This facile isotopic exchange reflects the enhanced acidity of the aromatic protons and the α-methyl (or methylene) protons of arene units when they are complexed to chromium(0) or chromium(I).

Recent ESR studies of the interaction of the π-bis(benzene)chromium cation with various strong bases [e.g., K in dimethoxyethane (DME) or K in tetrahydrofuran (THF), etc.] revealed that these bases abstract one proton from one of the arene rings complexed to chromium to give, in low yield, the neutral paramagnetic species **1**, Eq. (3-5). The data further indicate that

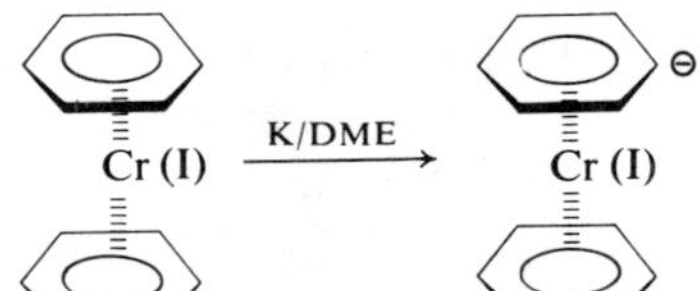

1

(3-5)

a rapid intramolecular and interannular proton exchange occurs in the neutral complex **1**; this proton exchange is presumed to proceed via a hydridochromium species, Eq. (3-6) [11].

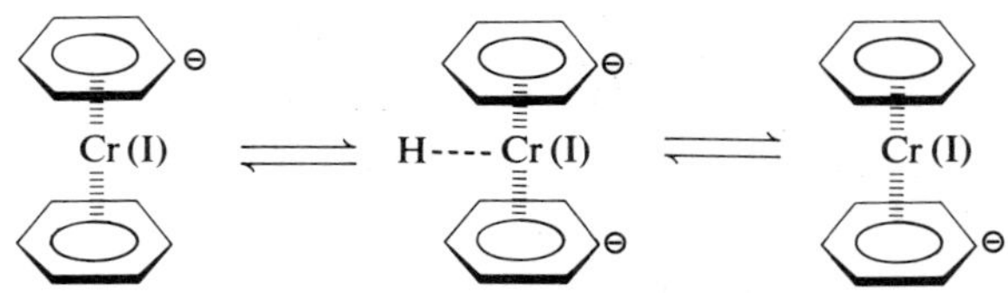

(3-6)

1

c. *Metalation*

The direct metalation of π-bis(benzene)chromium(0) apparently requires more drastic conditions than those required for the simple isotopic exchange reaction. Thus, Fischer and his colleagues found that whereas π-bis(benzene)-chromium(0) is inert toward *n*-butyllithium in benzene or diethyl ether solution [4] it does react with amylsodium in hexane solution [12–14]. The product from this reaction is not, however, one specific sodium derivative of π-bis(benzene)chromium(0), since subsequent methylation or carbonation leads to a complex mixture of products. Thus, when the mixture of alkyl-bis(arene)chromium π-complexes, obtained by reacting the metalated π-bis(benzene)chromium(0) with methyl iodide, is decomposed ($Na_2Cr_2O_7$), the resulting hydrocarbons consist of a mixture of benzene and alkylated

$$\pi\text{-}(C_6H_6)_2Cr(0) \xrightarrow[(2)\ CH_3I]{(1)\ NaC_5H_{11}} [(CH_3)_nC_6H_{6-n}]_2Cr(0) \xrightarrow{[O]}$$
$$C_6H_6,\ CH_3C_6H_5,\ m + p\text{-}(CH_3)_2C_6H_4,\ \text{isomeric-}(CH_3)_3C_6H_3,\ CH_3(CH_3CH_2)C_6H_4 \qquad (3\text{-}7)$$

benzenes [13], Eq. (3-7). Similarly, carbonation of metalated π-bis(benzene)-chromium(0) furnishes a complex mixture of carboxy-bis(arene)chromium π-complexes. Esterification of this mixture [$(CH_3)_2SO_4$] and subsequent column chromatography of the product permitted the isolation of the pure mono- and bis(methoxycarbonylbenzene)chromium(0) π-complexes **2** and **3**, Eq. (3-8) in 4% and 5% yield, respectively [13,14].* The other products [Eq. (3-8)] consisted of unreacted π-bis(benzene)chromium(0) and tris- and tetrakis(methoxycarbonylbenzene)chromium(0) π-complexes. The arene components in the latter π-complexes consisted of methylbenzoate and the *m*- and *p*-bis(methoxycarbonyl)benzenes (i.e, the dimethyl esters of isophthalic acid and terephthalic acids) [14].

* The tetraphenylboron salt of π-(benzoic acid)-π-(biphenyl)chromium(I) has also been prepared by the carbonation of the intermediate in the Hein–Grignard syntheses [15], which is discussed further in Section F,3,c,i of this chapter.

These results from the methylation and carbonation of the metalated π-bis(benzene)chromium(0) show that the product is a complex mixture containing mono-, bis-, tris-, and even tetrakis(sodium)arene derivatives.

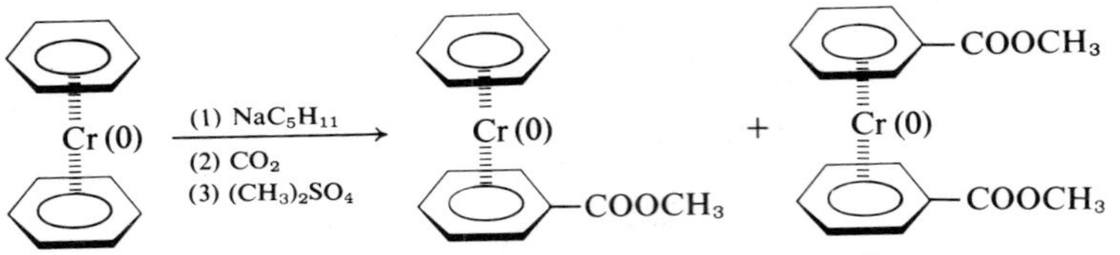

2 **3**

+ other products (3-8)

Other transformations of the "sodium derivative of π-bis(benzene)-chromium(0)" have been effected and again mixtures of products were obtained. However, discrete π-bis(arene)chromium(0) complexes bearing one functional group in one of the arene rings could be isolated by column chromatography (C.C.) of the crude reaction mixtures (see Scheme 3-1). Aliphatic aldehydes (paraformaldehyde and paraldehyde) react with the sodium derivative of π-bis(benzene)chromium(0) to give the corresponding carbinols while benzaldehyde reacts with concomitant hydrogen abstraction to give the corresponding ketone. The yields of pure products obtained in these reactions are low (4% to 40%); however, the functional groups in the mono-substituted arenechromium π-complexes can be transformed chemically (e.g., reduction, oxidation) to give high yields of products, this without disrupture of the bis(arene)chromium π-complex (see Scheme 3-1). Hence a variety of bis(arene)chromium π-complexes containing an aryl ester, a primary, secondary, or tertiary aryl alcohol, an aryl ketone, or an aryl aldehyde complexed to chromium have been prepared. Elschenbroich found that π-bis(benzene)chromium(0) is readily metalated by the *N,N,N',N'*-tetramethylethylenediamine [TMEDA] complex of *n*-butyllithium in cyclohexane solution [16]. The major product formed in this reaction is, in contrast to the product of metalation with amylsodium, the π-bis(lithium benzene)-chromium(0) complex **4**, Eq. (3-9), since subsequent reaction of the metalated complex **4**, Eq. (3-9), with D_2O or trimethylchlorosilane gives the corresponding bis(substituted arene)chromium(0) complexes as major products [16,17], Eq. (3-9) (TMEDA = *N,N,N',N'*-tetramethylethylenediamine).

When π-bis(benzene)chromium(0) and *n*-butyllithium/TMEDA complexes are allowed to interact in stoichiometric ratios of 1:10, the major product is still π-bis(lithium benzene)chromium(0) (**4**) since methylation (CH_3I) of the metalated complexes gives substantial quantities of π-bis(toluene)chromium-(0) [18]. Higher ratios of reagents (1:20) lead to polymetalation of the π-bis(benzene)chromium(0) since when the metalated complex is methylated

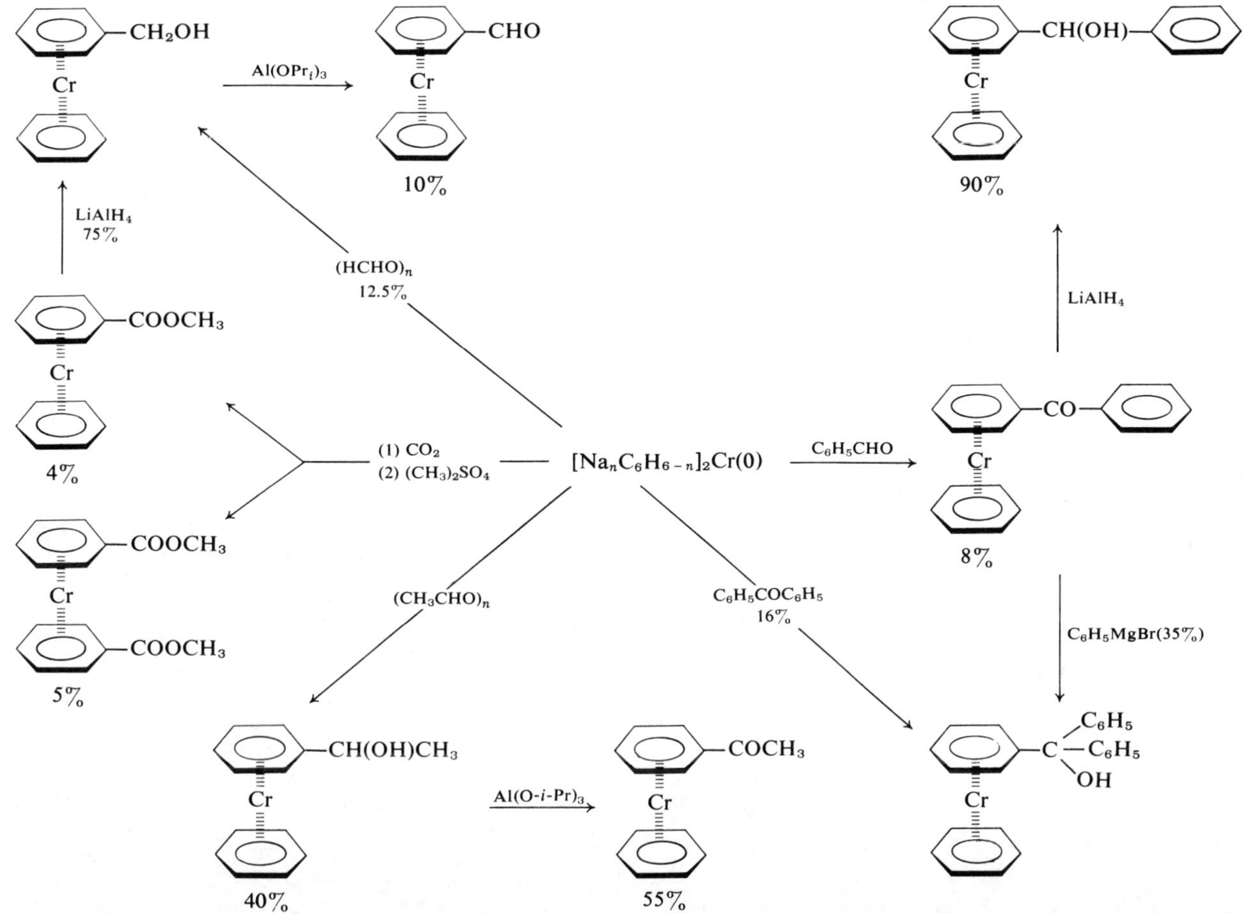

Scheme 3-1. Reactions of the sodium derivative of π-bis(benzene)chromium(0) and subsequent transformations of the products. The percentage figures are the yields of pure products [13,14].

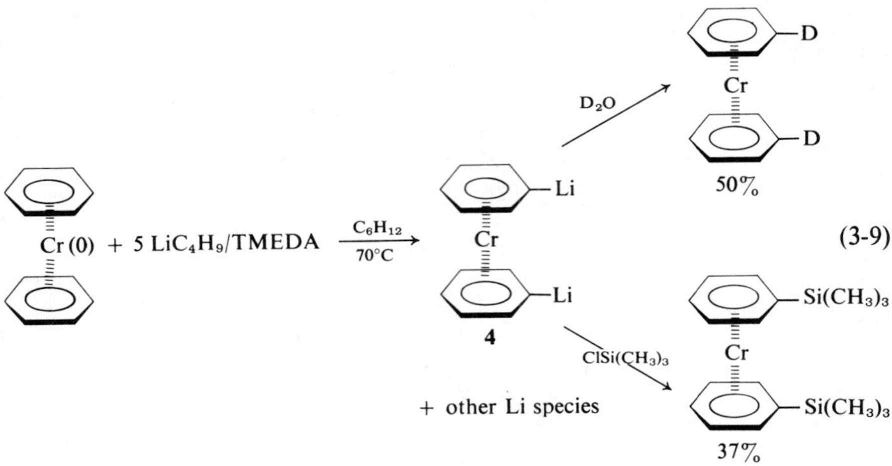

(3-9)

(by reaction with CH_3I) the hydrocarbons that are formed from the oxidative decomposition of the resulting methylbenzenechromium π-complexes now consist of benzene, toluene, *m*- and *p*-xylene (some *o*-xylene), and isomeric trimethylbenzenes [18], Eq. (3-10). The metalation of π-bis(toluene)chro-

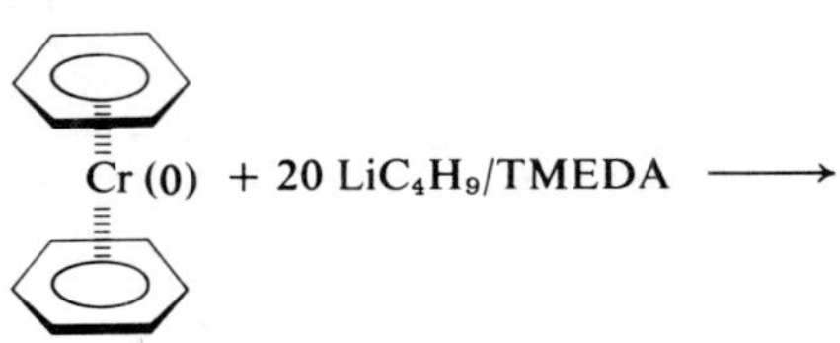

$$[Li_nC_6H_{6-n}]_2Cr(0) \xrightarrow[(2)\ Na_2Cr_2O_7]{(1)\ CH_3I} C_6H_6 + CH_3C_6H_5 + (CH_3)_2C_6H_4 + (CH_3)_3C_6H_3 \quad (3\text{-}10)$$

mium(0) and π-bis(*m*-xylene)chromium(0) with *n*-butyllithium/TMEDA, in stoichiometric ratios of 1:10, has also been studied. In the former case, metalation of both the aromatic rings and of the methyl substituent occur, since the hydrocarbon mixture, obtained after methylation and cleavage of the metalated complex, contain *m*- and *p*-xylene together with ethylbenzene [18]. In the second case [i.e., π-bis(*m*-xylene)chromium(0)], the aromatic ring is the main site of metalation, since the final hydrocarbon mixture, obtained after methylation and cleavage of the metalated complex consists mainly of 1,3,5-trimethylbenzene [18].

The mechanisms of these metalation reactions are not known and, in particular, the precise role of TMEDA in promoting the metalations with *n*-butyllithium remains obscure. It is known that TMEDA forms a strong complex with the lithium ion, thereby promoting the carbanion character of the *n*-butyl group in the butyllithium/TMEDA complex; this, in turn, will promote the abstraction of a proton from the complexed arene ring by the reagent, thereby facilitating the overall metalation process. The data available from the limited studies carried out to date permit the following conclusions concerning the metalation of π-bis(arene)chromium(0) complexes:

(i) Amylsodium metalates π-bis(arene)chromium(0) compounds to give a mixture of metalated complexes; the TMEDA complex of butyllithium, on the other hand, metalates π-bis(arene)chromium(0) complexes to give mainly the π-bis(lithium arene)chromium(0) complex.

(ii) When benzene is complexed to chromium(0), it is metalated more readily than free benzene; in the case of the *n*-butyllithium/TMEDA complex there is a factor of 10 in the relative rates of metalation.

(iii) The introduction of one metal atom into the arene complex facilitates further metalation (by amylsodium or *n*-butyllithium/TMEDA complex) to give the arene complex metalated in both rings.

(iv) On polymetalation, an arene ring that is already metalated with Na or Li is further metalated in the meta or the para positions and only slightly, if at all, in the ortho one.

(v) The arene rings of methyl-substituted bis(arene)chromium π-complexes are metalated by *n*-butyllithium/TMEDA, meta and para to the methyl group; though some metalation of the methyl substituent may also occur.

The present metalation reactions of π-bis(arene)chromium(0) compounds, particularly that with the butyllithium/TMEDA complex, promise to be convenient reactions for the preparation of otherwise inaccessible substituted bis(arene)chromium π-complexes.

2. Reactions Involving the Chromium Center

Bis(arene)chromium(0) and chromium(I) π-complexes undergo a variety of oxidation/reduction reactions in which the chromium center is involved. In several of these reactions the arene units are in part lost since the reactions are in effect disproportionations; in others, the arene units are retained and the reactions are stoichiometric.

a. Reduction Reactions

The thermal decomposition of several π-bis(arene)chromium(I) salts proceeds via a disproportionation reaction, in which part of the π-complexes is reduced to the chromium(0) π-complexes and the remainder undergoes disrupture to give the free arenes and chromium(III) or (II) salts, Eqs. (3-11) and (3-12) [AcAc = $CH_3COCH{=}C(O^{\ominus})CH_3$] [19,20; and also

Chapter 2, Section B,1 and 2]. The latter reaction [i.e., Eq. (3-12)], when carried out in the presence of copper acetylacetonate, leads to the total disrupture of the π-(arene)chromium complex, Eq. (3-13) [21].

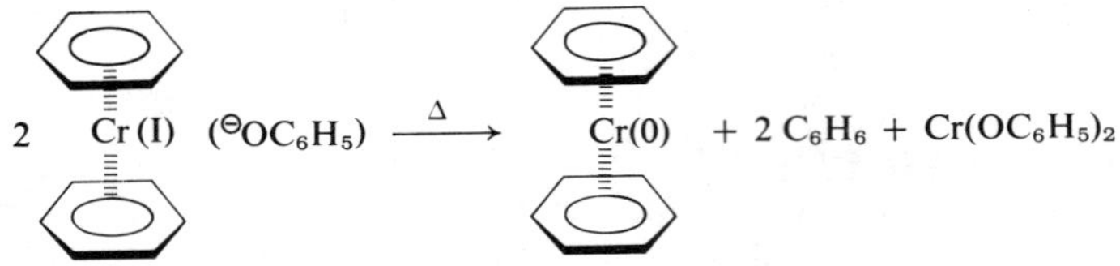

(3-11)

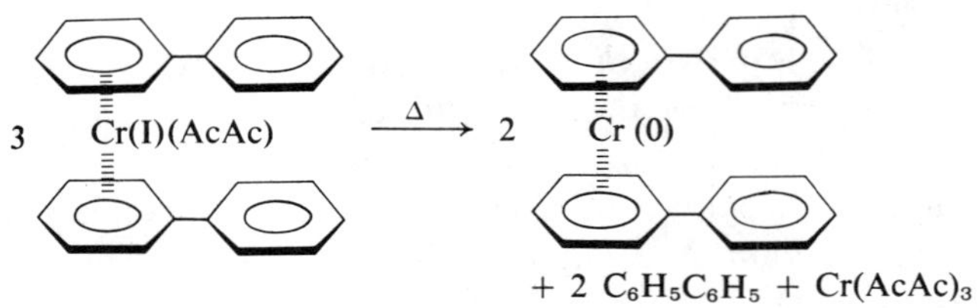

(3-12)

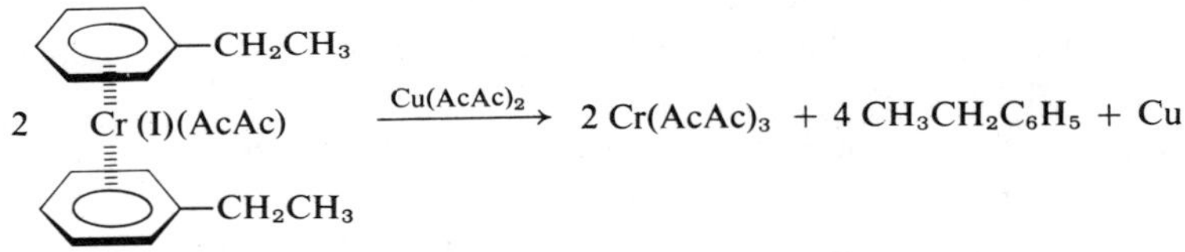

(3-13)

The quantitative reduction of π-bis(arene)chromium(I) hydroxides can be effected by a variety of reducing agents, e.g., alkaline sodium dithionite [7].

b. Oxidation Reactions

π-Bis(arene)chromium(0) compounds are readily oxidized to the corresponding π-bis(arene)chromium(I) compounds by molecular oxygen [22]. A more detailed investigation of this oxidation showed that alkylsubstituents in the arene rings complexed to chromium increase the rate of oxidation and that the products of oxidation are the π-bis(arene)chromium(I) hydroxide and hydrogen peroxide according to the stoichiometry outlined in Eq. (3-14) [23].

$$2\pi\text{-(arene)}_2\text{Cr(0)} + 2\text{H}_2\text{O} + \text{O}_2 \longrightarrow 2\ \text{(arene)}_2\text{Cr(I)(OH)} + \text{H}_2\text{O}_2 \qquad (3\text{-}14)$$

π-Bis(biphenyl)chromium(0) and π-(benzene)-(biphenyl)chromium(0) are oxidized by gaseous hydrogen chloride to give by oxidation and disproportionation reactions the Cr(I) chlorochromates and gaseous hydrogen, Eq. (3-15) (arene = $C_6H_5C_6H_5$ or C_6H_6 and $C_6H_5C_6H_5$, respectively [24].

$$2\ \pi\text{-(arene)}_2\text{Cr(0)} \xrightarrow{\text{HCl}} [\pi\text{-(arene)}_2\text{Cr(I)}][\text{Cr(II)Cl}_3] + \tfrac{3}{2}\text{H}_2 + 2\ \text{arene} \qquad (3\text{-}15)$$

π-Bis(arene)chromium(0) compounds are oxidized, quantitatively, to the corresponding Cr(I) halides by a variety of organic halides [Eq. (3-16) (Table 3.1)] [25,26]. The sequence of reactivities of the alkyl halides is iodide >

$$\pi\text{-(arene)}_2\text{Cr(0)} + \text{RX} \longrightarrow \pi\text{-(arene)}_2\text{Cr(I)X} + \text{R}\cdot \qquad (3\text{-}16)$$

bromide > chloride and $CH_3 > C_2H_5 > n\text{-}C_3H_7 > n\text{-}C_4H_9 \geqslant i\text{-}C_5H_{11}$. The organic products that are formed in these oxidation reactions depend upon the organic halide used (see Table 3.1) [26,27]. Thus, with alkyl halides, the products are hydrocarbons and/or olefins formed from the radical R· by hydrogen abstraction or by disproportionation, and with allyl or aryl halides the products from the dimerization of the radical R· are formed.

TABLE 3.1

Reaction of π-bis(ethylbenzene)chromium(0) Complexes with Organic Halides [26,27]

ORGANIC HALIDE	ORGANIC PRODUCTS	REMARKS
C_2H_5I	$CH_3{-}CH_3$ (48%), $CH_2{=}CH_2$ (37%)	No C_4H_{10}
$n\text{-}C_4H_9I$	C_4H_{10}	No olefin or dimer
$i\text{-}C_5H_{11}I$	$i\text{-}C_5H_{12}$	No olefin or dimer
C_6H_5I	C_6H_6 and $C_6H_5C_6H_5$	—
$CH_2{=}CH{-}CH_2I$	$CH_2{=}CH(CH_2)_2CH{=}CH_2$	—
$C_6H_5CH_2Cl$	$C_6H_5CH_2CH_2C_6H_5$	Complex partially destroyed to give $CH_3CH_2C_6H_5$
$(C_6H_5)_2CHCl$	$(C_6H_5)_2CHCH(C_6H_5)_2$	—
$(C_6H_5)_2CCl_2$	$(C_6H_5)_2C{=}C(C_6H_5)_2$	—

The reaction with benzyl chloride proceeds somewhat differently insofar as decomposition of the π-bis(arene)chromium complex also occurs, Eq. (3-17) [26].

$$3\pi\text{-}(CH_3CH_2C_6H_5)_2\text{Cr(0)} + 4\,C_6H_5CH_2Cl \longrightarrow 2\,\pi\text{-}(CH_3CH_2C_6H_5)_2\text{Cr(I)Cl} + 2\,C_6H_5CH_2CH_2C_6H_5 + CrCl_2 + 2\,CH_3CH_2C_6H_5 \qquad (3\text{-}17)$$

The standard oxidation-reduction potentials $E°$ for the system π-(arene)$_2$Cr(I) + $e \rightleftharpoons$ π-(arene)Cr(0) are in the range of 0.6–1.0 V [28–33] and all lie close to that associated [34] with the system $RX + e \rightarrow X^{\ominus} + R\cdot$. This suggests that π-bis(arene)chromium(0) compounds should be more readily oxidized by halides whose $E°$ value is less than that of the π-complex; and indeed it is observed that methylene iodide and allyl iodide oxidize π-bis(arene)chromium(0) complexes more readily than do saturated alkyl halides [25,26].

The mechanism of the reaction of alkyl halides with π-bis(arene)chromium(0) compounds may involve either the oxidative cleavage of the R–X bond by zerovalent chromium, or the nucleophilic attack on the carbon atom of the C–X bond by zerovalent chromium.

The former mechanism, which is analogous to that proposed for the reaction of chromous salts with benzyl halides (Chapter 1, Section F,1,c), involves the attack of the halogen atom by Cr(0) to give directly the chromium(I) π-complex and the organic radical, which subsequently couples, disproportionates, or abstracts hydrogen to give finally the observed organic

$$\pi\text{-}(C_6H_6)_2Cr \quad X{-}R \longrightarrow \pi\text{-}(C_6H_6)_2Cr(I)X^{\ominus} + [R\cdot \longrightarrow \text{products}] \qquad (3\text{-}18)$$

products, Eq. (3-18). The second mechanism involves nucleophilic substitution at the carbon bonded to halogen by the zerovalent chromium to give a σ-bonded (organo)-π-bis(arene)chromium complex [**5**, Eq. (3-19)]. This

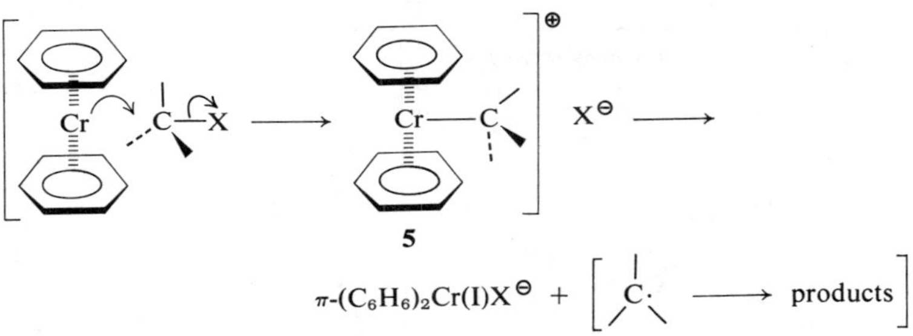

(3-19)

intermediate (**5**) may undergo fragmentation to give the π-bis(arene)chromium(I) salt and the organic radical R$\cdot$; the latter being converted subsequently into the observed organic products [25,26].

π-Bis(arene)chromium(I) halides catalyze the reduction of alkyl halides by sodium dithionite ($Na_2S_2O_4$) or by aminoiminomethane sulfinic acid

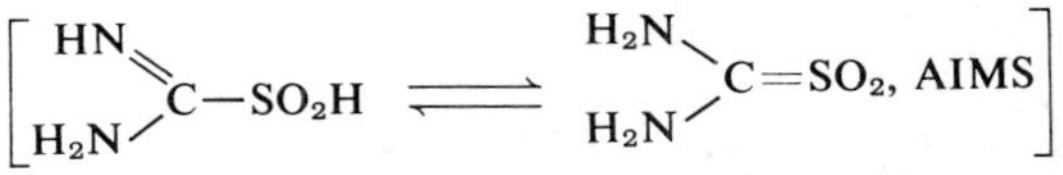

[27,35]. The products of these reductions are the hydrocarbons derived from the alkyl radical R$\cdot$ (see Table 3.2) [27,35]. The initial step in these reactions is believed to be the reduction of the Cr(I) complex to the Cr(0) complex,

$$\pi\text{-}(CH_3CH_2C_6H_5)_2Cr(I)I \xrightarrow{AIMS^{\ominus}} AIMS\cdot + \pi\text{-}(CH_3CH_2C_6H_5)_2Cr(0) \qquad (3\text{-}20)$$

Eq. (3-20). Both the radical produced from the reducing agent (AIMS$\cdot$) and the zerovalent chromium complex may now react with the alkyl halide to give the organic radical R$\cdot$, which in turn may couple, or abstract hydrogen, or undergo disproportionation to give the observed organic products, Eq.

TABLE 3.2

Catalytic Reduction of Alkyl Halides with π-Bis(ethylbenzene)chromium(I) Iodide [27] *and AIMS*[a] *or* $Na_2S_2O_4$

REDUCING AGENT	ALKYL HALIDE	PRODUCT
AIMS	$(C_6H_5)_2CHCl$	$(C_6H_5)_2CHCH(C_6H_5)_2$ (98%)
AIMS	$(C_6H_5)_2CCl_2$	$(C_6H_5)_2C{=}C(C_6H_5)_2$ (93%)
AIMS	$C_6H_5CH_2Cl$	$C_6H_5CH_2CH_2C_6H_5$ (30%) + solid m.p.150°
AIMS	$C_6H_5CCl_3$	$C_6H_5C(Cl)_2C(Cl)_2C_6H_5$ (20%)
$Na_2S_2O_4$	$(C_6H_5)_3CCl$	$(C_6H_5)_3CH$ (84%)
AIMS	$CH_2{=}CHCH_2Br$	$CH_2{=}CH(CH_2)_2CH{=}CH_2$ (17%)
AIMS	CH_3I	CH_4 (35%)
AIMS	CH_3CH_2I	$CH_3CH_3 + CH_2{=}CH_2$ (48%)

[a] AIMS represents $(H_2N)(HN{=})C{-}SO_2H$, aminoiminomethane sulfinic acid.

(3-21) [27,35]. The presence of excess reducing agent AIMS$^{\ominus}$ and of the radical derived therefrom (AIMS·) ensures the continual reduction of the π-bis(arene)chromium(I) complex to the zerovalent complex [27,35], thereby making the system a catalytic one.

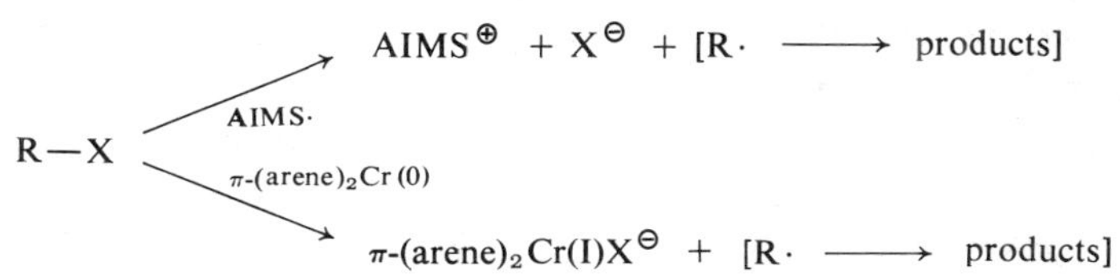

(3-21)

3. Catalytic Reactions

There are many reports concerning the use of π-bis(arene)chromium compounds modified either chemically [e.g., 36–40] or thermally [e.g., 36, 39,41], or in combination with other organometallic compounds [e.g., 37, 42–44], as catalysts for the polymerization, hydrogenation, or isomerization of olefins or dienes. However, there are very few reports concerning the use of discrete π-bis(arene)chromium compounds or simple identifiable derivatives thereof as active catalysts in polymerization, hydrogenation, or isomerization processes. Thus, whereas π-bis(benzene)chromium(0) does not catalyze the polymerization of ethylene at 80°–160°C [39], it does catalyze the polymerization of ethylene when heated above its decomposition temperature (i.e., ca. 250°C) [41], or when heated in combination with aluminosilicates at 80°–160°C [39] or when used in combination with tris(isobutyl)-aluminum [44].

Domrachev and Razuvaev studied the interaction of π-bis(benzene) chromium(0) with a variety of monomers and reported that whereas the pure complex does not catalyze the polymerization of styrene, α-methylstyrene, methyl methacrylate, vinyl chloride, vinyl acetate, or ethylene oxide, it does catalyze the polymerization of acrylonitrile at 20°C under nitrogen. This polymerization is promoted by the zerovalent arene chromium complexes (arene = benzene, toluene, ethylbenzene, xylene, mesitylene, and hexamethylbenzene) but not by π-bis(arene)chromium(I) salts or hydroxides. The composition of the polymer, after exposure to air, indicates the presence of several acrylonitrile units and also of the π-bis(arene)chromium(I) hydroxide hydrate. Further experiments showed that the chromium π-complex is bound to the acrylonitrile polymer and that during the polymerization process the bis(arene)chromium(0) remains zerovalent. The oxidation to Cr(I) only occurs when the polymer is exposed to air. The authors concluded, therefore, that the polymerization process is not initiated by the transfer of an electron from the zerovalent Cr complex to acrylonitrile in a radical-initiated reaction, but rather that the process is initiated by chemical inter-

$$\pi\text{-(arene)}_2\text{Cr(0)} + \text{AN} \longrightarrow \pi\text{-(arene)}_2\text{Cr(I)} + \text{AN}^{\overline{\cdot}} \qquad (3\text{-}22)$$

action between the zerovalent chromium π-complex and acrylonitrile, e.g., cyanoethylation of the arene rings complexed to chromium [45].

Hagihara and his colleagues studied the catalytic activity of π-bis(arene)-chromium(0) compounds in the polymerization of various monomers. They found that whereas the zerovalent complexes were inactive in the polymerization of styrene and ethylene, they were active when used in conjunction with certain oxidizing agents, e.g., *tert*-butyl peroxide or oxygen [15,46–52].

The polymerization of styrene by π-bis(benzene)chromium(0) in the presence of *tert*-butyl peroxide is occasioned by the *tert*-butyl radicals generated in the oxidation–reduction reaction outlined in Eq. (3-23) [46,52].

$$\pi\text{-}(C_6H_6)_2\text{Cr(0)} + [(CH_3)_3CO]_2 \longrightarrow 2\,C_6H_6 + [(CH_3)_3CO]_4\text{Cr(IV)} + (CH_3)_3CO\cdot \qquad (3\text{-}23)$$

Chromium(IV)-tetrakis(*tert*-butoxide) also effects the polymerization of methyl acrylate and styrene in the presence of alcohols, the active species is again the alkoxy radical generated in the oxidation–reduction reaction outlined in Eq. (3-24) [47].

$$[(CH_3)_3CO]_4\text{Cr(IV)} \xrightarrow{\text{ROH}} (CH_3)_3COH + \text{Cr(III)}[OR]_3 + RO\cdot \qquad (3\text{-}24)$$

The polymerization of ethylene by π-bis(benzene)chromium(0) after activation by molecular oxygen gives a highly linear polyethylene (mol wt 1,660,000; m.p. 137°–140°C) [48]. The active catalyst in this reaction was identified as a chromate salt $[\pi\text{-}(C_6H_6)_2Cr(I)]_2Cr(VI)O_4$. Recent kinetic studies have shown that the autoxidation of $\pi\text{-}(C_6H_6)_2Cr$ and $\pi\text{-}(CH_3CH_2\text{-}C_6H_5)_2Cr$ are bimolecular heterolytic reactions and that the initial products are the corresponding arenes and chromates $[\pi\text{-(arene)}_2Cr]_2CrO_4$ [53].

The dichromate salt $[\pi\text{-}(C_6H_6)_2Cr(I)]_2Cr_2O_7$ is also an effective catalyst in the polymerization of ethylene [40,48,52]. The same chromate and dichromate salts are also effective catalysts in the hydrogenation of olefins, dienes, and acetylenes [54].

The foregoing evidence would tend to suggest that the bis(arene)chromium(0) complexes are themselves not active catalysts. The redox potential for the system $Cr(0) \overset{e}{\rightleftharpoons} Cr(I)$ is such that the bis(arene)chromium(0) complexes should act as reducing agents (in stoichiometric processes) or as initiators in certain radical-promoted catalytic processes.

C. π-(Cyclopentadienyl)chromium Compounds

The chemistry of the π-(cyclopentadienyl)chromium compounds has not been studied extensively [55] and this may be due, in part, to the general difficulties inherent to their preparation and to the experimental difficulties often encountered in the handling, purification, and characterization of some of the more reactive compounds. These compounds undergo a variety of reactions involving the ligands on chromium and/or the chromium center (e.g., ligand exchange, ligand transformation, oxidation-reduction reactions, and in some instances catalytic reactions). In the ensuing discussion of all these reactions, the π-(cyclopentadienyl)chromium compounds have been divided for convenience into the following main categories: (1) π-bis(cyclopentadienyl)chromium(II) and the related chromium(III) complexes; (2) π-(cyclopentadienyl)$Cr(CO)_3$ dimer and the related hydride and sodium salt; (3) $\pi\text{-}(C_5H_5)Cr(CO)_2NO$; (4) $\pi\text{-}(C_5H_5)Cr(NO)_2Cl$; (5) π-(cyclopentadienyl)chromium compounds containing one cyclopentadienyl ring and one other alicyclic unit with delocalized electrons [e.g., $\pi\text{-}(C_5H_5)\text{-}\pi\text{-}(C_6H_6)Cr$, $\pi\text{-}(C_5H_5)\text{-}\pi\text{-}(C_7H_7)Cr$, etc.).

1. π-Bis(cyclopentadienyl)chromium(II) and Related Compounds

π-Bis(cyclopentadienyl)chromium(II) (chromocene) is a highly reactive compound, which though unreactive toward water, is cleaved by mineral acids [56]. Most of the reactions of chromocene involve either ligand exchange reactions or oxidation reactions with or without the loss of one cyclopentadienyl unit. The compound has also been used to effect the polymerization of olefins.

a. Ligand Replacement Reactions

The cyclopentadienyl units of π-bis(cyclopentadienyl)chromium(II) may be transferred from chromium to iron(II) by reaction with $FeCl_2$ in tetrahydrofuran, Eq. (3-25) [56]. The lability of the cyclopentadienyl rings of

$$\pi\text{-}(C_5H_5)_2Cr(II) + FeCl_2 \xrightarrow{THF} \underset{70\%}{\pi\text{-}(C_5H_5)_2Fe} + CrCl_2(THF)_2\,(?) \qquad (3\text{-}25)$$

chromocene has been illustrated by the study of the exchange reactions between LiC_5D_5 or π-$(C_5D_5)_2Ni$ and chromocene in tetrahydrofuran or heptane as solvents. Mass spectrometric data of the nickelocene recovered from the exchange reaction with chromocene, proved that the cyclopentadienyl units exchanged as entities, Eq. (3-26) [57].

$$\pi\text{-}(C_5D_5)_2Ni + \pi\text{-}(C_5H_5)_2Cr \overset{\Delta}{\rightleftharpoons} \pi\text{-}(C_5D_5)_n\text{-}\pi\text{-}(C_5H_5)_{2-n}Ni(II) + \pi\text{-}(C_5H_5)_n\text{-}\pi\text{-}(C_5D_5)_{2-n}Cr(II) \qquad (3\text{-}26)$$

Chromocene reacts with carbon monoxide, both in the presence and in the absence of hydrogen, to give a variety of products in which one cyclopentadienyl unit is replaced by three CO units. These reactions, which have been discussed in Chapter 1, constitute a method of preparation for the tricarbonyl dimer **6** and the hydridochromium compound **7**, Scheme 3-2.

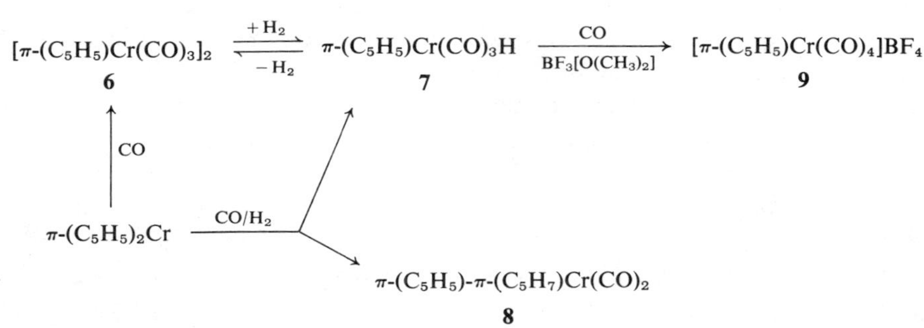

Scheme 3-2. Reactions of chromocene with CO.

The dicarbonylchromium complex **8**, which contains one cyclopentadienyl unit (bonded as π-C_5H_5) and a cyclopentenyl unit (bonded as a π-allyl unit), is a by-product in the reaction of chromocene with CO and H_2 [58].

The hydridochromium compound **7**, which is conveniently prepared by the oxidative addition of cyclopentadiene to tris(acetonitrile)$Cr(CO)_3$ [59], and the carbonyl dimer **6** are interconvertible by the reversible loss or gain of hydrogen [60]. The compound **7** reacts further with CO, in the presence of boron trifluoride etherate, to give the cationic tetracarbonyl complex **9**, Scheme 3-2 [61].

b. Oxidative Reactions

Chromocene is a powerful reducing agent and is oxidized by a variety of organic halogeno compounds to π-(cyclopentadienyl)chromium(III) complexes, Scheme 3-3. Some of these reactions, in particular the oxidation to π-(cyclopentadienyl)dichlorochromium(III) (**10**) and its solvates (**11**, **12**, Scheme 3-3), represent satisfactory methods of preparation and have already

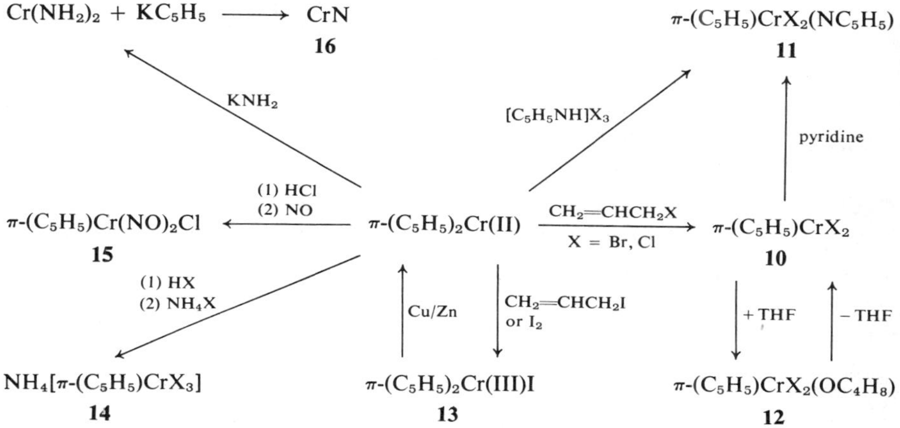

Scheme 3-3. Some reactions of π-bis(cyclopentadienyl)chromium(II).

been discussed in Chapter 1. The most unusual feature of these oxidations is that in some cases both cyclopentadienyl rings are retained and in others one of the rings is lost. Thus, oxidation of chromocene with allyl iodide gives the bis(cyclopentadienyl)chromium(III) cation **13**, while oxidation with allyl bromide (or chloride) gives, by loss of one C_5H_5 unit, monocyclopentadienyl dichlorochromium(III) (**10**, Scheme 3-3) [62,63]. The mechanism of these oxidations is not clear, and the loss of the C_5H_5 units may be associated with their lability referred to above in Section C,1,a. The allyl groups in the oxidations with allyl halides, couple to give, as product, 1,5-hexadiene. This may suggest, in analogy with the oxidation of π-bis(arene)chromium(0) compounds [Eqs. (3-18) and (3-19)], that an allyl radical is formed at some stage in the process, either by oxidative cleavage of the allyl–halogen bond, or by homolysis of an allyl–chromium bond in an intermediate of the type $[\pi\text{-}(C_5H_5)_2\text{-}\sigma\text{-}(C_3H_5)CrX]^{\ominus}$, cf. Eq. (3-19). π-Bis(cyclopentadiene)chromium-(III) iodide is reduced to chromocene by zinc-copper powder at 180°–190°C [63].

The precise products formed in the interaction of chromocene and mineral acids depend critically upon the reaction conditions. Thus, Fischer, Ulm, and Kuzel found that chromocene reacted with concentrated HCl (HBr) in the cold to give a blue solution containing the π-$(C_5H_5)CrCl_3$ anion with cyclopentadiene and hydrogen as by-products. The anion could be isolated in the form of its ammonium salt (**14**, Scheme 3-3) by treatment with NH_4X (X = Cl or Br) [62].

c. Miscellaneous Reactions

Chromocene reacts successively with gaseous HCl and NO to give, by replacement of one cyclopentadienyl ring, the crystalline π-(cyclopentadienyl)-bis(nitric oxide)dichlorochromium (**15**, Scheme 3-3), in good yield (see Chapter 1) [64]. The latter compound is the starting material for the preparation of many bridged chromium compounds and σ-bonded organochromium compounds (see Sections C,4,a and c).

Chromocene dissolves in liquid ammonia to give ammoniates [π-$(C_5H_5)_2Cr \cdot 9NH_3$ and π-$(C_5H_5)_2Cr \cdot 4NH_3$]. These are stable only at low temperatures since when they are warmed, ammonia is lost, in stages, to give finally unchanged chromocene [65]. With potassium amide, however, the cyclopentadienyl units are exchanged for $NH_2^{\ominus}$ to give a chromium amide that subsequently loses hydrogen to give a chromium nitride **16**, Scheme 3-3 [65]. It would be of interest to know whether the cyclopentadienyl units of chromocene may be exchanged by other anionic species.

Werner and Richards have studied the reaction of chromocene with organodiazo compounds and found that, in contrast to nickelocene, the chromium compound does not catalyze the rapid decomposition of diazomethane, and no polymethylene and/or volatile hydrocarbons are formed. Chromocene, however, does react with ethyldiazoacetate in the presence of cyclohexene to give the products formulated in Eq. (3-27) in the ratios 1.5: 0.6:1.0 [66].

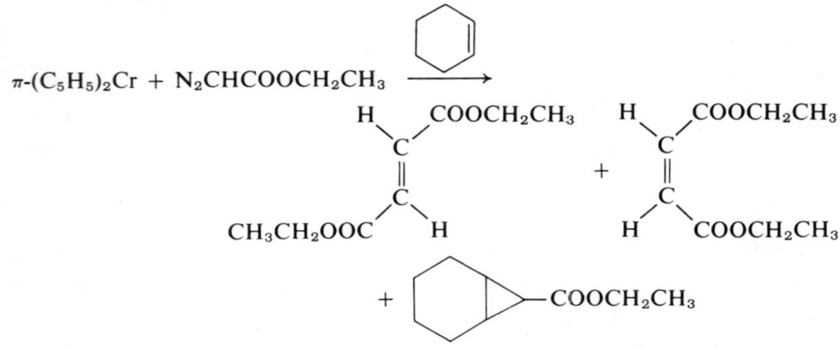

(3-27)

d. Catalytic Reactions

π-Bis(cyclopentadienyl)chromium(II), like bis(arene)chromium(0) compounds, is itself not an effective catalyst in the polymerization of ethylene

[67]. However, when the compound is modified chemically (e.g., by reaction with O_2 [37] or by interaction with alumina or silica, etc. [67–69]) or when it is used in conjunction with other organometallic compounds [e.g., $Al(CH_2CH_3)_3$] [37], effective catalyst systems for the hydrogenation, polymerization, and isomerization of olefins are obtained. Carrick and his associates have studied the polymerization of ethylene by the chromocene/silica catalyst [67]. They have found that the formation of the catalyst, by the interaction of chromocene and silica, is accompanied by the liberation of cyclopentadiene and that in the active catalyst the chromium remains bivalent. They have suggested, therefore, that, since silica contains weakly acidic surface hydroxyl groups, the active catalyst is formed by the displacement of one cyclopentadienyl unit of chromocene by a surface OH group to give a chromium(II)-siloxy-π-C_5H_5 complex, e.g., Eq. (3-28). The polymeriza-

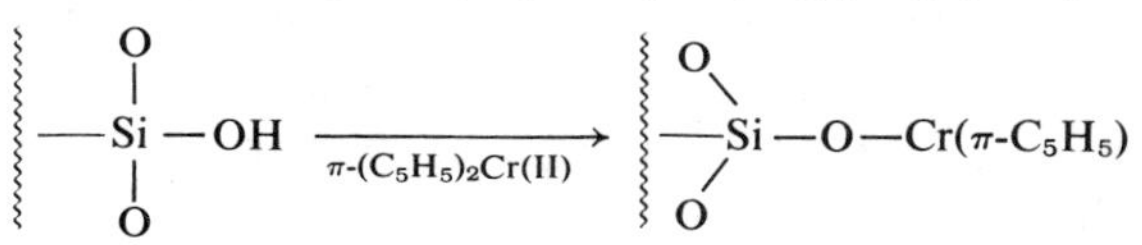

(3-28)

tion of the ethylene is assumed to occur on the coordinatively unsaturated chromium centers via an anionic mechanism. The initiation of the polymerization remains obscure; however, the authors have suggested that it involves insertion of ethylene into a Cr–R bond (R = O, H, alkyl, etc.), but it is not clear whether this "insertion" proceeds via a four-center intermediate or whether it proceeds by the stepwise transfer of R and Cr to coordinated ethylene, Eq. (3-29a and b). The latter mechanism would require the presence of at least one free coordination site on the chromium center.

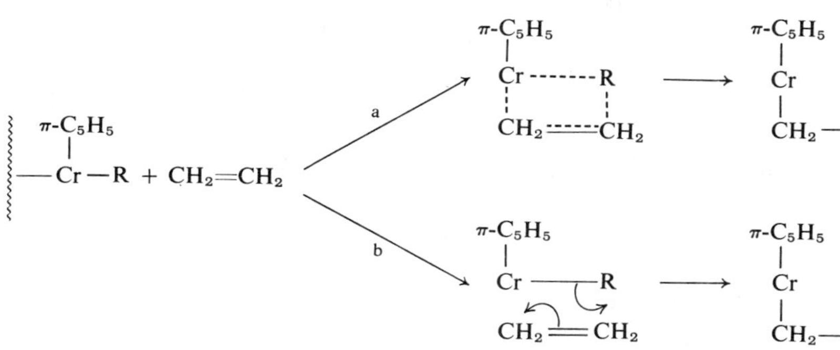

(3-29)

2. π-(CYCLOPENTADIENYL)CHROMIUM TRICARBONYL DIMER AND THE RELATED π-$(C_5H_5)Cr(CO)_3H$ AND π-$(C_5H_5)Cr(CO)_3Na$

The chemistry of the title compounds **17**, **18**, **19**, Scheme 3-4, has not been extensively studied, and as will be seen from the ensuing pages the compounds have mainly been used for the preparation of complexes containing chromium-to-metal bonds and catalysts in the hydrogenation of conjugated dienes. When considering the reactions of this group of compounds, it is important to bear in mind that the hydridochromium compound **18** is on the one hand a weak acid and so gives rise to sodium and potassium salts **19**, and on the other hand, it is thermally unstable and readily eliminates hydrogen to give the dimer **17**. The dimer **17** possesses a relatively weak Cr–Cr bond ($-E_{1/2} = 1.3$ V on Hg and 0.8 V on Pt [70,71]; Cr–Cr bond length 3.281(1) Å [71a]) and can thus act as a source of radicals π-$(C_5H_5)(CO)_3Cr\cdot$ and may therefore participate in redox reactions and radical initiated processes.

a. Ligand Replacement Reactions

There are few reports concerning ligand replacement reactions within the present group of compounds. The best known examples are: the replacement of one CO group of $[\pi\text{-}(C_5H_5)Cr(CO)_3]_2$ by NO in the synthesis of π-$(C_5H_5)Cr(CO)_2NO$ (**20**, Scheme 3-4) [72], the replacement of one CO group in **17**, Scheme 3-4, by a π-crotyl group in the synthesis of π-(C_5H_5)-π-(crotyl)$Cr(CO)_2$, (**21**, Scheme 3-4 [73,74], Chapter 1, Section 4,b,iv), and the displacement of all the CO groups of **17**, by reaction with allylic halides, to give π-cyclopentadienyl dibromochromium, **22**. This last reaction has been mentioned in Chapter 1, Section C,1,b [62]. There is a report that $[\pi\text{-}(C_5H_5)Cr(CO)_3]_2$ reacts with CH_3SSCH_3 to give by replacement of all the CO groups a compound formulated as π-$(C_5H_5)_2Cr_2(CH_3S)_3$ [75]. One of the carbonyl groups of the mercury-bridged complex **23** may be replaced by certain phosphines to give moderate yields (20%–50%) of the unstable complexes, Eq. (3-30) (R = C_6H_5 and OCH_3) [76].

$$[\pi\text{-}(C_5H_5)Cr(CO)_3]_2Hg \xrightarrow{PR_3/\Delta} [\pi\text{-}(C_5H_5)Cr(CO)_2PR_3]_2Hg \qquad (3\text{-}30)$$

Manning and his colleagues have observed that two of the carbonyl groups of the dimer $[\pi\text{-}(C_5H_5)Cr(CO)_3]_2$ may be replaced by phosphines to give crystalline dimers $[\pi\text{-}(C_5H_5)Cr(CO)_2PR_3]_2$ [76a]. It is pertinent to mention here that the attempted ligand exchange of the CO groups in π-(cyclopentadienyl)methylchromium tricarbonyl by certain triarylphosphines led to migration of the methyl group (from Cr to CO) and introduction of the phosphine, to give an acylchromium complex, Eq. (3-31) [R = C_6H_5, 4-$CH_3OC_6H_4$, and $C_6H_3(CH_3)_2$] [77].

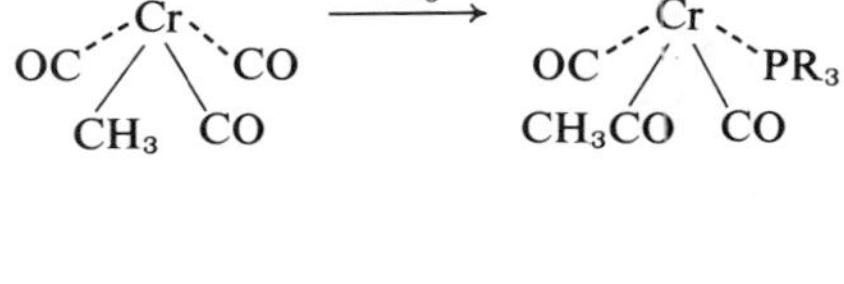

(3-31)

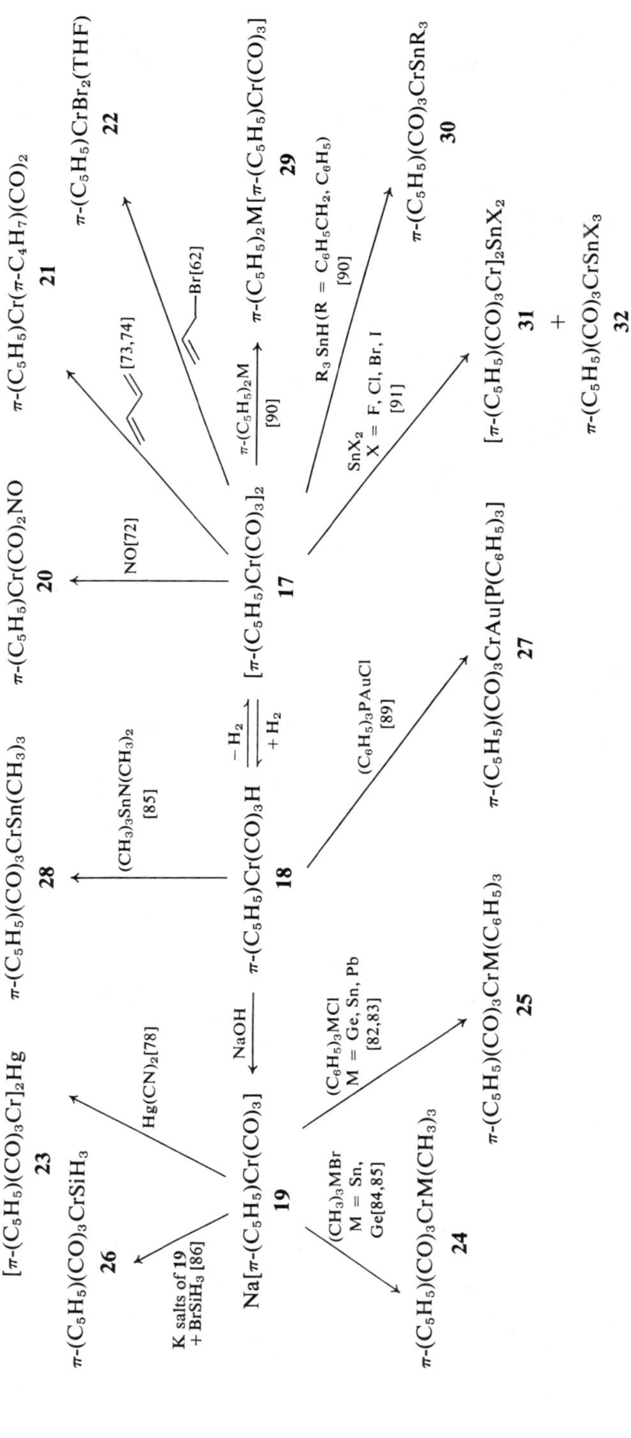

Scheme 3-4. Some reactions of $[\pi\text{-}(C_5H_5)Cr(CO)_3]_2$, $\pi\text{-}(C_5H_5)Cr(CO)_3H$, and $[\pi\text{-}(C_5H_5)Cr(CO)_3]Na$.

b. Formation of Compounds with Chromium-to-Metal Bonds

The hydridochromium compound **18** and the chromium carbonyl dimer **17**, or simple derivatives thereof have been used as the starting material in the preparation, by the routes outlined below, of several compounds containing bonds between chromium and other metals (e.g., Hg, Zn, Cd, Si, Ge, Sn, Pb, and Au) (see Scheme 3-4). The sodium salt **19**, Scheme 3-4, reacts with $Hg(CN)_2$ to give the mercury-bridged [Cr–Hg–Cr] complex **23** [78,79] which, in turn, reacts with metallic zinc or cadmium in tetrahydrofuran to give the corresponding zinc- and cadmium-bridged complexes, Eq. (3-32) (M = Zn or Cd) [80]. The mercury complex **23** also undergoes a ligand-redis-

$$[\pi\text{-}(C_5H_5)Cr(CO)_3]_2Hg \xrightarrow{\text{Zn or Cd/THF}} [\pi\text{-}(C_5H_5)Cr(CO)_3]_2M \tag{3-32}$$

tribution reaction with bis[tris(ethyl)germyl]mercury to give a new complex with a chromium–germanium bond, Eq. (3-33) [81]. The sodium salt **19** reacts,

$$[\pi\text{-}(C_5H_5)(CO)_3Cr]_2Hg + [(CH_3CH_2)_3Ge]_2Hg \longrightarrow \pi\text{-}(C_5H_5)(CO)_3CrGe(CH_2CH_3)_3 \tag{3-33}$$

in diglyme, with various organotin, germanium, or lead halides to give, by elimination of NaX, new complexes with chromium-to-tin, germanium, and lead bonds, i.e., **24** and **25**, Scheme 3-4 [82–85]. The corresponding silicon compound could not be prepared by this metal-halide elimination reaction in solution [83] but the successful preparation of a silylchromium derivative **26** has been effected by the interaction of the potassium salt (**19**, K in place of Na) and silyl bromide in the absence of solvent (see Scheme 3-4) [86,87]. This silylchromium compound is sensitive to air and moisture and is readily cleaved by gaseous HCl according to Eq. (3-34) [86]. The presence of a

$$\underset{\mathbf{26}}{\pi\text{-}(C_5H_5)(CO)_3CrSiH_3} \xrightarrow{\text{HCl}} \pi\text{-}(C_5H_5)Cr(CO)_3H + H_3SiCl \tag{3-34}$$

metal-to-metal bond in this group of compounds has been confirmed by an x-ray structure analysis of π-$(C_5H_5)(CO)_3CrSn(C_6H_5)_3$; the chromium-to-tin bond length is 2.85 Å, which is somewhat larger than the sum of the atomic radii (2.65 Å) [88].

The hydridochromium compound **18** may also be used as a starting material for compounds containing a chromium-to-metal bond. Thus, reaction of **18** with triphenylphosphine gold chloride gives the complex **27**, Scheme 3-4, with a chromium-to-gold bond [89]. Compound **18** reacts with dimethylaminotrimethylstannane to give, by cleavage of the tin–amide bond, the complex **28** with a chromium-to-tin bond, Scheme 3-4 [85].

The π-cyclopentadienylchromium carbonyl dimer **17**, undergoes a reduction–oxidation reaction with other π-bis(cyclopentadienyl) metal complexes, tris(organo)tin hydrides, and tin halides to give complexes with new chromium to metal bonds, e.g., compounds **29–31**, Scheme 3-4 [90,91]. The products obtained by reaction with π-$(C_5H_5)_2M$, i.e., the complexes **29**, Scheme 3-4, are readily cleaved by interaction with $NaB(C_6H_5)_4$ or H_2O thereby regenerating the sodium and hydrido-(π-cyclopentadienyl)chromium tricarbonyls, respectively, Eq. (3-35) [90]. The redox reaction between the

$$\pi\text{-}(C_5H_5)_2M[Cr(CO)_3(\pi\text{-}C_5H_5)] \begin{cases} \xrightarrow{NaB(C_6H_5)_4} \pi\text{-}(C_5H_5)_2M[B(C_6H_5)_4] + [\pi\text{-}(C_5H_5)Cr(CO)_3]Na \\ \xrightarrow{H_2O} [\pi\text{-}(C_5H_5)Cr(CO)_3]H \end{cases} \quad (3\text{-}35)$$

dimer **17** and organotin hydrides is accompanied by the evolution of hydrogen, and the products are the complexes with chromium-to-tin bonds **30** ($R = C_6H_5$ or $C_6H_5CH_2$), Scheme 3-4 [90]. The same compounds may also be prepared by the salt-elimination reaction discussed above.

The interaction of the tricarbonyl chromium dimer **17** and tin halides (SnX_2) may be brought about either by heating or by ultraviolet irradiation and the products formed depend not only upon the mode of reaction but also upon the stoichiometry of the reagents [91]. Thus, when equimolar quantities of reagents are heated in butanol, the main product formed by the direct "insertion" of SnX_2 into the Cr–Cr bond (or by the oxidative addition of the dimer to SnX_2?) is the dichromium–tin complex **31**, Scheme 3-4, and when the dimer to tin halide ratio is 1:10, a mixture of dichromium-tin and monochromium-tin compounds (**31** and **32**, Scheme 3-4), is obtained. The relative amounts of the dichromium compound formed depend upon the tin halide used in the reaction and fall in the series $F > Cl > Br \geqslant I$. It has been suggested that these reactions do not involve direct insertion into the Cr–Cr bond, but that they proceed via the monochromium-tin trihalide π-(C_5H_5)-$(CO)_3CrSnX_3$ as an intermediate. The interaction of the dimer **17** and tin halides gives, when promoted by UV irradiation, mixtures of mono- and dichromium complexes **31** and **32**, Scheme 3-4 [91]. π-Cyclopentadienyl-(trimethylstannyl)chromium tricarbonyl can also be prepared by the interesting ligand redistribution reaction outlined in Eq. (3-36) [59].

$$(CH_3CN)_3Cr(CO)_3 + (CH_3)_3SnC_5H_5 \longrightarrow \pi\text{-}(C_5H_5)(CO)_3CrSn(CH_3)_3 + 3\,CH_3CN \quad (3\text{-}36)$$

c. *Catalytic Reactions*

All three types of π-(cyclopentadienyl)chromium tricarbonyl compounds, i.e., the dimer $[\pi\text{-}(C_5H_5)Cr(CO)_3]_2$, the hydride $[\pi\text{-}(C_5H_5)Cr(CO)_3H]$, and the salt $[\pi\text{-}(C_5H_5)_2M][\pi\text{-}(C_5H_5)Cr(CO)_3]$ ($M = Cr$ and V), are active catalysts in the hydrogenation of conjugated dienes and of ene-ynes [92–96]. Miyake and Kondo found that the hydridochromium compound **18** reacted quantitatively with isoprene according to Eq. (3-37). The dimer **17** can be

$$\underset{\mathbf{18}}{2\,[\pi\text{-}(C_5H_5)Cr(CO)_3H]} + CH_2{=}C(CH_3){-}CH{=}CH_2 \longrightarrow \underset{\mathbf{17}}{[\pi\text{-}(C_5H_5)Cr(CO)_3]_2} + CH_3{-}C(CH_3){=}CHCH_3 \qquad (\mathbf{17} \xrightarrow{H_2} \mathbf{18}) \tag{3-37}$$

reconverted to the hydridochromium compound by interaction with hydrogen and, therefore, a catalytic cycle can be set up for the hydrogenation of dienes using either **17** or **18** in the presence of hydrogen. Various conjugated linear and cyclic dienes, trienes, and tetraenes have been hydrogenated with $[\pi\text{-}(C_5H_5)Cr(CO)_3]_2$ and the molybdenum and tungsten analogs, and of the Cr, Mo, and W triad the chromium compound is reported to be the most active [92,93].

The published results concerning the hydrogenation of various olefins indicate that (i) only conjugated dienes are reduced, i.e., compounds with isolated double bonds are not reduced; and (ii) the hydrogen is added to the terminal carbon atoms of the conjugated systems. Exceptions would seem to be sterically encumbered dienes, e.g.,

$$CH_2{=}CH_2{-}CH{=}C(CH_3)_2$$

and cyclic dienes. Thus the reduction of 6,6-dimethylfulvene gives the mixture of products formulated in Eq. (3-38) [92]. The present catalyst

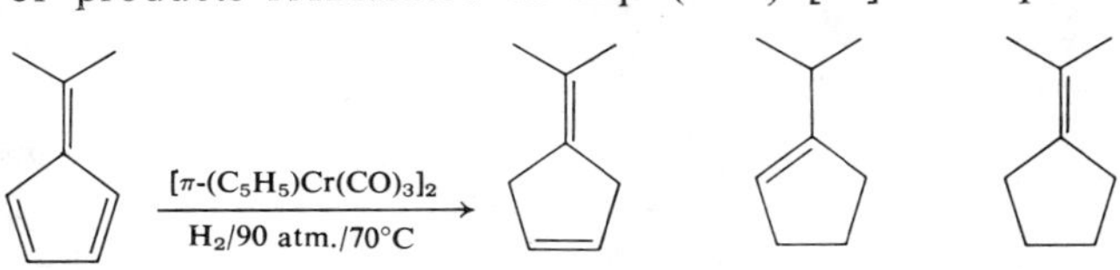

20% 29% 51% (3-38)

systems would seem to lack the stereoselectivity associated with the π-(arene) chromium tricarbonyl catalyst, see Section D,1,c of this chapter.

3. π-(Cyclopentadienyl)Cr(CO)$_2$NO

The title compound may be prepared either as outlined in Scheme 3-4 above or by the interaction of sodium $\pi\text{-}(C_5H_5)Cr(CO)_3$ with certain nitroso compounds, Eq. (3-39) [97–99]. The chemistry of this diamagnetic compound has not been extensively investigated; however, from the scant information that is available, it would seem that it is more reactive than the corresponding tricarbonyl complex π-$(C_5H_5)Cr(CO)_3$. Thus, as will be seen from Scheme 3-5, not only may the CO ligands be exchanged more readily, but also the cyclopentadienyl and carbonyl ligands may be transformed by electrophilic reagents.

$$\pi\text{-}(C_5H_5)Cr(CO)_3Na \xrightarrow[CH_3\cdot C_6H_4SO_2NCH_3(NO)]{\text{amyl nitrite}} \underset{\mathbf{20}}{\pi\text{-}(C_5H_5)Cr(CO)_2NO} \quad (3\text{-}39)$$

a. Ligand Replacement Reactions

The products formed in the thermal interaction of π-$(C_5H_5)Cr(CO)_2NO$ (**20**) and triphenylphosphine depend critically upon the temperature of the reaction. Thus, in a triphenylphosphine melt one of the carbonyl groups is replaced by the phosphine to give π-$(C_5H_5)Cr(CO)(NO)P(C_6H_5)_3$ (**33**). However, the reaction is of restricted value as a method of preparation since the desired compound is difficult to separate from the accompanying decomposition product and excess reagents [97].

The replacement of one CO group of **20** by phosphine, olefin or acetylene may be readily effected by the photocatalyzed interaction of the nitrosocarbonyl **20** with the appropriate substrate [97,99,100]. Thus, the photocatalyzed interaction of π-$(C_5H_5)Cr(CO)_2NO$ and cyclooctene gives the new complex **34**,

L = 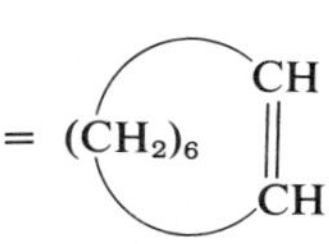

Scheme 3-5, in which one of the CO ligands is replaced by cyclooctene. The cyclooctene in this compound may in turn be replaced by a variety of other olefins or acetylenes:

$CH_2{=}CH_2$, $CH{\equiv}CH$, $CH_3OOCC{\equiv}CCOOCH_3$, 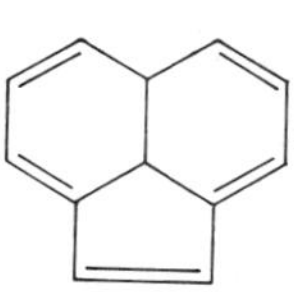, , 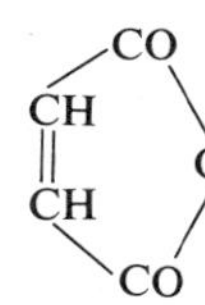

There are thus two synthetic routes to olefin or acetylene complexes of the type π-$(C_5H_5)Cr(CO)(NO)(L)$ (**34**): the one is the photocatalyzed interaction of the parent dicarbonyl **20** with the unsaturated substrate; the other is ligand exchange between the cyclooctene complex **34** (L = cyclooctene) and the unsaturated substrate [100].

The photoinduced interaction of the π-acetylcyclopentadienyl complex **35**, Scheme 3-5, with triphenylphosphine also leads to the replacement of one CO

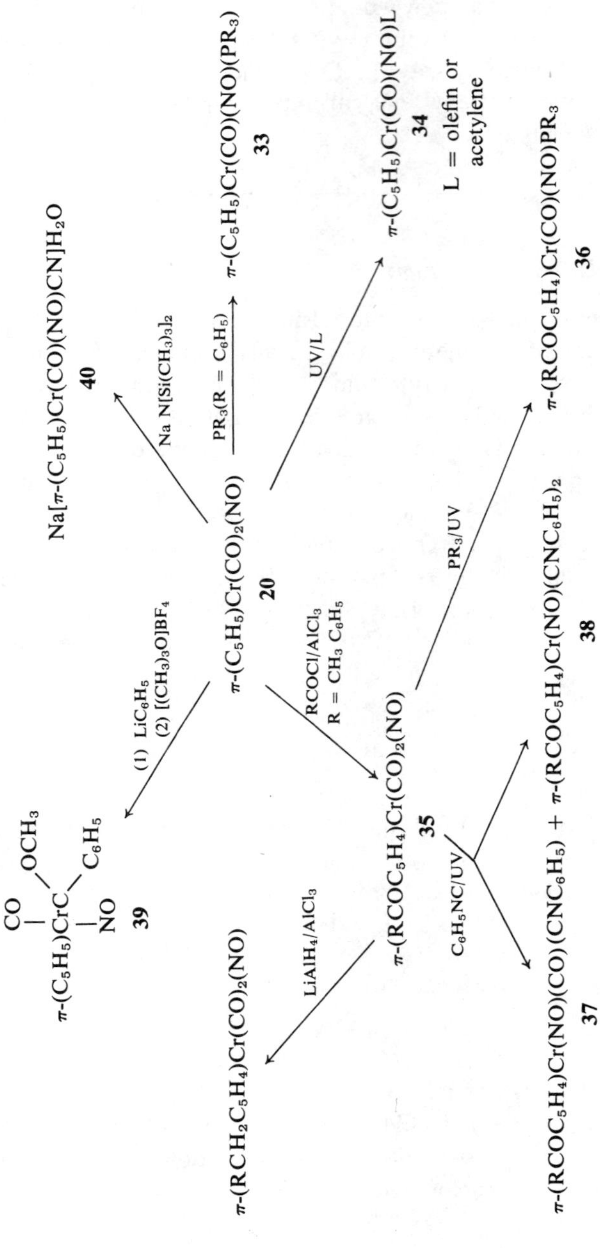

Scheme 3-5. Some reactions of π-$(C_5H_5)Cr(CO)_2(NO)$.

group by $P(C_6H_5)_3$ to give **36**, Scheme 3-5. The same reaction with phenylisocyanide, however, leads to the replacement of not only one CO group, but also of a second CO group to give a mixture of the mono- and the bisisocyanide complexes **37** and **38**, respectively, Scheme 3-5. The monoderivative **37** may undergo subsequent disproportionation to π-$(CH_3COC_5H_4)Cr(CO)_2(NO)$, (**20**) and π-$(CH_3COC_5H_4)Cr(NO)(CNC_6H_5)_2$, (**38**) [99]. The compounds derived from π-$(RC_5H_4)Cr(CO)_2NO$ by the replacement of one CO group by a ligand L (i.e., compounds **33**, **34**, **36**, **37**, Scheme 3-5) or by the chemical transformation of one of the CO ligands (i.e., compounds **39**, **40**, Scheme 3-5) all possess an asymmetric center at chromium, and it should therefore be possible to resolve them into optical antipodes (e.g., **41** and **42**). To date, however, there are no reports of the successful resolution of such asymmetric chromium compounds, though asymmetric manganese compounds have been resolved [101].

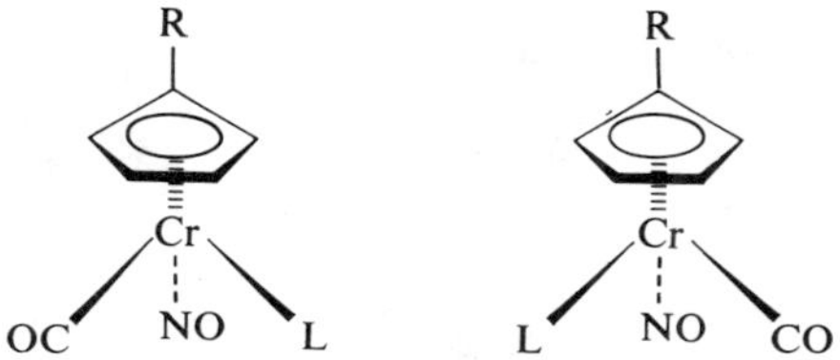

41 **42**

b. Chemical Transformations of the Ligands

The cyclopentadienyl ring of π-$(C_5H_5)Cr(CO)_2NO$ (**20**) undergoes electrophilic substitution upon interaction with acetyl and benzoyl tetrachloraluminates to give the ketones **35**, Scheme 3-5 ($R = CH_3$ and C_6H_5, respectively) [72,99,102]. The results of the competitive Friedel–Craft acylation of mixtures of **20** and other substrates have established that the organochromium compound is much less reactive than is ferrocene, anisole, or π-$(C_5H_5)Mn(CO)_3$ and is about as reactive as π-$(C_5H_5)V(CO)_3$ and benzene [102].

One of the carbonyl groups of π-$(C_5H_5)Cr(CO)_2(NO)$ undergoes nucleophilic substitution upon interaction with phenyllithium or sodium bis(trimethylsilyl)amide to give finally the carbenoid and cyano compounds **39** and **40**, respectively, Scheme 3-5 [103,104].

The carbenoid compound **39** is obtained, as in the case of other carbenoid complexes (Chapter 1), by the reaction of the phenyllithium adduct **43**, with trimethyloxonium tetrafluoroborate, Eq. (3-40).

The conversion of one of the carbonyl groups of **20** to a cyanide unit proceeds via nucleophilic attack of the silylamide anion to give the intermediate **44**, Eq. (3-41). The subsequent transformation of **44** may involve a migration of a trimethylsilyl unit from nitrogen to oxygen (to give **45**) followed by extrusion of the silyl ether unit to give the cyanide complex **40**, Eq. (3-41). This transformation has its analogy in the reaction of non-enolizable ketones with $NaN[Si(CH_3)_3]_2$ [105].

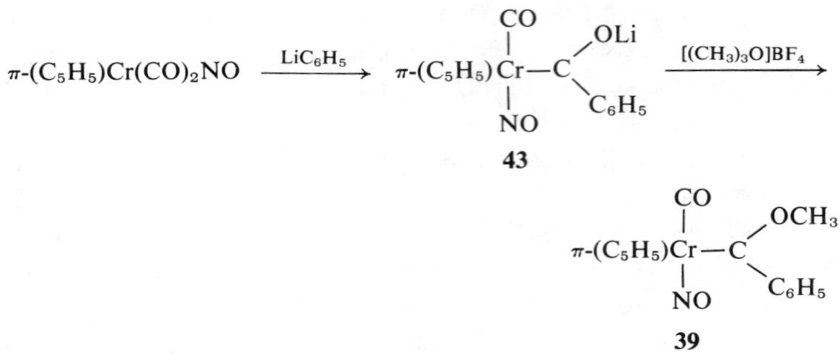

(3-40)

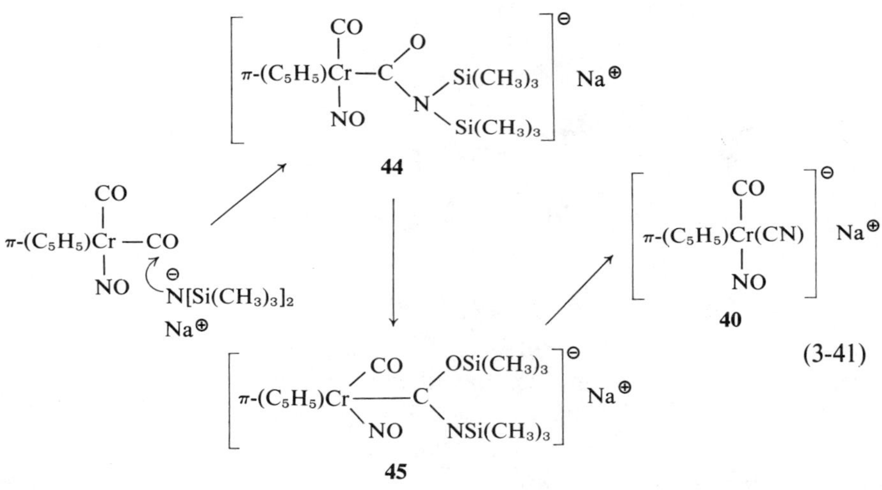

(3-41)

4. π-(CYCLOPENTADIENYL)$Cr(NO)_2Cl$

The chemistry of the title compound has not been explored extensively (for reviews, see [55,106]), and from the data available (see Scheme 3-6), it is evident that the compound undergoes metathetical exchange of Cl for various anions and ligand exchange of one NO group. Furthermore, the compounds may be transformed into new bridged dimers containing Cr–Cr bonds.

a. Metathetical Exchange Reactions

The chloride ion of the compound **15** dissociates in aqueous solution and may be exchanged for a variety of other anions (e.g., SCN [107] and NCSe [108]) first by the removal of the halide ion (by reaction with $AgNO_3$) and then by treatment of the aquated nitrosyl cation with NaX or KX, to give the new complexes π-$(C_5H_5)Cr(NO)_2X$ (**46**, Scheme 3-6).

π-(Cyclopentadienyl)dinitrosylchlorochromium has been used as the starting material for the synthesis of a variety of σ-bonded organochromium compounds (**47**, Scheme 3-6) (e.g., by reaction with RMgX, CH_2N_2, and NaC_5H_5, etc., [64]). These reactions have already been discussed in Chapter 1, Section C,1,d.

The σ-bonded benzylchromium compound **47** ($R = CH_2C_6H_5$) reacts with tetracyanoethylene (TCNE) to give products formed by the "insertion" of the TCNE unit into the $Cr–CH_2C_6H_5$ bond, e.g., Eq. (3-42) [109]. This

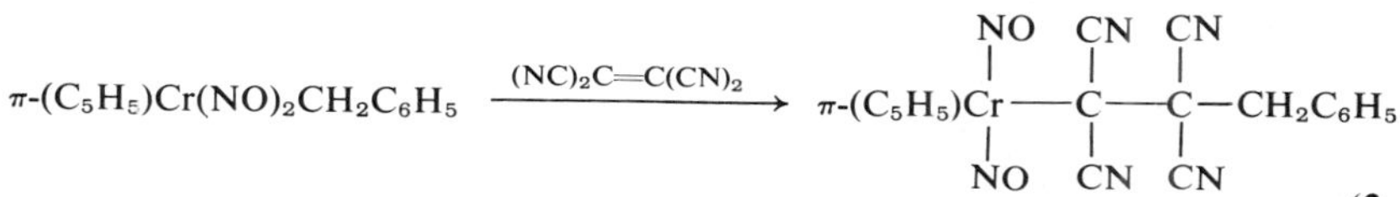

(3-42)

insertion reaction is analogous to the reaction of disubstituted acetylenes with π-bonded tris(organo)chromium(III) compounds, where it has been suggested that the initial process is the formation of a vinylchromium compound by "insertion" of the acetylenic unit into the carbon-to-chromium σ-bond (see Section F,3,a,ii below).

b. Ligand Replacement Reactions

One of the nitrosyl groups of the diamagnetic π-$(C_5H_5)Cr(NO)_2Cl$ may be exchanged for various amines [e.g., piperidine, piperazine, pyridine, γ-picoline], $P(C_6H_5)_3$, $P(n\text{-}C_5H_9)_3$, $As(C_6H_5)_3$, and cyclohexylisonitrile to give the corresponding paramagnetic mononitrosyl compounds **48**, Scheme 3-6. In the case of reaction with piperazine, a paramagnetic bridged dimer is obtained [110].

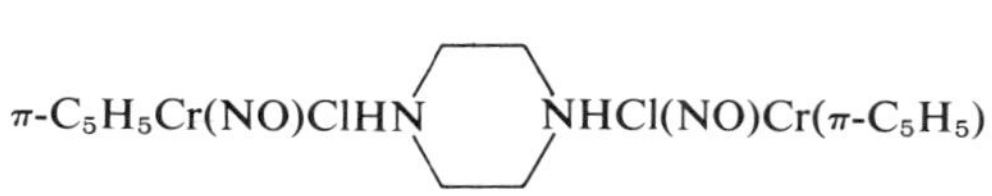

The chlorine of **15** may be replaced by CO, to give the cationic complex **49** [111].

c. Bridged Dimeric Chromium Compounds

The reduction of the title compound **15** with sodium borohydride leads to a mixture of dimeric products **50** and **51**, Scheme 3-6 [112–114]. The initial stage of this reduction may be the formation of the hydridochromium complex π-$(C_5H_5)Cr(NO)_2H$, which subsequently dimerizes to the nitroso-bridged compound **50** with concomitant extrusion of hydrogen [c.f., the thermal dimerization of π-$(C_5H_5)Cr(CO)_3H$, Scheme 3-4]. Reduction of one of the bridging nitroso groups of **50** gives the bridged amide **51** [112,113].

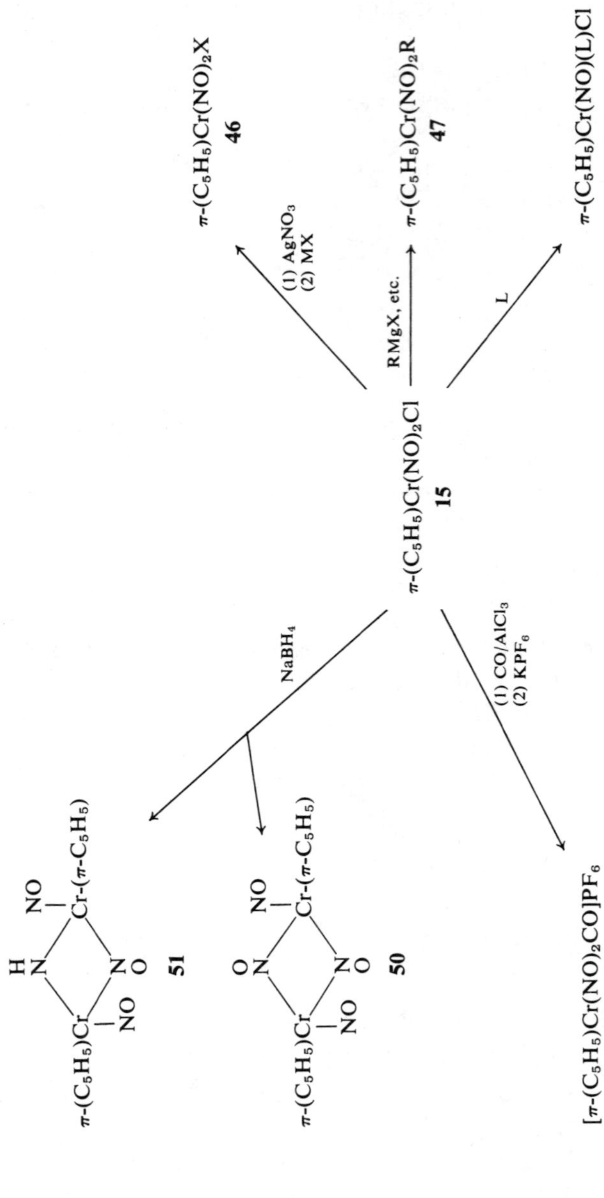

Scheme 3-6. Reactions of π-$(C_5H_5)Cr(NO)_2Cl$.

Complete x-ray structure analyses of **50** and **51** confirm that they are dimers with the two nitrosyl groups trans to each other (see Chapter 2, Table 2.3). Other bridged dimeric chromium compounds, analogous to **50** and **51**, are known (see Chapter 2, Table 2.3) and the sulfur-bridged complex **52** is prepared by heating the monomer **53**, Eq. (3-43) [106,115].

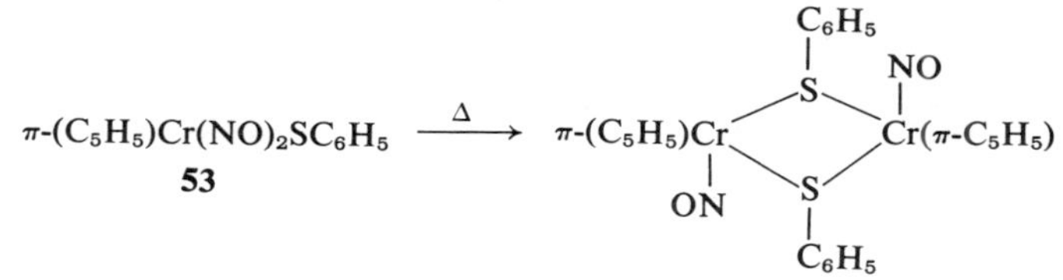

(3-43)

52

5. OTHER π-(CYCLOPENTADIENYL)CHROMIUM COMPOUNDS

This group includes compounds consisting of one cyclopentadienyl unit and one other acyclic unit with delocalized electrons complexed to chromium e.g., π-(C_5H_5)-π-(C_6H_6)Cr, π-(C_5H_5)-π-(C_7H_7)Cr, etc. The information about the chemistry of these compounds is very sparse, and this may be due to the difficulty of their preparation and the modest yields obtained (see Chapter 1, Sections B,3 and C,3).

Broadly speaking, this class of compound is quite reactive and the ligands may be exchanged; they may be metalated and substituted by nucleophiles and as a rule the complexes undergo redox processes at chromium (see Scheme 3-7).

a. Ligand Replacement Reactions

The interaction of π-(C_6H_6)-π-(C_5H_5)Cr(I) (**54**) and CO under pressure leads to the replacement of the arene ring to give π-$(C_5H_5)Cr(CO)_3$ dimer [73]. The π-bonded arene unit of **54** may also be displaced by alicyclic trienes (e.g., cycloheptatriene, 1-phenyl- and 1-methylcycloheptatriene, azulene, 4,6,8-trimethyl azulene) under the action of Lewis acids (Scheme 3-7) [116,117]. The initial products (already discussed in Chapter 1) are the chromium(I) cations **55** and **56**, which may be isolated as their crystalline salts **57** (X = PF_6, I, etc.,). The π-(phenylcycloheptatrienyl)-π-(cyclopentadienyl)chromium(I) cation may also be prepared by the rather unusual ring expansion reaction that results from the interaction of acid halides and $AlCl_3$ and **54** (see Chapter 1, Section C,3,b) [118].

b. Oxidation-Reduction Reactions

The cationic chromium(I) complexes **55–57** are readily reduced to the corresponding Cr(0) compounds **58**, which in turn may be reoxidized to

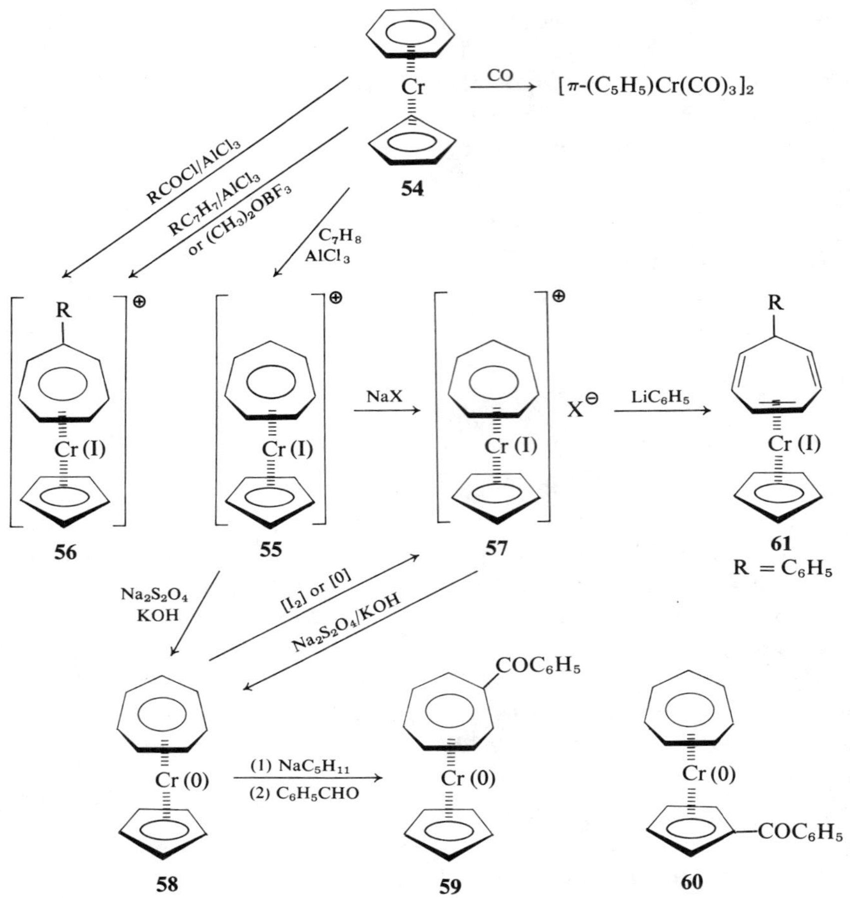

Scheme 3-7. Some reactions of π-$(C_5H_5)Cr(\pi$-$C_6H_6)$ and π-(C_5H_5)-π-$(C_7H_7)Cr$ [71,116,117].

the Cr(I) compounds [116,119,120], Scheme 3-7. The redox potential for the system

$$[\pi\text{-}(C_5H_5)\text{-}\pi\text{-}(C_7H_7)Cr(I)]^{\oplus} \underset{-e}{\overset{+e}{\rightleftharpoons}} \pi\text{-}(C_5H_5)\text{-}\pi\text{-}(C_7H_7)Cr(0)]$$

is $-E_{1/2} = 0.7$ V [116,121] and this value is similar to that for the π-bis(arene)-Cr(I)/Cr(0) system. There are no reports, however, of the oxidation of the π-(C_5H_5)-π-$(C_7H_7)Cr(0)$ compounds with organic halides.

c. Substitution of the π-Complexed Rings

π-(Cyclopentadienyl)-π-(cycloheptatrienyl)chromium(0) (**58**) is unaffected by phenyllithium but may be metalated by reaction with sodium amyl.

Reaction of the resulting sodium derivative with benzaldehyde gives poor yields of two ketonic products, **59** and **60**, Scheme 3-7. The formation of ketones rather than the anticipated alcohols has also been observed in the reaction of metalated π-bis(benzene)chromium(0) with benzaldehyde and reflects the ease of oxidation of such carbinols when complexed to chromium. The formation of products with substituents in the C_7 and the C_5 rings indicates that apparently there is no selectivity in the metalation of either of these rings [116]. There are no reports of attempts at the metalation of **58** with the more powerful butyllithium/TMEDA complex or of attempts to effect the isotopic exchange of the hydrogen atoms in the C_7 and C_5 rings.

The cationic chromium(I) complex **57** ($X = I^{\ominus}$), in contrast to the chromium(0) complex, undergoes nucleophilic substitution when treated with phenyllithium to give the π-(cycloheptatriene)chromium(I) complex, **61**, Scheme 3-7. The stereochemistry of this compound has not been defined, though the fact that the compound cannot be dehydrogenated to the chromium(0) complex (by Raney nickel [116]) might suggest that the hydrogen is on the same side as the chromium center, i.e., that it is π-(*exo*-7-phenylcycloheptatriene)-π-(cyclopentadienyl)chromium(I).

D. Carbonylchromium π-Complexes

There are two main classes of carbonylchromium π-complexes, the tricarbonylchromium π-complexes [e.g., π-(arene)$Cr(CO)_3$ and π-(triene)$Cr(CO)_3$] and the tetracarbonylchromium π-complexes [e.g., π-(norbornadiene)-$Cr(CO)_4$]. The information concerning the chemistry of the latter group is scant and the ensuing discussion will be restricted, therefore, to the reactions of tricarbonylchromium π-complexes. π-(Arene)- and π-(cycloheptatriene)-chromium tricarbonyls are relatively stable to air and protic solvents; they are readily prepared, purified (chromatography) and characterized (NMR spectroscopy), and the combination of all these factors makes the study of their chemistry a very fruitful field.

1. π-(Arene)chromium Tricarbonyls

π-(Arene)chromium tricarbonyl compounds undergo a variety of reactions in which the ligands around the chromium center are either exchanged or transformed chemically by reaction with electrophiles and nucleophiles; substituents in the arene rings complexed to chromium may also be transformed by various chemical reactions and these transformations may proceed, at times, with a high degree of stereospecificity. The π-(arene) chromium carbonyl compounds have also been used as catalysts in the hydrogenation of olefins and, in a modified form, in the polymerization of olefins.

The electron distribution within the π-(arene)chromium tricarbonyls will play a significant role in determining their chemical reactions. These compounds have an inert gas configuration and have formally three unshared pairs of electrons associated with the chromium center. Such an arrangement does not favor direct nucleophilic attack at chromium but it should favor the attack at chromium by electrophiles (e.g., strong Lewis acids like $AlCl_3$ and CF_3COOH). The electron distribution and the bonding within π-(arene)-chromium tricarbonyls has been the subject of much interest [1,122–139]. The bonding of arenes (or trienes) to a metal center involves, in principle, a forward donation of electrons to the metal center and a back-donation from the metal (*d* orbitals) to the vacant orbitals (antibonding) of the arene (or triene), features that have been likened to the $-R$ and $+R$ effects of a substituent in an arene ring [124]. An analogous situation obtains in the $Cr(CO)_3$ unit, namely, donation from the carbonyl unit to chromium and back-donation from chromium to the CO unit. The accumulated evidence indicates that the $Cr(CO)_3$ unit not only exerts an overall electron-withdrawing inductive effect on the arene (or triene) ligand but may also supply electrons to an adjacent electron-deficient center (e.g., in stabilizing benzylic carbonium ions) [124,126–129]. Furthermore, work by Gubin and his colleagues suggests that the $Cr(CO)_3$ unit may also increase the positive charge in the carbon σ orbitals of the arene unit [134,136]. The combination of these factors account satisfactorily for several of the observed reactions of π-(arene)chromium carbonyls, e.g., the facts that: π-(aniline)$Cr(CO)_3$ is a weaker base than aniline [140]; that the arene units of π-(arene)chromium tricarbonyls are more susceptible to nucleophilic attack than the free arenes and are apparently slightly less susceptible to electrophilic attack than are the free arenes [123,130,131].

a. Ligand Replacement Reactions

Both the arene and the carbonyl ligands of π-(arene)chromium tricarbonyl compounds may be exchanged for other π- or σ-donor ligands by either thermal or photochemical reactions; some individual aspects of these ligand exchange reactions have been reviewed [e.g., 1,122,141,142].

The arene ring in π-(arene)$Cr(CO)_3$ may be completely displaced from the chromium by a thermal reaction with strong σ-donor ligands, e.g., NH_3 [143], amines [123], phosphines, or phosphites [123,144], to give the corresponding *cis*-tris(ligand)chromium tricarbonyls, **62**, Scheme 3-8. The analogous reaction of π-$(C_6H_6)Cr(CO)_3$ with bidentate ligands LL (2,2′-bipyridyl, 2,2′,2″-tripyridyl and 1,10-phenanthroline) gives by displacement

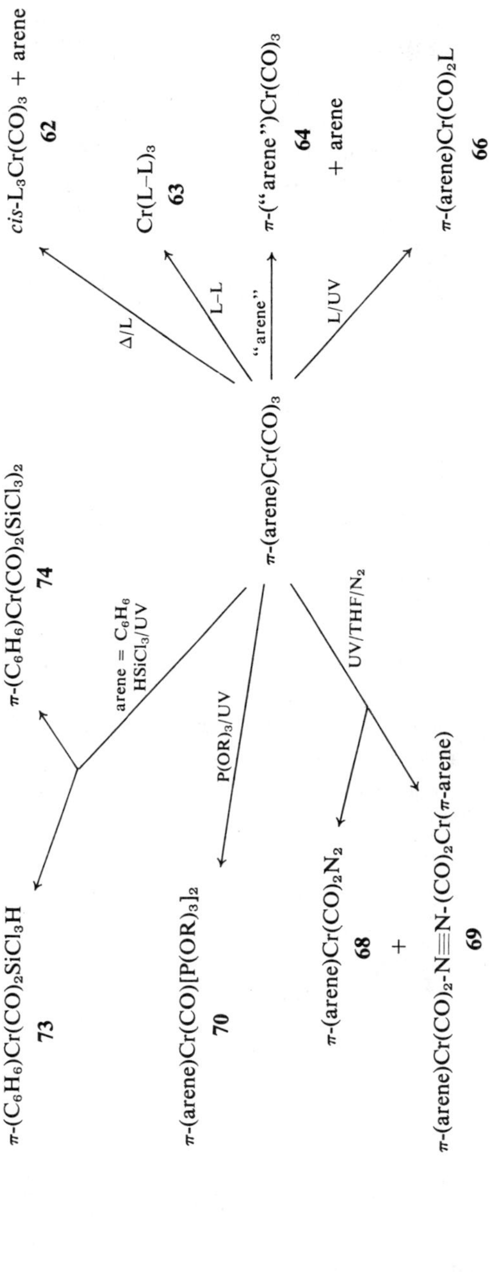

Scheme 3-8. Some ligand-exchange reactions of π-(arene)chromium tricarbonyls.

of the arene unit and the three CO units the tris(LL)Cr(0) complex **63**, Scheme 3-8 [143].

The arene units of π-(arene)$Cr(CO)_3$ compounds may be exchanged thermally for other arene units giving a new π-(arene)$Cr(CO)_3$ complex **64**, Scheme 3-8; the exchange process is enhanced by ultraviolet irradiation. Strohmeier and his colleagues have studied in detail the mechanism of this reaction (e.g., [145,146], and for reviews, [122,141]) and found that the photocatalyzed exchange of arene ligands in the system "arene"/π-(arene)Cr-$(CO)_3$ involved two fundamental processes, Eqs. (3-44) and (3-45). The first process involves the formation of a charge-transfer type complex **65**, which rapidly fragments to benzene, and an active $Cr(CO)_3$ fragment; the latter reacts with the second "arene" to give the observed products, Eq. (3-44).

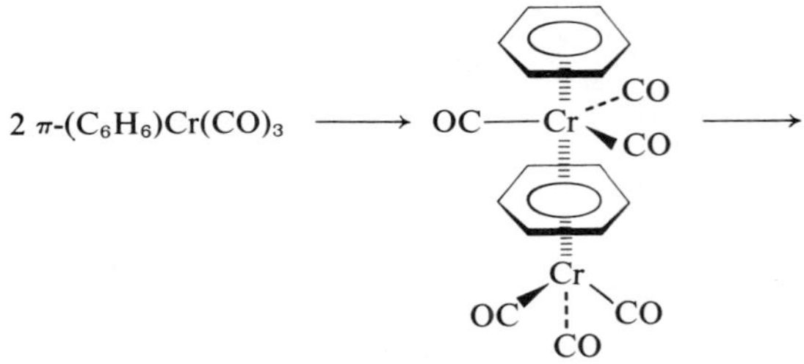

65

(3-44)

This type of mechanism is supported by the variation in reaction rates observed with variation of the substituents in the arene unit, electron-releasing substituents increase the rate. The second, slower process involves the direct displacement of the π-C_6H_6 unit by an arene unit 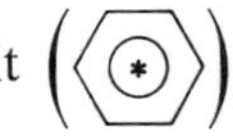 in an S_N2 type reaction, i.e., an association process, Eq. (3-45).

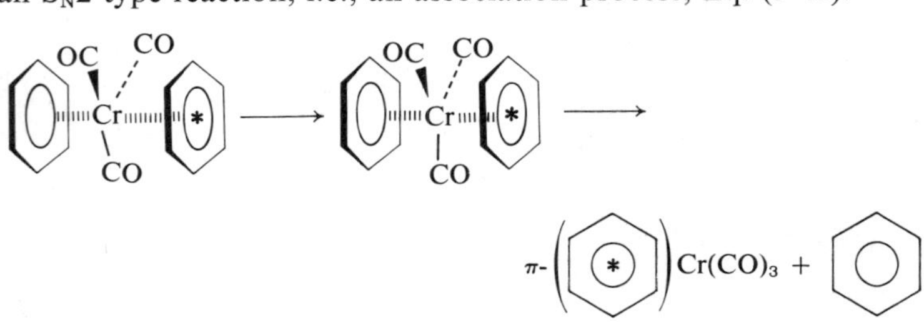
(3-45)

One of the carbonyl groups of π-(arene)chromium tricarbonyls may be exchanged for a variety of σ- and π-donor ligands to give compounds of the type π-(arene)$Cr(CO)_2L$ (**66**, Scheme 3-8, L = phosphines) [e.g., 137], olefins, or acetylenes [e.g., 125,147,148], pyridine or piperidine [e.g., 135], and nitriles [e.g., 138]. The precise mechanisms for these exchange processes are not known and may involve an associative mechanism analogous to that formulated by Basolo and his colleagues for ligand-exchange reactions in π-(arene)molybdenum tricarbonyls [149], or it may involve a dissociative mechanism as outlined in Eq. (3-46). An analogous dissociative mechanism

$$\pi\text{-(arene)Cr(CO)}_3 \xrightarrow{\text{UV}} [\pi\text{-(arene)Cr(CO)}_2] \xrightarrow{\text{L}} \pi\text{-(arene)Cr(CO)}_2\text{L} \qquad (3\text{-}46)$$

has been put forward in order to account for the kinetic data obtained for the ligand-exchange reaction outlined in Eq. (3-47) [147]. A third mechanism

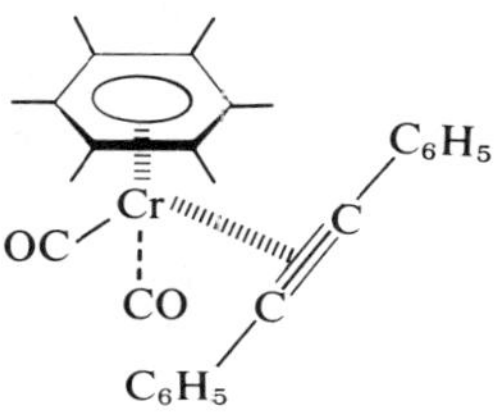

$$\xrightarrow{\text{PR}_3} \pi\text{-[(CH}_3)_6\text{C}_6\text{]Cr(CO)}_2\text{PR}_3 + \text{C}_6\text{H}_5\text{C}{\equiv}\text{CC}_6\text{H}_5 \qquad (3\text{-}47)$$

could involve what is formally a nucleophilic attack at the ligands bonded to chromium (the arene part or more likely the CO). Thus, the CO group when bonded to transition metal centers is susceptible to nucleophilic attack (e.g., the formation of carbenoid-chromium compounds). Certain strong donor ligands might, therefore, attack the CO unit in π-(arene)$Cr(CO)_3$ to give the intermediate **67**, Eq. (3-48); migration of the ligand from carbon to chromium with concomitant extrusion of a CO would then terminate the ligand-exchange process, Eq. (3-48). The precise mechanism of these exchange processes will

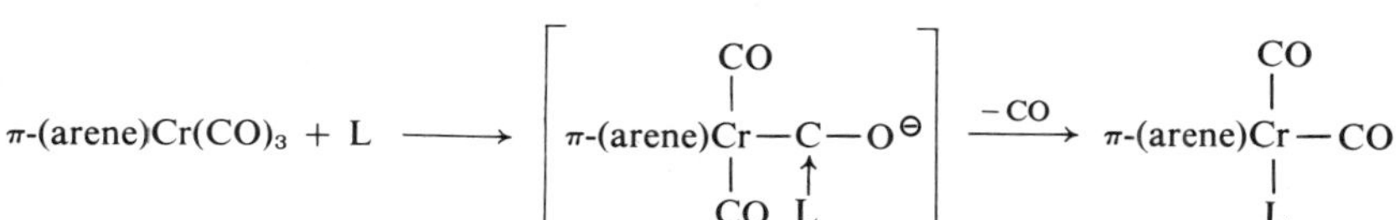

67 (3-48)

clearly depend upon the nature of the entering ligand and upon the solvent used in the reaction, since it is known that UV irradiation of π-(arene)$Cr(CO)_3$ compounds in tetrahydrofuran (THF) solution results in the formation of π-(arene)$Cr(CO)_2$THF, and the tetrahydrofuran in this complex may now be readily exchanged for another ligand (e.g., $CH_2{=}CHCN$) in a thermal reaction [e.g., 138].

The CO group of some π-(arene)$Cr(CO)_3$ compounds may be exchanged in a photochemical reaction for dinitrogen to give the monomeric complexes **68** and the dinitrogen bridged dimer **69**, Scheme 3-8 (arene = mesitylene and hexamethylbenzene). This ligand exchange is presumed to occur via the tetrahydrofuran complex: π-(arene)$Cr(CO)_2$(THF) [150]. π-(Mesitylene)Cr-$(CO)_2N_2$ is rather unstable and undergoes disproportionation to give the dinitrogen bridged dimer **69** and π-(mesitylene)$Cr(CO)_3$. The monomer **68** (arene = mesitylene or benzene) can, however, be prepared by the controlled oxidation of the hydrazine complexes as outlined in Eq. (3-49) [150,151].

$$\pi\text{-(arene)}Cr(CO)_2N_2H_4 + 2\,H_2O_2 \xrightarrow{Cu^{2\oplus}/THF,\,-70^\circ C} \underset{\mathbf{68}}{\pi\text{-(arene)}Cr(CO)_2N_2} + H_2O \qquad (3\text{-}49)$$

Whereas the photocatalyzed interaction of π-(arene)chromium tricarbonyls and *phosphines* leads to the replacement of one CO group (to give **66**, Scheme 3-8), the same reaction with *phosphites* leads to the replacement of not only one CO group [to give **66**, Scheme 3-8, arene = C_6H_6 and $CH_3OOCC_6H_5$, L = $P(OCH_2CH_3)_3$ or $P(OCH_3)_3$] but also of a second CO group to give the bis(phosphite) complexes **70**, Scheme 3-8 [152]. Furthermore, one of the CO ligands of the monophosphite complex **66** may in turn be exchanged for a second phosphite to give, among other products, the mixed bis(phosphite) complex, e.g.,

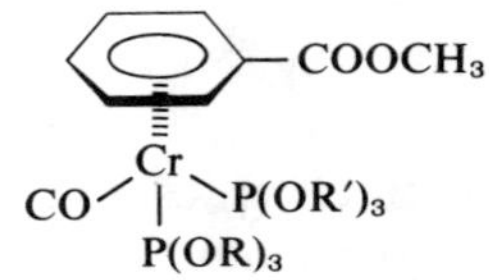

R = CH_3 and R′ = CH_2CH_3 [152]. Such complexes, like the cyclopentadienyl complexes π-(C_5H_5)Cr(CO)(NO)PR_3 prepared by Knox and Robertson

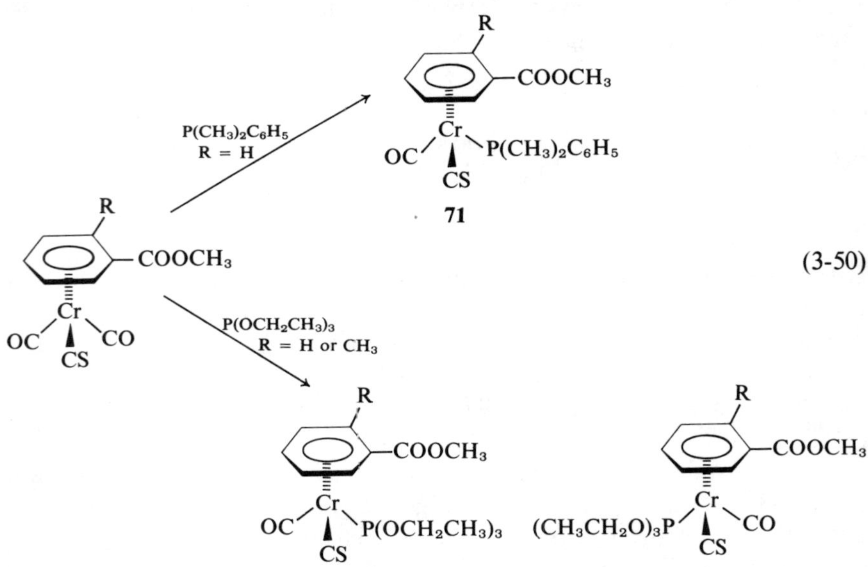

71 **72a** **72b** (3-50)

[99] (see Chapter 3, Section C,3,a [44]) possess an asymmetric center at chromium. Other π-(arene)chromium complexes, with an asymmetric center at chromium, have been prepared from π-(arene)Cr$(CO)_2$CS, e.g., **71** and **72**, Eq. (3-50) [153,154]. The asymmetry of the chromium center in the compounds **71** and **72** a and b (R = CH_3) has been established by NMR spectroscopic methods. Thus, in **71**, the methyl groups attached to phosphorus are magnetically nonequivalent (δ 1.85 and 1.97, respectively), and for the diastereoisomeric pair **72** a and b (R = CH_3), two pairs of resonance signals were observed for the aromatic- and ester-methyl groups $\delta_{aryl-CH_3}$ 2.47 and 2.50, $\delta_{ester-CH_3}$ 3.87 and 3.86 [154].

All the ligand-exchange reactions discussed up to this point have not involved an overall change in the coordination number at chromium. Jetz and Graham have reported, however, that ultraviolet irradiation of π-$(C_6H_6)Cr(CO)_3$ and trichlorsilane in hexane results in a mixture of two hepta-coordinate compounds **73** and **74**, Scheme 3-8 [155].

b. Chemical Transformation of the Ligands on Chromium

i. Isotopic H/D *Exchange.* The protons of the arene ring in π-(arene)-chromium tricarbonyls and the protons on the carbon atom α to the arene ring in π-(alkyl-substituted arene)chromium tricarbonyls are readily exchanged for deuterium when the appropriate complexes are treated with $NaOCH_2CH_3/DOCH_2CH_3$ or K(*t*-BuO) and perdeuteriodimethyl sulfoxide (DMSO-d_6) [156–163]. Setkina and her colleagues found that whereas benzene, anisole, toluene, ethyl benzoate, and *N,N*-dimethylaniline were not affected by $NaOCH_2CH_3/DOCH_2CH_3$, the corresponding π-(arene)$Cr(CO)_3$ complexes underwent rapid H/D exchange when treated with these reagents. A comparison of the relative reaction rate constants [π-$(CH_3OC_6H_5)Cr(CO)_3$, 2.8×10^{-5} sec^{-1}; π-$(C_6H_6)Cr(CO)_3$, 1.1×10^{-5} sec^{-1}; π-$(CH_3C_6H_5)Cr(CO)_3$, 0.8×10^{-5} sec^{-1}; π-$(NaOOCC_6H_5)Cr(CO)_3$, 0.7×10^{-5} sec^{-1}; and π-$[(CH_3)_2NC_6H_5]Cr(CO)_3$, 0.3×10^{-5} sec^{-1}] [159,161] indicated that the presence of electron-donating or electron-withdrawing substituents in the arene ring had little effect on the overall rate constant of the isotopic exchange [156–159,161]. These isotopic-exchange reactions are, however, accompanied by extensive decomposition of the π-(arene)chromium complex; thus with π-(anisole)$Cr(CO)_3$, π-(ethylbenzoate)$Cr(CO)_3$, and π-(dimethylaniline)$Cr(CO)_3$, 14%, 7%, and 18% of the deuterated complexes were recovered. Furthermore, mass spectrometric analysis of these deuterated complexes indicated that they were rather complex mixtures (e.g., for anisole D_0, 25%, D_1, 38%; D_2, 24%; D_3, 8%; and for ethylbenzoate D_0, 6.8%; D_1, 15.0%; D_2, 27.0%; D_3, 29%; D_4, 17.0%; D_5, 6.0% [160]). These isotopic compositions show that though an electron-withdrawing substituent does not influence apparently the rate constant of isotopic exchange, it does influence the extent of isotopic exchange, *viz.* the presence of D_4 and D_5 compounds in π-(deuterioethyl benzoate)$Cr(CO)_3$.

The protons of the CH_3 group in π-(acetophenone)$Cr(CO)_3$ may be exchanged for deuterium, by reaction with $KOH/DOCH_2CH_3$; under these reaction conditions the arene protons are evidently not exchanged [162].

Trahanovsky and Card found that the benzylic protons of π-(alkylbenzene)-chromium tricarbonyls could be exchanged stereoselectively and in high yield by the reaction of the complex with K(t-BuO)/DMSO-d_6 [163]. Both the benzylic protons of π-(1,4-diphenylbutane)$Cr(CO)_3$ are selectively replaced by deuterium, in the reaction outlined in Eq. (3-51) [R = $(CH_2)_3$-C_6H_5], to give 1,1-dideuterio-1,4-diphenylbutane in 70% yield and 96%

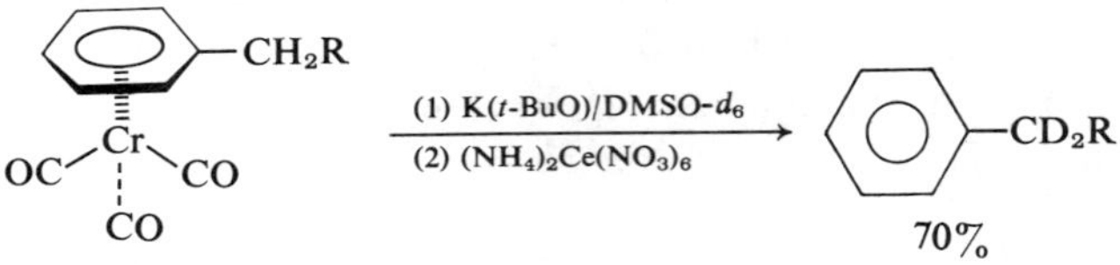

(3-51)

isotopic purity. Similarly, the benzylic protons anti (i.e., on the face opposite) to the $Cr(CO)_3$ unit in the compounds **75** to **77** are selectively replaced by deuterium when the complexes are treated with K(t-BuO)/DMSO-d_6 [163].

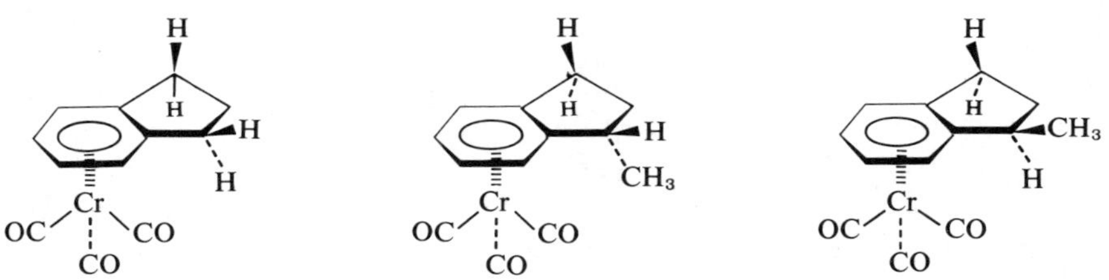

75 **76** **77**

The fact that the benzylic protons and not the arene protons are replaced in the above reactions is perhaps surprising in view of the result of the H/D exchange reactions discussed earlier. One possible explanation is that it is a kinetic phenomenon, associated with the relative acidities of the benzylic and the arene protons. However, the high stereospecificity of the exchange processes (observed with **75**–**77**) suggests that the chromium center exerts a direct influence in stabilizing the negative charge on the benzylic carbon atom.

The protons of π-(arene)chromium carbonyl complexes may also be exchanged for deuterium by the reaction of the complexes either with $BF_3/D_2O/CF_3COOD/D_2SO_4$ or with CF_3COOD [164–167]. Setkina and her colleagues found that isotopic exchange of the protons of the arene units of π-(arene)$Cr(CO)_3$ compounds occurs in highly acidic solutions. The exchange reaction for the complexed arene proceeds, however, much more slowly (by a factor of 7 in the order of the rate constant) than for free benzene [i.e., relative rate constants are benzene, 10 sec^{-1}; π-$(C_6H_6)Cr(CO)_3$, $1.2 \times 10^{-7} sec^{-1}$; π-$(CH_3C_6H_5)Cr(CO)_3$, $0.9 \times 10^{-7} sec^{-1}$; π-(1,3,5-$(CH_3)_3$-$C_6H_3)Cr(CO)_3$, $1.2 \times 10^{-7} sec^{-1}$; π-$(CH_3OC_6H_5)Cr(CO)_3$, $3.4 \times 10^{-7} sec^{-1}$] [164–166]. The replacement of one CO ligand by PPh_3, to give complexes of type π-(arene)$Cr(CO)_2PPh_3$, greatly facilitates the isotopic exchange process that now proceeds in the less acid system, *viz.* CF_3COOD; the rate constants of isotopic exchange for various complexes are: π-(C_6H_6), $5.0 \times 10^{-7} sec^{-1}$; π-[1,3,5-$(CH_3)_3C_6H_3$], $1.9 \times 10^{-7} sec^{-1}$; π-$(CH_3OC_6H_5)$, $1.9 \times 10^{-7} sec^{-1}$ [165], π-$(CH_3OOCC_6H_5)$, $4.7 \times 10^{-6} sec^{-1}$ [166]. It is of interest that in these isotopic exchange processes the protons in the phosphine ligands and in the substituents in the arene ligands [e.g., the CH_3 of π-$(CH_3OOCC_6H_5)$] are not exchanged under the conditions of the reaction [165,166]. The exact mechanism of this isotopic exchange process is not known, and it is unlikely that it is a simple electrophilic substitution at the arene carbon atoms, since it is known from spectroscopic evidence (IR and NMR) that several π-(arene)chromium tricarbonyls *protonate at chromium* in $BF_3/H_2O/CF_3COOH$ [167–170] {δ for Cr–H between 4.0 and 5.0 ppm upfield of [$(CH_3)_4Si$]} and that several complexes of the type π-(arene)$Cr(CO)_2PPh_3$ *protonate at chromium* in CF_3COOH [δ for Cr–H again between 4.00 and 5.0 ppm upfield from $(CH_3)_4Si$] [168,169]. The initial step in the H/D exchange is, therefore, probably the formation of a chromium deuteride of the type

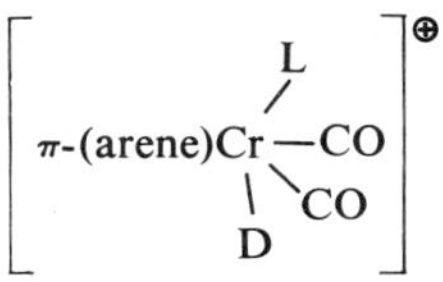

The actual introduction of the deuterium into the arene can proceed either by an intramolecular process (the proton being removed by a base [171]) or by an intermolecular process with a second molecule of deuterio acid.

ii. Metalation of π-(Arene)$Cr(CO)_3$. The direct metalation of π-(arene)-chromium tricarbonyls would appear to require more drastic conditions than those required for the simple isotopic exchange reactions discussed under Section D,1,b, above. Furthermore, from the scant information available, it would appear that the precise products formed in the reaction of the title compound with organolithium compounds depend critically upon several factors (e.g., the solvent, the nature of the organolithium compound, and the nature of the arene unit in the π-complex). Thus π-(benzene)$Cr(CO)_3$ is metalated in the arene ring by reaction with butyllithium in tetrahydrofuran, since subsequent carbonation gives the π-benzoic acid complex **78**, Scheme 3-9 [172]. In contrast, various π-(arene)chromium tricarbonyls react with phenyllithium in diethyl ether to give, by nucleophilic attack on one CO ligand, the lithium acylmetalates, which in turn react with trimethyloxonium fluoroborate to give the π-(arene)carbenoid-chromium dicarbonyls (**79**). Arene = C_6H_6, $CH_3C_6H_5$, 1,4-$(CH_3)_2C_6H_4$ and 1,3,5-$(CH_3)_3C_6H_3$ (Scheme 3-9) [173,174].

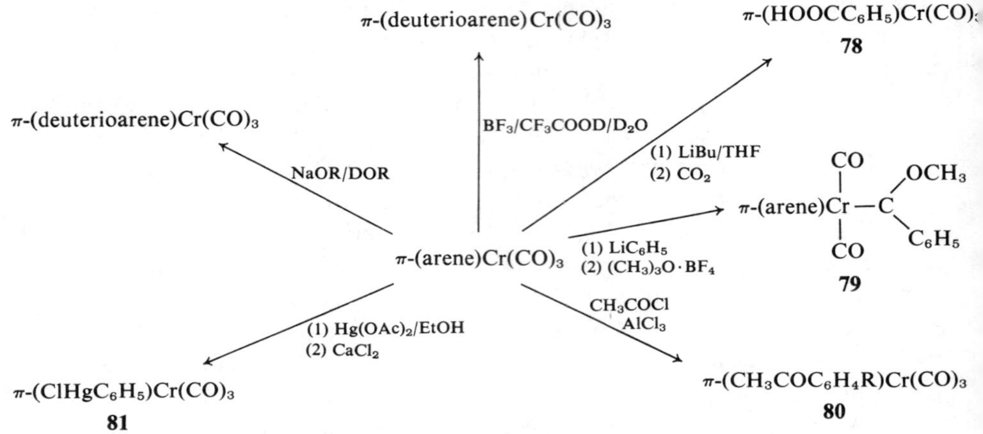

Scheme 3-9. Some reactions of π-(arene)$Cr(CO)_3$ compounds.

iii. Electrophilic Substitution at Carbon. Simple theoretical considerations suggest that the presence of an electron-withdrawing unit [i.e., $Cr(CO)_3$] complexed to an arene ring should reduce the reactivity of the carbon centers in this ring toward electrophiles. However, the instability of π-(arene)chromium tricarbonyls toward strong Lewis acids is such that these compounds are destroyed by many of the classical electrophilic reagents and therefore make it rather difficult to determine whether an arene ring in π-(arene)$Cr(CO)_3$ is indeed less reactive toward electrophiles than is the free arene. The Friedel–Crafts acylation of several π-(arene)$Cr(CO)_3$ compounds has been effected, and the products, which are formed in relatively low yields, are the corresponding π-monoacylarene complexes **80**, Scheme 3-9 [R = CH_3, H, CH_3CH_2, CH_3O, $(CH_3)_2CH$, $(CH_3)_3C$] [131,132,175–177]. However, it does not follow that these reactions involve simple electrophilic substitution processes at the carbon atom of the arene ring in the π-(arene)$Cr(CO)_3$ compounds; indeed, Brown and his associates have found that the π-(arene)-$Cr(CO)_3$ compounds react with $AlCl_3$ in CS_2 to give a chromium(II) complex Cr(II)$(AlCl_3)_2(CS_2)_2$ [132]. Therefore, it is possible that the electrophilic substitution actually takes place in a complex [of the type π-$(C_6H_6)Cr(CO)_3$-$(AlCl_3)_n$] in which the electron-withdrawing effect of the $Cr(CO)_3$ group (on the arene unit) is modified by coordination to $AlCl_3$.

The mercuration of π-$(C_6H_6)Cr(CO)_3$ with $Hg(OAc)_2$ probably involves the direct electrophilic substitution of an arene ring complexed to chromium. This reaction, which requires more severe conditions than the mercuration of benzene gives initially the acetoxymercuriarene complex, which reacts with $CaCl_2$ to give finally the chloromercuribenzene complex **81**, Scheme 3-9, in 43% yield [178]. The same chloromercuribenzene complex may be prepared in 80% yield from bis(phenyltricarbonylchromium)mercury and $HgCl_2$ as outlined in Eq. (3-52) [179]. However, it cannot be prepared by the reaction

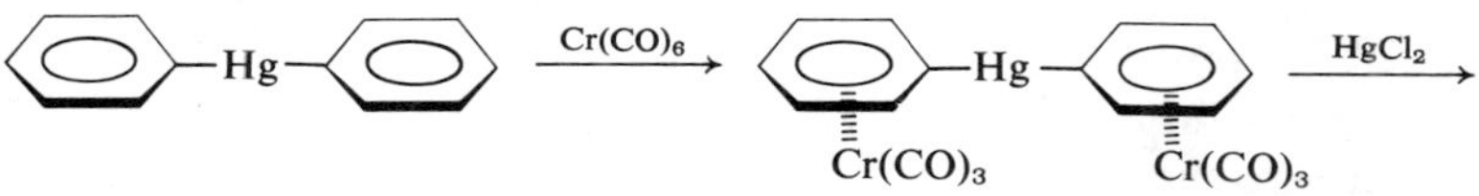

π-$(ClHgC_6H_5)Cr(CO)_3$ (3-52)

of the corresponding π-(triphenylstannylbenzene)chromium tricarbonyl and $HgCl_2$ [180].

iv. Substitution at Carbon in or Adjacent to the π-Arene Ring. Nicholls and Whiting found that whereas chlorobenzene is relatively inert toward sodium methoxide, the corresponding π-complex, i.e., π-$(ClC_6H_5)Cr(CO)_3$ reacts readily with sodium methoxide to give π-$(CH_3OC_6H_5)Cr(CO)_3$ in high yield. The authors ascribed this reactivity of π-bonded arene rings toward nucleophiles to the capacity of the $Cr(CO)_3$ unit to withdraw electrons from the arene ring [123]. The kinetics of the reaction of $NaOCH_3$ and $NaNH_2$ with various π-(fluoroarene)chromium tricarbonyls (including optically active complexes) not only confirmed the reactivity of π-bonded arene rings toward nucleophiles, but also showed that the reaction is a bimolecular nucleophilic process (S_N2) and does not involve a benzyne- (aryne-) type intermediate [131,181,182]. The experiments with the optically active compounds also showed that whereas the meta compounds π-$[3\text{-}FC_6H_4COOH(CH_3)]Cr(CO)_3$ underwent nucleophilic substitution without racemization, the ortho compounds π-$[2\text{-}FC_6H_4COOH(CH_3)]Cr(CO)_3$ underwent partial racemization in the process. This racemization is associated in some undefined way with the nucleophilic process, since the products $\{\pi$-$[2\text{-}CH_3OC_6H_4COOH(CH_3)]Cr(CO)_3\}$ did not racemize under the conditions of the reaction [181,182].

Pettit and his collaborators have found that the rate of solvolysis of π-$(C_6H_5CH_2Cl)Cr(CO)_3$ is greater than that of noncomplexed benzyl chloride and ascribed this enhancement in rate to the capacity of the chromium of the $Cr(CO)_3$ unit to *donate* electrons to the arene ligand (i.e., +R effect) thereby stabilizing the benzylic cation. This effect can proceed either from chromium via the arene ring to the benzylic cation, or directly from chromium to the benzylic carbocation. The authors tried, without success, to prepare stable π-benzylchromium tricarbonyl cations by the interaction of π-$(C_6H_5CH_2OH)$-$Cr(CO)_3$ and acids [124]. Wells and Trahanovsky obtained spectroscopic evidence for the existence of π-($C_6H_5CH_2^{\oplus}$)$Cr(CO)_3$ and related cations derived from other π-(arene)$Cr(CO)_3$ complexes [arene = C_6H_5CHO, $C_6H_5COCH_3$, $C_6H_5COC_6H_5$, C_6H_5COOH, $C_6H_5CON(CH_3)_2$] in solutions of the π-(arene)chromium carbonyls in 82.6% H_2SO_4 [126,127]. Further evidence of the capacity of the $Cr(CO)_3$ unit to stabilize carbonium ions adjacent to or in proximity of a π-complexed arene ring has been obtained by comparing the rates of solvolysis π-(cumyl chloride)$Cr(CO)_3$ [183] and of acetolysis of several substituted aryl sulfonates both free and complexed to $Cr(CO)_3$ [128,129,184,185]. This is illustrated by the acetolysis of π-($C_6H_5C(CH_3)_2CH_2OMes$)$Cr(CO)_3$ (Mes = methane sulfonate) in which the major products are the acetate and hydrocarbons formed by the migration of the aryl group (see Scheme 3-10). The mechanism proposed for the overall reaction involves the stabilization of the initially formed carbonium ion by direct interaction with the surplus *d*-electrons on the chromium atom (see Scheme 3-10) [184]. Two other examples of the capacity of $Cr(CO)_3$ and $Cr(CO)_2PPh_3$ ligands to stabilize carbonium ions adjacent to the arene ring in π-(arene)chromium compounds are (i) the rapid isomerization of π-(benzylthiocyanate)$Cr(CO)_3$ (**82**) to the isothiocyanate complex **83**, Eq. (3-53) [186], and (ii) the rapid and quantitative reduction of the ester group

$$\underset{\mathbf{82}}{\pi\text{-}(C_6H_5CH_2SCN)Cr(CO)_3} \xrightarrow{\Delta/\text{acetone}} \underset{\mathbf{83}}{\pi\text{-}(C_6H_5CH_2NCS)Cr(CO)_3} \qquad (3\text{-}53)$$

$$\underset{\mathbf{84}}{\pi\text{-}[4\text{-}(CH_3)_3CC_6H_4COOCH_3]Cr(CO)_2PPh_3} \xrightarrow{LiAlH_4/AlCl_3} \pi\text{-}[4\text{-}(CH_3)_3CC_6H_4CH_3]Cr(CO)_2PPh_3 \qquad (3\text{-}54)$$

in **84** (to CH_3) by $LiAlH_4/AlCl_3$, Eq. (3-54) [187]. In the reduction, Eq. (3-54), the corresponding chromium tricarbonyl is reduced much more slowly and incompletely; this difference in the ease of reduction of the two compounds has been interpreted as an indication that the $Cr(CO)_2PPh_3$ unit has an even greater capacity than the $Cr(CO)_3$ unit for the back-donation of electrons to an adjacent carbonium ion [187].

There are many other examples of the chemical transformation of arene, and substituted arene ligands, in π-(arene)$Cr(CO)_3$ complexes, and the major limiting feature in the chemistry of these complexes (as compared with that of the free arene) is their instability toward certain reagents (e.g., strong acids, oxidizing agents, UV light, and in some instances oxygen). Two highly favorable features of the chemistry of these complexes are (i) the electron-donating capacity of the $Cr(CO)_3$ unit, which (as is evident from the foregoing discussion) has the effect of promoting and/or facilitating reactions involving the generation of carbonium ions adjacent either to the arene ring or the chromium center; (ii) the steric bulk of the $Cr(CO)_3$ unit, which (as will be

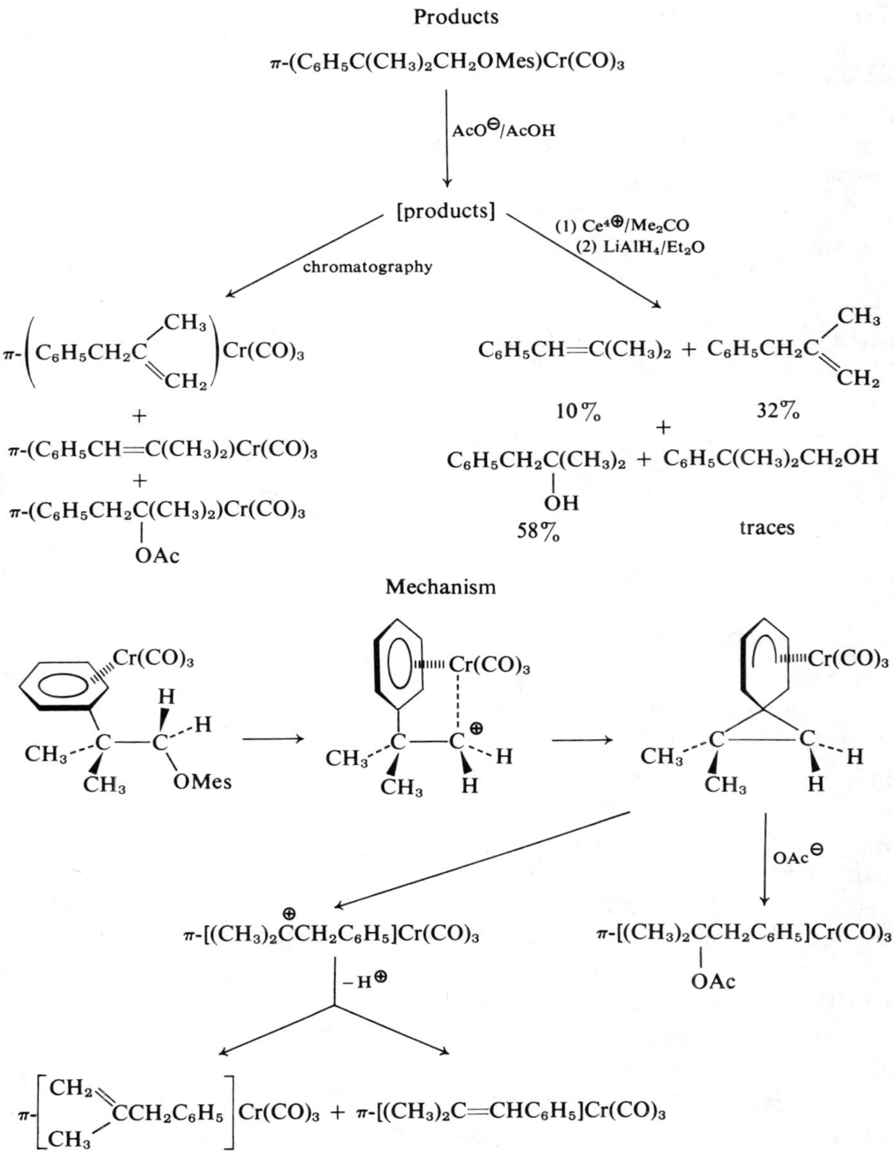

Scheme 3-10. Products and mechanism of solvolysis of π-($C_6H_5C(CH_3)_2CH_2OMes$)-$Cr(CO)_3$.

explained directly) has the effect of promoting stereoselective reactions at carbon atoms adjacent to the chromium center.

The π-(arene)chromium tricarbonyls have a "piano-stool" configuration depicted in **85** for π-(1-indanone)$Cr(CO)_3$ and the bulk of the $Cr(CO)_3$ unit is such that it impedes the approach of a reagent to the underside of the arene unit (provided of course that the reagent does not initially form a bond with the chromium center). The effect, therefore, of the presence of the $Cr(CO)_3$ unit will be to favor the approach of the reagent, in particular nucleophiles, to the upper face of the arene unit [i.e., the side opposite to $Cr(CO)_3$] and therefore to impose a stereospecificity on the overall chemical reaction. This is illustrated by the stereospecific reductions of π-(indanone)-$Cr(CO)_3$ and π-(tetralone)$Cr(CO)_3$ with $LiAlH_4$ or $NaBH_4$ to the endo-(cis) alcohols [188–191], e.g., **85–86**, Eq. (3-55) [189] (n = 1 or 2). Such a

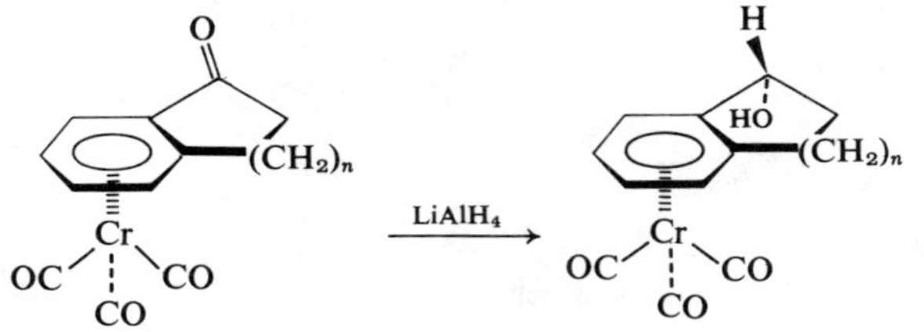

85 **86**

(3-55)

stereospecific reduction of an optically active ketone (complexed to chromium) will lead eventually in an overall process of asymmetric induction to diastereoisomeric alcohols, as is illustrated in Scheme 3-11 [190]. The initial interaction of the optically active arene with $Cr(CO)_3$ can possibly give two isomers **87** and **88**, Scheme 3-11. Reduction of both ketones **87** and **88** gives stereoselectively the complexes with the OH group on the side of the $Cr(CO)_3$ unit; these compounds may be decomposed, by UV irradiation, to the diastereoisomeric alcohols, Scheme 3-11 [190]. Other asymmetric inductions, occasioned by the bulk of the $Cr(CO)_3$ unit, in the reactions (e.g., reductions and reactions with alkyl Grignards) of a variety of optically active π-(arene)-$Cr(CO)_3$ compounds has been extensively investigated [e.g., 192–194].

The foregoing description of the electronic and steric influences occasioned by the presence of the $Cr(CO)_3$ unit in π-(arene)$Cr(CO)_3$ compounds illustrates the importance of such organometallic compounds in classical organic chemistry. Thus, by the simple process of forming a π-(arene)chromium tricarbonyl complex, it becomes possible to effect chemical transformations with a facility and a stereoselectivity that is not possible in the free arene molecule.

One example of the unexpected influence that a $Cr(CO)_3$ unit can have on a complexed arene, is found in the observation that π-(styrene)$Cr(CO)_3$,

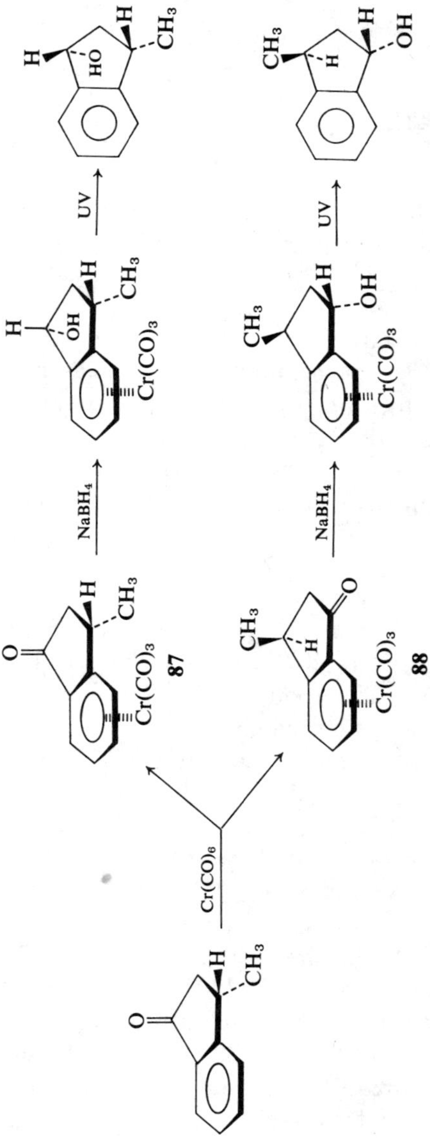

Scheme 3-11. Stereoselective reduction of isomeric π-(optically-active 3-methylindan-1-one)Cr(CO)$_3$.

in contrast to free styrene, is not homopolymerized by various radical and anionic initiators [139,195–197]. The reason for the resistance of π-(styrene)-$Cr(CO)_3$ to homopolymerization remains obscure, since it will undergo the Diels–Alder reaction with cyclopentadiene as outlined in Eq. (3-56) [195]

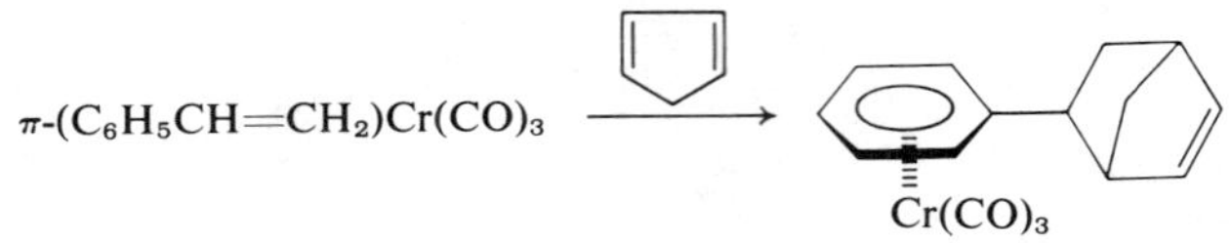

(3-56)

and may be copolymerized with a variety of monomers [e.g., styrene, methyl acrylate, and π-(vinylcyclopentadienyl)$Mn(CO)_3$] [139].

c. Catalytic Reactions

π-(Arene)chromium tricarbonyl compounds have been found to be efficient catalysts in a variety of chemical reactions. Thus, it has been reported that π-(arene)chromium tricarbonyls are catalysts in the alkylation, acylation, and sulfonylation of aromatic compounds [198], and, when they are deposited on aluminosilicates, they are effective catalysts in the polymerization of ethylene [199]. Furthermore, diphenylacetylene, which reacts with $(CH_3CN)_3$-$Cr(CO)_3$ to give tetraphenylcyclopentadienone [200], is trimerized (to hexaphenylbenzene) when heated (270°C) in the presence of a catalytic quantity of a π-(arene)chromium tricarbonyl; the activity of various π-(arene)-chromium tricarbonyls falls in the sequence cycloheptatriene > C_6H_6 > $(CH_3)_6C_6$ > 1,4-$(CH_3O)_2C_6H_4$ [201]. However, the most interesting reaction that is catalyzed by π-(arene)chromium tricarbonyls is possibly the stereo-specific hydrogenation of conjugated dienes [202–214].

Brown and his associates observed that moist chromium hexacarbonyl and acid solutions of π-(arene)chromium tricarbonyls will reduce a variety of unsaturated substrates (e.g., *p*-benzoquinone, benzil, and azobenzene) [215]. Cais, Frankel, and Rejoan found, however, that π-(arene)chromium tri-carbonyl compounds are very efficient catalysts in the stereoselective hydro-genation of conjugated and certain nonconjugated dienes, dienoic and trienoic esters, e.g., Eqs. (3-57)–(3-59) [arene = $(CH_3OOCC_6H_5)$; R = $(CH_2)_7COOCH_3$ and R′ = $CH_3(CH_2)_4$] [202–207]. These authors also found that the course of these hydrogenations is profoundly influenced by the nature of the π-(arene)chromium tricarbonyl compound, the temperature of the reaction, and the nature of the unsaturated substrate.

Thus the rate, but not the stereoselectivity, of these hydrogenations depends upon the nature of the substituents in the arene unit of the π-(arene)$Cr(CO)_3$ complex; and the rate decreases for the series phenanthrene, cycloheptatriene, ClC_6H_5, 3-$HOCC_6H_4COOCH_3$, $(C_6H_5CH{=}CH)_2$, $CH_3OOCC_6H_5$, C_6H_6, $CH_3C_6H_5$, $(CH_3)_3C_6H_3$, $(CH_3)_6C_6$ [202,205,206]. These differences in catalytic activities may be associated with changes in the bonding of the various arene units to the chromium center. Electron-withdrawing substituents should facilitate the displacement of the arene unit from the chromium center; also in certain π-(polycyclic arene)chromium tricarbonyls and π-(cycloheptatriene) $Cr(CO)_3$ the chromium is not equidistant from the six carbon atoms of the arene (triene) unit, which is complexed to chromium (see Chapter 2, Section D,5) and this may also favor the displacement of the arene (or triene) unit from the carbonyl complex.

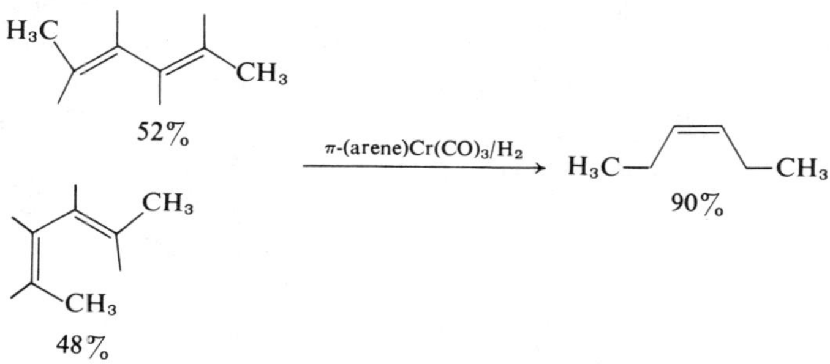

(3-57)

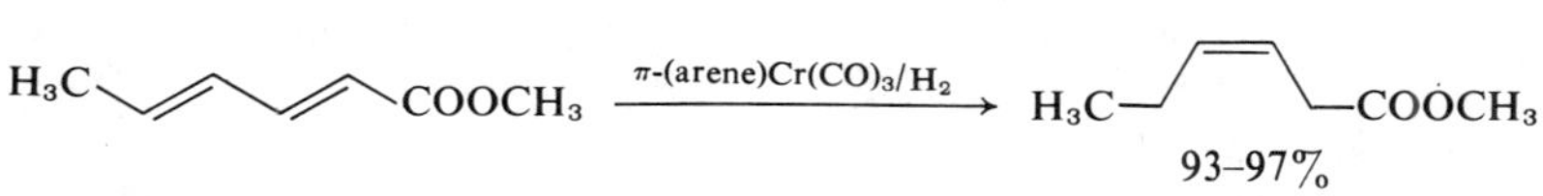

(3-58)

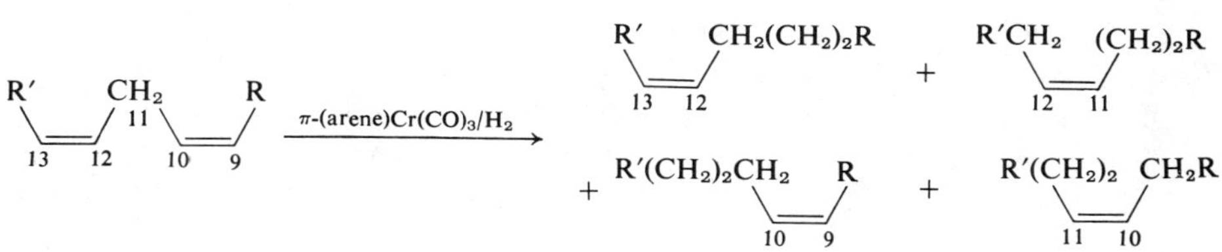

(3-59)

The thermal stability of the π-(arene)$Cr(CO)_3$ complexes is also of importance in these hydrogenations since they will determine the maximum operating temperature for the reaction [205]. Thus, π-(cycloheptatriene)$Cr(CO)_3$ is a more active hydrogenation catalyst than is π-$(CH_3OOCC_6H_5)Cr(CO)_3$; however, because of its relative thermal instability, it can only be used in the temperature range of 100°–125°C (i.e., in the hydrogenation of conjugated 1,3-dienes), whereas π-$(CH_3OOCC_6H_5)Cr(CO)_3$, which is more stable thermally, can be used both at 120°–160°C for the reduction of 1,3-dienes and at 160°–175°C for the reduction of nonconjugated 1,4-dienes [205, 207,209].

The nature of the unsaturated substrate, in particular the location and steric environment of the double bonds, is also of particular importance in these hydrogenations. It was found originally that only conjugated dienes underwent rapid and stereoselective hydrogenation at 120°–160°C to the *cis*-monoolefin [e.g., Eq. (3-57)], but later it was observed that the more thermally stable π-(arene)chromium tricarbonyl compounds were capable of isomerizing nonconjugated 1,4-dienes (but not 1,5- and 1,6-dienes) to the conjugated 1,3-dienes at temperatures of 160°—175°C, and of reducing the latter stereoselectively to the *cis*-monoolefins, e.g., Eq. (3-59) [204,205,207, 209]. The influence that the steric environment of the double bonds has on these hydrogenations is seen in the "reductions" of the alkyl-substituted butadienes, of the isomeric cyclohexadienes, and of norbornadiene. Substituents on C_2 and C_3 of the butadiene system do not greatly influence either the rate or the course of the reduction; thus 2,3-dimethylbutadiene is reduced (**89**, R = H, $R_1 = CH_3$) to the *sym*-tetramethylethylene (**90**, R = H, $R_1 = CH_3$), Eq. (3-60). Substituents on the terminal carbon atoms C_1 and C_4, however, not only slow down the reduction but may also change its

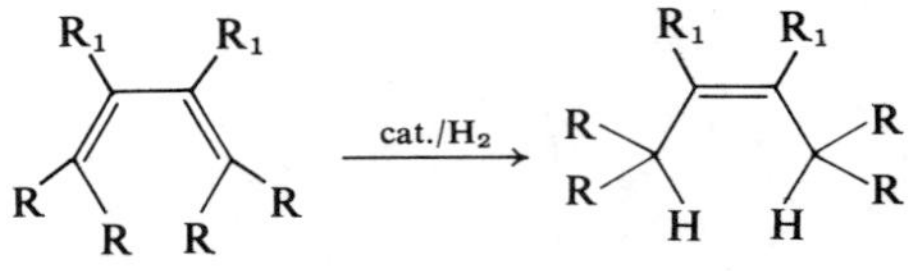

89 **90** (3-60)

course, e.g., 1,1,4,4-tetramethylbutadiene **89** (R = CH_3, R_1 = H) is only slowly and incompletely reduced to *sym*-tetramethylbutene-2 (**90**, R = CH_3, R_1 = H), Eq. (3-60); and 1,1-dimethylbutadiene (**91**) is reduced to a mixture of olefins containing, as main product, the 1,1-dimethylbutene-1 [**92**, Eq. (3-61)]. The formation of **92** is assumed to proceed via a thermal isomerization of **91** to the less sterically encumbered diene **93** and subsequent 1,4 addition

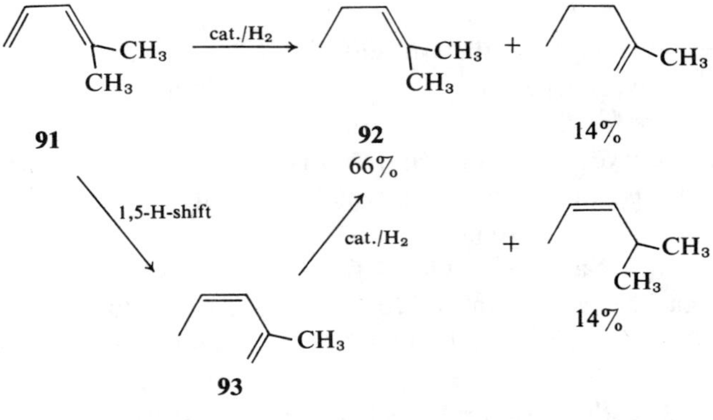

(3-61)

91 **92** **93**

of hydrogen to the latter, Eq. (3-61) [205]. The steric effect of substituents in these dienes probably interferes with the ability of the dienes to complex with the chromium center, in particular with the ability of the dienes to adopt the cisoid configuration:

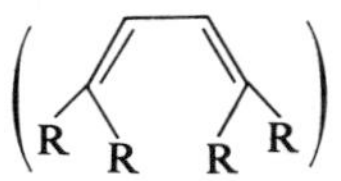

Both 1,3- and 1,4-cyclohexadiene are hydrogenated to cyclohexene [205], again the reduction of the 1,4-isomer is assumed to proceed via an initial isomerization to the 1,3-isomer [205,212] followed by 1,4 addition of hydrogen to the latter, rather than by the direct 1,2 addition of hydrogen to an isolated double bond in 1,4-cyclohexadiene. Norbornadiene is reduced by π-(arene)chromium tricarbonyl and hydrogen to a mixture of nortricyclene and norbornene, Eq. (3-62) [206,211].

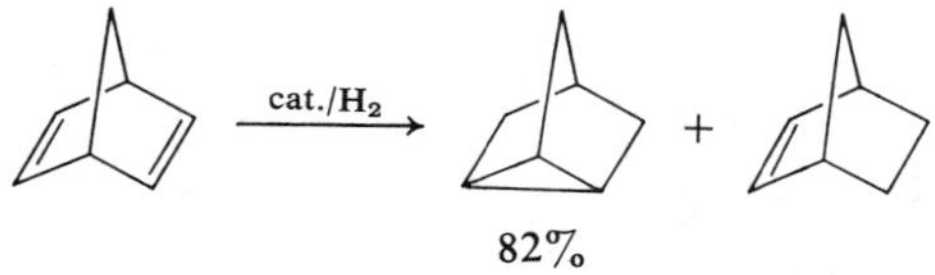

(3-62)

It is therefore evident that in these hydrogenation reactions the diene substrate must be a conjugated diene or a diene that may be isomerized to a conjugated *cisoid*-diene, or a diene in which the double bonds are so disposed sterically as to permit ready interaction with the chromium center (or centers), and that the overall reduction process may involve a combination of isomerization and hydrogenation reactions.

The actual mechanism of the hydrogenation of dienes by π-(arene)$Cr(CO)_3$ compounds is still not very clear. However, from the extensive studies of Cais, Frankel, and their colleagues and particularly from studies with D_2 gas, it is evident that the hydrogenation is a highly stereospecific process and that the hydrogen (deuterium) is transferred to the diene substrate by a 1,4- or 1,2-addition from the chromium side of a *cisoid*-(diene)chromium complex [203–205,210,211]. Thus the reduction of methyl sorbate, with π-$(CH_3OOCC_6H_5)Cr(CO)_3/D_2$, gives a high yield of methyl 1,4-dideuterio-hex-3-enoate (Eq. 3-63) along with traces of tri- and tetradeuterio compounds [203] and presumably deuterated methyl hex-2-enoate [202]. Norbornadiene

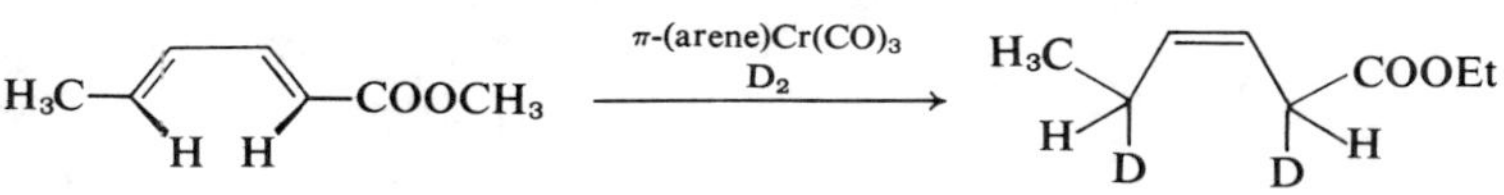

(3-63)

is reduced by π-(arene)$Cr(CO)_3$ and D_2 to gve *endo-endo*-dideuterio-nortricyclene and norbornene, Eq. (3-64) [211]. The reduction of nonconjugated dienoic esters (e.g., methyl linoleate) with π-(arene)$Cr(CO)_3$ and D_2 leads to a very complex mixture of deuterated products thereby indicating that deuterium is incorporated into these molecules not only in the hydrogenation step but also in the isomerization step [204].

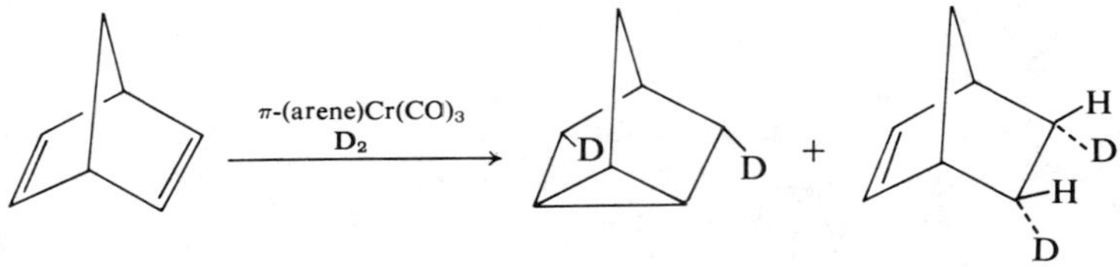

(3-64)

The precise nature of the actual active hydrogenation catalyst is not known. There is an induction period at the start of most of the hydrogenations with π-(arene)Cr(CO)$_3$ compounds and the nature of the arene unit bonded to chromium influences not only the length of this induction period but also, as was discussed above, the rate of the hydrogenation. A considerable proportion of the π-(arene)Cr(CO)$_3$ complex (up to 75%) may be recovered once the hydrogenation is terminated, but the reaction products also consistently contain the free arene, a green chromium compound and some Cr(CO)$_6$ [202,203,206].

Two types of mechanism have been advanced for the hydrogenation of dienes by π-(arene)chromium tricarbonyl.

The first is a dissociative mechanism, in which the π-(arene) complex undergoes dissociation, in a rate-determining step, to an active Cr(CO)$_3$ unit, this with $H_2(D_2)$ gives a dihydride that reacts with a *cisoid*-diene to give by the 1,4 addition of hydrogen, the monoene. This is represented schematically in Eq. (3-65) [203]. The second is an associative mechanism in

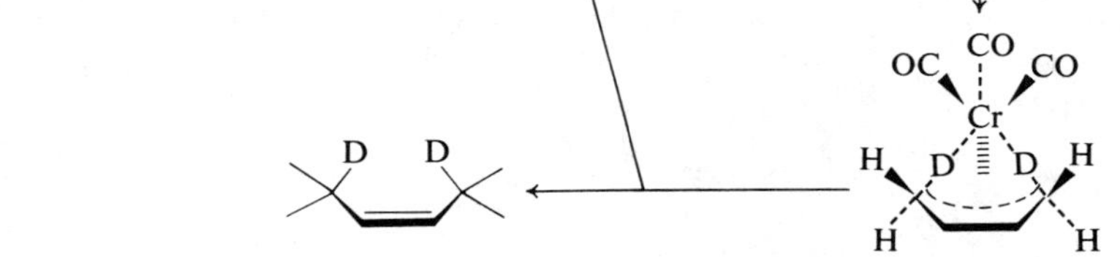

(3-65)

which the diene and the π-(arene) complex combine to give a *cisoid*-diene-(arene)chromium tricarbonyl (**94**) in which the diene and the arene are, respectively, bidentate and monodentate ligands, Eq. (3-66). It is proposed that **94** can now react with hydrogen to give either, with retention of the arene unit, the hydrogen species **95** or, by loss of the arene unit, the dihydride

96, Eq. (3-67), analogous to that postulated by Frankel [see Eq. (3-65)]. The subsequent transfer of hydrogen within these complexes is assumed to be stepwise giving first the π-(allyl)monohydrides and, subsequently, by a rapid transfer of hydrogen to either end of the π-(allyl) system, the monoene products, Eq. (3-67) [206].

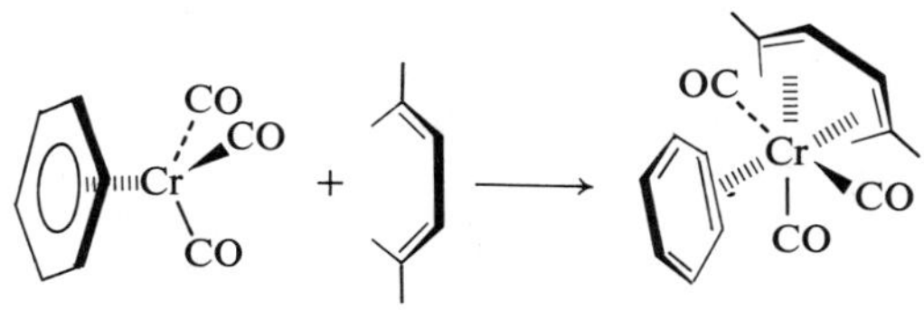

(3-66)

94

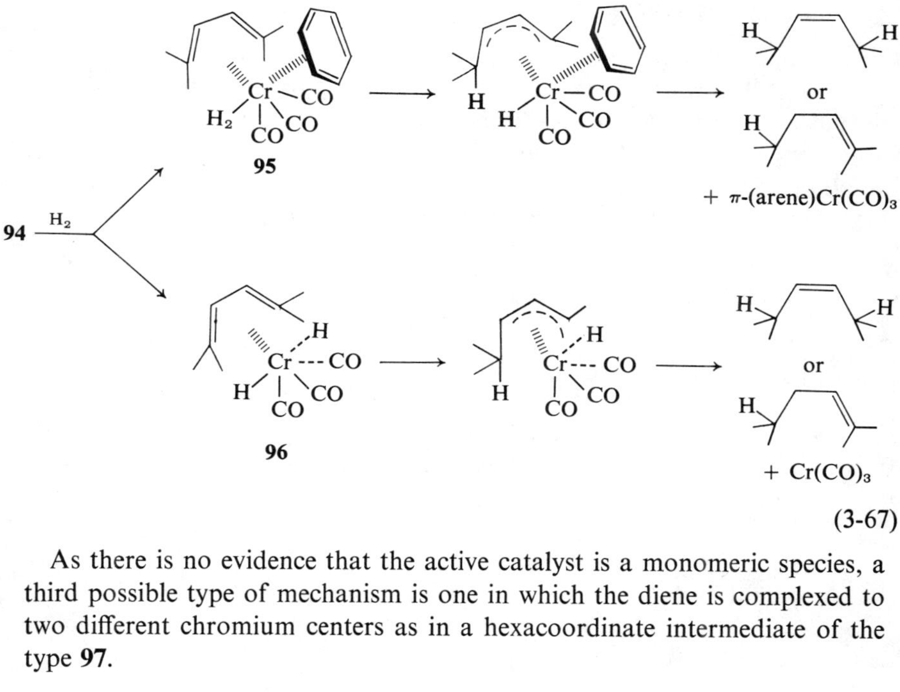

(3-67)

As there is no evidence that the active catalyst is a monomeric species, a third possible type of mechanism is one in which the diene is complexed to two different chromium centers as in a hexacoordinate intermediate of the type **97**.

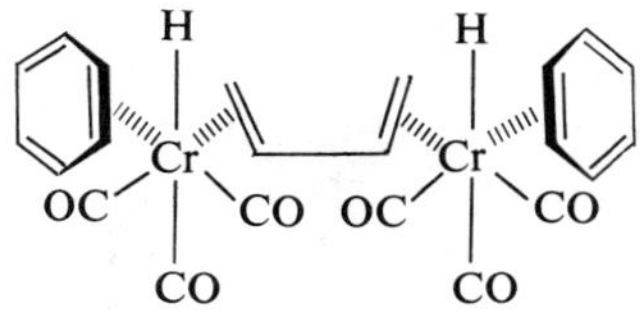

97

Conjugated dienes may also be reduced stereospecifically to the corresponding *cis*-monoenes by the photoassisted interaction of the diene,

$Cr(CO)_6$, and hydrogen [216,217]. Here again it is probable that the reaction proceeds via a *cisoid*-diene complex, since in competitive reductions with a mixture of *trans*-1,3-pentadiene and *cis,cis*-2,4-hexadiene, only the former is reduced, Eq. (3-68).

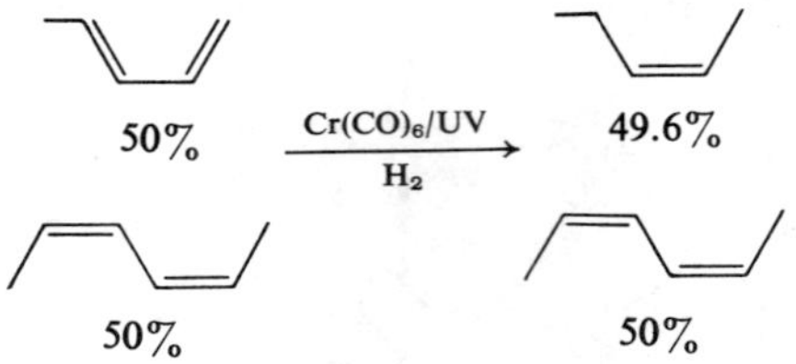

(3-68)

2. π-(Cycloheptatriene)chromium Tricarbonyls

There are two main types of cyclic C_7-triene chromium tricarbonyls, the π-(cycloheptatriene)chromium tricarbonyls **98** and the cationic π-(cycloheptatrienyl), or π-(tropylium)chromium tricarbonyls **99**. The substituted cycloheptatriene complexes may exist either as the exo or the endo isomers (**100** and **101**), whereas the substituted cycloheptatrienyl complex exists in only one form, since the C_7-ring in these cationic complexes is planar.

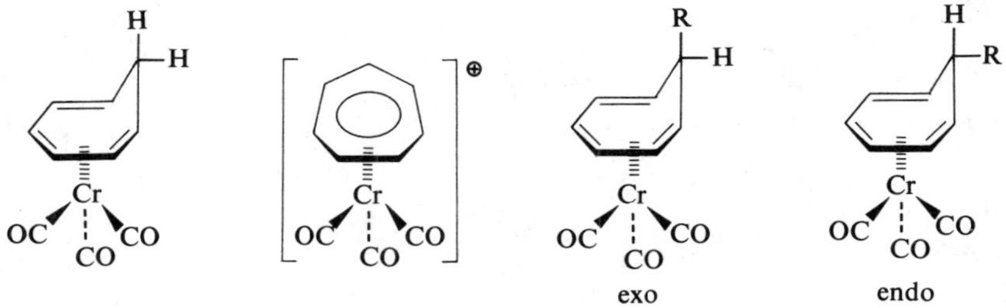

98 **99** **100** **101**

The parent cycloheptatriene and cycloheptatrienyl complexes **98** and **99**, are interconvertible by the abstraction or addition of a hydride. The hydride abstraction may be effected by the reaction of **98** with $(C_6H_5)_3CBF_4$ [e.g., 218] or with $(CH_3CH_2)_3OBF_4$ [219]; the hydride addition may be effected by the reaction of **99** with $NaBH_4$ [218,219].

The abstraction of the hydride is stereospecific, and it is the hydrogen that is farthest from the chromium center (i.e., the exo hydrogen) that is removed. This is illustrated by the fact that the 7-endo isomers (**101**, R = CH_3, OCH_3, OCH_2CH_3) react with $(C_6H_5)_3CBF_4$ to give the corresponding substituted cycloheptatrienyl chromium tricarbonyls [220,221], whereas the corresponding 7-exo isomers (**100**, R = CH_3, CH_3O) are either inert toward $(C_6H_5)_3$-CBF_4 (e.g., 7-*exo*-methylcycloheptatriene compound) or lose the alkoxy group when treated with $(C_6H_5)_3CBF_4$ to give [π-(cycloheptatrienyl)Cr-$(CO)_3]^{\oplus}BF_4^{\ominus}$ [220].

The stereochemistry of the hydride addition to cationic π-(cycloheptatrienyl)$Cr(CO)_3$ compounds is not so clear-cut since the reaction of substituted cycloheptatrienyl compounds with $NaBH_4$ gives 1- and 2-substituted cycloheptatriene compounds as the main products, e.g., Eqs. (3-69), (3-70) (R = CH_3 or CH_3CH_2) [220,222]. However, the fact that some π-(7-*endo*-

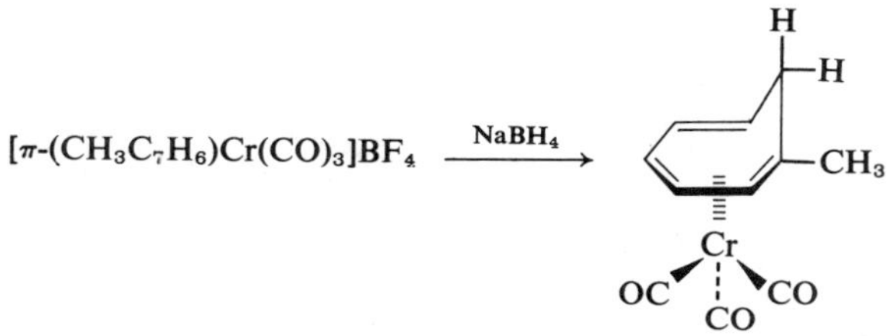

(3-69)

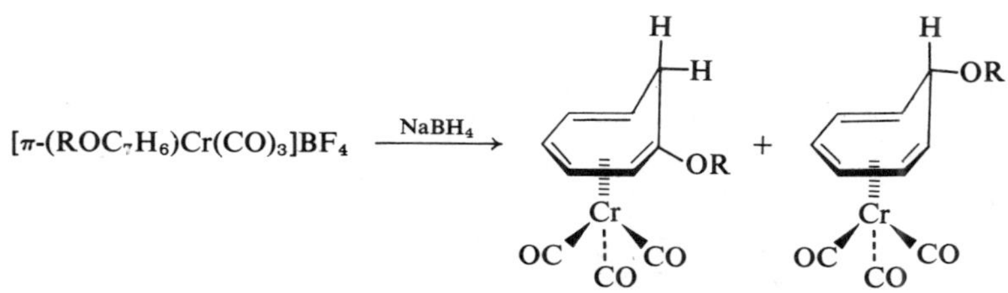

(3-70)

methoxycycloheptatriene)$Cr(CO)_3$ was obtained in the reaction of [π-methoxycycloheptatrienyl)$Cr(CO)_3]^{\oplus}BF_4^{\ominus}$ with $NaBH_4$ [Eq. (3-70), R = CH_3] [222] indicates that the hydride addition takes place on the face of the C_7-ring, which is opposite to the chromium center.

Finally, it may be mentioned that π-(cycloheptatriene)chromium tricarbonyl has been used as a catalyst in the thermal trimerization of diphenylacetylene [201] and in the stereospecific hydrogenation of conjugated dienes (see Section D,1,c of this chapter). The main chemistry of the present group of compounds, which will now be discussed, falls under the two headings of ligand-replacement reactions and chemical transformations of the C_7-ring (there are no reports of the chemical transformation of the CO groups bonded to chromium).

a. Ligand Replacement Reactions

The cycloheptatriene unit and/or one or more of the carbonyl groups of π-(cycloheptatriene)$Cr(CO)_3$ may be replaced by a variety of π- or σ-donor ligands (see Scheme 3-12) [218,223–233].

The C_7-ring may be replaced by an arene unit (e.g., mesitylene) to give π-(mesitylene)$Cr(CO)_3$ [218] or by various monodentate ligands (L) (e.g., phosphines, phosphites, arsines, nitriles, and isonitriles) in thermally assisted

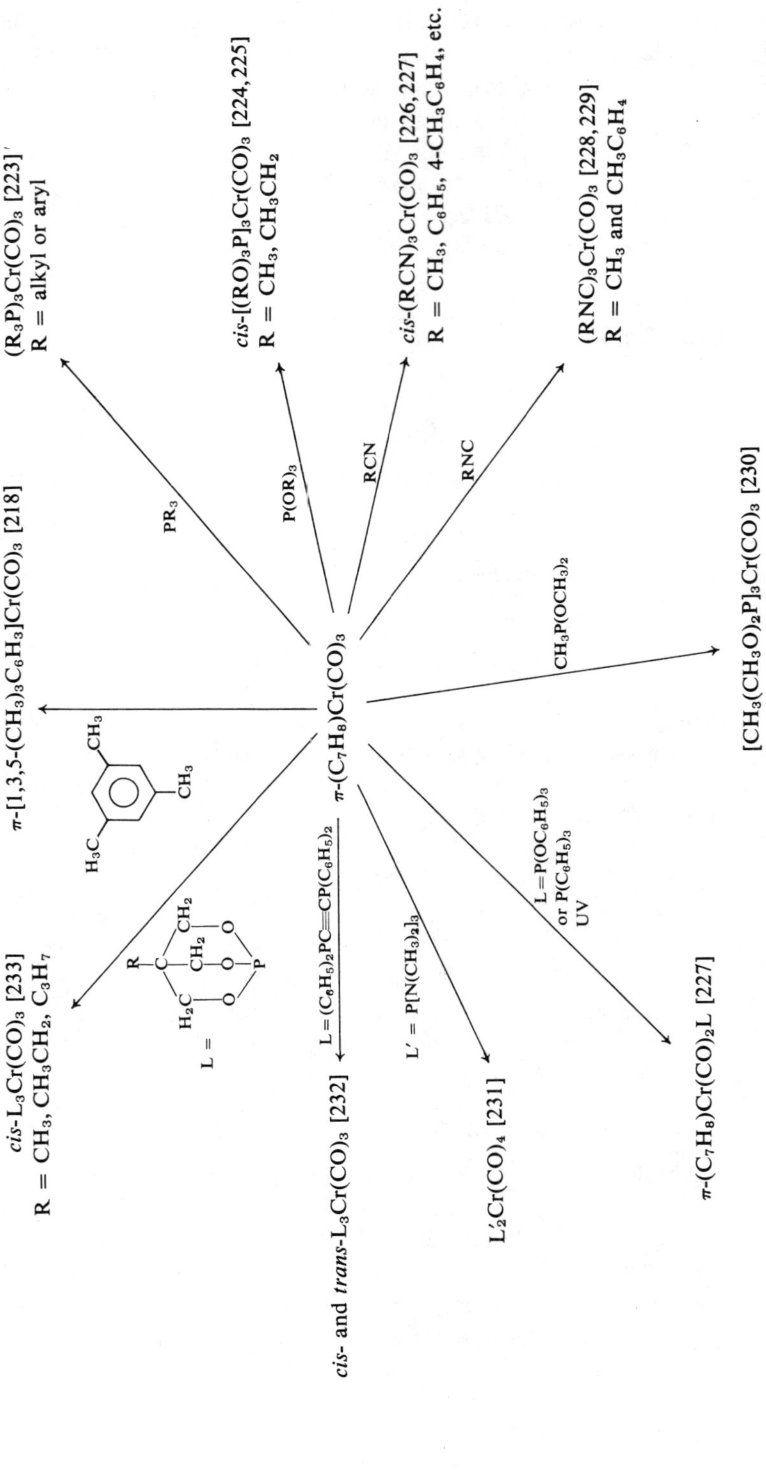

Scheme 3-12. Some ligand-replacement reactions of π-(C_7H_8)Cr(CO)$_3$.

reactions to give complexes of the type $L_3Cr(CO)_3$, e.g., Scheme 3-12 [223–232]. An exception would seem to be the reaction with tris(dimethylamino)-phosphine (L′), when the tetracarbonyl complex $L_2'Cr(CO)_4$ is formed [231]. However, $[(CH_3)_2N]_3P$ is an unusual phosphine because though it is a strong donor ligand; the presence of the nitrogen atoms adjacent to phosphorus makes it a poor acceptor ligand.

The photoassisted interaction of π-$(C_7H_8)Cr(CO)_3$ with some phosphines and phosphites, on the other hand, leads, by replacement of one CO ligand, to the dicarbonyl complexes π-$(C_7H_8)Cr(CO)_2L$ [$L = P(OC_6H_5)_3$ and $P(C_6H_5)_3$] [227]. Attempts to effect the analogous photoassisted replacement of one CO group by cyclooctene were, however, not successful [227].

The products formed by the reaction of π-$(C_7H_8)Cr(CO)_3$ and bidentate σ-donor ligands depend upon the ligand and, at times, upon the polarity of the solvent. Thus, with the bidentate phosphine ligand $[(C_6H_5)_2PCH_2]_2$[L–L] the bridged tricarbonyl complex **102** is obtained, Eq. (3-71) [234]. The reaction with the bidentate arsenic analog $[(C_6H_5)_2AsCH_2]_2$ [L′–L′] results in a mixture of the di- and tetracarbonyls, Eq. (3-72) [235]. The reaction of π-$(C_7H_8)Cr(CO)_3$ with 2,2-bipyrididyl (bipy) or 1,10-phenanthroline (phen) in the polar solvent, methanol, gives the tri- and dicarbonyl complex **103** and **104**, respectively, Eq. (3-73) [236]. In contrast, the same reaction in the less polar solvent, tetrahydrofuran, gives a mixture of the tetra- and dicarbonyl complexes, Eq. (3-74) [236]. Treatment of **103** with benzene leads

π-$(C_7H_8)Cr(CO)_3$ $\xrightarrow{(C_6H_5)_2PCH_2CH_2P(C_6H_5)_2}$ 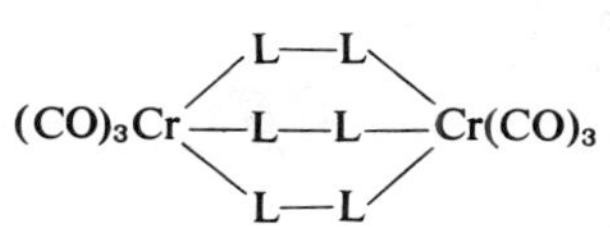 (3-71)

102

$$\pi\text{-}(C_7H_8)Cr(CO)_3 \xrightarrow{(C_6H_5)_2AsCH_2CH_2As(C_6H_5)_2} (CO)_2Cr(L'\text{–}L')_2 + (CO)_4Cr(L'L') \quad (3\text{-}72)$$

$$\pi\text{-}(C_7H_8)Cr(CO)_3 \xrightarrow[60^\circ\text{–}80^\circ C]{bipy/MeOH} (CO)_3Cr(bipy)(CH_3OH) \ \textbf{103}$$

$$\pi\text{-}(C_7H_8)Cr(CO)_3 \xrightarrow[60^\circ\text{–}80^\circ C]{phen/MeOH} (CO)_2Cr(phen)_2 \ \textbf{104} \quad (3\text{-}73)$$

$$2\text{-}\pi\text{-}(C_7H_8)Cr(CO)_3 \xrightarrow[60^\circ\text{–}120^\circ C]{bipy/THF,\ 60^\circ\text{–}120^\circ C,\ or\ phen/THF} Cr(CO)_4(bipy) + Cr(CO)_2(bipy)_2 + 2C_7H_8 \ \text{or} \ Cr(CO)_4(phen) + Cr(CO)_2(phen)_2 + 2C_7H_8 \quad (3\text{-}74)$$

to the formation of the carbonyl-bridged complex **105**, Eq. (3-75), admixed with $Cr(CO)_4(bipy)$ and $Cr(CO)_2(bipy)_2$ [236].

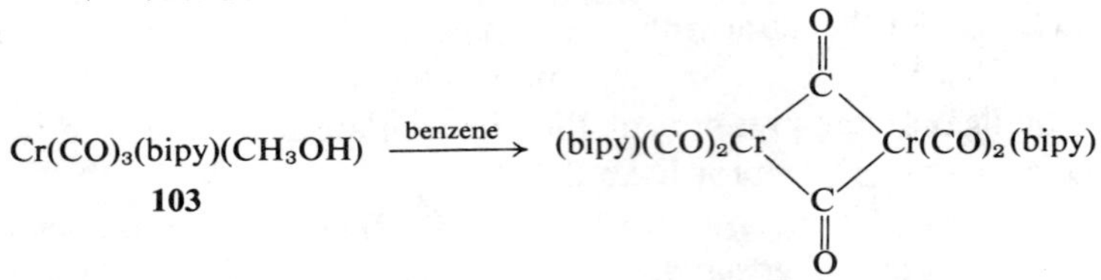

(3-75)

The tridentate ligand, pentamethyldiethylenetriamine $[(CH_3)_2NCH_2CH_2]_2$-NCH_3 [L–L–L] displaces the C_7-ring of π-$(C_7H_8)Cr(CO)_3$ to give the tricarbonyl complex [227].

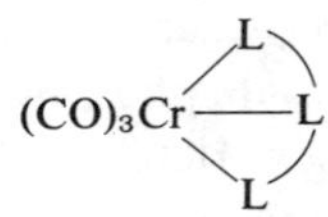

b. Chemical Transformations of the C_7*-Ligand*

The abstraction of a hydride from π-(cycloheptatriene)$Cr(CO)_3$ to give the corresponding π-(cycloheptatrienyl) complex has been discussed in the introductory section. An elegant utilization of this reaction in the preparation of highly reactive heptafulvene compounds is outlined in Scheme 3-13. The 1-substituted-cycloheptatriene compounds **106** are readily accessible from the free triene and $(CH_3CN)_3Cr(CO)_3$, and undergo hydride abstraction to give the π-(substituted-cycloheptatrienyl) complexes **107**. Abstraction of the

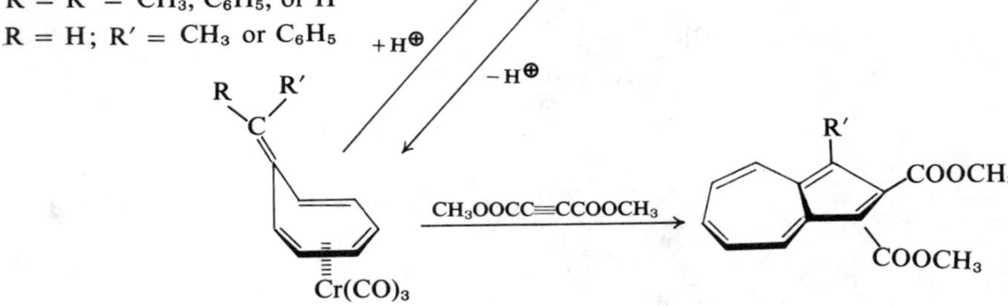

Scheme 3-13. Synthesis of heptafulvene $Cr(CO)_3$ compounds.

proton on the carbon adjacent to the tropylium ring in **107** [by reaction with 1,8-bis(dimethylamino)naphthalene] now gives the heptafulvene complex **108**. The free heptafulvenes, although very unstable, are quite stable in the form of their complexes with $Cr(CO)_3$, i.e., **108**. Protonation of the heptafulvene complex **108** with HPF_6 regenerates the cycloheptatrienyl complex **107**. The heptafulvene unit of **108** (R = H, R′ = H, C_6H_5, or CH_3) may be displaced from the chromium center in an appropriate arene solvent in the presence of a trapping agent; thus, reaction with $CH_3OOCC{\equiv}CCOOCH_3$ gives, interestingly enough, with loss of two hydrogens, the azulenes **109** (R′ = H, C_6H_5, or CH_3) [237].

Pauson and his colleagues found that several π-(7-substituted cycloheptatriene)chromium tricarbonyls undergo thermal rearrangement to give finally the π-(2-substituted cycloheptatriene) complexes [238,239]. A detailed investigation of this reaction showed that it proceeded via a series of stereospecific 1,5-hydrogen shifts involving the ψ-equatorial endo hydrogen. Thus, whereas the 7-*endo*-phenyl complex **101** is thermally stable up to its decomposition temperature, the 7-*exo*-complexes **100** (R = C_6H_5, CH_3, 4-CH_3-C_6H_4) all underwent smooth and rapid isomerizations when their solutions were heated. Furthermore, the authors established that the isomerization proceeded stepwise to give successively the 3-isomer **110**, the 1-isomer **111**,

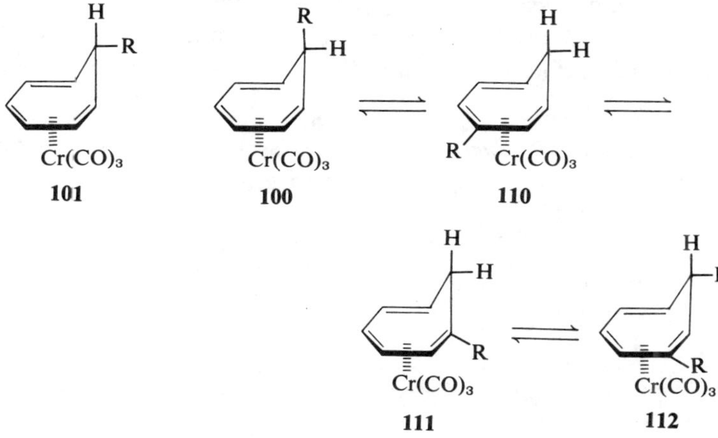

(3-76)

and finally the 2-isomer, Eq. (3-76). The complexed ligands rearrange more readily than the free ligands and the activation energy for the rearrangement of the ligand complexed to chromium (e.g., for **100**, R = CH_3, $E_A = 24.2$ kcal mole^{-1} [239] is lower than that for the free ligand ($E_A = 33.25$ kcal mole^{-1} [240]). The foregoing evidence shows that the isomerization process involves a series of metal-assisted 1,5-hydrogen shifts [239].

Pauson and his colleagues have investigated extensively the chemistry of "π-(cycloheptatriene)chromium tricarbonyl compounds" and the majority of the reactions involve the action of various nucleophiles on π-(cycloheptatrienyl) and π-(cycloheptatriene) complexes to give new π-(cycloheptatriene) complexes. Many of these reactions are stereospecific, and before discussing them it is pertinent to recall that most 7-substituted cycloheptatrienes react with $Cr(CO)_6$ [or $(pyridine)_3Cr(CO)_3/BF_3$] to give the 7-endo

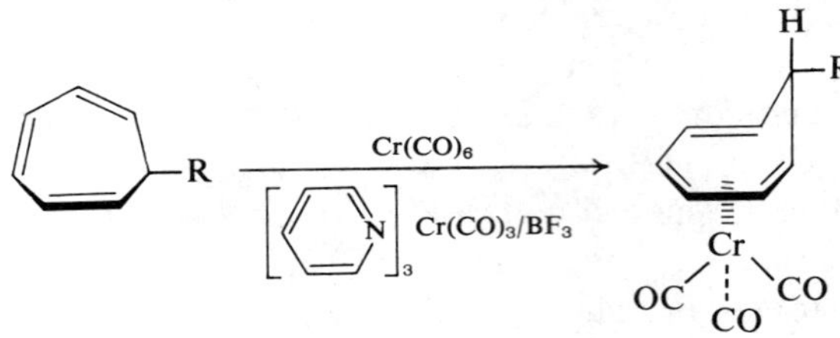

(3-77)

endo isomer

R = CH_3, C_6H_5, CN, $CH_2COOCH_2CH_3$, $CH(COOCH_2CH_3)_2$, $CH_2C{\equiv}CH$ [241], OCH_3, OCH_2CH_3 [221]

complexes, Eq. (3-77). The reaction of nucleophiles with salts of π-(cycloheptatrienyl)chromium tricarbonyl may result in the formation of the following three types of products: (1) Normal substitution products, (2) π-(arene)chromium tricarbonyl compounds, and (3) π-(ditropyl)chromium tricarbonyls.

i. Normal Substitution Products. Salts of π-(cycloheptatrienyl)$Cr(CO)_3$ react with a variety of nucleophiles (anions) to give by nucleophilic sub-

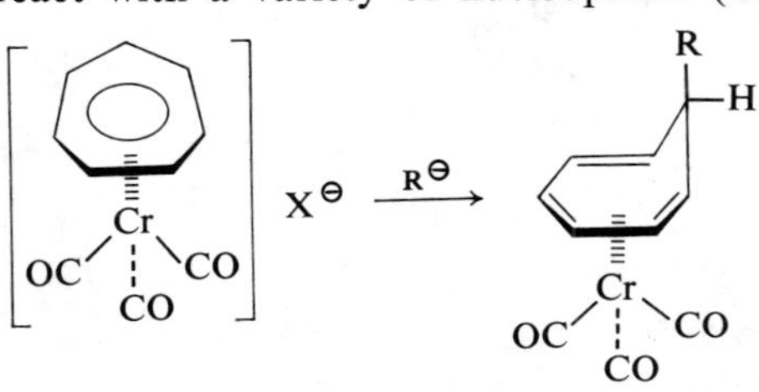

(3-78)

100 7-exo isomer

R = H, OCH_3, SH, $C_5H_4C(CH_3)_3$, $C(CH_3)(COOCH_2CH_3)_2$, $CH(COOCH_2CH_3)_2$, CN, CH_3, C_6H_5,

$C_6H_5C{\equiv}C$,

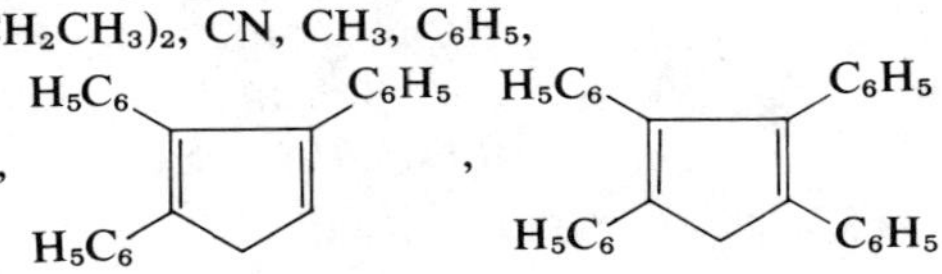

$(CH_3)_2N$

stitution the corresponding π-(7-*exo*-substituted cycloheptatriene)chromium tricarbonyls [**100**, Eq. (3-78)] [218,220,242,243]. The proof for the assignment of the exo configurations to all these products is based on an x-ray structure analysis of **100** ($R = C_6H_5$) and NMR spectroscopy (see Chapter 2, Sections D,2 and D,5,b). The fact that all the products of the normal nucleophilic substitution have the exo configurations indicates that the nucleophiles attack the organic ligand on the face opposite to chromium.

π-(7-Exo-substituted cycloheptatriene)chromium tricarbonyls may also be prepared by the reaction of π-(7-*exo*-methoxycycloheptatriene)$Cr(CO)_3$ with various anions, Eq. (3-79) [220]. This nucleophilic substitution (S_N1 ?) is

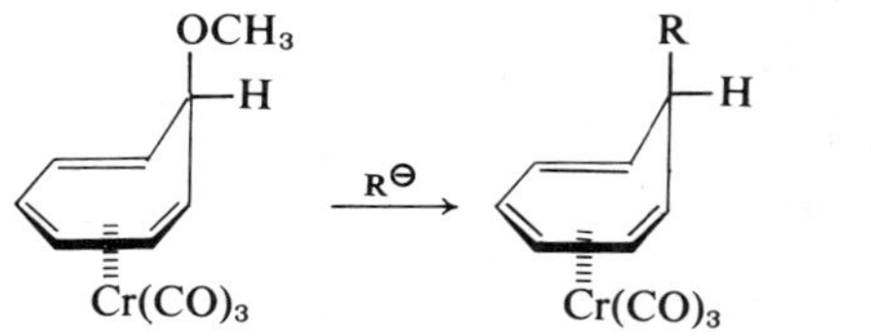

(3-79)

R = $CH(COOCH_2CH_3)_2$,
CH_3, C_6H_5,
4-CH_3–C_6H_4S,
$CH_2C{\equiv}CH$
$CH_2COOCH_2CH_3$

effected with an overall retention of the configuration and is not usually accompanied by the side reactions associated with nucleophilic substitution on the cycloheptatrienyl salt [220].

ii. π-(Arene)chromium Compounds. The interaction of π-(cycloheptatrienyl)chromium tricarbonyl perchlorate with excess sodium cyclopentadienide or sodium malonic ester results in an unusual ring contraction to give, as product, π-(benzene)$Cr(CO)_3$, Eq. (3-80) [$R = C_5H_5$ or $CH(COOCH_2CH_3)_2$]. Pauson and his colleagues were able to show: (i) that the

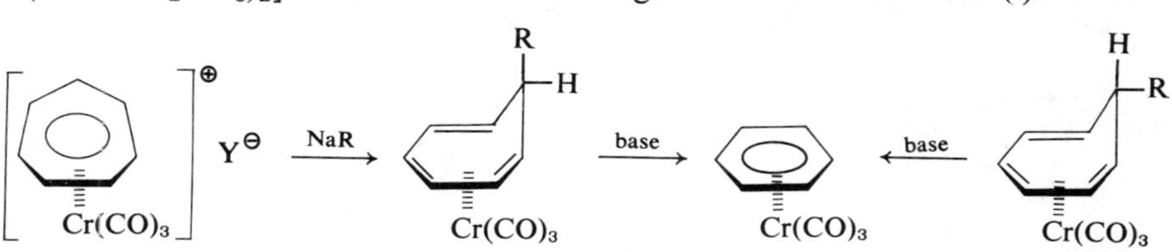

113 **114**

(3-80)

six carbon atoms of the arene ring originated in the cycloheptatrienyl unit, (ii) the product from the normal nucleophilic substitution [i.e., **113**, Eq. (3-80)] was an intermediate in the overall rearrangement, (iii) the hydrogen on the carbon adjacent to the C_7-ring was essential to the rearrangement, since **113** [$R = CCH_3(COOCH_2CH_3)_2$] did not rearrange [243], (iv) the

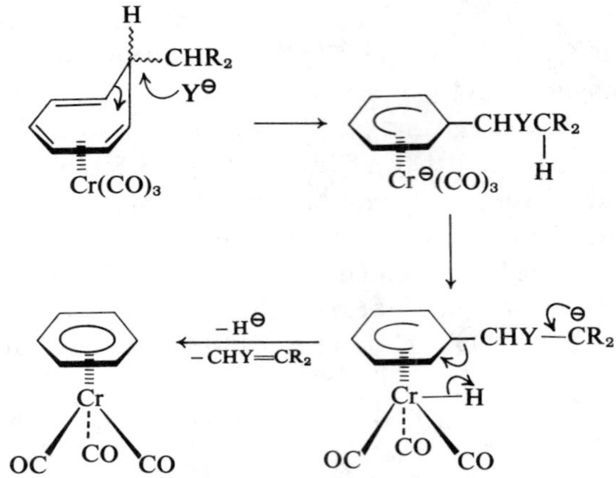

Scheme 3-14. Ring contraction of π-(cycloheptatriene)Cr(CO)$_3$ compounds.

7-endo isomer of **113**, [R = $CH(COOCH_2CH_3)_2$] (i.e., compound **114**) also rearranged to π-(benzene)Cr(CO)$_3$ when treated with base [241].

The mechanism, outlined in Scheme 3-14, has been proposed for the rearrangement of the π-cycloheptatrienyl malonic esters ($Y^\ominus$ = attacking anion and R = $COOCH_2CH_3$). The key step is the nucleophilic attack on the ring, which initiates the ring contraction; the terminating step is the extrusion of vinylic species containing the attacking nucleophile $Y^\ominus$ [241].

The ring contractions mentioned above and the ring contraction of π-(tropone)Cr(CO)$_3$ to π-(C_6H_6)Cr(CO)$_3$ (promoted by organomagnesium and organolithium compounds) [221] involve the extrusion of the C_7-ring carbon atom together with the attached substituents. However, π-(methyl 7-*exo*-methoxycyclohepta-1,3,5-triene-1-carboxylate)Cr(CO)$_3$ (**115**) undergoes base-catalyzed ring contraction with retention of the $COOCH_3$ group [to give π-(methylbenzoate)Cr(CO)$_3$] on chromatography over alumina, Eq. (3-81) [244].

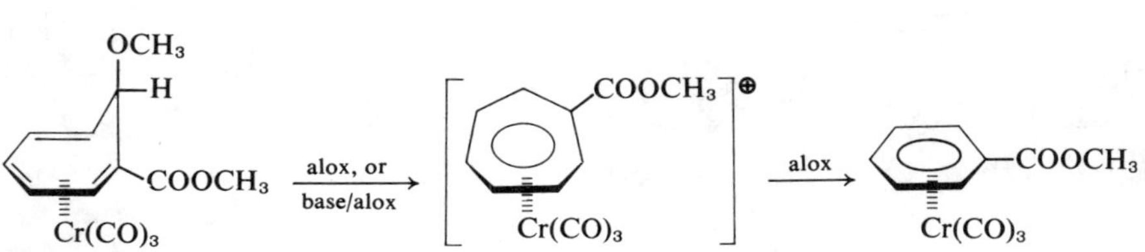

115 **116** (3-81)

The initial step in the reaction is methoxide elimination to give the π-(cycloheptatrienyl) complex **116**, Eq. (3-81). and this is confirmed by the observation that the salt **116** BF_4, when in contact with alumina, undergoes ring contraction to π-(methyl benzoate)Cr(CO)$_3$ [244]. A possible rationaliza-

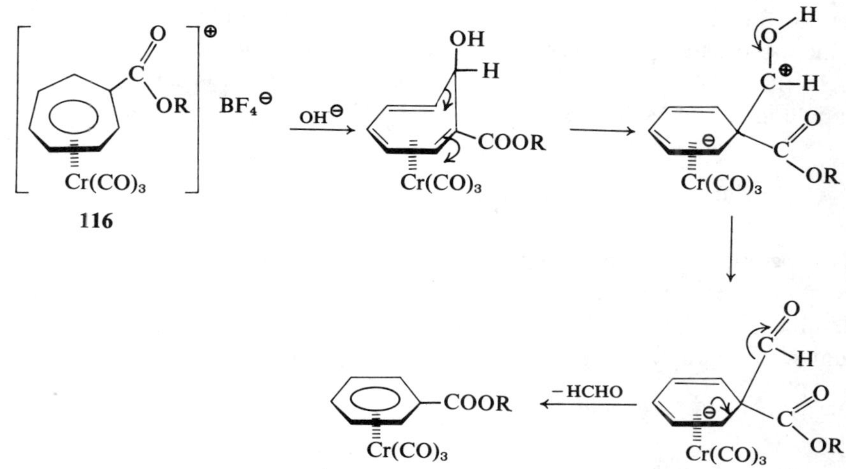

Scheme 3-15. Rearrangement of π-(methoxycarbonylcycloheptatrienyl)$Cr(CO)_3 \cdot BF_4$.

tion of the ring contraction of the salt **116** is outlined in Scheme 3-15 (R = CH_3). The initial attack of hydroxide on the C_7-ring in the salt would, in analogy to the attack of OCH_3, be expected to give the 7-*exo*-hydroxy-cyclohepta-1,3,5-triene-1-carboxylate complex. The rearrangement step is analogous to that outlined in the rearrangement mechanism in Scheme 3-14, and the final aromatization step is the simple extrusion of formaldehyde.

iii. π-(Ditropyl)chromium Compounds. Several anions [e.g., $CN^{\ominus}$, $OAc^{\ominus}$, $NH_2^{\ominus}$, $HCO_3^{\ominus}$, and $CH_3^{\ominus}$ (as $LiCH_3$)] cause the reductive coupling of π-(cycloheptatrienyl)$Cr(CO)_3$ salts and of π-(7-*exo*-CH_3O cycloheptatriene)-$Cr(CO)_3$ to give the ditropyl complexes **117, 118**, Eq. (3-82) [220,242]. The mechanism of these reductive dimerizations is not known, but the fact that

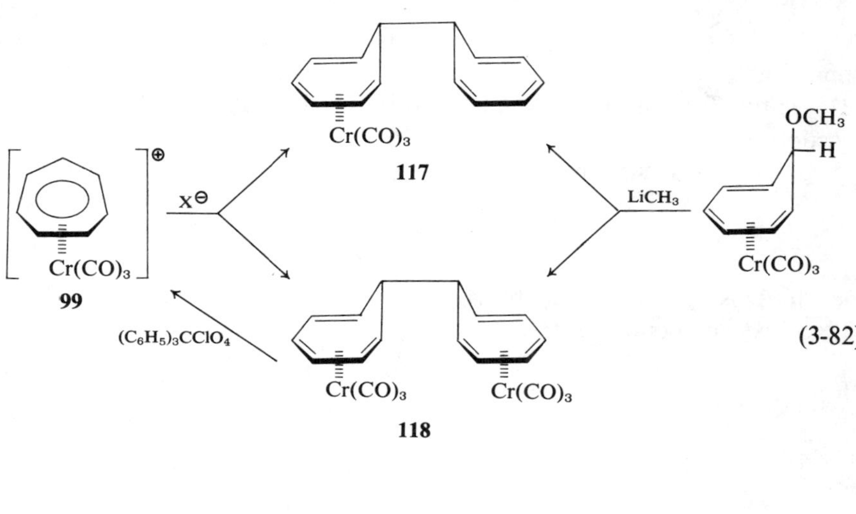

(3-82)

part of the chromium is lost may suggest that it might be involved in the reduction process [242]. The reaction of the π-(ditropyl) complex **(118)** with $(C_6H_5)_3C \cdot ClO_4$ did not result in hydride abstraction, instead the original tropylium salt **99** is regenerated, Eq. (3-82) [242].

E. Carbenoid-Chromium Compounds

Carbenoid-chromium compounds are relatively stable to both air and moisture, and may be readily purified by conventional techniques and identified by IR and NMR spectroscopies (see Chapter 2, Section E). The known compounds may be divided into the following four main types:

monocarbenoid compounds

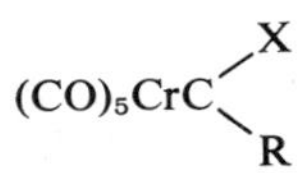

X = OR, NRR′, SR, etc.
R = CH_3, C_6H_5, etc.

biscarbenoid compounds

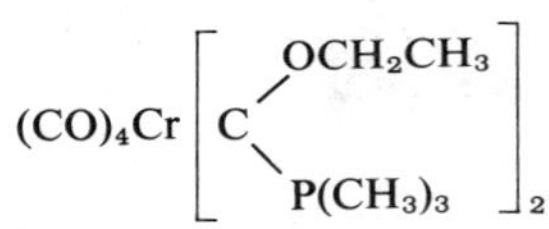

π-(arene)chromium carbenoid compounds

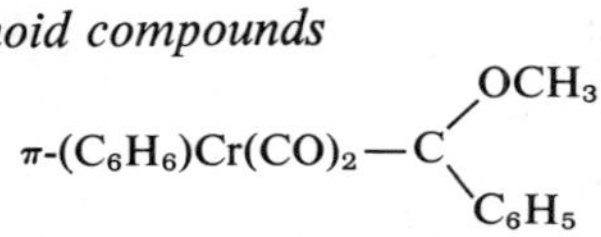

π-(cyclopentadienyl)chromium carbenoid compounds

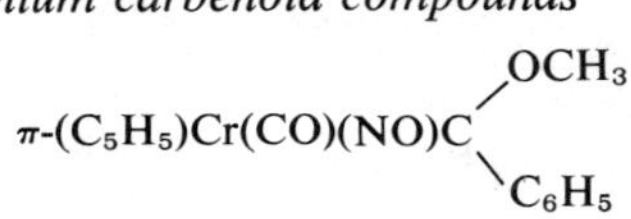

As yet, however, only the chemistry of the monocarbenoid-chromium compounds has been studied to any great detail (for earlier reviews, see [245–250]). The ensuing discussion will, therefore, be restricted to the reactions of:

(1) compounds of the type

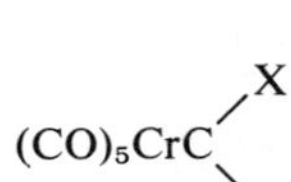

(X = OR, NR_2, SR, SeR, and SiR_3, R′ = an organic group), and

(2) acetoxy compounds of the type

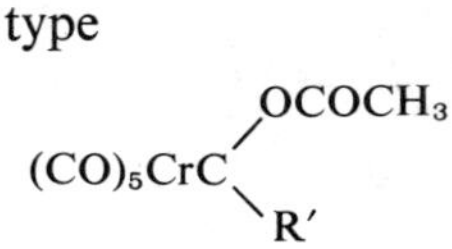

These reactions fall into three main categories: (a) those in which the carbonyl groups are displaced, (b) those in which the carbenoid unit is displaced, and (c) those in which the carbenoid unit is transformed.

1. Reactions of Alkoxy-, Amino-, Thio-, and Selenocarbenoid Complexes ($X = OR, NR_2, SR, SeR$ and SiR_3)

a. Displacement of the CO *Group*

The precise products formed in the reaction of pentacarbonylchromium carbenoid compounds and phosphorus compounds depend critically upon the nature of the phosphorus compound and the reaction conditions (see Scheme 3-16).

Tertiary alkyl- and arylphosphines react with alkoxycarbenoid-chromium compounds (**119**, $R = CH_3$ and C_6H_5), in various solvents (e.g., decane, cyclohexane, di-*n*-butylether, or dioxane) to give a mixture of the corresponding *cis*- and *trans*-phosphine carbenoid complexes (**120**, Scheme 3-16).

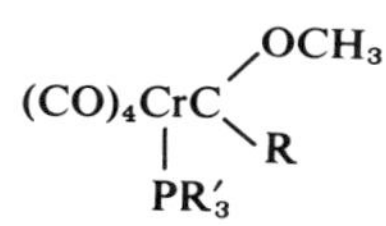

$R'_3 = (C_6H_{11})_3$, $(4\text{-}CH_3C_6H_4)_3$, $(C_6H_5)_3$, $(C_6H_5)_2C_2H_5$, $(C_6H_5)(C_2H_5)_2$, $(C_2H_5)_3$, $(n\text{-}C_4H_9)_3$ [251–253] and the *trans*-diphosphine chromium tetracarbonyl complexes $(CO)_4Cr(PR'_3)_2$ (**121**). The former compounds (i.e., **120**) are formed by the replacement of a carbonyl group of the original carbenoid compound by phosphine, while the latter (i.e., **121**) are formed by the displacement of the carbenoid unit in the phosphine carbenoid-chromium complexes (**120**) by the excess phosphine that is present in the reaction mixture [251,252]. The pure cis and trans isomers of the phosphine carbenoid-chromium compounds [**120**, $R'_3 = (C_2H_5)_3$, $(i\text{-}C_3H_7)_3$, $(n\text{-}C_4H_9)_3$, $(C_6H_{11})_3$ and $(n\text{-}C_8H_{17})_3$] may be isolated as crystalline solids, and they both isomerize when heated in benzene solution [253]. The cis isomer of **120** [$R = CH_3$ and $PR'_3 = P(C_6H_5)_3$] may also be prepared as outlined in Eq. (3-83) [254].

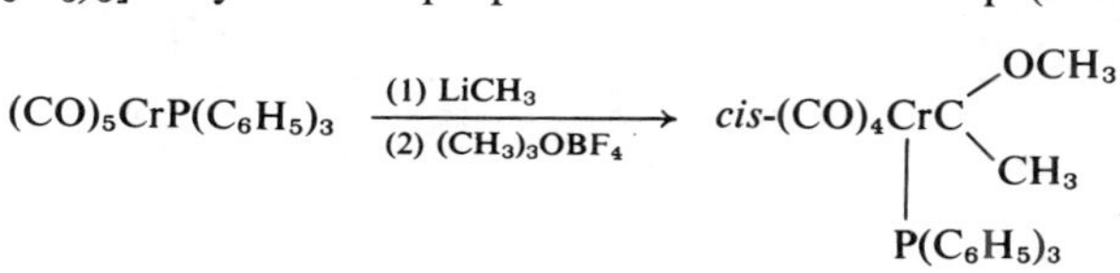

(3-83)

Secondary and tertiary phosphines [e.g., $HP(CH_3)_2$, $P(CH_3)_3$, $P(CH_2CH_3)_3$] and the alkoxycarbenoid compound **119** ($R = C_6H_5$) interact at low temperatures in hexane solution to give the corresponding 1:1 adducts **122** and **123** in which the phosphine is bonded to $C_{carbene}$ [255,256]. These adducts are stable only at low temperatures and the stability sequence is

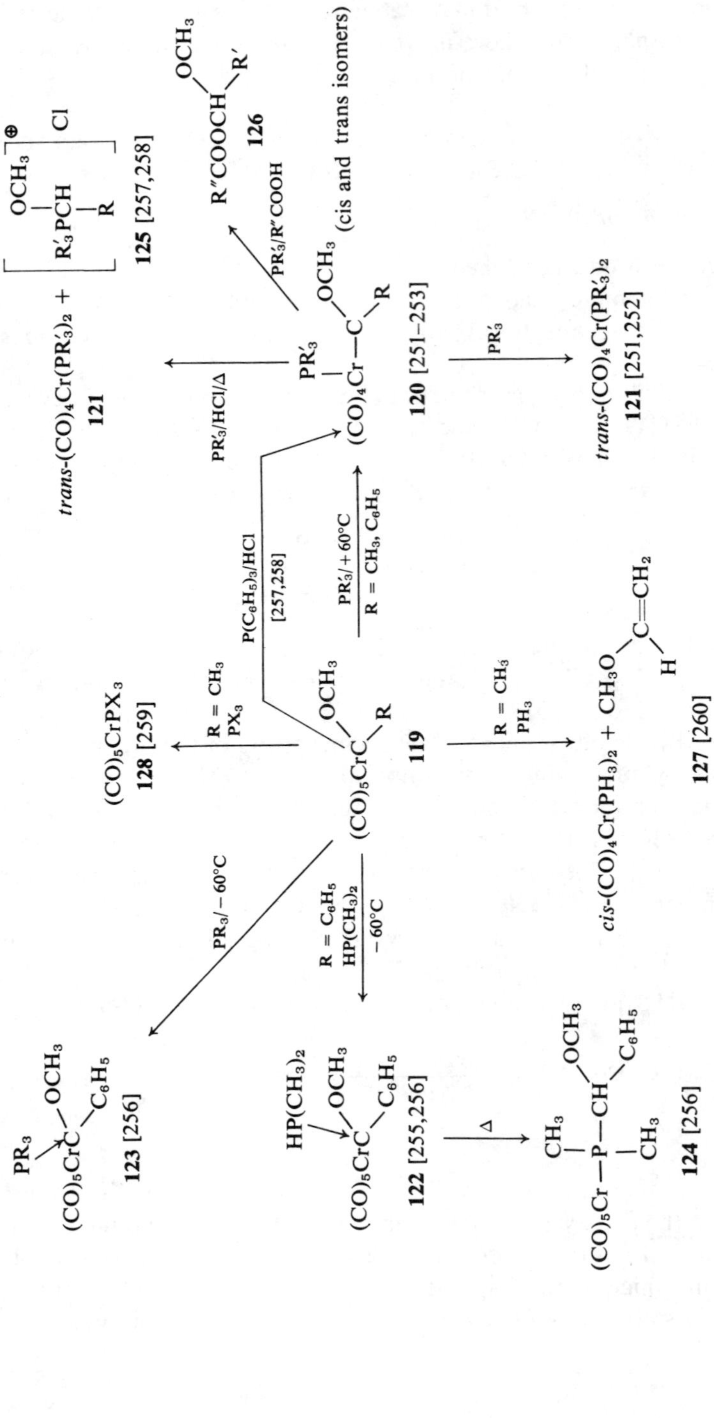

Scheme 3-16. Some reactions of alkoxy-carbenoid-chromium pentacarbonyls with some phosphorus compounds.

$P(CH_3)_3 > P(CH_2CH_3)_3 > HP(CH_3)_2$. When the 1:1 adduct with $HP(CH_3)_2$ **122**, dissolved in acetone, is allowed to warm to room temperature it is converted, in part, to the phosphine complex **124**, Scheme 3-16.

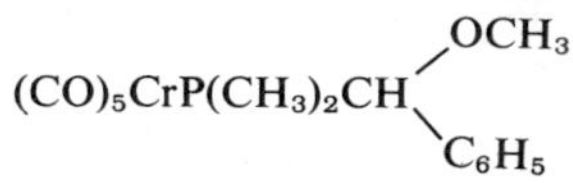

This "insertion reaction" involves the transfer of the hydrogen, which was present on the phosphorus atom, to $C_{carbene}$ and the transfer of the phosphorus unit from $C_{carbene}$ to chromium [255,256].

The carbenoid compounds **119** (R = CH_3 and C_6H_5) also react with $P(C_6H_5)_3$ in the presence of HCl to give, initially, the corresponding *cis*-phosphine carbenoid compounds **120**. These phosphine complexes, however, react further with $P(C_6H_5)_3$ and HCl to give finally the corresponding *trans*-bis(phosphine) complexes $[(CO)_4Cr(PR'_3)_2]$ **121** and the phosphonium salts **125** (R = CH_3 and C_6H_5, R′ = C_6H_5 and 4-$CH_3C_6H_4$) [257,258]. In these reactions the $C_{carbene}$–Cr bond is cleaved, presumably by the attack of the nucleophile $Cl^{\ominus}$ on $C_{carbene}$ to give a "carbenoid fragment" (chlorobenzylmethyl ether), which subsequently reacts with the excess phosphine, present in the reaction mixture, to give the phosphonium salt as it outlined in Eq. (3-84) (R = R′ = C_6H_5). The phosphine carbenoid compound **120**

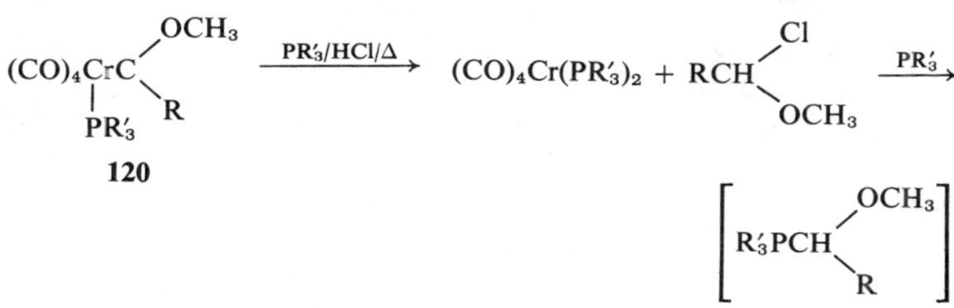

(3-84)

(R = CH_3 and C_6H_5) also reacts with carboxylic acids, in the presence of stoichiometric amounts of triphenylphosphine when the carbenoid units are again displaced and are finally converted to the corresponding methoxyesters **126** [Eq. (3-85), R = CH_3, R″ = C_6H_5; R = C_6H_5, R″ = CH_3; R = R″ = C_6H_5] [258].

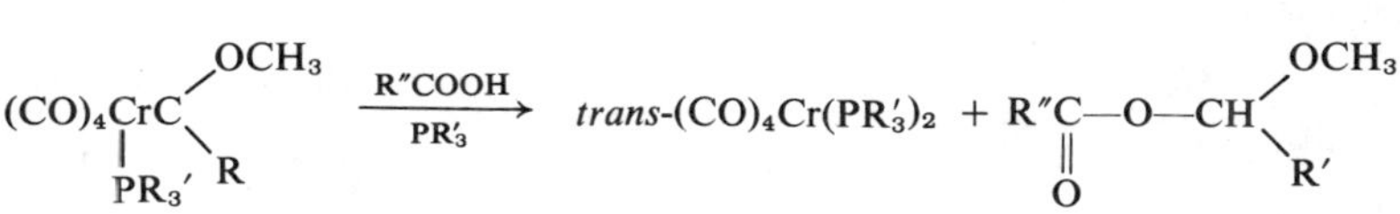

(3-85)

120 **126**

The carbenoid-chromium compound **119** [Scheme 3-16 (R = CH_3)] reacts with PH_3 [259] and PX_3 (X = I or Br) [259]; in both cases the carbenoid unit is displaced and found as methylvinyl ether [259]. The chromium products in these reactions are, however, not of the same type; thus, in the reaction with phosphine the *cis*-bis(phosphine) compound $(CO)_4Cr(PH_3)_2$ (**127**) is formed [260], while in the reactions with PX_3, the monophosphine complexes $(CO)_5CrPX_3$ (**128**, X = Br or I) are formed [259].

b. Displacement of the Carbenoid Unit

The carbenoid unit of alkoxy- and aminocarbenoid-chromium compounds may be displaced either thermally or by reaction with various chemical

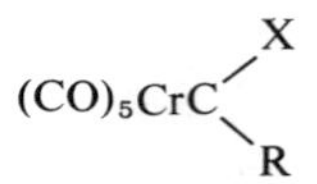

reagents. The fate of the carbenoid unit in these reactions is diverse; it may be recovered in an isomerized or dimerized form, or in the form of an adduct with the substrate or, finally, in a chemically modified form. The actual displacement of the carbenoid unit may involve the rupture of one of the three bonds in the carbenoid compound, i.e., the $C_{carbene}$–chromium bond, the $C_{carbene}$–C bond or the $C_{carbene}$–heteroatom bond.

The carbenoid unit of the alkoxycarbenoid compounds **129** and **130** may be displaced from the chromium center by a thermal reaction with pyridine or CO (under pressure) and is found in the products either as the "dimer," Eqs. (3-86) [261], Eq. (3-87) [262,263], or in an isomerized form, Eq. (3-88); this isomerization is assumed to proceed via a 1,2-hydride shift in the carbenoid unit promoted by the pyridine base [261,264].

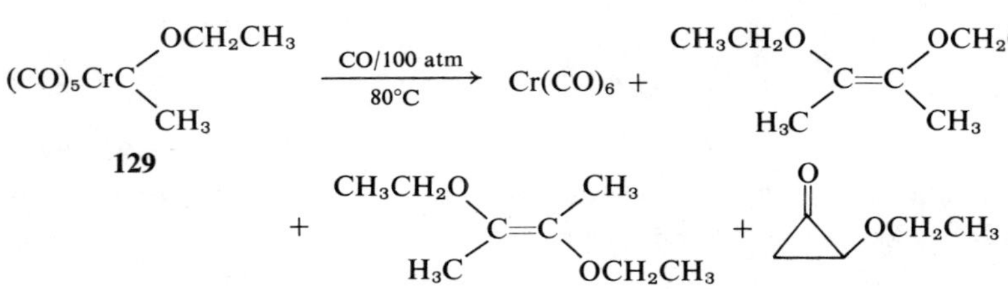

(3-86)

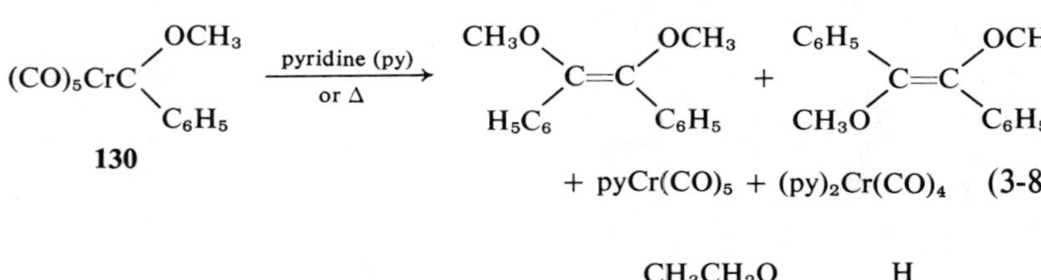

(3-87)

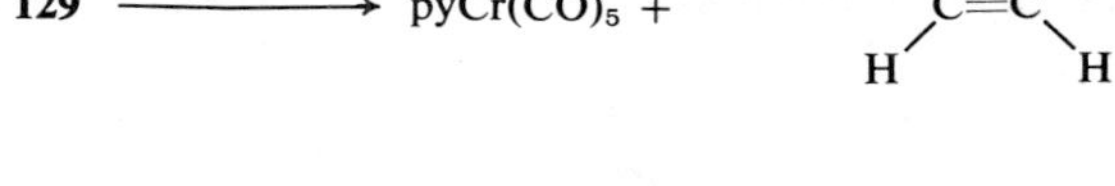

(3-88)

Attempts to trap the carbenoid unit, liberated in these reactions, in the form of their adducts with simple olefins, e.g., cyclohexene or dimethylbutene, were not successful. Fischer and his colleagues found, however, that these "carbenoid units" could be trapped in the form of their adducts **131** and **132** with activated olefins, as outlined in Eq. (3-89).

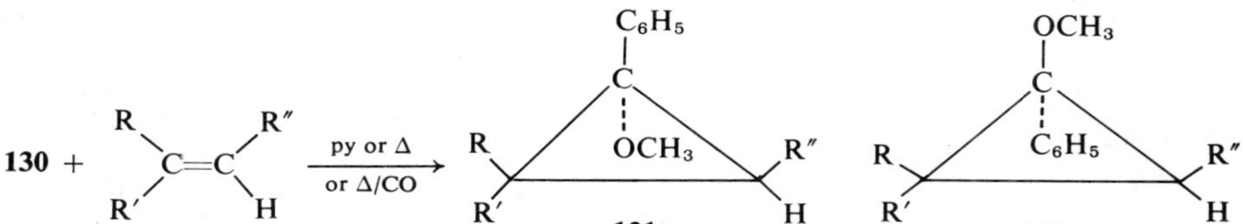

131 **132**

R = H, R′ = CH_3, R″ = $COOCH_3$ [265,266] (3-89)
R = H, R′ = C_6H_5, R″ = $COOCH_3$ [266]
R = H, R′ = $COOCH_2CH_3$, R″ = $COOCH_2CH_3$ [266]
R = $COOCH_2CH_3$, R′ = H, R″ = $COOCH_2CH_3$ [266]
R = R′ = H, R″ = OCH_2CH_3 [267]
R = R′ = H, R″ = $OCH_2(CH_3)_2$ [267]

The formation of the cyclopropyl compounds **131** and **132** is assumed to proceed by a mechanism involving four-center transition states of the type:

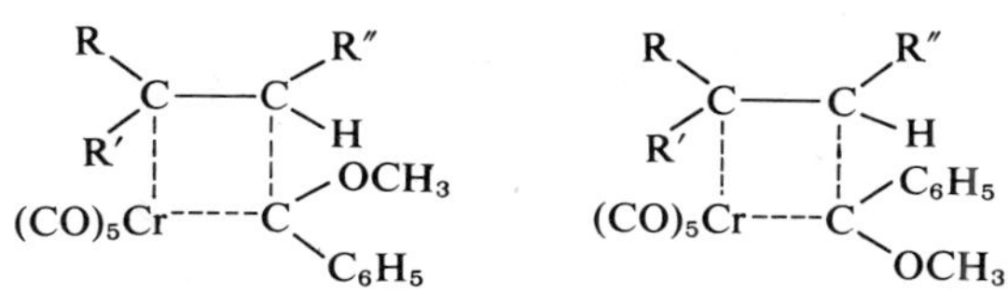

rather than by a mechanism involving the generation of a free carbene unit [266]. Fischer and Cooke have recently obtained an optically active cyclopropane derivative by the transfer of the carbenoid unit from a chromium compound containing an optically active phosphine ligand to diethyl fumarate Eq. (3-90). The asymmetric induction observed in this reaction shows that

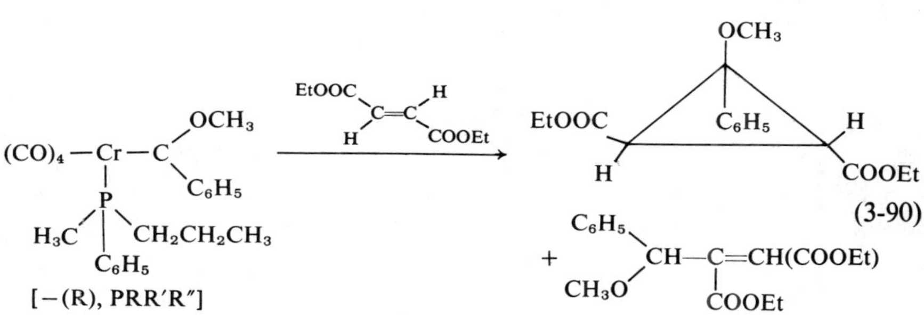

(3-90)

minor product

the transfer of the carbenoid unit to the activated olefin must occur within a metal–olefin complex [268].

Connor and his colleagues have also been able to trap the carbenoid units that are generated in the thermal reaction of alkoxy- and aminocarbenoid compounds with pyridine, by carrying out these reactions in the presence of group IV organometallic hydrides (R_3MH) [Eq. (3-91)].

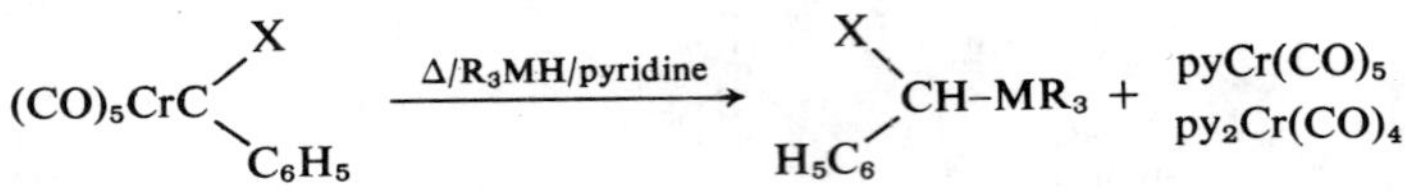

(3-91)

133

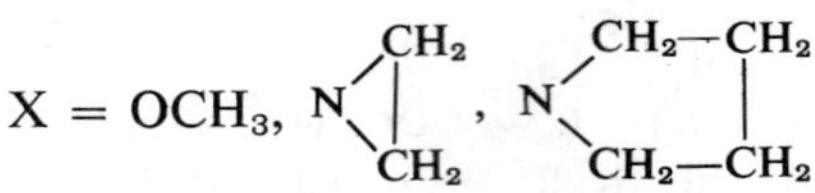

R = alkyl and aryl, M = Si, Ge, Sn [269,270].

The "carbenoid unit" inserts into the metal–hydrogen bond of R_3MH to give the tetraorganometallic compounds **133**, Eq. (3-91). A study of the kinetics of the reaction of the alkoxycarbenoid compound **130** with $(CH_3CH_2)_3SiH$, $(C_6H_5)_3GeH$, and $(C_6H_5)_3SnH$ (R_3MH) showed that the insertion reaction did not involve a free carbene species, but proceeded via

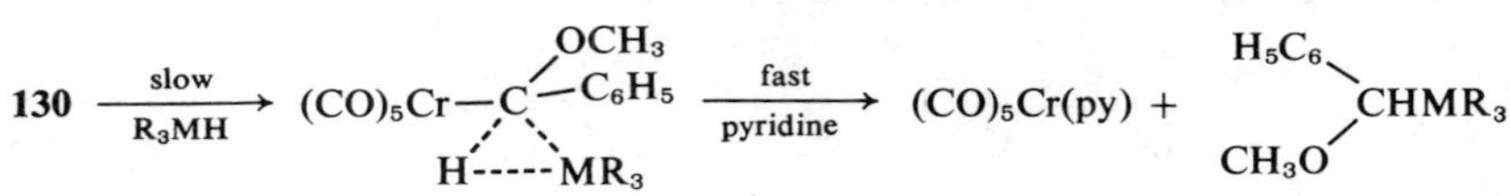

134 **133**

(3-92)

a three-center transition state complex of the type **134**, Eq. (3-92) [271]. The insertion of the carbenoid unit of

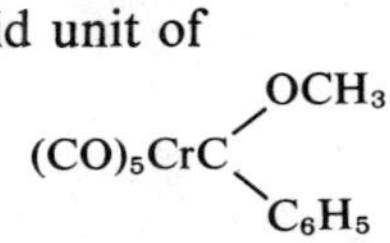

into the Si–H bond of $(C_6H_5)_2SiH_2$ to give

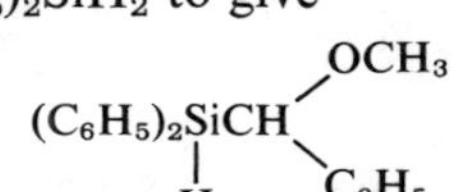

may be effected by warming the reagents together at 55°C [272].

The reaction of the alkoxy- and aminocarbenoid ferrocenyl compounds

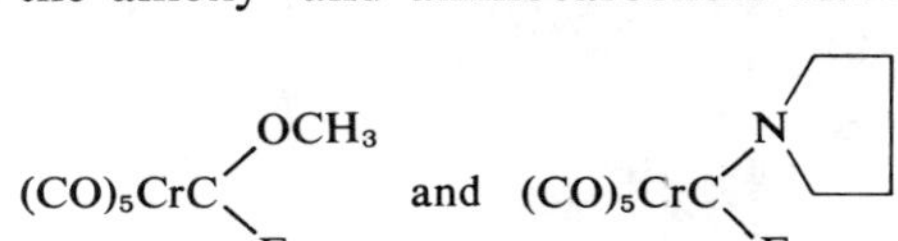

(Fc = ferrocenyl) with dimethylfumarate and 1,1-diphenylethylene have revealed some difference in the reactivity of the two types of carbenoid compounds [273].

The alkoxy compound reacts with dimethylfumarate, according to Eq. (3-93), but does not react with 1,1-diphenylethylene. The amino compound

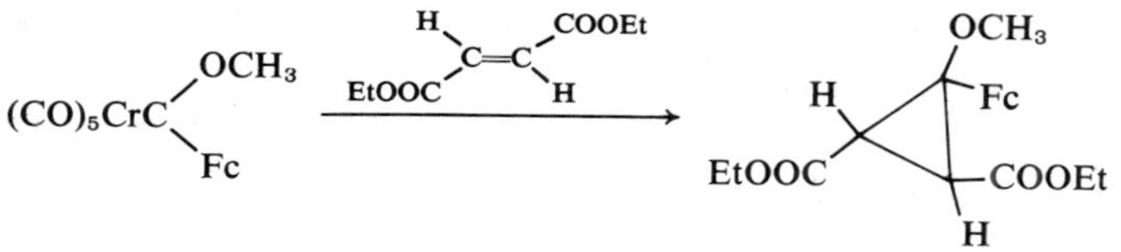
(3-93)

reacts with dimethylfumarate and 1,1-diphenylethylene to give compounds containing one carbenoid unit and two olefin units, i.e., **135** and **136**. Their formation is presumed to involve the transfer of the carbenoid unit to the olefin to give the cyclopropyl compounds **137** and **138**. These react with a second molecule of olefin substrate to give, after the appropriate rearrangements, the observed products, Eqs. (3-94) and (3-95) [in Eqs. (3-94) and (3-95), R =  Fc = ferrocenyl; in Eq. (3-94), R′ = $COOCH_2CH_3$, and in Eq. (3-95), R′ = C_6H_5] [273].

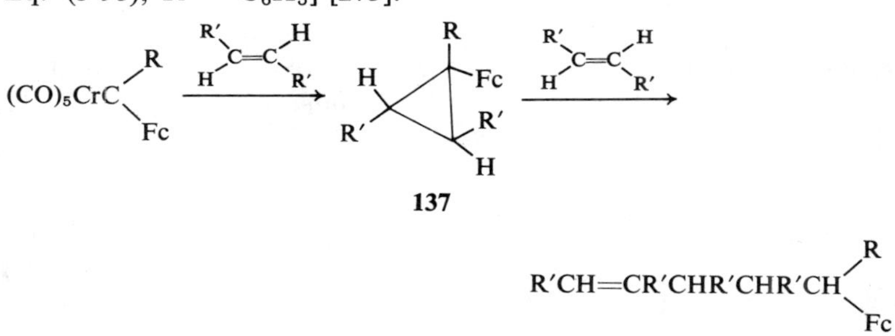
137

135
(3-94)

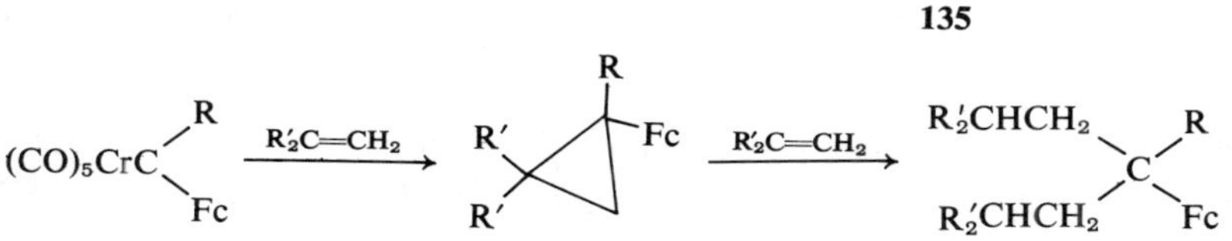
138 136
(3-95)

The carbenoid unit of alkoxycarbenoid-chromium compounds may also be displaced by a variety of nucleophiles, and some of these reactions are given in Scheme 3-17 [245,250,258,274]. The reaction of alkoxycarbenoid compounds with phosphorus compounds in the presence and absence of acids has been discussed in Section 1 of this chapter.

The products formed when the carbenoid unit of aminocarbenoid compounds is displaced from the chromium center (e.g., thermally or by reaction

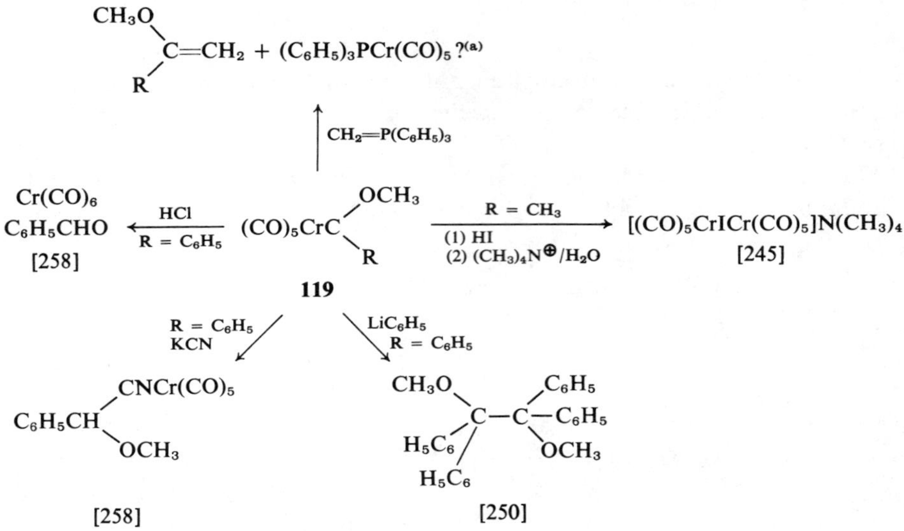

Scheme 3-17. Displacement of the carbenoid unit from $(CO)_5CrC(OCH_3)R$ compounds. (a) This reaction is known for the carbenoid-tungsten compound

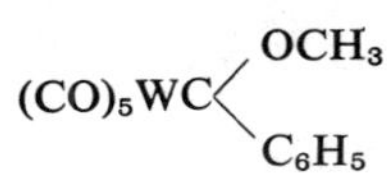

[274].

with pyridine or CO, etc.,) depend upon the nature of the amino unit, as is illustrated in Eqs. (3-96)–(3-101). Connor has pointed out [264] that when

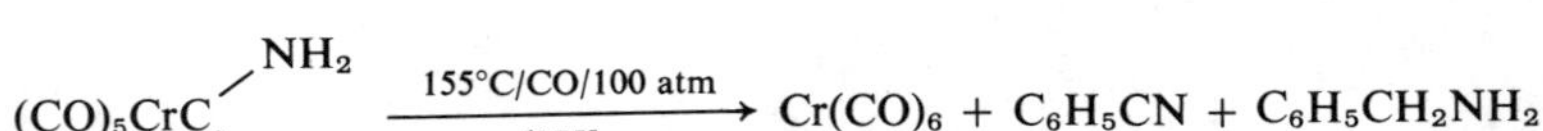

(3-96)

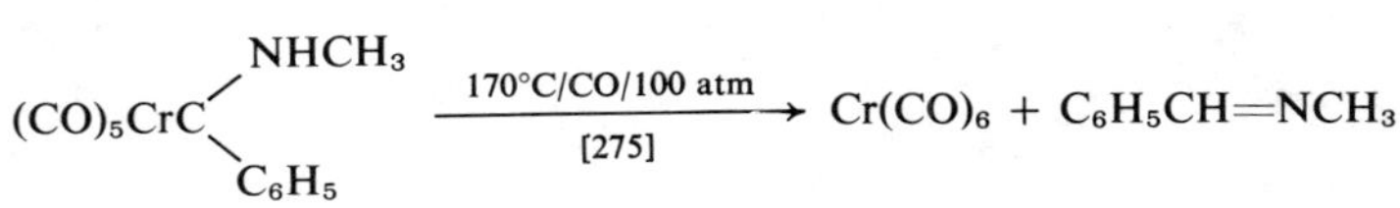

(3-97)

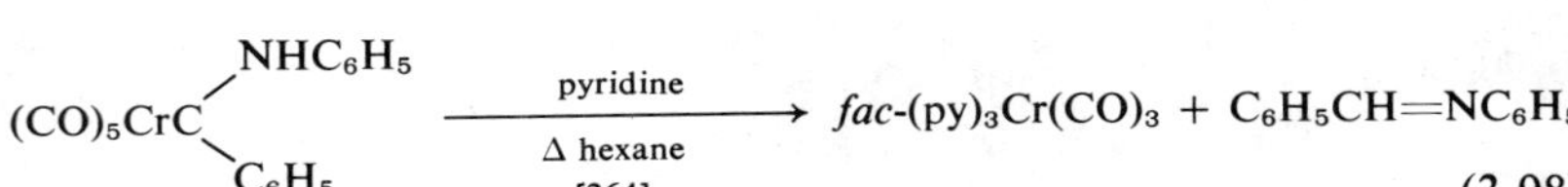

(3-98)

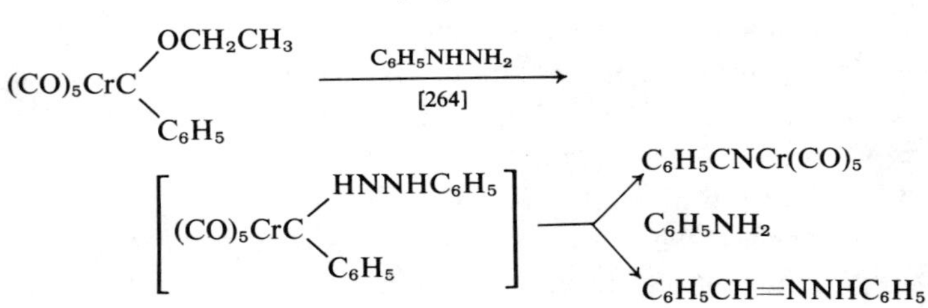

(3-99)

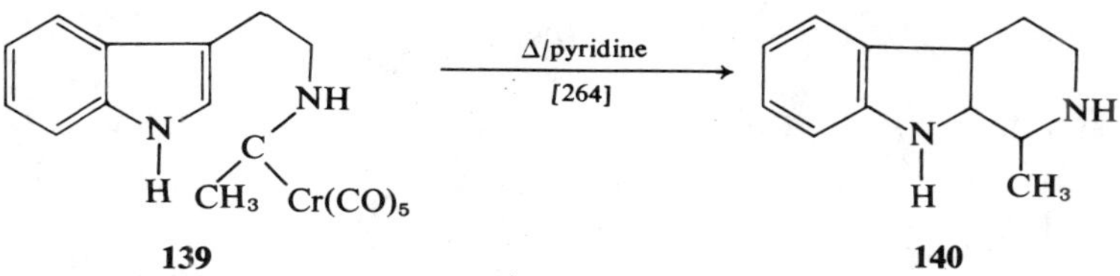

(3-100)

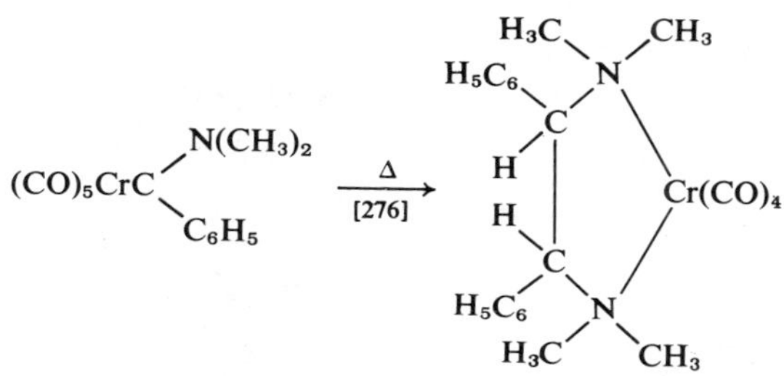

(3-101)

there are hydrogen atoms on the amino group bonded to $C_{carbene}$, this hydrogen migrates to $C_{carbene}$ during the displacement reaction to give imines and other related products illustrated in Eqs. (3-96)–(3-100). When there are no hydrogen atoms on the nitrogen adjacent to $C_{carbene}$, the carbenoid fragment will scavenge hydrogen to give a radical species that will couple to give vicinal diamine complexes, e.g., Eq. (3-101).

The thermal decomposition of the hydrazine intermediate in Eq. (3-99) proceeds by two paths, the one involving hydrogen transfer to give the phenylhydrazone, and the other involving cleavage of the N–N bond to give aniline and the pentacarbonyl nitrile compound. Other reactions in which the carbenoid unit is displaced in a chemically modified form will be discussed in Section c. The formation of tetrahydroharman **140** in the thermal displacement of the carbenoid unit of **139**, Eq. (3-100), is presumed to involve the transformations outlined in Eq. (3-102); again, a 1,2-hydride shift in the

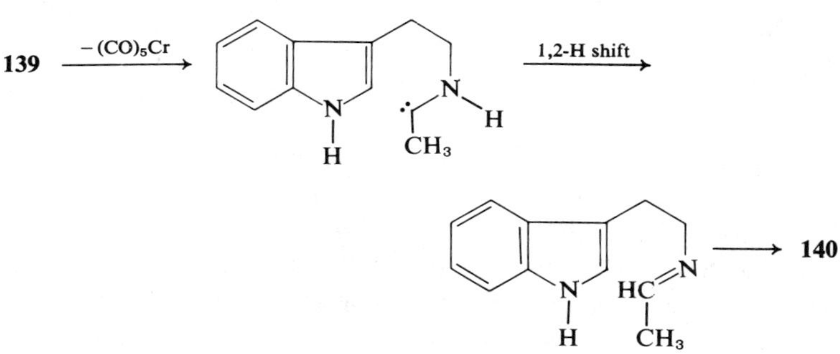

(3-102)

carbenoid fragment is presumed to lead to the imine, which then cyclizes to product [264].

The carbenoid ligand of the trimethylsilylmethyl carbenoid compound **141** may be displaced by reaction with iodine or acetyl chloride, Eq. (3-103) (the corresponding acetoxy compound **142** is not isolated from the latter reaction, c.f., Chapter 1, Section E). The fate of the carbenoid unit in these reactions could not be ascertained [277].

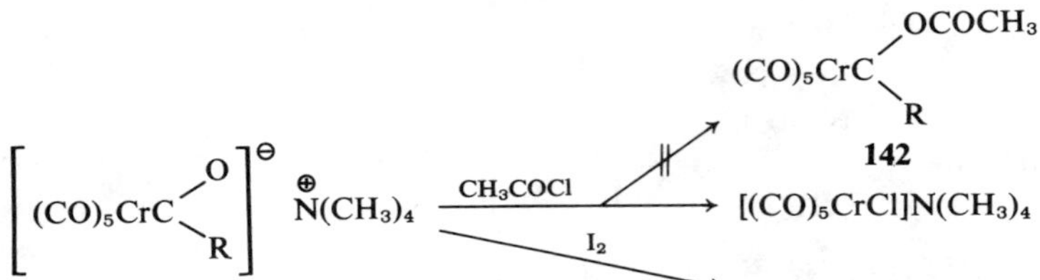

(3-103)

141

$R = CH_2Si(CH_3)_3$

The displacements that have been discussed up to this point have all involved the rupture of the $Cr{-}C_{carbene}$ bond. Connor and his colleagues [277] have found that the thermolytic displacement of the carbenoid unit in certain tetraalkylammonium carbenoid salts involves the cleavage of the $C_{carbene}{-}C$ bond. Thus, thermal decomposition of the salts of **141** and of the cyclopropyl-carbenoid complex **143** leads to the zwitterionic complex **144** together with, respectively, tetramethylsilane and cyclopropane, Eq. (3-104).

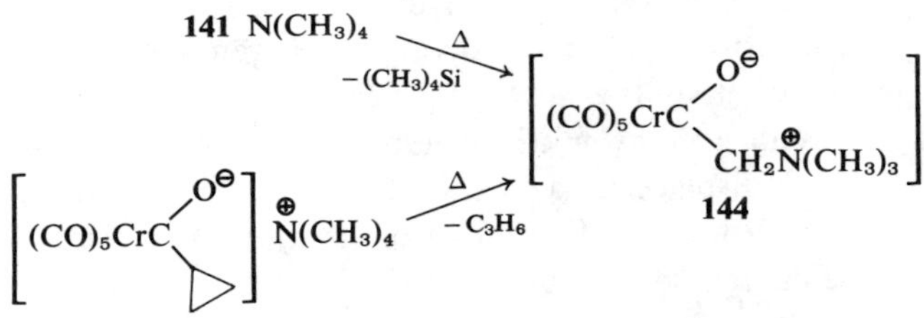

(3-104)

143

Pyrolysis of the tetraethylammonium salt of **141** gave the homologous zwitterionic complex

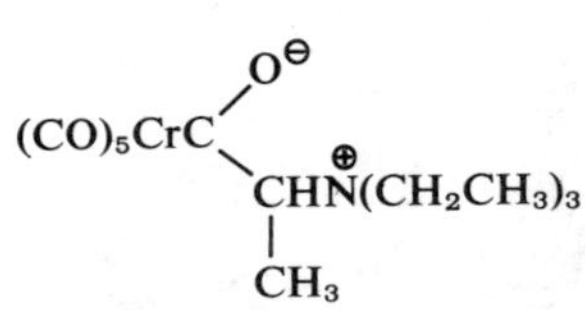

and tetramethylsilane, thereby indicating that it is indeed the $C_{carbene}$–carbon bond that is cleaved in these thermolytic reactions [277].

Fischer and his colleagues found that the displacement of the carbenoid unit of some thiocarbenoid compounds may involve rupture of the $C_{carbene}$–heteroatom bond. Thus, reaction of

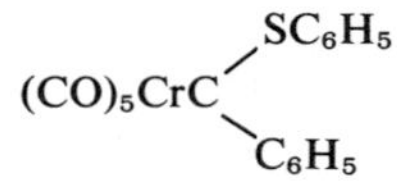

with CO in benzene gives C_6H_5S–SC_6H_5 [278], and reaction of the methyl analog of **145** with an *N*-acylimine gives by cleavage of the $C_{carbene}$–S bond, the thioether amide [Eq. (3-105), R = C_6H_5, 4-FC_6H_4]. The formation of

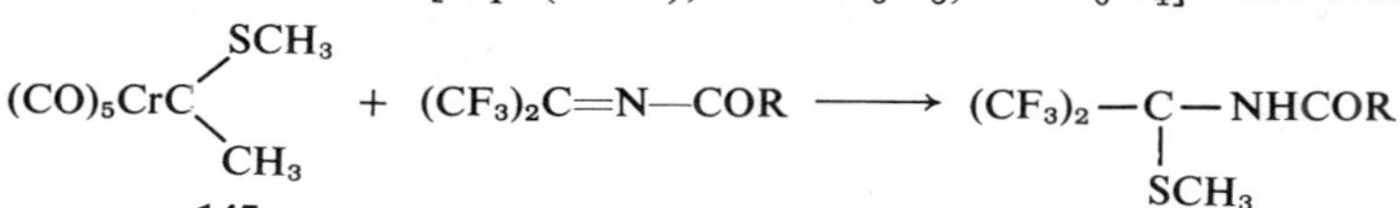 (3-105)

145

the thioether amide is presumed to involve the nucleophilic attack of S–CH_3 on the C=C of the imine followed by the transfer of hydrogen from either the C–CH_3 fragment or from solvent [278].

c. Reactions in Which the Carbenoid Unit is Transformed

Fischer and his colleagues found that alkoxycarbenoid-chromium pentacarbonyls react with a variety of nucleophiles (e.g., amines, thiols, selenols, etc.) to give products in which the carbenoid unit, though modified, is retained bonded to chromium. As will be evident from the ensuing discussion, the alkoxycarbenoid unit behaves, in many instances, as a carboxylic ester unit in which the carbenoid carbon has a pronounced cationic character.

i. Reactions with Ammonia and Amines. The alkoxy- and trimethylsiloxycarbenoid compounds of the type **146** [R = CH_3 or $Si(CH_3)_3$] [279] react with ammonia to give the corresponding aminocarbenoid complexes **147**, Scheme 3-18. The alkoxytrimethylsilylmethyl carbenoid compound **148**

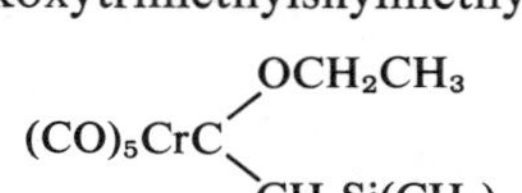

148

reacts with ammonia to give, by replacement of the OCH_2CH_3 group and cleavage of the CH_2–Si bond, the amide [277];

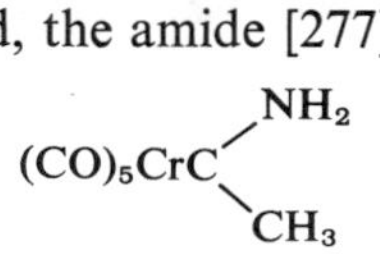

analogous cleavage of the CH_2–Si bond has been observed in the hydrolysis of **148** to give [277]:

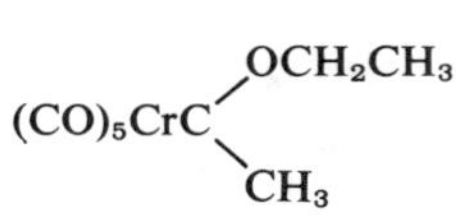

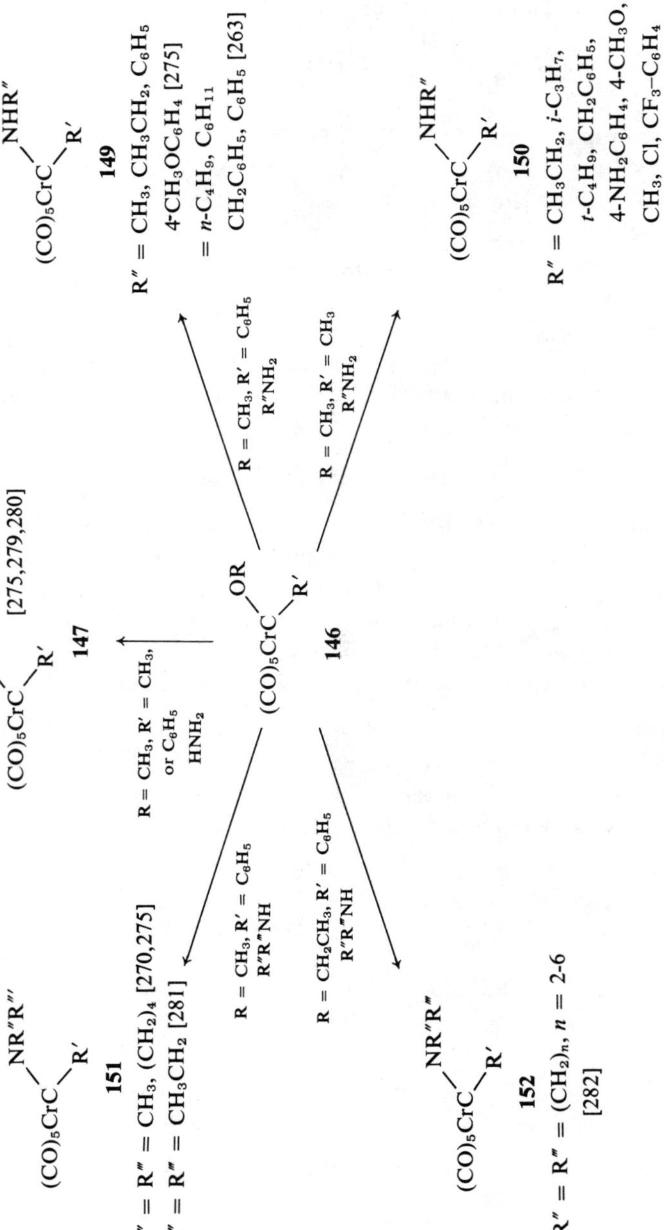

Scheme 3-18. Some reactions of alkoxycarbenoid-chromium complexes with amines.

The alkoxycarbenoid compounds of the type **146** also react with a variety of primary amines and certain secondary amines to give, by ammonolysis of the alkoxy group, the corresponding amides **149–152**, Scheme 3-18 [275, 279–282]. The products formed in the reaction with secondary amines depend upon the nature (steric bulk?) of the amine. Thus, with some amines {e.g., $HN(C_6H_5)_2$, $HN(CH_3)C_6H_5$, $HN[Si(CH_3)_3]_2$} no reaction takes place, while with others some unusual and ill-defined fragmentations of the amine are involved in the formation of the products, e.g., Eqs. (3-106) and (3-107)

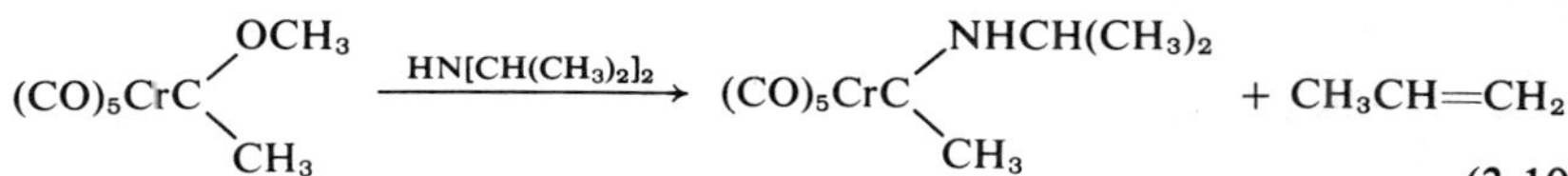

(3-106)

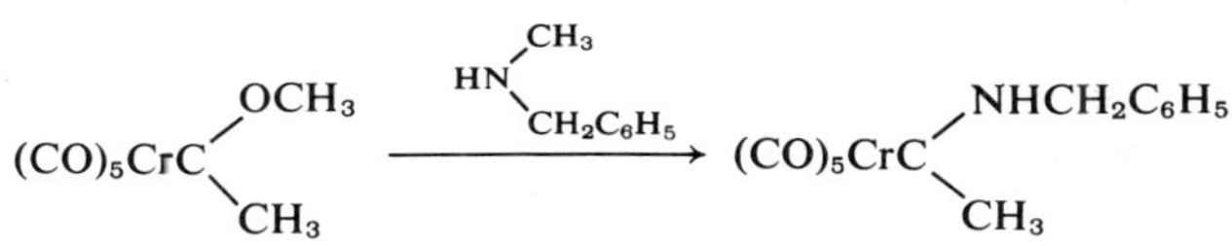

(3-107)

[282,283]. The preparation of the substituted amide derivatives of the type

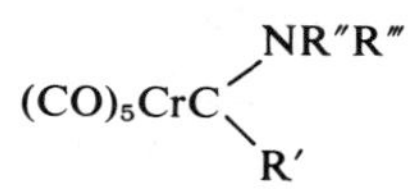

can, however, be effected by the reaction of the acetoxycarbenoid compound with various secondary amines [284]; see Scheme 3-20.

A detailed kinetic study of the reaction of

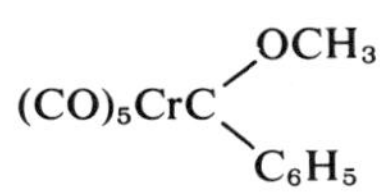

with various primary amines indicated that the replacement of the alkoxy group by the amino group was not a straightforward nucleophilic substitution process. The formation of the aminocarbenoid compound followed a rather complex (fourth-order) rate law [285,286] and the push-pull mechanism outlined in Eq. (3-108) was proposed in order to explain the kinetic

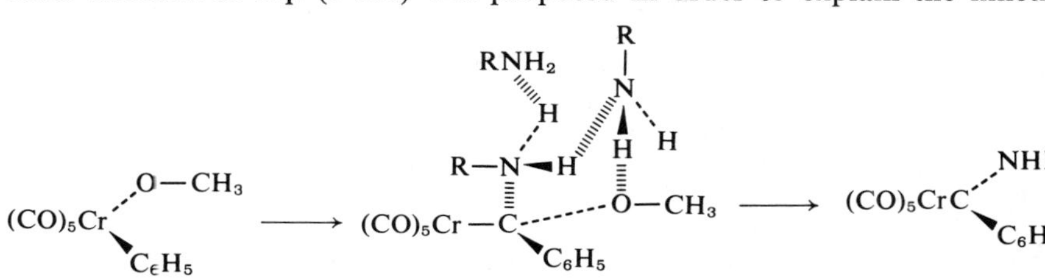

(3-108)

and spectroscopic data [286]. In nonpolar solvents (e.g., decane), the amine must act as a proton donor (to oxygen) and a proton acceptor (from amine), while in donor solvents (e.g., dioxane), the solvent may act as the proton acceptor. The formation of a nitrogen–ylide complex in this reaction is supported by the recent observation that a tertiary amine reacts with **130** to give a 1:1 adduct formulated as the nitrogen ylide **153**, Eq. (3-109). The ylide may be hydrolyzed to the hydrocarbenoid compound **154**, Eq. (3-109) [287].

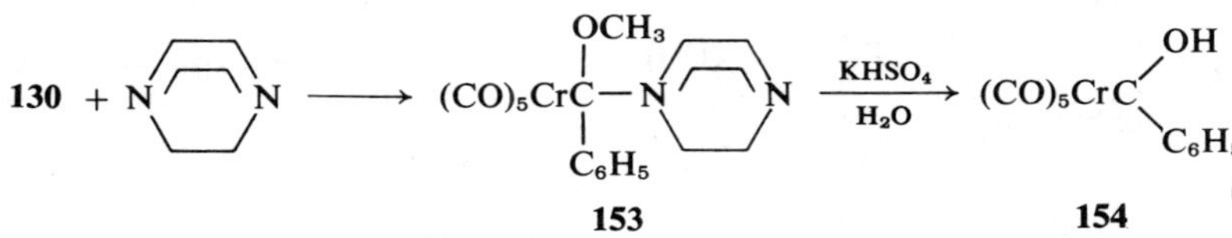

(3-109)

Aliphatic diamines of the type $H_2N–R'''–NH_2$ react with **146** ($R = R' = CH_3$) to give bridged aminocarbenoid compounds, Eq. (3-110) [288]. Aromatic diamines, on the other hand, react with **146** ($R = R' = CH_3$) to give the monoamidocarbenoid complexes, Eq. (3-111) [288].

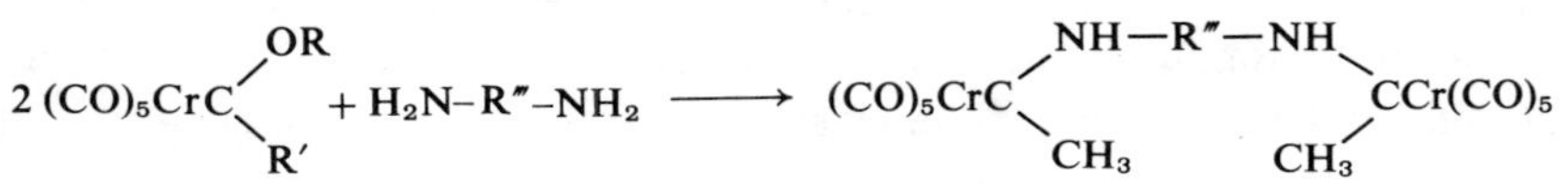

146 (3-110)

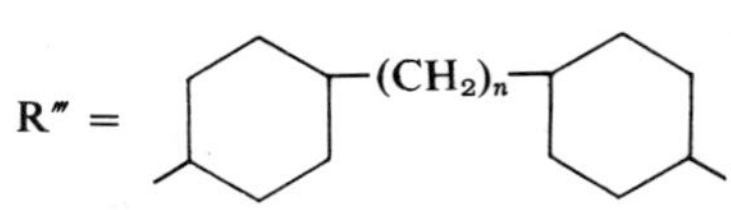

$n = 3$ to 6 and 10

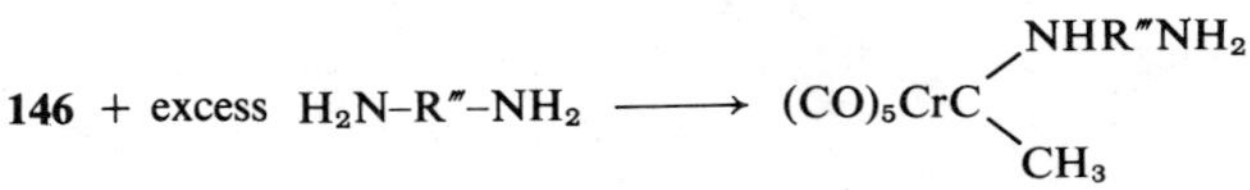

(3-111)

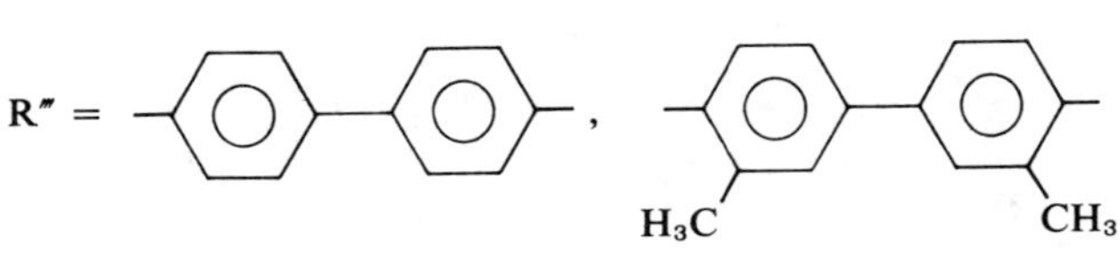

ii. Reactions with Other Nitrogen-Containing Compounds. Fischer and his colleagues have shown that the alkoxycarbenoid-chromium compounds do not react with acetamide and thioacetamide [281], while they do react with a variety of oximes, imines, hydrazines, isonitriles, etc. (see Scheme 3-19).

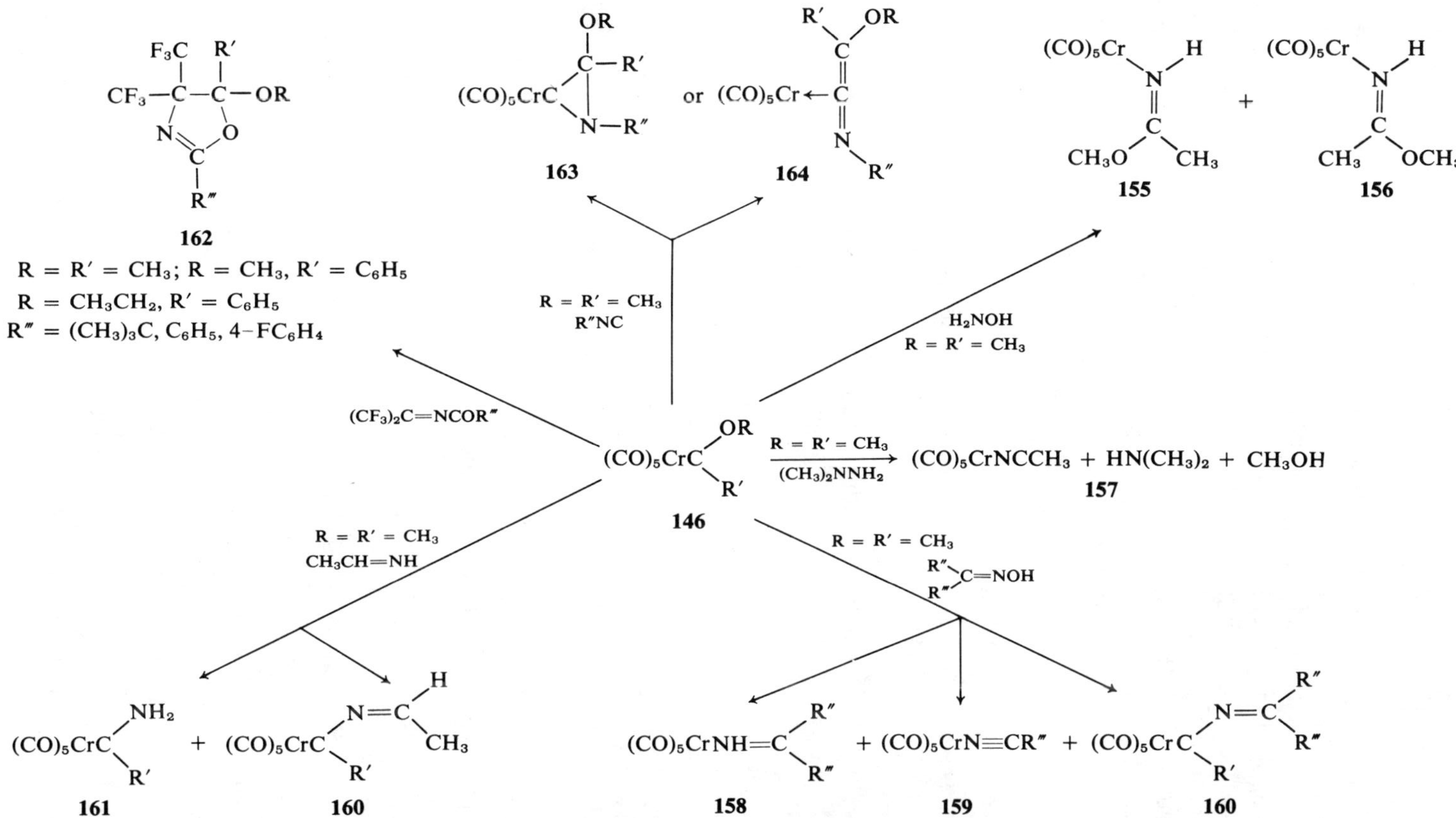

Scheme 3-19. Some reactions of alkoxycarbenoid-chromium pentacarbonyl complexes with oximes, imines, and isonitriles.

Anhydrous hydroxylamine reacts with the alkoxycarbenoid compound **146** to give the cis-trans isomeric compounds **155** and **156**, in which the carbenoid unit, though displaced from the chromium center, is retained within the final product (a type of insertion reaction) [289].

Substituted hydrazines [e.g., $(CH_3)_2NNH_2$] react with methylmethoxycarbenoid chromium (**146**, R = R′ = CH_3) to give the acetonitrile complex $(CO)_5CrNCCH_3$ (**157**) as main product together with methanol and dimethylamine [289]. The nitrile **157** is probably formed by the cleavage of the N–N bond in the hydrazine complex

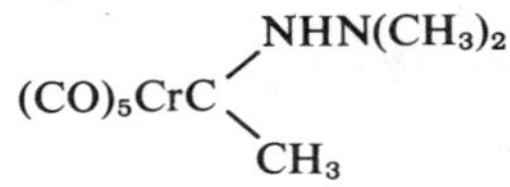

cf. Eq. (3-99).

Oximes react with **146** to give a variety of products in which the carbenoid unit is lost, e.g., **158** and **159**, or is retained in a modified form, e.g., **160**, see Scheme 3-19.

Benzaldehyde oxime reacts with the methoxycarbenoid compound to give all three types of products; the aldimine compound, **158**,

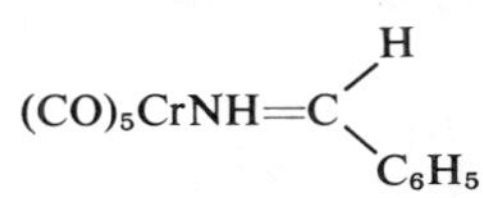

158

the nitrile complex $(CO)_5CrNCC_6H_5$ (**159**), and the benzylidenamino compound **160**:

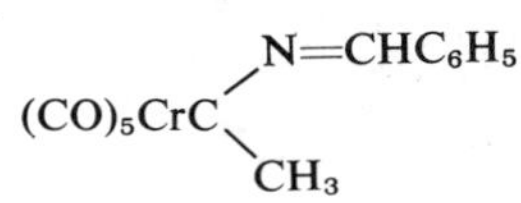

160

The phenyl analog of this last compound, i.e.,

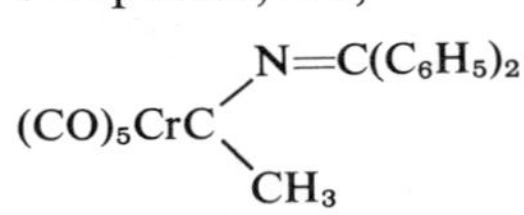

may be prepared, in low yield (21%) by the reaction of **146** (R = R′ = CH_3) with benzophenonimine [290].

Aliphatic and aromatic oximes react with **146** (R = R′ = CH_3) to give, in low yields (5%–20%) mainly the ketimine compounds **158**:

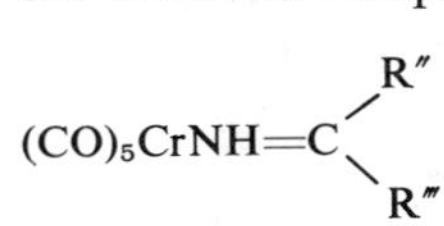

(R″ = R‴ = CH_3, C_2H_5, C_6H_5; R″ = CH_3; and R‴ = C_2H_5 or C_6H_5; R″ = R‴ = $(CH_2)_n$, n = 4 or 5) [291]; in the reaction with acetophenone oxime, some of the methylbenzylidene amino compound **160** (R′ = R″ = CH_3, R‴ = C_6H_5) was also formed [290].

Acetaldehyde imine (or the ammonia adduct of acetaldehyde) reacts with **146** (R = CH_3, R′ = C_6H_5) to give, in 6% yield, the ethylideneamino-carbenoid complex **160** (R′ = C_6H_5, R″ = H, R‴ = CH_3) together with the simple ammonolysis product (**161**) [292].

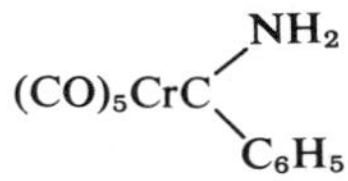

161

The N-acylimines of hexafluoracetone react with the alkoxycarbenoid compounds **146** (R′ = CH_3 and C_6H_5) and with

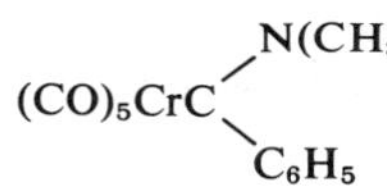

to give the oxazolines **162** [and the corresponding oxazoline with $N(CH_3)_2$ in place of OR and R′ = C_6H_5]. The corresponding thiocarbenoid compounds react with cleavage of the $C_{carbene}$–S bond, see Eq. (3–105) [278].

Cyclohexyl- and methylisonitrile react with the alkoxycarbenoid compound **146** (R = R′ = CH_3) to give a 1:1 adduct, formulated as being either the arizinylidene complex **163** or the ketinimine complex **164**, Scheme 3-19. The 1:1 adduct with cyclohexylisonitrile may be converted into three new aminocarbenoid complexes **165**–**167** by respective treatment with acids,

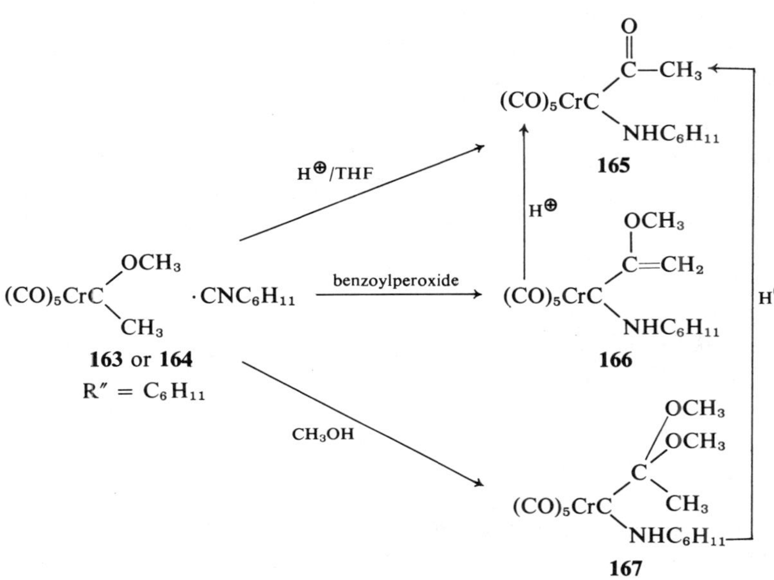

(3-112)

benzoylperoxide, or absolute methanol, Eq. (3-112). The vinyl compound **166** and the ketal **167** are converted, by treatment with acid, into the acetyl compound **165** [293]. The crystal structure of the complex **166** has been discussed in Chapter 2, Table 2.5 [294].

An interesting example of the application of the aminolysis of alkoxy-carbenoid-chromium compounds is in the synthesis of peptides. Thus, reaction of **130** with L-alanine methyl ester gives the cis and trans isomers of the amino acid ester-carbenoid compound **168**, Eq. (3-113). The amino acid unit of **168** may be built up into a polypeptide unit by first hydrolyzing the ester to the acid and subsequently treating the acid with an amino acid ester in the presence of *N*-hydroxysuccinimide (NHS) and dicyclohexylcarbodi-imide (DCCD), Eq. (3-113). The peptide carbenoid compounds with 1-,

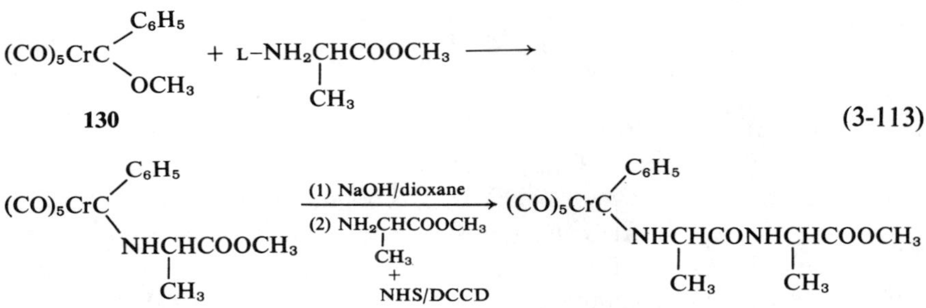

130 **168** (3-113)

2- and 3-alanine units are readily soluble in polar organic solvents but are insoluble in water and hydrocarbon solvents. The "peptide unit" may be cleaved from the chromium center by reaction with trifluoracetacetic acid (or with 80% acetic acid); the carbenoid residue is converted to benzaldehyde [295].

iii. Reactions with Thiols and Selenols. The alkoxycarbenoid compound **146** (R′ = CH_3 and C_6H_5, R = CH_3) reacts with both aliphatic and aromatic thiols to give the corresponding thiocarbenoid compound **169**, Eq. (3-114) [280,296]. A more direct route to thiocarbenoid complexes of chromium

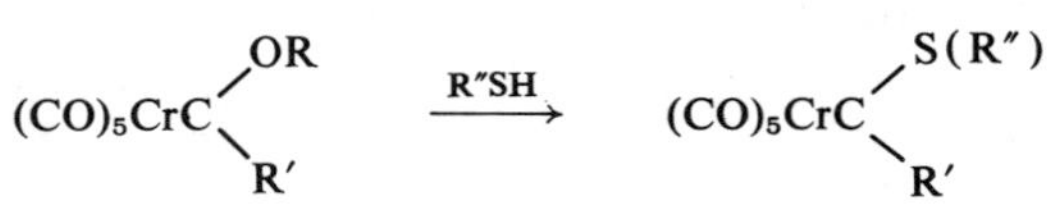 (3-114)

146 **169**

R = CH_3 R′ = CH_3, C_6H_5

R′ = CH_3 and C_6H_5 R″ = CH_3, CH_3CH_2, C_6H_5

might be by the interaction of thiocarbonylchromium pentacarbonyl [$(CO)_5$-$CrCS$] [297] with nucleophiles. The tungsten analog reacts with secondary amines (e.g., dimethylamine and piperidine) to give the corresponding carbenoid compounds, Eq. (3-115) [297]. The products formed in the reaction

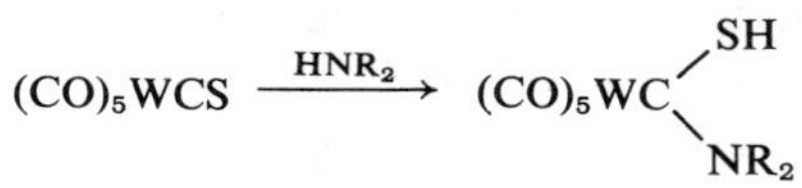

(3-115)

of alkoxycarbenoid compounds with selenols would appear to depend upon the nature of the selenol. Thus, with selenophenol the product is the seleno ether compound **170**, Eq. (3-116) ($R = R' = CH_3$) [298]. This compound is

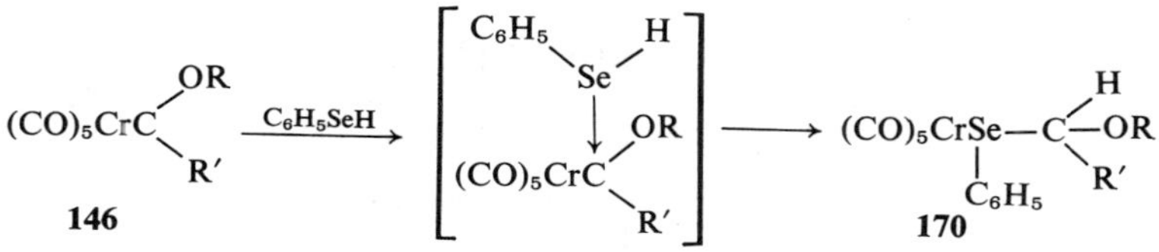

(3-116)

probably formed by the nucleophilic attack of the selenophenol at $C_{carbene}$ followed by transfer of chromium to selenium and hydrogen to $C_{carbene}$; an analogous "insertion reaction" has been observed in the thermal rearrangement of the dimethylphosphine adduct

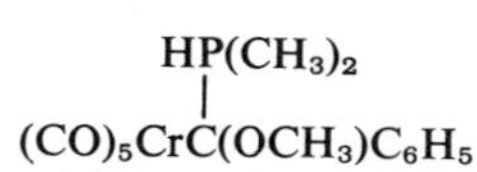

(see Scheme 3-16, compounds **122–124**). Methyl selenol reacts with the alkoxycarbenoid complex **146**, $R = R' = CH_3$, to give the expected selenocarbene compound [299].

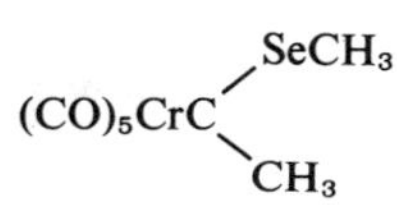

iv. Reactions with Acids. The methylthiocarbenoid-chromium compound **169** reacts with gaseous HBr to give, by nucleophilic addition of HBr to the $C_{carbene}$ and subsequent rearrangement of the adduct **171**, the dialkyl sulfide compound **172**, Eq. (3-117) [300]. Analogous rearrangements of 1:1 adducts

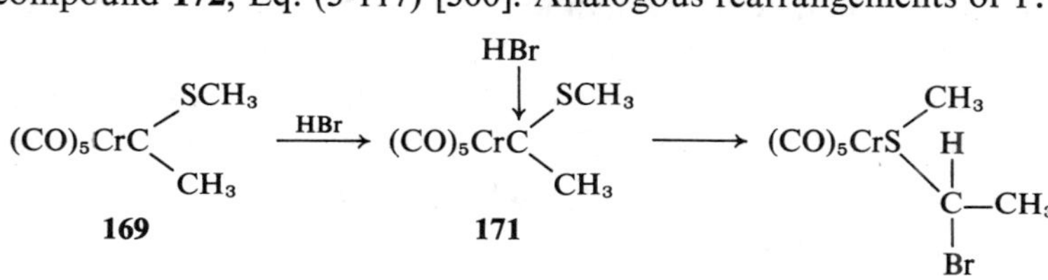

(3-117)

172

of carbenoid complexes with selenophenol and dimethylphosphine have already been discussed [Eq. (3-116) and Scheme 3-16].

The aminocarbenoid-chromium complexes **173** also react with gaseous HCl or HBr to give 1:1 adducts, which, on the basis of spectroscopic data, have been formulated as the iminium salts **174** (formed by rearrangement of the HX adduct **175**), Eq. (3-118) [301].

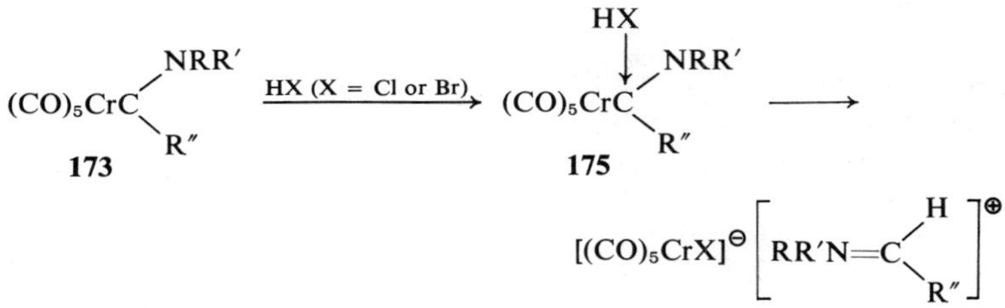

(3-118)

173 **175** **174**

$R'' = CH_3$ or C_6H_5
$R = H, R' = CH_3$;
$R = R' = CH_3, (CH_2)_4$

v. Miscellaneous Reactions. Lithium aluminum hydride reacts with various anilinocarbenoid compounds

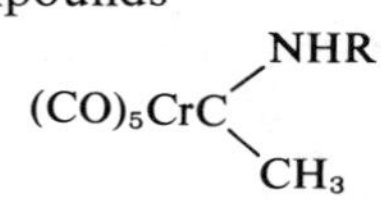

(R = 4-NH_2, CH_3, H, Cl, CF_3, CH_3O, C_6H_4) to give, by reductive cleavage of the carbenoid unit, the corresponding *N*-ethylanilines [283]. The methyl methoxycarbenoid compound

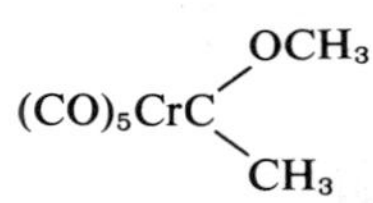

reacts with Li[AlH(*t*-BuO)$_3$] to give in low yield ($< 0.1\%$) the butadienyl complex [302].

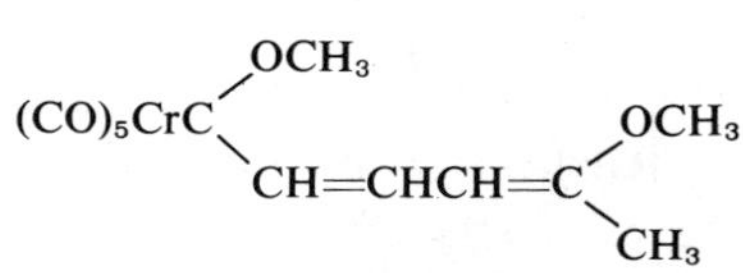

There are no reports to date of the addition of a hydride to alkoxycarbenoid chromium compounds to give the σ-bonded alkylchromium compounds.

Boron halides (BX_3, X = I, Br, Cl) react with the alkoxycarbenoid compounds **146** (R = CH_3, R′ = CH_3 and C_6H_5) to give novel carbyne-chromium compounds **176** (R′ = CH_3 and C_6H_5), Eq. (3-119) [303]. The

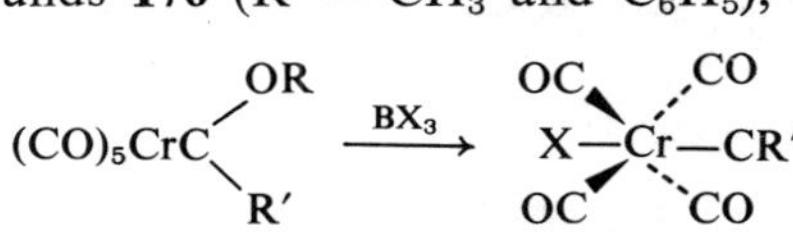

(3-119)

146 **176**

reaction involves the elimination of the alkoxy group and the replacement of one carbonyl group by a halogen atom. A single crystal x-ray structure analysis of the tungsten analog, $R' = C_6H_5$, $X = I$, showed that the halogen is trans to the "carbyne unit" [303].

Many of the reactions of alkoxycarbenoid compounds that have been discussed in the preceding pages imply that when such a unit is bonded to chromium it behaves like an organic ester group. This analogy is further illustrated in the observation that the hydrogens of the CH_3 group bonded to $C_{carbene}$ are acidic. The electron-withdrawing nature of the

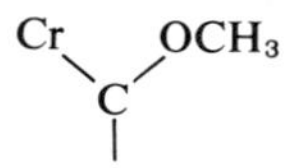

unit is such that the protons of the CH_3 group of **146** ($R = R' = CH_3$) are readily exchanged for deuterium when treated with $NaOCH_3/DOCH_3$, Eq. (3-120a) [304]. This observation implies, in turn, that methyl (or other

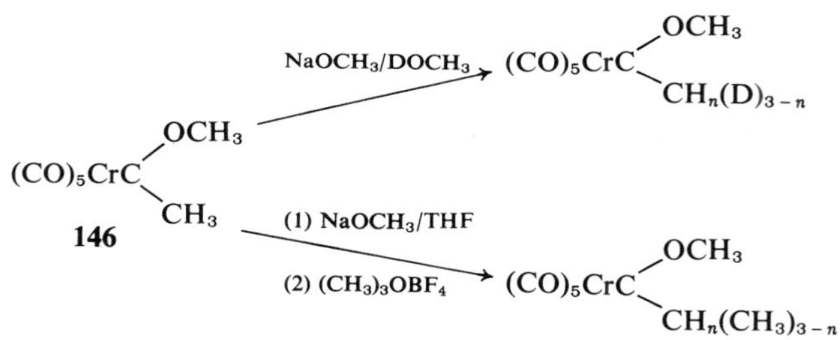

(3-120a)

(3-120b)

alkyl) alkoxycarbenoid-chromium compounds should undergo many of the base-catalyzed condensation reactions associated with CH_3–COOR or RCH_2COOR compounds, and it has been reported that the $C_{carbene}$-methyl group of **146** ($R = R' = CH_3$) may be methylated to give a mixture of C-methylated homologs, Eq. (3-120b) [245].

Several base-catalyzed condensation reactions involving the $C_{carbene}$-methyl group of the tungsten analog have been described, e.g., the methylation (to give 46% of the homologous ethyl complex) and condensation with benzaldehyde (to give the styryl complex), Eqs. (121a) and (121b) [305].

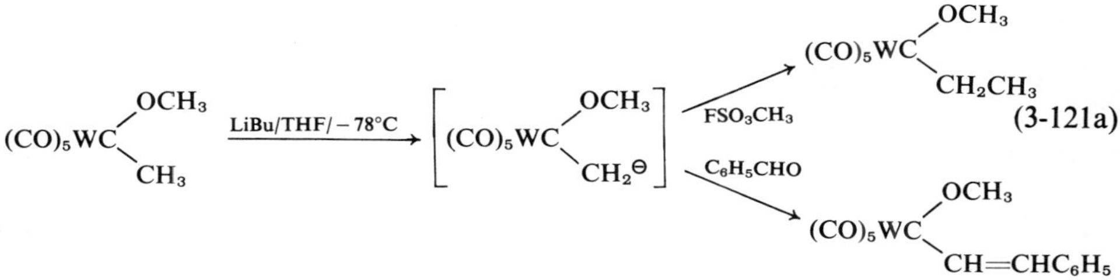

(3-121a)

(3-121b)

2. Reactions of Acetoxycarbenoid-Chromium Compounds

Connor and his colleagues have shown that the reactions of acetoxycarbenoid compounds of the type

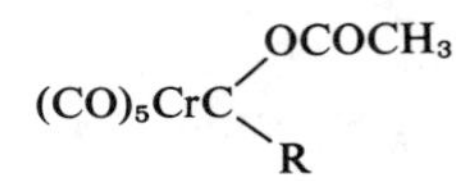

(**177**, Scheme 3-20) resemble those of an acid anhydride. Thus, reaction of **177** with alcohols, thiols, and amines gives the corresponding alkoxy-, thio-, and aminocarbenoid compounds, see Scheme 3-20. The reaction of **177** with hydrazoic acid gives the furanonitrile carbenoid complex **178**, Scheme 3-20; this reaction is assumed to proceed via the azido compound, which rearranges, with concomitant extrusion of N_2, to give **178** [284,306].

The corresponding cyclopropyl- and trimethylsilylmethylacetoxycarbenoid compounds could not be isolated from the reaction of the respective tetra-

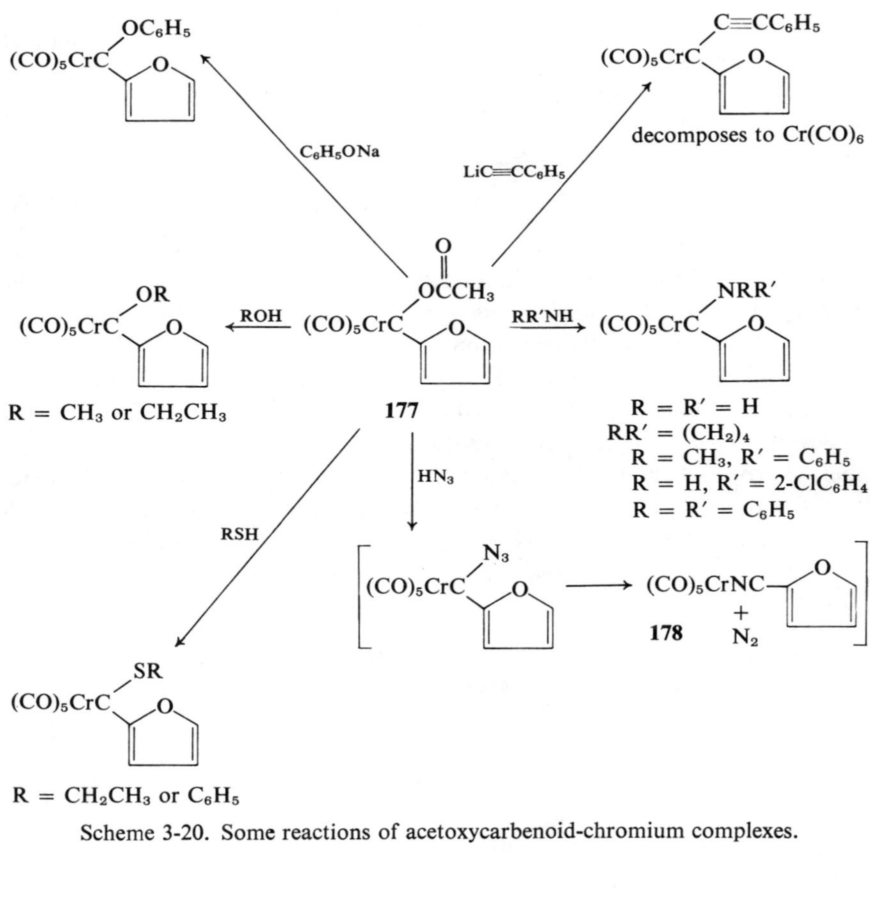

Scheme 3-20. Some reactions of acetoxycarbenoid-chromium complexes.

alkylammonium salts with acetyl chloride [c.f. Eq. (3-103)]. Their formation was inferred, however, from their further reaction with hydrazoic acid, which gives the corresponding nitrile and isonitrile complexes, Eqs. (3-122) and (3-123) [307]. The formation of these latter complexes is best interpreted

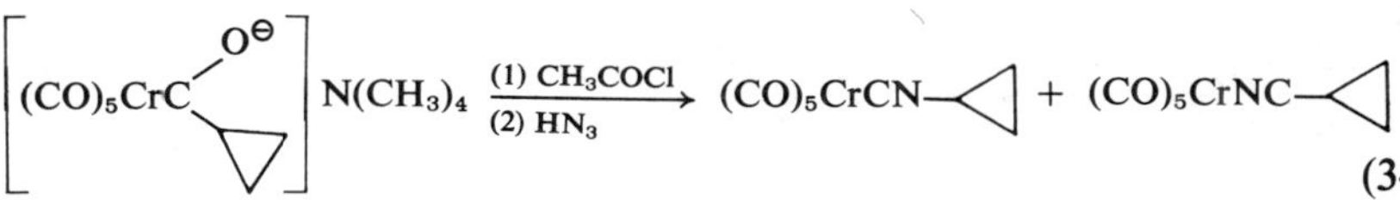

(3-122)

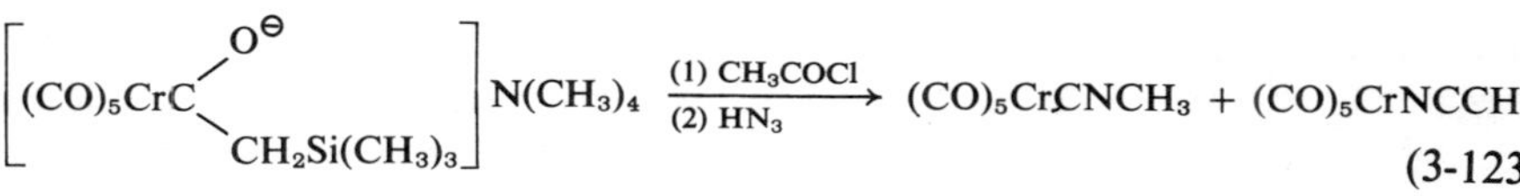

(3-123)

in terms of a mechanism whereby the tetraalkylammonium salts react with CH_3COCl to give the acetoxycarbenoid compounds (cf. Scheme 3-20). These react with HN_3 to give the azides, which in turn rearrange to the nitriles or isonitriles. In the case of the reaction of the trimethylsilylmethyl compound, the CH_2–Si bond is cleaved at some stage and the observed products are the methyl nitrile and isonitrile complexes instead of the corresponding $(CH_3)_3SiCH_2$ compounds.

The rearrangement of the azides **179** is believed to involve extrusion of N_2 to give the nitrenes **180**, which subsequently rearrange to either the nitriles or the isonitriles, Eq. (3-119):

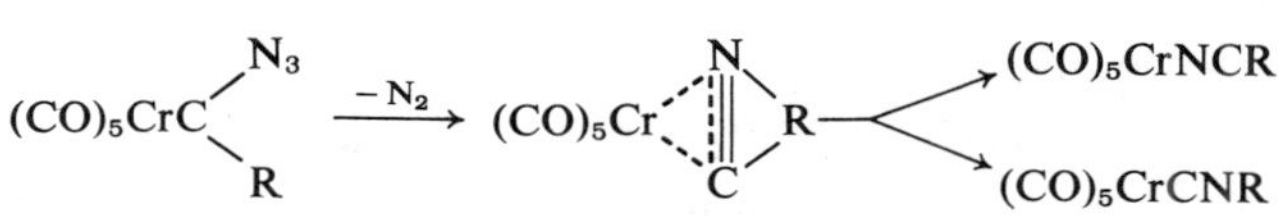

(3-124)

179 **180**

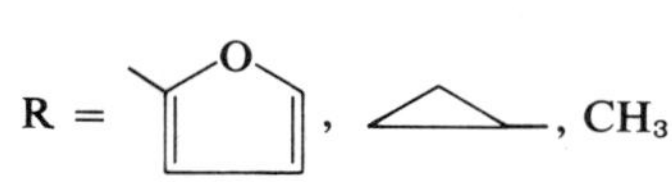

F. σ-Bonded Organochromium Compounds

There are five known main classes of stable σ-bonded organochromium compounds: mono(organo)chromium(III) compounds, bis(organo)chromium(III) compounds, tris(organo)chromium(III) compounds, mono(organo) chromium(II) compounds, and bis(organo)chromium(II) compounds (see Chapter 1, Tables 1.2, 1.3, 1.4, and 1.5). The stabilities and reactions (e.g., toward O_2 and H_2O) of these different types of compounds differ widely not only within a given type but also from one type to another.

The stability of an organochromium compound reflects its tendency to undergo fragmentation reactions, whereby the chromium-to-carbon bond is cleaved, and, therefore, it must be discussed in terms of the various pathways by which these reactions may occur. There are many such pathways and some of them have been discussed in review articles [308,309], but, as will be seen in the ensuing discussion, the pathways available for these fragmentations depend upon a number of factors, such as the following:

(i) Whether or not the fragmentation occurs in a mononuclear or a binuclear complex: the formation of a binuclear complex effectively brings the organic ligands on the individual chromium centers into a juxtaposition such that concerted hydrogen abstraction or dimerization processes may occur readily, e.g., Eqs. (3-128)–(3-130).

(ii) Changes in the geometry and in the coordination number of the chromium center occasioned by the acquisition or extrusion of a ligand; this may have the effect of either liberating or removing the coordination site necessary for certain concerted processes (e.g., metal-hydride elimination).

(iii) The relative ease with which the chromium center undergoes redox reactions [e.g., Cr(III) to Cr(II), (I), (0)]; this may have the effect of promoting a homolytic cleavage of a C–Cr bond since certain Cr(II) units [e.g., Cr(II)-Cl_2] are good "leaving groups," cf. Section 1,a of this section.

(iv) The nature of the organic group bonded to the chromium center; the presence or absence of hydrogen on the carbon atoms *in proximity to the chromium center* will determine whether the complex may undergo fragmentation by concerted metal-hydride elimination.

(v) The nature of the ligands associated with the chromium center; the presence of strong donor ligands will effectively block the coordination sites on the chromium that are required for concerted elimination reactions.

(vi) The propensity of chromium to abstract hydrogen from an organic ligand; this will have the effect of promoting fragmentation pathways that involve metal-hydride elimination or other hydrogen-transfer processes.

The chemistry of the five main classes of organochromium compounds has not been studied extensively and this may be due, in part, to the difficulties associated with their preparation and characterization. Furthermore, not all the reactions have been carried out with distinct crystalline compounds but have been effected with solutions of the organochromium compounds prepared *in situ*. However, it will be evident that many of the reactions that will be discussed reflect the tendency of organochromium compounds to act as sources of radicals, carbanions, and carbenoid units and others illustrate the capacity of chromium to act as a center for coordination syntheses and hydrogen transfer processes.

1. Reactions of σ-Bonded Mono(organo)chromium(III) Compounds

There are two main classes of σ-bonded mono(organo)chromium(III) compounds; the dichlorochromium(III) compounds $RCr(Cl)_2(S)_n$ and the cationic pentaaquochromium(III) complexes $[RCr(H_2O)_5]^{2\oplus}[X^{\ominus}]_2$. Both classes of compounds are sensitive to air (O_2), but only the former are sensitive to water. The chemical reactions that are discussed separately in the ensuing pages, reflect the tendency of the carbon-to-chromium bond to undergo both homolytic and heterolytic cleavage.

a. Mono(organo)chromium Dihalides

Several stable mono(organo)chromium(III) compounds of the type $RCrCl_2(THF)_n$ (n = 3 or 2) are known in which R = aryl, alkyl, benzyl, and diphenylmethyl (see Chapter 1, Table 1.2). The corresponding allyl compound has not been prepared and attempts to prepare the triphenylmethyl analog have led to the formation of solvated chromous chloride [310]. Similarly, attempts to prepare vinyl(dichloro)chromium(III) resulted, even at $-50°C$, in the formation of solvated $CrCl_2$, ethylene, acetylene, and butadiene [311]. The chemical reactions of these compounds have not been extensively investigated; thus, apart from their reactions with $HgCl_2$, H_2O, and I_2 (see Chapter 2, Section F) and the replacement of the THF ligands by pyridine (Chapter 1, Section F), the only other reaction that has been studied in any detail is their fragmentation.

All the known mono(organo)dichlorochromium(III) compounds undergo fragmentation on heating either as solids or as solutions in tetrahydrofuran to give chromous chloride and various organic products (Table 3.3). The latter depend upon the organic group originally bonded to the chromium center; thus when this is an aryl or an arylmethyl group, the major products are the biaryl or the bis(aryl)ethane (Nos. 5–8, Table 3.3); when it is an alkyl group (other than ethyl or methyl), the products are the corresponding alkane (70%) and alkene (30%) (Nos. 3, 4, 9, 10, Table 3.3); and when it is a methyl or ethyl group, the products are methane and ethane, together, with, however, appreciable quantities of the hydrocarbon dimer, see Table 3.3 [310,312–319].

The kinetics of the fragmentation reaction of all the compounds mentioned in Table 3.3, under comparable conditions, have not been studied. However the available data suggest that these reactions proceed *via* homolysis of the carbon-to-chromium bond and that the driving force is the formation of insoluble solvated $CrCl_2(THF)_2$. Zeiss and Sneeden found that the fragmentation of the specifically deuterated compound **181** proceeds as outlined in Eq. (3-125) [319].

Mass and NMR spectroscopic analysis of the organic products proved

TABLE 3.3

Thermal Decomposition of **$RCr(Cl)_2(THF)_n$**

	R	TEMPERATURE (°C)	PRODUCTS INORGANIC	PRODUCTS ORGANIC			REF.
1.	CH_3[a,b]	85	$CrCl_2$	CH_4 (80.6)	CH_3CH_3 (18.9)	$CH_2{=}CH_2$ (0.5)	312
2.	CH_3CH_2[a,b]	70	$CrCl_2$	CH_3CH_3 (48.1)	$CH_2{=}CH_2$ (8.5)	C_4H_{10} (43.4)	312
3.	$CH_3CH_2CH_2$[a]	40	$CrCl_2$	$CH_3CH_2CH_3$ (71.5)	$CH_3CH{=}CH_2$ (22.9)	C_6H_{14} (5.9)	312
4.	$(CH_3)_2CHCH_2$[a]	30	$CrCl_2$	i-C_4H_{10} (77.5)	i-C_4H_8 (22.3)	C_8H_{18} (trace)	312
5.	4-$CH_3C_6H_4$	130–190	$CrCl_2$	$(4\text{-}CH_3C_6H_4)_2$ + some $C_6H_5CH_3$			313
6.	C_6H_5[a,b]	150	$CrCl_2$	$(C_6H_5)_2$ + some C_6H_6			314
7.	$(C_6H_5)_2CH$[a,b]	110	$CrCl_2$	$[(C_6H_5)_2CH]_2$			310
8.	$C_6H_5CH_2$[c]	~20	$CrCl_2(THF)_2$	$(C_6H_5CH_2)_2$ + some $C_6H_5CH_3$			315–317
9.	$CH_3(CH_2)_8CH_2$[c]	~30	$CrCl_2(THF)_2$	$CH_3(CH_2)_8CH_3$ (75)	$CH_3(CH_2)_7CH{=}CH_2$ (24)		318
10.	$C_6H_5(CH_2)_2CD_2CH_2$[c]	~30	$CrCl_2(THF)_2$	$C_6H_5(CH_2)_2CD_2CH_3(D)$ (73)	$C_6H_5(CH_2)_2CD{=}CH_2$ (22)		319

	R	TEMPERATURE (°C)	INORGANIC PRODUCTS	RATE CONSTANTS	REF.
11.	CH_3[c]	22.5	$CrCl_2$	$k \times 10^{-4} = 0.22$	312
12.	CH_3CH_2[c]	7.0	$CrCl_2$	$k \times 10^{-4} = 0.098$	312
13.	$CH_3CH_2CH_2$[c]	22.5	$CrCl_2$	$k \times 10^{-4} = 1.22$	312

[a] Thermal decomposition of the solid. Figures in parentheses represent relative percentages.
[b] These compounds may be dissolved in tetrahydrofuran at 50°–60°C for the purpose of recrystallization.
[c] Thermal decomposition of solutions of the compounds in tetrahydrofuran.

that all of the deuterium lost in the formation of the olefin is found as trideuterioalkane [319], furthermore no hydrogen or chromium hydride was, formed either in the present or in analogous fragmentation reactions [312, 318,319]. The authors interpreted their results in terms of a reaction whereby the carbon-to-chromium bond is cleaved, homolytically, and the resulting "radical" acquires hydrogen from solvent or from the β-position of another alkyl group.

$$\underset{\mathbf{181}}{C_6H_5(CH_2)_2CD_2CH_2Cr(Cl)_2S_n} \xrightarrow[(2)\ H_2O]{(1)\ \Delta/THF} \begin{array}{ll} C_6H_5(CH_2)_2CD_2CH_3 & (50\%) \\ C_6H_5(CH_2)_2CD_2CH_2D & (22\%) \\ C_6H_5(CH_2)_2CD=CH_2 & (22\%) \end{array} \quad (3\text{-}125)$$

The data available concerning the fragmentation of mono(organo)-chromium(III) compounds can be rationalized in terms of the homolysis of the C–Cr bond either in a mononuclear complex, Eq. (3-126) (R = aryl or benzyl) and Eq. (3-127), or in a dinuclear complex, Eq. (3-128) (R = aryl or benzyl) and Eq. (3-129). The results with the specifically deuterated

$$RCr(III)Cl_2(THF)_3 \longrightarrow [R\cdot Cr(II)Cl_2(THF)_2] \begin{cases} \rightarrow \text{coupling} \rightarrow RR \\ \rightarrow \text{H-abstraction} \rightarrow RH \end{cases} \quad (3\text{-}126)$$

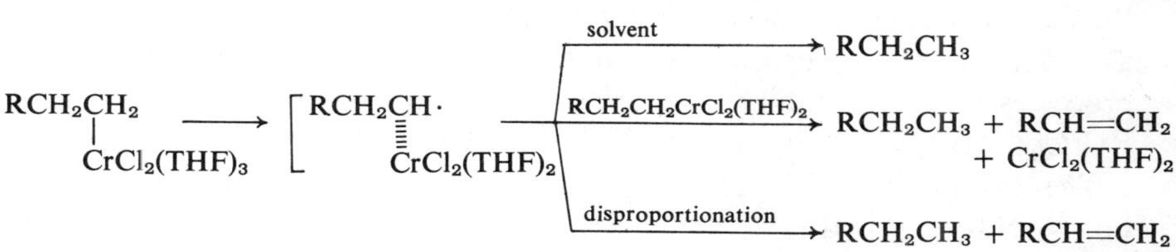

(3-127)

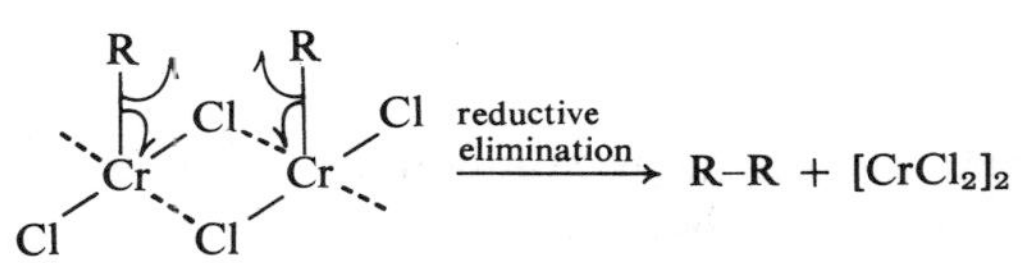

(3-128)

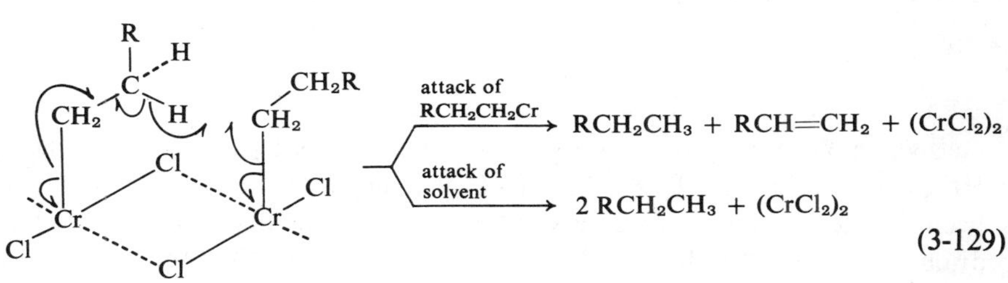

(3-129)

compound **181**, outlined in Eq. (3-125), show that in this case the fragmentation proceeds via the abstraction of the β-hydrogen of the alkyl group bonded to chromium (to give the alkane and alkene directly). However, in other cases the fragmentation could proceed by the abstraction of the α-hydrogen to give the alkane and a "carbene," which may rearrange to the olefin or in the absence of a β-hydrogen, may dimerize to the alkene, e.g., Eq. (3-130). Such a mechanism could account for the formation of ethylene in the fragmentation of $CH_3Cr(Cl)_2(THF)_3$. The fragmentation reactions in the bi-

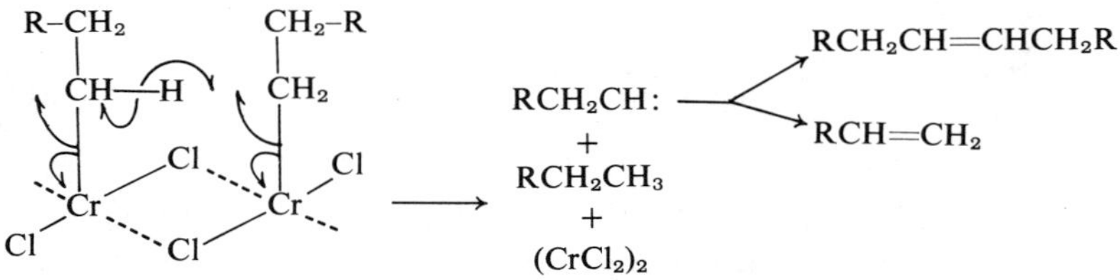

(3-130)

nuclear complexes outlined in Eqs. (3-128)–(3-130) are all concerted processes and therefore would be expected to proceed more readily than the simple homolytic reactions in the mononuclear compounds outlined in Eqs. (3-126) and (3-127).

It is not possible to decide from the evidence available whether the fragmentation of all the compounds of the type $RCrCl_2(THF)_3$ proceed in mononuclear compounds or via the binuclear complexes. However, were the fragmentation to proceed via the simple homolysis of the C–Cr bond in a mononuclear complex then in analogy with other unimolecular homolytic processes [320] the stability sequence expected for compounds of the type $RCrCl_2(THF)_3$ would be aryl > vinyl > methyl > alkyl > benzyl ≥ allyl.

The mono(alkyl)-dichloro compounds alone and in combination with Lewis acids (e.g., $AlBr_3$ or $SnCl_4$) are reported to effect the polymerization of ethylene [312,321].

b. Mono(organo)chromium(III) Pentaaquates

There are several organochromium pentaaquo compounds of the general type $[RCr(H_2O)_5]^{2\oplus}$ known where R is benzyl or substituted benzyl {e.g., $C_6H_5CH_2$; $C_6H_5CH(R)$; $XC_6H_5CH_2$ [322,323]}, pyridiomethyl [324–327],

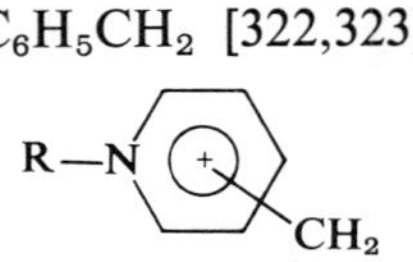

methyl, and substituted methyl {e.g., CH_3, X_nCH_{3-n} (X = halogen), HOCRR, $ROCHCH_3$ [328–330]}. However, none of these compounds has been isolated in the form of a crystalline salt and, therefore, the ensuing discussion refers to the stabilities and reactions of solutions (usually in $HClO_4$) of these compounds.

The majority of the known mono(organo)chromium pentaaquates are moderately stable, in the absence of halide ion, to fragmentation and hydrolysis reactions but they all react rapidly with oxygen. The chemical reactions of these compounds (see Scheme 3-21) reflect their capacity to act as sources of organic radicals or anions.

i. Hydrolysis. Kochi and Buchanan have studied the decomposition of solutions of $[C_6H_5CH_2Cr(H_2O)_5]^{2\oplus}$ and find this reaction to be highly sensitive to the presence of oxygen and of certain anions [331]. The protonolysis of benzylchromium under anaerobic conditions, in perchloric acid, and in the presence of acetate buffer follows first-order kinetics (activation parameters $\Delta H^* = 16.2$ kcal/mole and ΔS^* -13.9 e.u.). These authors have proposed a two-step mechanism for this reaction. The first step, is a preequilibrium reaction with a nucleophile $X^{\ominus}$ (e.g., acetate) and leads to an intermediate complex **182**, Eq. (3-131), in which the effective positive charge on the metal is reduced and the benzyl group is rendered more labile by the presence of the anion $X^{\ominus}$. The second step is rate determining and is the protonolysis of the benzyl group, Eq. (3-131) [331]. This mechanism explains

$$C_6H_5CH_2Cr^{2\oplus} \underset{}{\overset{X^{\ominus}}{\rightleftharpoons}} \underset{\textbf{182}}{[C_6H_5CH_2CrX]^{\oplus}} \xrightarrow{H^{\oplus}} C_6H_5CH_3 + CrX^{2\oplus} \qquad (3\text{-}131)$$

Anet and Leblanc's observation that whereas benzyl chloride reacts with $Cr(ClO_4)_2/HClO_4$ to give benzylchromium, it reacts with $CrCl_2/HCl$ to give toluene since, in the latter reaction the protonolysis (or "decomposition") of the organochromium compound is promoted by the presence of the chloride anion in the complex $[C_6H_5CH_2CrCl(OH_2)_4]^{\oplus}$ [322].

Schmidt, Swinehart, and Taube have studied the kinetics of the protonolysis of several substituted methylchromium pentaaquates of the type $[(H_2O)_5\text{-}CrC(OH)RR']^{2\oplus}$ in perchloric acid. The reactions follow first-order kinetics (activation parameters ΔH^*, 17.1 kcal/mole and ΔS^*, -11 e.u.) and are assumed to proceed via a solvate **183**, Eq. (3-132) in which the proton and the carbon atom bonded to chromium are so disposed as to facilitate the extrusion of the neutral organic fragment Eq. (3-132) [330].

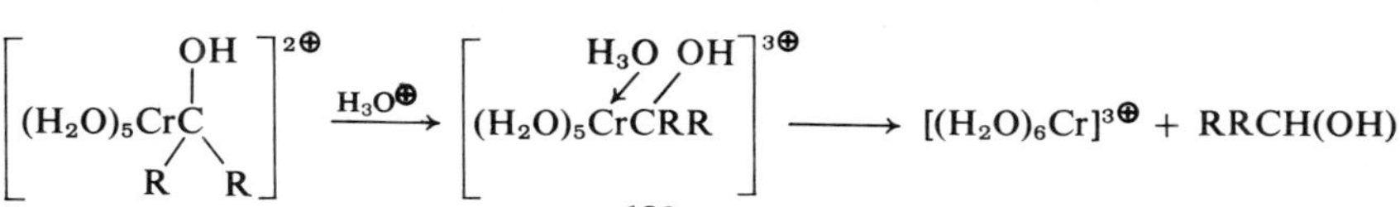

183 (3-132)

Coombes and Johnson have studied the kinetics of the protonolysis of the isomeric pyridiomethylchromium pentaaquates in perchloric acid under aerobic conditions. These reactions also followed first-order kinetics (activation parameters ΔH^*, 30–40 kcal/mole; ΔS^*, up to 37 e.u.), the products, however, contained the corresponding pyridine aldehydes, pyridines, and sometimes the 1,2-bis(pyridyl)ethanes. The overall reactions are, therefore,

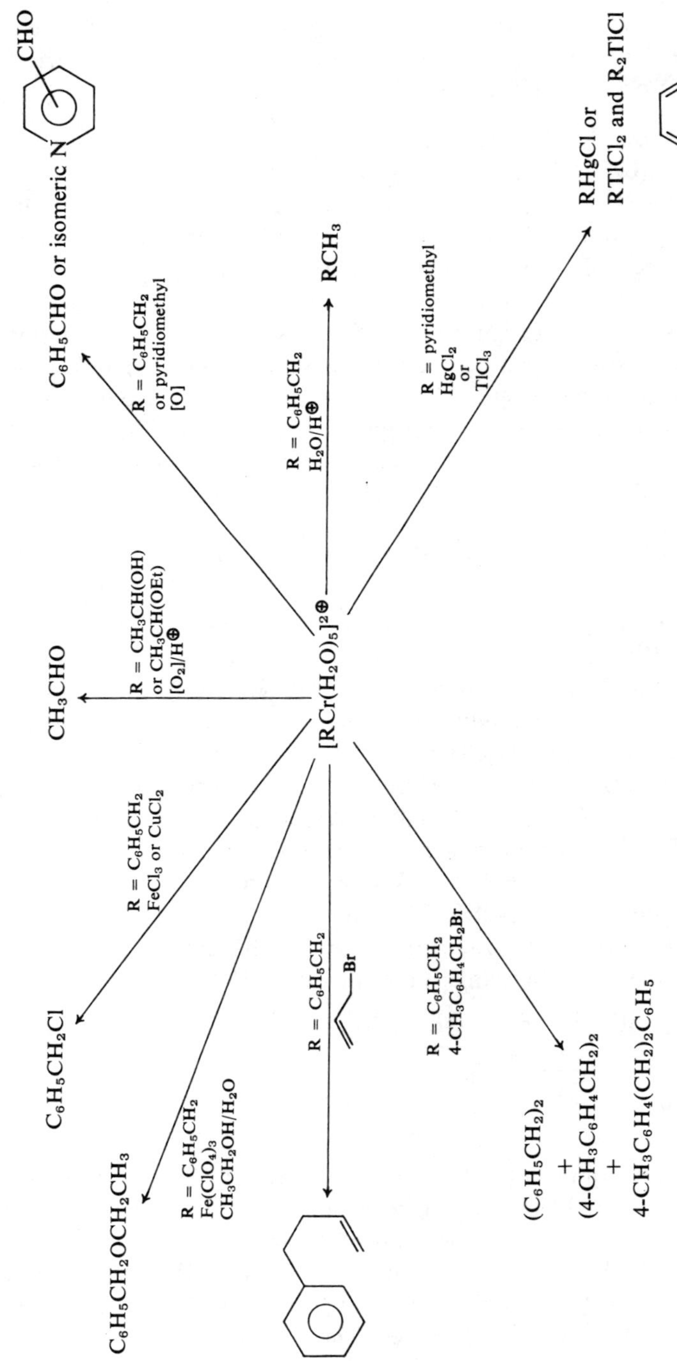

Scheme 3-21. Some reactions of mono(organo)chromium(III) pentaaquates. Pyridiomethyl represents the isomers of 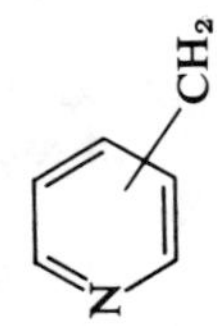

rather complex; however, the authors have suggested that they proceeded via the homolysis of the carbon-to-chromium bond and the radical thus engendered abstracts hydrogen (from coordinated H_2O), reacts with O_2, or attacks a second molecule of organochromium compound as outlined in Eq. (3-133) [327].

$$[pyCH_2Cr(H_2O)_5]^{2\oplus} \longrightarrow pyCH_2\cdot \; \cdot Cr(II)(H_2O)_5 \quad \begin{cases} \xrightarrow{H_2O} pyCH_3 \\ \xrightarrow{[O]} pyCHO \\ \xrightarrow{[pyCH_2Cr]} pyCH_2CH_2py + Cr(II) \end{cases} \quad (3\text{-}133)$$

py = 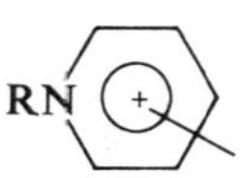

ii. Reactions with $HgCl_2$ *and* $TlCl_3$. Coombes and Johnson have found that the kinetics of the reaction of $HgCl_2$ with the isomeric pyridiomethyl-chromium compounds are in accord with an S_E2 mechanism in which Cr(III) is displaced at carbon by Hg(II) {see Chapter 2, Eq. (2-23) [226]}.

The analogous reaction with thallium(III) chloride is more complex since at high reagent concentrations the product is the bis(organo)thallium halide while at low reagent concentrations the product is the mono(organo)-thallium halide. Johnson and his collaborators have found that, at low reagent concentrations, the reactions proceed by a bimolecular displacement pathway (S_E2) analogous to that already outlined for the reaction with $HgCl_2$, i.e., Eq. (3-134) [332].

$$[RCH_2Cr(H_2O)_5]^{2\oplus} + TlCl_2^{\oplus} \longrightarrow \begin{matrix} Cl_2Tl^{\oplus} \cdots CH_2 \cdots Cr(OH_2)_2 \\ | \\ R \end{matrix} \longrightarrow \begin{matrix} Cl_2TlCH_2R \\ + \\ [Cr(H_2O_5]^{3\oplus} \end{matrix} \quad (3\text{-}134)$$

R = 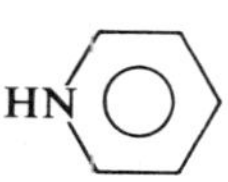

iii. Oxidation Reactions. All the organochromium pentaaquates react with oxygen to give by oxidative cleavage of the C–Cr bond, the corresponding aldehyde, Scheme 3-21 [316,322,327,330,333]. The oxidation of the ethanol and diethyl ether chromium compounds **184**, R = CH_3, R′ = H or R = CH_3, R′ = CH_2CH_3, may proceed via the hydrate or hemiacetal **185**, R = CH_3, R′ = H, or R = CH_3, R′ = CH_2CH_3, as outlined in Eq. (3-135) [330].

$$\underset{\mathbf{184}}{\left[\begin{matrix} R{-}CHOR' \\ | \\ Cr(OH_2)_5 \end{matrix}\right]^{2\oplus}} \xrightarrow{[O_2]} \underset{\mathbf{185}}{RCH(OR')(OH)} + [HOCr(OH_2)_5]^{2\oplus} \xrightarrow{H^{\oplus}} RCHO \quad (3\text{-}135)$$

Benzylchromium pentaaquate reacts with a variety of metal salt oxidants (e.g., $FeCl_3$, $CuCl_2$, $CuSO_4$) and the products depend upon the solvent(s) and metal salt used. Thus, with $FeCl_3$ and $CuCl_2$ in methanol/water the organic product is mainly benzyl chloride together with some bibenzyl [316, 331] and with $Fe(ClO_4)_3$ and $CuSO_4$ in ethanol/water the product is benzyl ethyl ether [331]. A two-step mechanism has been proposed for these oxidations, i.e., Eq. (3-136) ($R = C_6H_5CH_2$). The reaction of $FeCl_3$, in which

$$[RCr(OH_2)_5]^{2\oplus} \xrightarrow{FeCl_3} R\cdot + [CrCl]^{2\oplus} + FeCl_2 \xrightarrow{FeCl_3} RCl + [CrCl]^{2\oplus} + 2\,FeCl_2 \qquad (3\text{-}136)$$

precisely two equivalents of $FeCl_2$ are produced from one equivalent of organochromium compound, constitutes an excellent method for the quantitative estimation of the benzylchromium(III) pentaaquate cation. It is of interest to note that the σ-bonded arylchromium(III) dichlorides [e.g., $4\text{-}CH_3C_6H_4CrCl_2(THF)_3$] do not reduce $FeCl_3$ [313].

iv. Reactions with Organic Halides and Halogens. Kochi and Buchanan have found that benzylchromium reacts with a variety of benzylic halides to give mixtures of toluenes, symmetrical bibenzyls, and unsymmetrical bibenzyls, e.g., for the reaction with *p*-xylyl bromide, Eq. (3-137). A similar

$$[C_6H_5CH_2Cr(OH_2)_5]^{2\oplus} \xrightarrow{4\text{-}CH_3C_6H_4CH_2Br} C_6H_5CH_3 + 4\text{-}CH_3C_6H_4CH_3 + (C_6H_5CH_2)_2 + 4\text{-}CH_3C_6H_4(CH_2)_2C_6H_5 + (4\text{-}CH_3C_6H_4CH_2)_2 \qquad (3\text{-}137)$$

distribution of products results from the interaction of the individual benzylchromium compounds. The overall reaction of benzylic halides with benzylchromium compounds is therefore composed of two processes: The first is a ligand-exchange process whereby a benzylic halide reacts with benzylchromium to give a new benzylchromium complex, e.g., Eq. (3-138) and the

$$4\text{-}CH_3C_6H_4CH_2Br + [C_6H_5CH_2Cr(OH_2)_5]^{2\oplus} \rightleftharpoons [4\text{-}CH_3C_6H_4CH_2Cr(OH_2)_5]^{2\oplus} + C_6H_5CH_2Br \qquad (3\text{-}138)$$

second is the interaction of a benzylic halide with the organochromium compound to give the bibenzyl, e.g., Eq. (3-139), $R = H$ or CH_3, $R' = H$ or CH_3 (when $R = R' = H$ or CH_3, symmetrical products are formed; when $R = H$ or CH_3 and $R' = CH_3$ or H, unsymmetrical products are formed [331]).

$$[RC_6H_4CH_2Cr(OH_2)_5]^{2\oplus} + R'C_6H_4CH_2Br \longrightarrow RC_6H_4CH_2CH_2C_6H_4R' + [CrBr]^{2\oplus} \qquad (3\text{-}139)$$

Slaugh and Raley have found that allylic halides react with benzylchromium to give 4-phenylbutene-1, without the formation of 1,5-hexadiene or bibenzyl [334].

Espenson and Chang report that benzylchromium pentaaquate reacts with bromine or iodine to give the corresponding benzyl halide. The kinetics of this reaction are in accord with an S_E2 mechanism in which the halogen electrophile attacks the carbon bonded to the chromium [335].

v. Polymerization Reactions. Lunk and Youngman have found that benzylchromium in perchloric acid polymerizes styrene and methylmethacrylate [336]. These polymerizations are probably radical initiated processes, since benzylchromium reacts with acrylonitrile (to give γ-phenylbutyronitrile) and with butadiene (to give phenylpentene-1) [323]. The related benzyldichloro-tris(pyridine)chromium(III) reacts with 1,3- or 1,4-cyclohexadiene to give toluene and benzene [316], and again this reaction is presumed to be a radical initiated process in which a benzyl radical abstracts H· from the substrate to give the cyclohexadienyl radical.

2. Reactions of Bis(organo)chromium(III) Compounds

The few known bis(organo)chromium(III) compounds (see Chapter 1, Table 1.3) are either neutral compounds of the type $R_2CrCl(S)_n$ or cationic complexes of the type $[R_2Cr(bipy)_2]^{\oplus}X^{\ominus}$. There are claims that bis(alkyl)- [337] and bis(benzyl)- [315] chlorochromium compounds have been prepared in solution; however, the evidence available to date suggests that simple solvated bis(organo)chromium(III) compounds are unstable and, in the absence of strong donor ligands, undergo either disproportionation [to $RCrCl_2(S)_n$ and $R_3Cr(S)_n$] or fragmentation reactions. The known stable and crystalline bis(organo)chromium(III) compounds either have strong donor ligands (e.g., 1,2-dimethoxyethane or 2,2′-bipyridyl) associated with the chromium center or are dimeric complexes {e.g., $[(CH_2{=}CHCH_2)_2CrX]_2$} [338].

The chemistry of these compounds has not been explored extensively and the little published information available concerns their reactions with $HgCl_2$, I_2, and protic solvents, and their use either alone or in combination with aluminum compounds in the oligomerization of olefins and dienes.

a. *Reaction with* $HgCl_2$ *and* H_2O

Both the neutral and the cationic compounds react rapidly with $HgCl_2$ (to give the corresponding organomercurial); in contrast, only the neutral compounds $(R_2CrCl(S)_x)$ react with H_2O to give, by cleavage of the carbon-to-chromium bond, the arene [339–341]. These differences in reactivities may be readily explained in terms of the mechanisms of the reactions and of the nature of the ligands associated with the chromium center [340].

The reaction of $HgCl_2$ with an organochromium compound is an electrophilic substitution at the carbon bonded to the chromium center (Chapter 2, Section F,2,b,i) and therefore the nature of the ligands associated with the chromium center (in particular the strength of the Cr ligand bond) will not usually inhibit this reaction. The hydrolysis of organochromium compounds, on the other hand, is a two-step process. The first step consists of the replacement of all or part of the solvating ligands attached to chromium by water, and the second step is the transfer of hydrogen either from coordinated H_2O or from a second molecule of H_2O to the organic ligand bonded to chromium, e.g., Eq. (3-140). This transfer is facilitated by the increased lability of the hydrogen in a water molecule that is coordinated to the chromium center.

$$R_nCrCl_{3-n}(S)_3 \xrightarrow{H_2O} R_nCrCl_{3-n}(H_2O)_3 \longrightarrow n\,RH + (HO)_nCrCl_{3-n}(H_2O)_{3-n} \tag{3-140}$$

The nature of the ligands associated with the chromium center will have a profound influence on the first step of the hydrolysis process. Thus, relatively labile ligands [e.g., the 1,2-dimethoxyethane in the neutral complexes $R_2CrCl(DME)$] will be readily replaced by water, and hydrolysis will occur, whereas the less labile ligands (e.g., 2,2′-bipyridyl in the cationic complexes) are not displaced by water and so hydrolysis will not occur.

b. Polymerization Reactions

There are reports that bis(alkyl)chromium(III) chlorides in combination with organoaluminum compounds or $AlBr_3$ polymerize ethylene to a polyethylene of molecular weight 200,000 to 800,000 [337] and that bis(allyl)-chromium iodide dimer acts as a catalyst in the cyclotrimerization of butadiene (to give cyclododecatriene) and that in combination with organoaluminum compounds it polymerizes ethylene [338]. The precise nature of the active catalyst in these reactions has not been established, however.

3. Reactions of Tris(organo)chromium(III) Compounds and Systems

Not all the tris(organo)chromium(III) compounds [$R_3Cr(THF)_n$] listed in Table 1.4, Chapter 1, have been isolated as crystalline solids, and those that have been isolated in a crystalline state (i.e., where R = aryl, allyl, and norcamphyl) will be referred to in the ensuing discussion as tris(organo)-chromium compounds, while those that have only been prepared in solution (i.e., where R = vinyl, benzyl, and alkyl) will be referred to as tris(organo) chromium systems. Both organochromium(III) compounds and systems react rapidly with $HgCl_2$ and I_2 (see Chapter 2, Section F,2,b,i) and most of them react readily with water and oxygen; the exceptions are compounds of the type $R_3Cr(III)$ with bidentate organic groups, i.e.,

R = 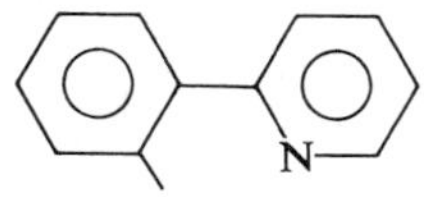 [342] 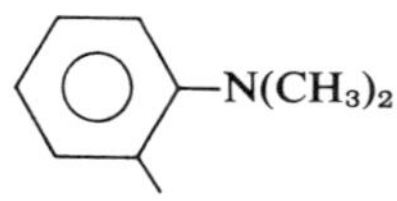 [343]

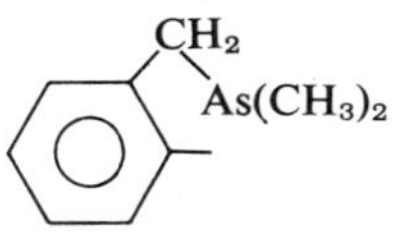 [344] 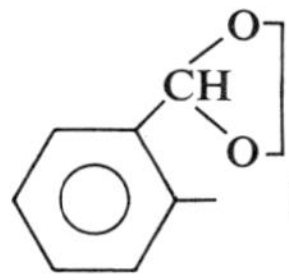 [345]

Curiously, the compound where

R = 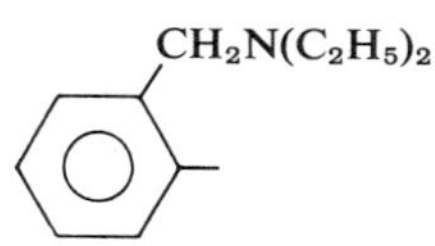

(which is monomeric in benzene) is reported to be highly reactive toward both air and moisture [346].

The chemistry of σ-bonded tris(organo)chromium compounds and systems has been extensively investigated and in particular their reactions with unsaturated substrates (e.g., acetylenes, dienes, CO, esters, etc.) and their fragmentation reactions. The results of these investigations have clarified considerably the many pathways whereby organochromium compounds may undergo fragmentation, and they also have revealed the capacity of chromium to act as a center for coordination syntheses and hydrogen-transfer reactions.

a. Reactions with Acetylenes

Zeiss and Herwig found that $(C_6H_5)_3Cr(THF)_3$ reacted with 2-butyne to give 1,2,3,4-tetramethylnaphthalene, hexamethylbenzene and π-bis(hexamethylbenzene)chromium [347,348]. This reaction was subsequently extended to other organochromium compounds of the type $R_3Cr(THF)_n$ (R = aryl, allyl, alkyl, and vinyl) and other acetylenes (see Table 3.4) [311,347,348, 350–361]. Most of these reactions have been effected with organochromium compounds that have been prepared *in situ*. However, in the case of the reactions of tris(phenyl)- tris(4-methylphenyl)- and tris(2-methoxyphenyl)-chromium with 2-butyne [348–350], control experiments have established that the chromium compound that had been prepared *in situ* gave the same products as did the crystalline tris(organo)chromium compounds, thereby proving that these products are indeed formed by coordination syntheses on the chromium center.

TABLE 3.4

***Products Formed in the Reaction of Tris(organo)chromium Compounds and Systems* (R_3Cr) *with Acetylenes* $R'C{\equiv}CR'$**

R in $R_3Cr(S)_x$	R′ in $R'C{\equiv}CR'$	Products	Ref.
1. C_6H_5	H	*trans*-$C_6H_5(CH{=}CH)_nC_6H_5$ $n = 1, 2, 3$ RR	351
2. C_6H_5	$COOCH_3$	*cis*-$C_6H_5(R')C{=}C(R')C_6H_5$	352
3. C_6H_5	CF_3	polymer	353
4. C_6H_5	CH_3	$(CH_3)_6C_6$ 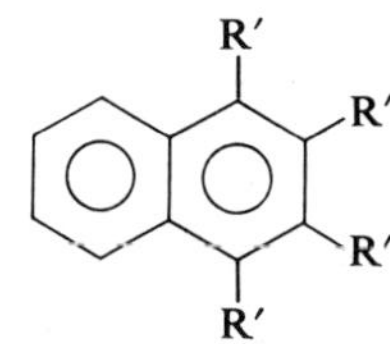 $\pi[(CH_3)_6C_6]_2Cr$	347 348 354
		cis-$C_6H_5(CH_3)C{=}CH(CH_3)$	
5. C_6H_5	C_6H_5	$(C_6H_5)_6C_6$ 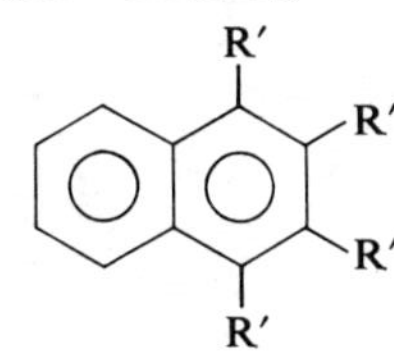	348 355
		$C_6H_5(R')C{=}C(R')C_6H_5$ 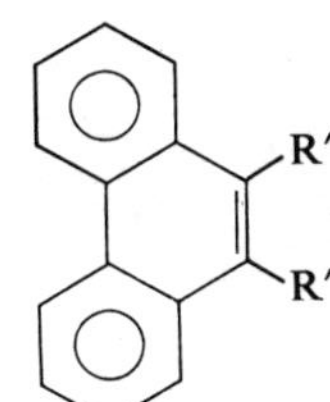	

6. 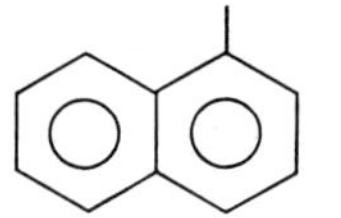	CH_3	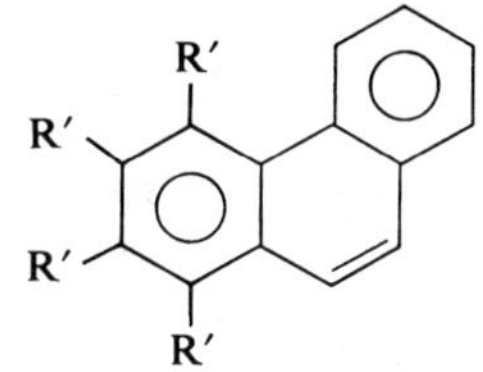	348
7. 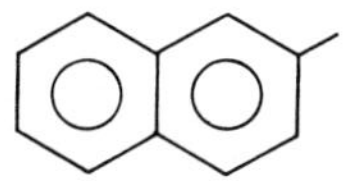	CH_3	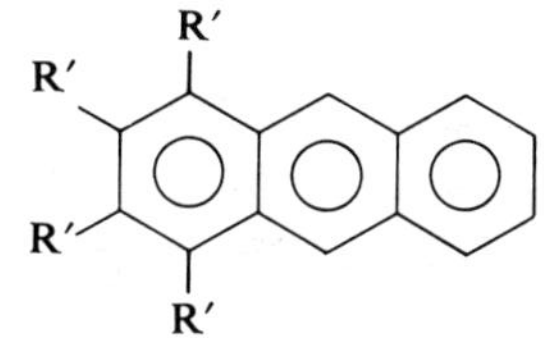 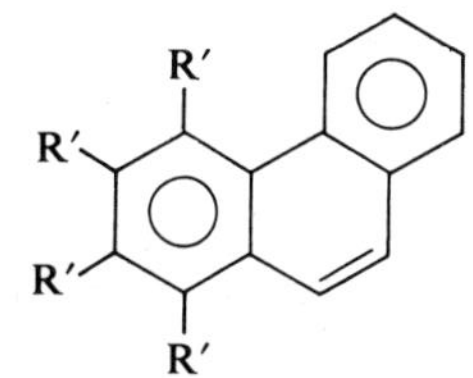	348
8. $4\text{-}CH_3C_6H_4$	CH_3	$(CH_3)_6C_6$ 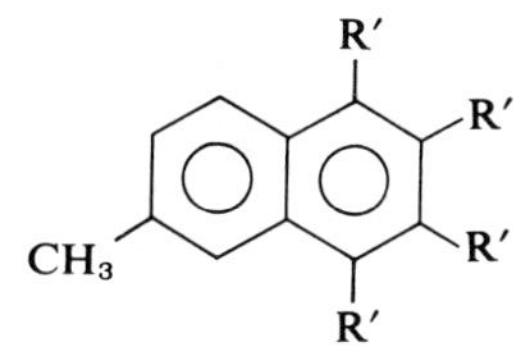 $(4\text{-}CH_3C_6H_4)_2$	349 354
9. $3\text{-}CH_3C_6H_4$	CH_3	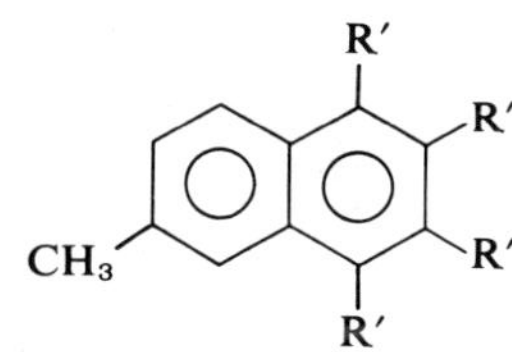 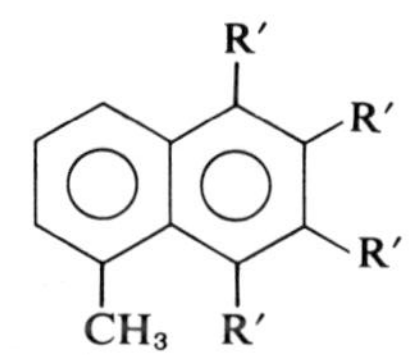	354
10. $2\text{-}CH_3C_6H_4$	CH_3	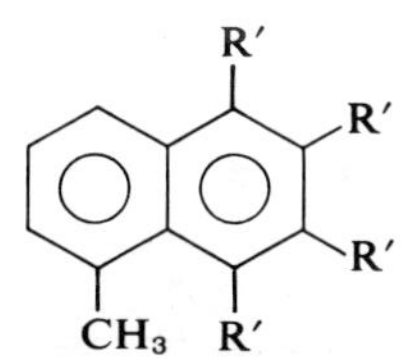	354

(continued)

TABLE 3.4 (*continued*)

	R in $R_3Cr(S)_x$	R′ in R′C≡CR′	Products	Ref.
11.	2-$CH_3C_6H_4$	C_6H_5	R(R′)C═C(R′)R	355
12.	4-$CH_3OC_6H_4$	CH_3	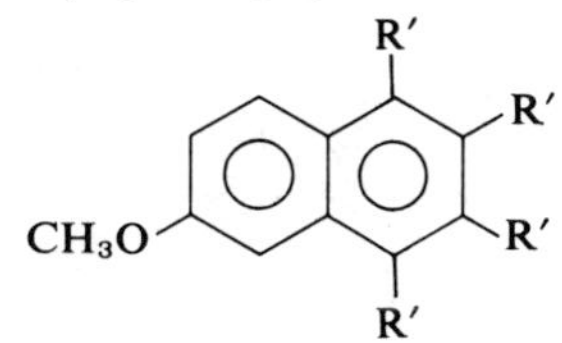 *cis*- + *trans*-R(R′)C═CH(R′) $(CH_3)_6C_6$ *cis*- + *trans*-R(R′)C═C(R′)R + RR	350
13.	3-$CH_3OC_6H_4$	CH_3	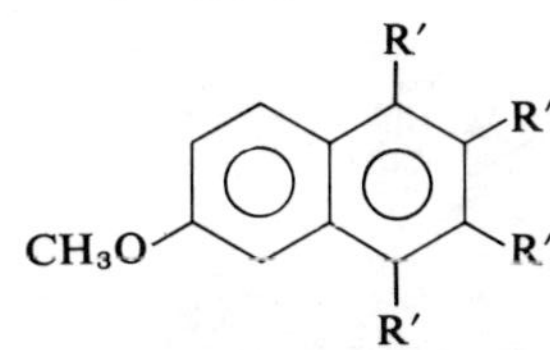 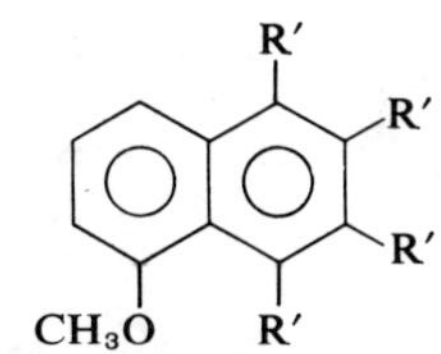 *cis*- + *trans*-R(R′)C═CH(R′) + *cis*- + *trans*-R(R′)C═C(R′)R + RR	350
14.	2-$CH_3OC_6H_4$	CH_3	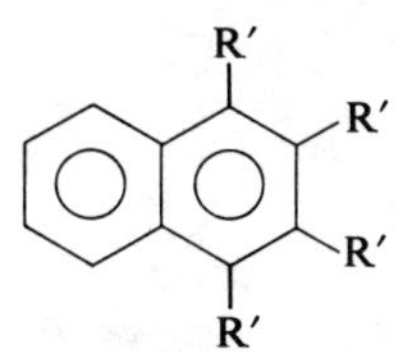 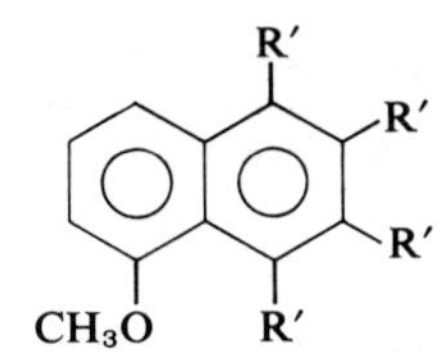 *cis*-R(R′)C═CH(R′) + *cis*-R(R′)C═C(R′)R + RR + RRCr (II)	350 356

15. 2-Cl-C_6H_4	C_6H_5	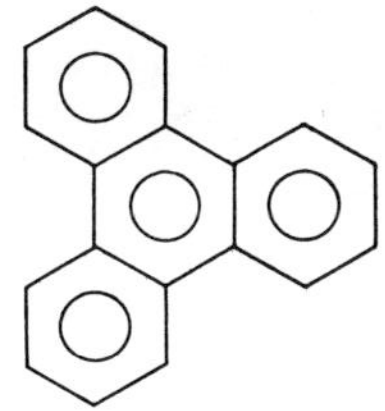	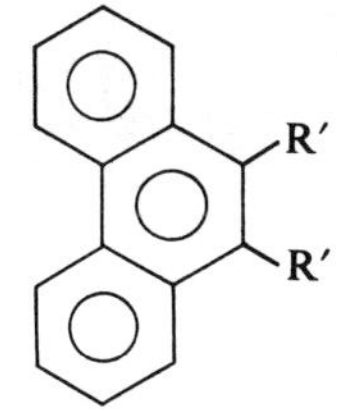	357
	C_6H_5, CH_3	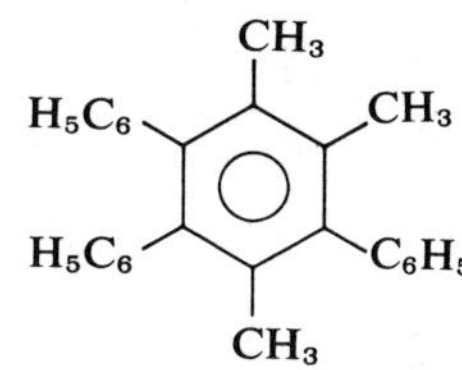	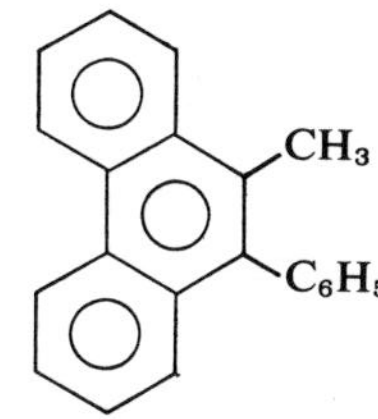	357
16. 2,4,6-$(CH_3)_3C_6H_2$	CH_3	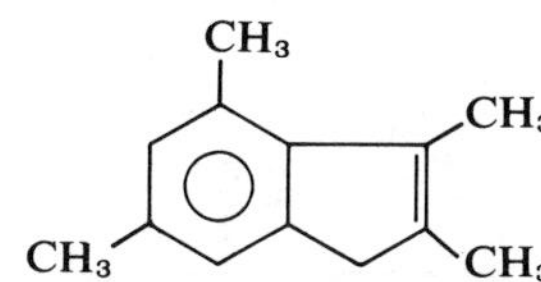	*cis*-R(R′)C=CH(R′)	352 354
16a. $(Mes)_2Cr(II)S_x$ [a]	C_6H_5	R'_6C_6 R′CH=CHR′		358
17. $C_6H_5CH{=}CH$	C_6H_5, H	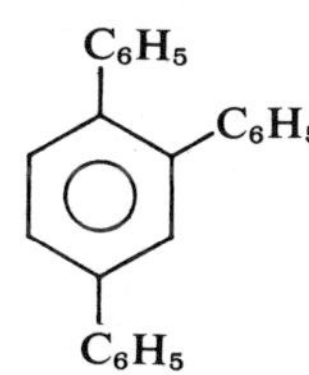	*trans*, *trans*-RR	359

(*continued*)

TABLE 3.4 (*continued*)

R in $R_3Cr(S)_x$	R′ in $R'C{\equiv}CR'$	Products		Ref.
18. $C_6H_5CH{=}CH$	C_6H_5, CH_3	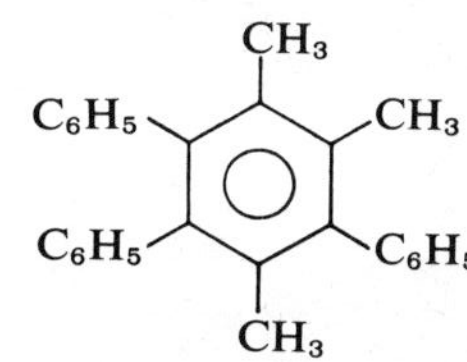	*trans*, *trans*-RR	359
19. $(C_6H_5)_2C{=}CH$	CH_3	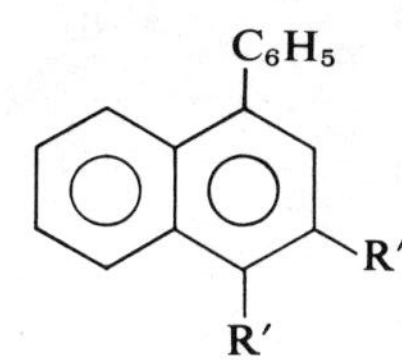	RR	354
20. $CH_2{=}CH$	CH_3 or C_6H_5	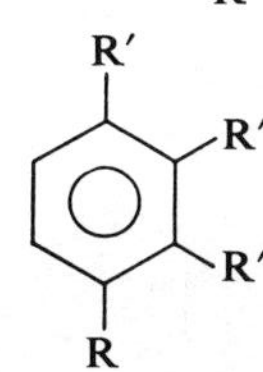	R'_6C_6 + RR and in case $R' = C_6H_5$ *trans*-$R'CH{=}CHR'$	311

21.	$CH_2{=}CHCH_2$	CH_3	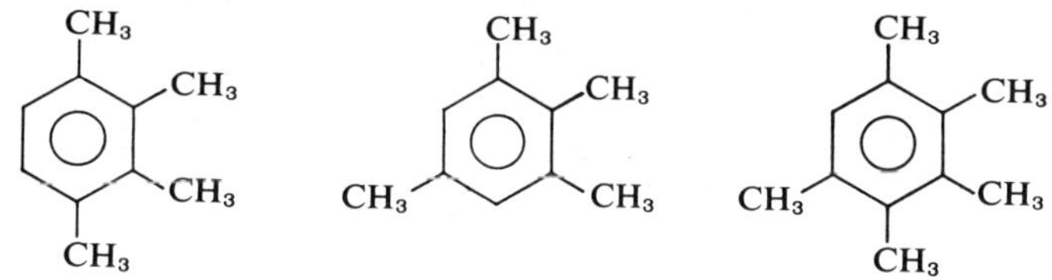 $R'_6C_6 + RR$	360
22.	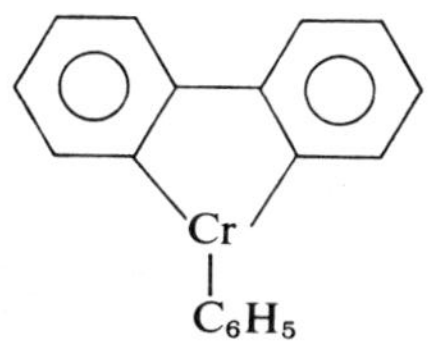	CH_3	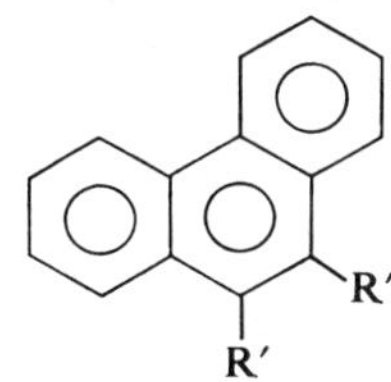	354
23.	CH_3	C_6H_5	see Table 3.5	
24.	CH_3CH_2	CH_3	R'_6C_6 + other organic products	347, 348
		C_6H_5	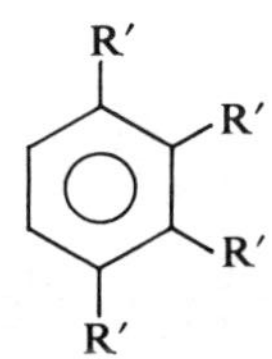 R'_6C_6	361
25.	$CH_3(CH_2)_3$	$CH_3(CH_2)_2$	R'_6C_6 + *cis*- and *trans*-4-octene	349

[a] $(Mes)_2Cr(II)S_x = [2,4,6\text{-}(CH_3)_3C_6H_2]_2Cr(II)S_x$.

TABLE 3.5

Products[a] Formed in the Reaction of the** $(CH_3)_3Cr$ **and** $(CD_3)_3Cr$ **Systems with $C_6H_5C{\equiv}CC_6H_5$ [361–367].

ORIGIN OF PRODUCT	PRODUCTS
1. Reaction gases	H_2, CH_4 [HD + CD_4, 0.2%; CD_3H, 85.3%; CD_2H_2, 14.5%]
2. Hydrolysis gases	H_2, CH_4 [HD + CD_4, 4.3%; CD_3H, 94.3%; CD_2H_2, 1.38%]
3. Transfer of "H" to one acetylene	*cis*- and *trans*-$C_6H_5CH{=}CHC_6H_5$ [$C_6H_5CD{=}CDC_6H_5$] $C_6H_5CH_2CH_2C_6H_5$
4. Transfer of "CH_3" to one acetylene	*cis*- and *trans*-$C_6H_5C(CH_3){=}CHC_6H_5$ [$C_6H_5(CD_3)C{=}CDC_6H_5$] $C_6H_5C({=}CH_2)CH_2C_6H_5$ [$C_6H_5C({=}CD_2){-}CD_2C_6H_5$] $C_6H_5CH(CH_3)CH_2C_6H_5$
5. Transfer of 2 "CH_3" to one acetylene	*meso*-$C_6H_5CH(CH_3)CH(CH_3)C_6H_5$
6. Transfer of "H" to two acetylenes	*cis*-, *cis*-$C_6H_5CH{=}C(C_6H_5)C(C_6H_5){=}CH(C_6H_5)$ 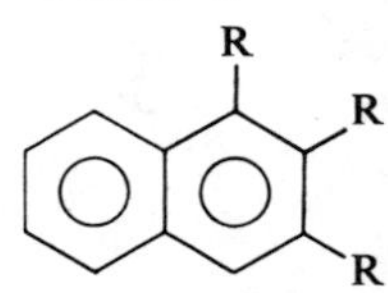
7. Transfer of "CH_2" or "CH_3" to two acetylenes	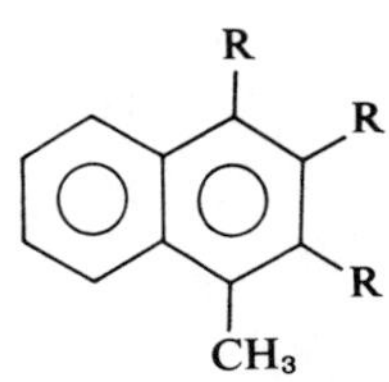 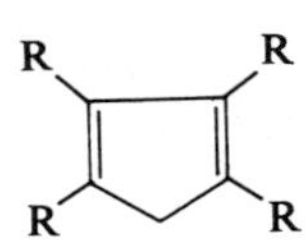
8. Transfer of 2 "CH_2" to two acetylenes	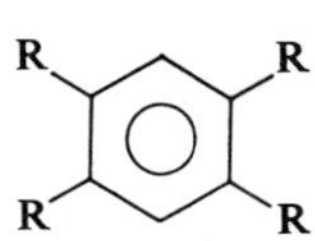
9. Trimerization of acetylene	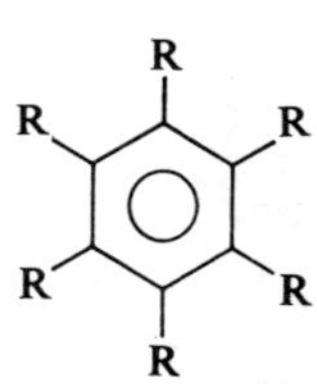

[a] In the formulae R = C_6H_5 and the products in brackets were isolated from reactions with the $(CD_3)_3Cr$ system.

i. Products. The precise products formed in the reaction of tris(organo)-chromium compounds and acetylenes depend upon the nature of the organic group bonded to the chromium center (see Tables 3.4 and 3.5); however, they are all, broadly speaking, of the following three main types:

(i) Those formed from one acetylenic unit and one or two of the organic groups bonded to chromium (or fragmentation products thereof), e.g., the mono- and bis(aryl)butenes (Table 3.4, numbers 2,4,5, 11–14, and 16 and the products 3–5) (Table 3.5) [361–367].

(ii) Those formed from two acetylenic units and one or two of the organic groups bonded to chromium (or fragmentation products thereof), e.g., the naphthalenes and phenanthrenes in Table 3.4, numbers 4–14 and the products 7 and 8 in Table 3.5.

(iii) Those formed from three acetylenic units with or without the organic groups originally bonded to chromium.

The structures of the various compounds, of types (i) and (ii), have supplied valuable information concerning the processes by which they are formed. Thus aryl- and 3- and 4-substituted arylchromium compounds all react with acetylenes to give products whose formation does not involve any rearrangement of the original organochromium compound (Table 3.4, numbers 6–14), thereby indicating that the carbon atom bonded originally to the chromium center is involved in the initiating step of the reaction. Certain ortho-substituted arylchromium compounds react with acetylenes to give products whose formation involves the extrusion or modification of this ortho substituent (e.g., numbers 14–16, Table 3.4), thereby indicating that the elimination of a Cr–X fragment is involved either in some intermediate stage or in the terminating step of the reaction. The tris(methyl)chromium system reacts with diphenylacetylene to give, among other products, stilbenes, methylstilbenes, and methylene bibenzyl (see Table 3.5), thereby indicating that a methyl group, when bonded to chromium, may act as a source of "H," "CH_3," or "CH_2."

ii. Mechanism. The manifold products formed in the reaction of organochromium compounds with acetylenes preclude the possibility of establishing a satisfactory overall material balance and of carrying out any meaningful kinetic studies. However, though the precise mechanisms of these reactions obscure, many rationales [348,350,354,368] have been proposed to account for the formation of the various products.

The formation of products derived from one acetylenic unit [i.e., of type (i)] may be explained in terms of an isomerization within an organochromium–acetylene complex in which one of the organic groups (or a fragmentation product thereof) originally bonded to chromium is transferred to the

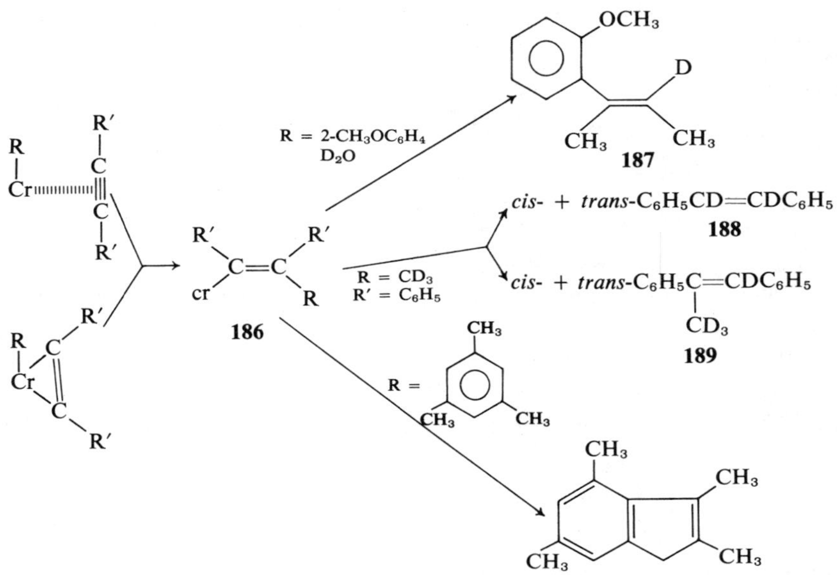

Scheme 3-22. Formation and some reactions of vinylchromium compounds.

acetylene to give a bonded vinylchromium compound **186**, Scheme 3-22, (R = H, aryl, alkyl, or other organic groups; R′ = CH_3 or C_6H_5, cr = chromium and associated ligands). There is no evidence available that would indicate whether the vinylchromium compound **186** is formed by the concerted or stepwise addition of R–cr to the acetylene or whether it is formed by the transfer of a group R from chromium to carbon in the chromocycle; there is, however, proof of the formation of such an intermediate. Thus, Zeiss and Sneeden have found that substantial quantities of isotopically pure *cis*-2-(2-methoxyphenyl)-3-deuterio-but-2-ene (**187**, Scheme 3-22) are formed when the products of the reaction of crystalline tris(2-methoxyphenyl)-chromium and 2-butyne were treated with D_2O [350]. Zeiss and his colleagues also have found that deuteriostilbenes and deuteriomethylstilbenes (**188** and **189**, Scheme 3-22) are obtained by the interaction of the $(CD_3)_3Cr$ system and diphenylacetylene [362,364]. Furthermore, Whitesides and Ehmann have found [354] that tris(mesityl)chromium reacts with 2-butyne to give, among other products, the indene **190**, Scheme 3-22, this compound being formed presumably by the elimination of cr–H in a vinylchromium compound. It is pertinent to point out at this stage that the elimination of Cr–H, and other Cr–X units may take place either directly (i.e., by the extrusion of Cr–H or Cr–X) or indirectly via chromocycles. In the latter cases, the chromocycles are formed either by oxidative addition of C–H(X) to cr or by the elimination of RH(X); reductive elimination of the chromium fragment from these chromocycles then gives the product [354]. These three processes are outlined for the case of the conversion of mesityl-vinylchromium to the indene **190** in Eq. (3-141).

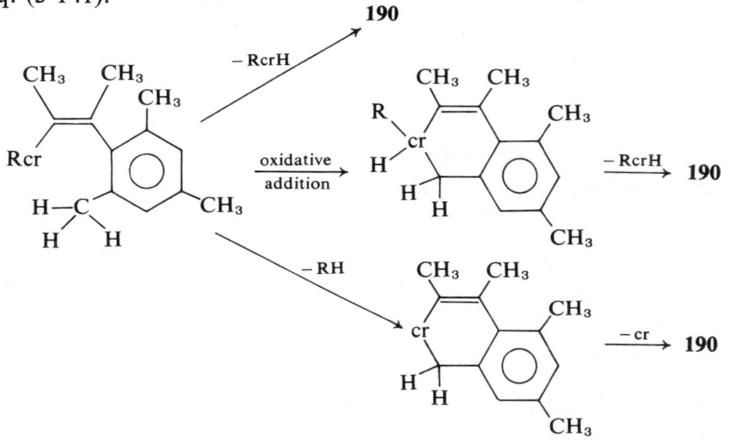

(3-141)

The formation of the other products of type (i) (i.e., those derived from one acetylene) (Tables 3.4 and 3.5) may be interpreted in terms of *cis*- and *trans*-vinylchromium compounds (**186**, **191**, Scheme 3-23) as intermediates.

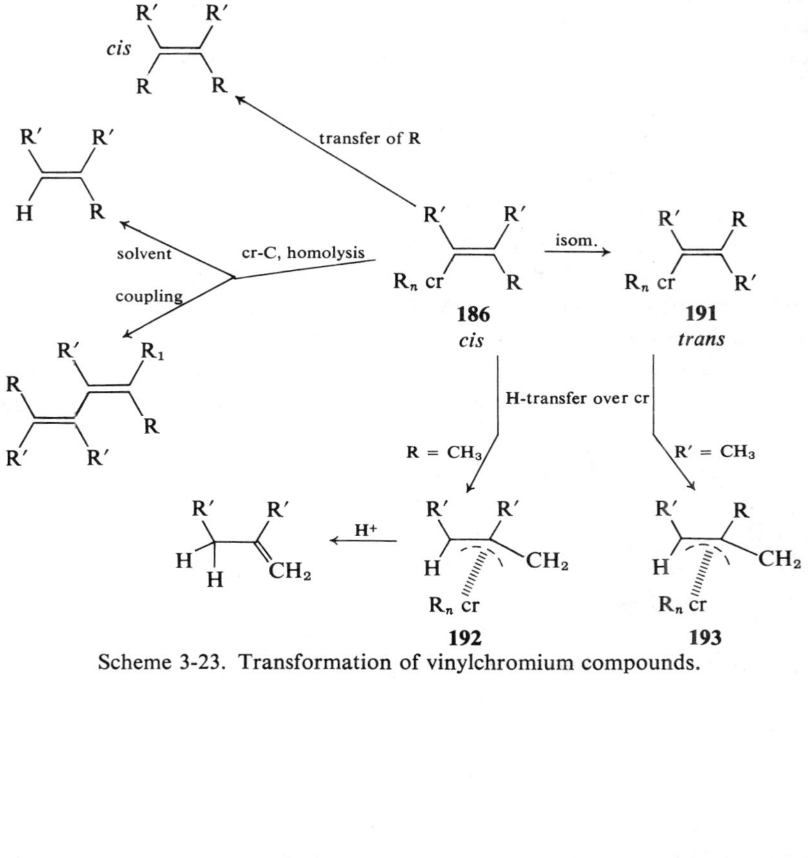

Scheme 3-23. Transformation of vinylchromium compounds.

The cis-monosubstituted butenes (or stilbenes) and disubstituted butadienes are formed by the homolysis of the carbon–chromium bond in **186** and the disubstituted *cis*-butenes (or stilbenes) by the transfer of a second organic group (or a fragmentation product thereof) from chromium to carbon in **186** (see Scheme 3-23). The trans-, mono-, and disubstituted butenes (or stilbenes) are obtained analogously from the *trans*-vinylchromium compound **191**.

The isomerization of the *cis*-vinylchromium compound (**186**) to the trans isomer (**191**) is facilitated because the chromium that is bonded to the vinyl carbon atom drains off sufficient electron density from the C═C to permit rotation about a quasisingle bond; the isomerization is, however, impeded by the presence within the molecule of groups capable of coordinating to chromium. Thus, in the reaction of the isomeric anisylchromium compounds with 2-butyne, only the ortho isomer gives high yields of the cis-substituted butenes, the other two isomers give modest yields of both cis- and trans-substituted butenes [350].

Zeiss and Sneeden have suggested that the formation of benzylstyrene $[C_6H_5C({=}CH_2)CH_2C_6H_5]$ in the reaction of the $(CH_3)_3Cr$ system and tolane is the result of the isomerization of the vinylchromium compound **186** to the allyl compound **192.** This isomerization requires that a methyl (or other alkyl group) be adjacent to the chromium center [thus **191**, R′ = CH_3, can, in principle, isomerize to **193** (Scheme 3-23)] and may proceed via the oxidative addition of C–H to the chromium, as outlined in Eq. (3-142) [350].

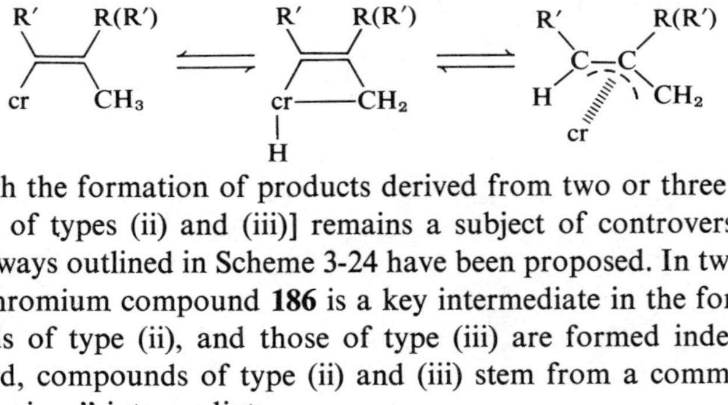

(3-142)

Although the formation of products derived from two or three acetylenic units [i.e., of types (ii) and (iii)] remains a subject of controversy, yet the three pathways outlined in Scheme 3-24 have been proposed. In two of these, the vinylchromium compound **186** is a key intermediate in the formation of compounds of type (ii), and those of type (iii) are formed independently, in the third, compounds of type (ii) and (iii) stem from a common "butadiene-chromium" intermediate.

Whitesides and Ehmann have proposed a pathway in which the vinylchromium compound **186** cyclizes (by extrusion of RH or RX) to give the chromium heterocycle **194**, and this compound subsequently reacts with a molecule of acetylene to give the naphthalenes **195** and **196** [i.e., compounds of types (ii)] [354].

Zeiss and his colleagues, on the other hand, have proposed two other pathways for the formation of compounds of the type (ii) [350,365,366]. In

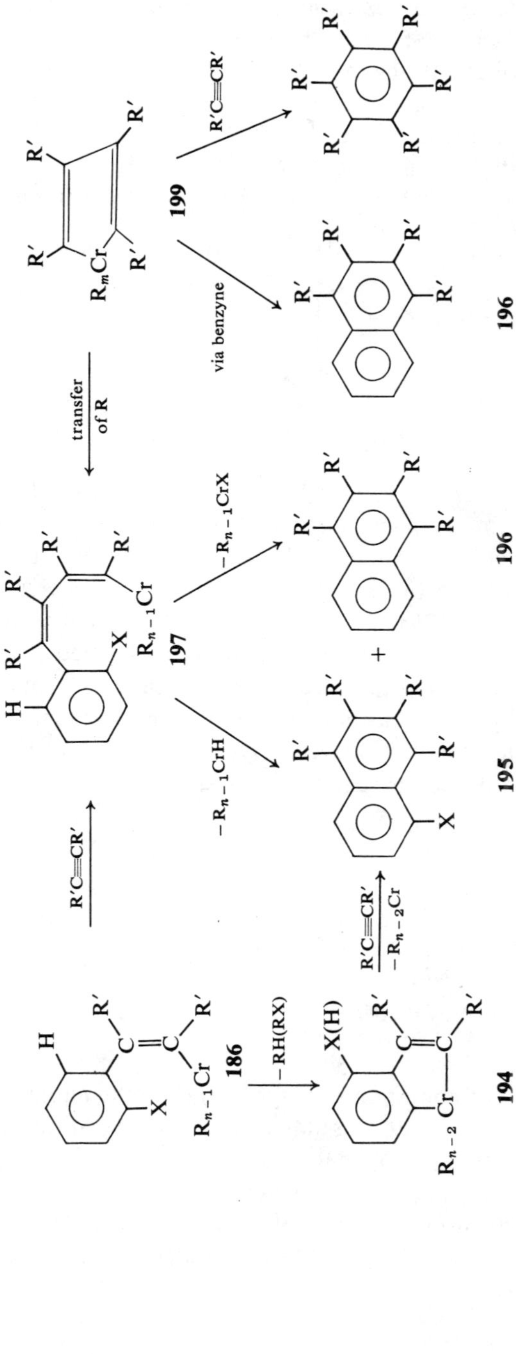

Scheme 3-24. Pathways for the formation of products derived from two and three acetylenic units.

one of these, the vinylchromium compound **186** reacts with a second molecule of acetylene giving the σ-bonded butadienylchromium compound **197**, which subsequently cyclizes, by extrusion of Cr–H or Cr–X [see Eq. (3-141)] to give the naphthalenes **195** and **196** directly. Some support for this sequence is found in the observations that: (1) tris(2-methoxyphenyl)chromium reacts with 2-butyne to give a mixture of 5-methoxy-1,2,3,4-tetramethylnaphthalene and 1,2,3,4-tetramethylnaphthalene [350], (2) tris(2,2-diphenylvinyl)chromium reacts with 2-butyne to give the naphthalene **198**, Eq. (3-143) [R = $(C_6H_5)_2$-C=CH] [354]. In the other and third pathway, the compounds of type (ii)

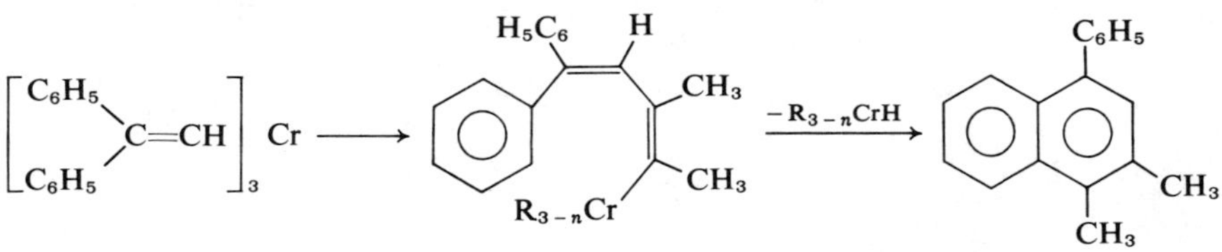

198

(3-143)

and (iii) are formed, independently of those of type (i), from the common "butadiene chromium" intermediate **199**, Scheme 3-24 [349,350,365,366, 368]. This intermediate, which could be formed by the direct interaction of two molecules of acetylene and a chromium species, has been variously formulated, e.g., **199–201**, Eq. (3-144). However, Whitesides and Ehmann have shown, from their studies with isotopically labeled 2-butyne, that such a "butadiene chromium" intermediate cannot have the symmetrical structure **201** [354,369,370]. In the ensuing discussion, therefore, this intermediate will be formulated as the chromocycle **199**, though the valence tautomer, the diradical **200**, could be the reactive species.

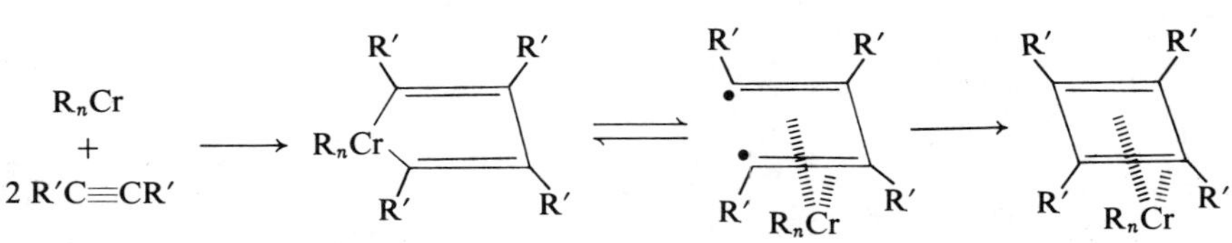

199 **200** **201**

(3-144)

The formation of compounds of type (ii) (i.e., the naphthalenes) from the chromium heterocycle (**199**) could proceed via the transfer of one of the organic groups (or a fragmentation product thereof) bonded originally to chromium to the complexed butadiene unit. The transfer, *per se*, of one of the organic groups would give the σ-bonded butadienylchromium compound (**197**, Scheme 3-24) and thence, the substituted naphthalenes **195** and **196**.

Alternatively, the transfer of a fragmentation product (e.g., when R = aryl, a benzyne unit) would give directly the substituted naphthalene, Scheme 3-24.

The hexasubstituted benzenes [i.e., compounds of type (iii)] could be formed by the direct interaction of the chromocycle **199** and a molecule of acetylene (Scheme 3-24); an analogous reaction, Eq. (3-145) has been described by Whitesides and Ehmann [354].

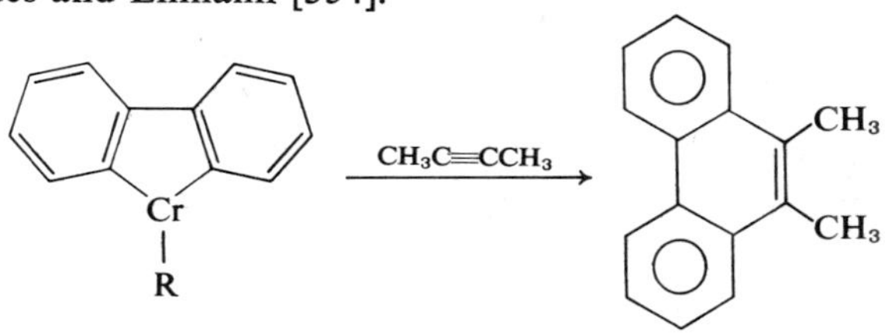

(3-145)

The available chemical evidence relating to the reactions of tris(aryl)- and tris(vinyl)chromium compounds can be interpreted in terms of any of these three mechanisms. Furthermore, there is x-ray structural evidence for the existence of both a σ-bonded butadienyl-Ru compound [371], and a metallocyclopentadiene complex of rhodium [372].

The interpretation of the results from reactions involving the tris(methyl)-chromium system and diphenylacetylene is rendered difficult by the fact that a methyl group bonded to chromium can act as a source of methyl "CH_2" units or hydrogen [362–367]. However, the formation of the products derived from one, two, and three acetylenic units may be readily explained in terms of the mechanisms already outlined in the preceding pages. Thus, the transfer of H or CH_3 to complexed acetylene would give the *cis*-vinylchromium compounds **202** and **203**, which may isomerize to the corresponding trans compounds **204** and **205**, Scheme 3-25.

The compounds derived from one acetylenic unit would then be formed by the fragmentation of these *cis*- and *trans*-vinylchromium compounds as outlined in Scheme 3-23, and those derived from two acetylenic units as outlined in Scheme 3-25.

Reaction of the vinylchromium compounds **202–205** with a molecule of acetylene would give the σ-bonded butadienylchromium compounds **206–209**. Those derived from the *trans*-vinylchromium compounds (i.e., **206** and **208**, Scheme 3-25) would have the aryl ring and the chromium center in juxtaposition, thereby, favoring metal-hydride elimination to give, respectively, 1,2,3-triphenylnaphthalene **210** and 1,2,3-triphenyl-4-methylnaphthalene **211**. In the compounds derived from the *cis*-vinylchromium compounds (i.e., **207** and **209**), the latter **209** has a methyl group contiguous to the chromium center and metal-hydride elimination would produce 1,2,3,4-tetraphenylcyclopenta-1,3-diene (**212**), and the former (**207**) has an aryl ring contiguous to the chromium center and metal-hydride elimination

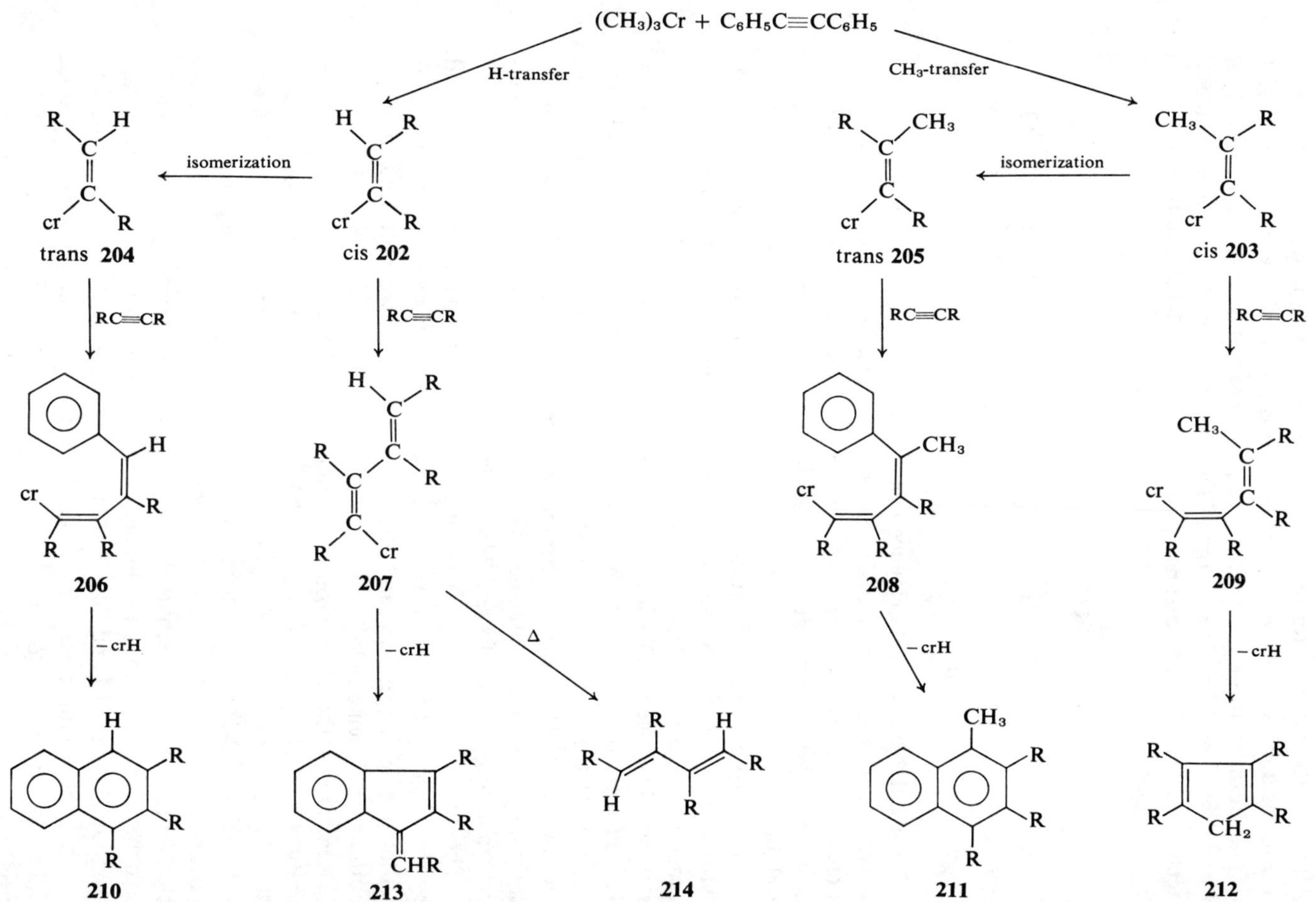

Scheme 3-25. Reaction of the tris(methyl)chromium system and diphenylacetylene, cr = chromium and associated ligands, R = C_6H_5.

would give 1-benzylidene-2,3-diphenylindene (**213**), whereas homolytic cleavage of the C–Cr bond would give the butadiene (**214**). The indene (**213**) has been isolated from the products formed in the reaction of the methylvanadium system and diphenylacetylene [366], and the butadiene (**214**) from products of the methylchromium system and diphenylacetylene [363].

The tetraphenylcyclopentadiene **212** could also be formed from the chromium heterocycle **199** as outlined in Eq. (3-146) ($R' = CH_3$, $R = C_6H_5$).

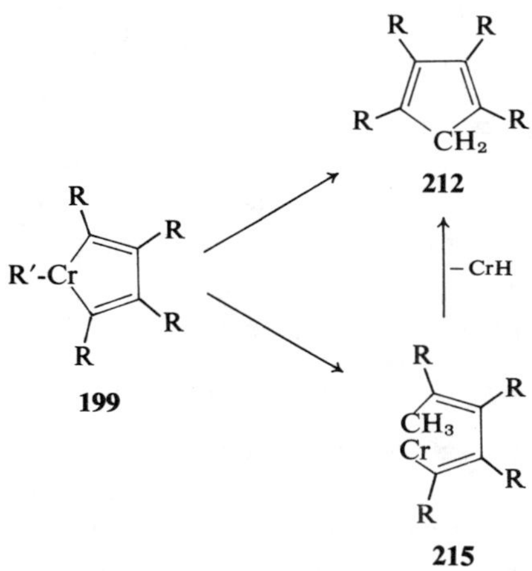

(3-146)

Insertion of a carbenoid unit would give the tetraphenylcyclopentadiene directly [365,366]; alternatively, transfer of a methyl group would result in the σ-bonded butadienylchromium compound [**215**, Eq. (3-146)], which may cyclize (by metal-hydride elimination) to the cyclopentadiene.

The formation of all the compounds of types (i), (ii), and (iii) that have been isolated to date from the reactions of tris(organo)chromium compounds (or systems) with disubstituted acetylenes may be rationalized in terms of the various pathways outlined above. Thus, the compounds derived from one acetylenic unit are almost certainly formed via a vinylchromium compound. Those derived from two acetylenic units may be formed by three routes: (a) from a chromium heterocycle derived from the vinylchromium compound, (b) from a σ-bonded butadienylchromium compound also derived from the vinylchromium compound, or (c) from a chromium heterocycle containing two acetylenic units. Those derived from three acetylenic units are probably formed via a chromium heterocycle containing two acetylenic units.

The products from the reactions of certain tris(aryl)chromium compounds also contain arenechromium-π-complexes [348]; however, there is no information available that might indicate how these products are formed. In this connection it is of interest to note that Wilke and his collaborators have found that when 2-butyne is allowed to react with $CrCl_3$ and an alkylaluminum compound in benzene solution, it is transformed into π-$[(CH_3)_6C_6]$-π-$[(CH_3)_5C_5]Cr(I)$ [Eq. (3-147) $R' = CH_3$, CH_2CH_3, and $CH_2CH_2CH_3$; $R = CH_3$] [373]. The authors have suggested that the C_5 unit may be

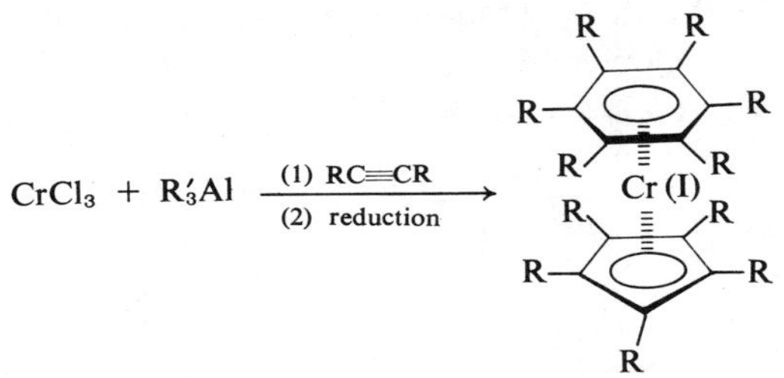

(3-147)

formed by the interaction of two acetylenic units and a "CH_3C" unit formed by the dichotomy of a molecule of butyne (the disproportionation of acetylenes may be effected with molybdenum catalysts [374]) rather than by the rearrangement of hexa(methyl)-Dewar-benzene [e.g., 375].

b. Reactions with Carbonyl Compounds and Olefins

i. CO. Job and Cassal found that a diversified mixture of organic compounds and $Cr(CO)_6$ were formed when carbon monoxide was bubbled through a reacting mixture of $CrCl_3$ and C_6H_5MgBr in diethyl ether [376, 377]. Zeiss and his colleagues found, however, that when carbon monoxide is bubbled through a solution of pure tris(phenyl)chromium(III) in tetrahydrofuran, the sole organic product was benzopinacol **216**, Scheme 3-26 [378]. Other authors have reported that they have isolated benzophenone and other ketonic products from this reaction [379,380]. As it is known that benzopinacol decomposes thermally to benzophenone [381], the apparent discrepancy in the results reported can be ascribed to the procedures used for the isolation of the products (i.e., crystallization as against distillation).

ii. Ketones. Tris(phenyl)- and tris(ethyl)chromium react with enolizable ketones (e.g., diethyl ketone and cyclohexanone) to give two main types of products [378]. The first are aryl or alkyl carbinols (**217**, Scheme 3-26) and are formed by the transfer of one of the organic groups bonded to chromium to complexed ketone, e.g., Eq. (3-148) ($R = C_6H_5$ or CH_2CH_3, $R' = CH_3$). Spectroscopic evidence for this type of mechanism has been obtained in the case of the reaction of tris(allyl)chromium(III) and acetone [382]. The second type of product (**218**, Scheme 3-26) is formed from one of the

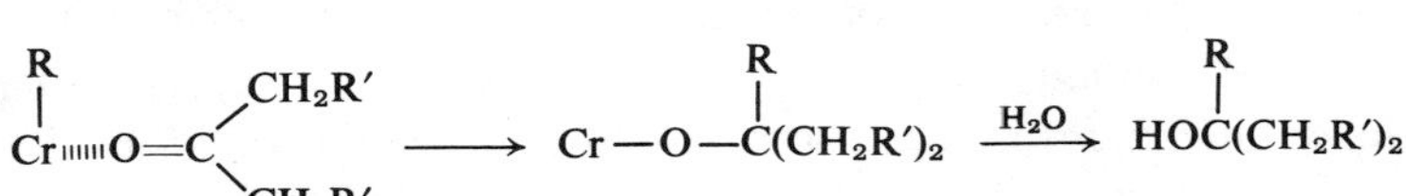

(3-148)

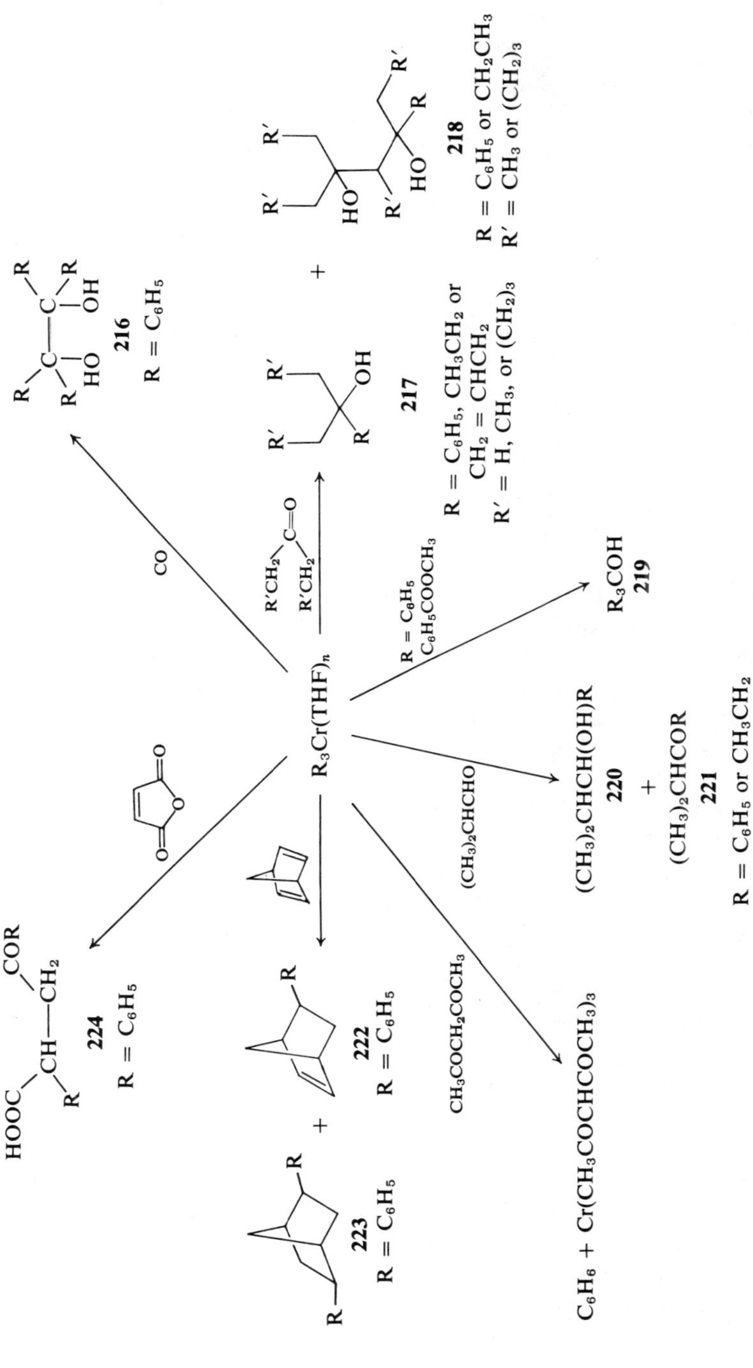

Scheme 3-26. Some reactions of tris(organo)chromium compounds with carbonyl compounds and olefins.

groups bonded to chromium and two molecules of ketone. The isolation of 2-(cyclohex-1-ene-1-yl)cyclohexanone from the reaction of $(C_6H_5)_3Cr$ and cyclohexanone suggests that the initial step of the reaction is the condensation of two ketone molecules on the chromium center followed by the transfer of one of the organic groups bonded to chromium as outlined in Eq. (3-149).

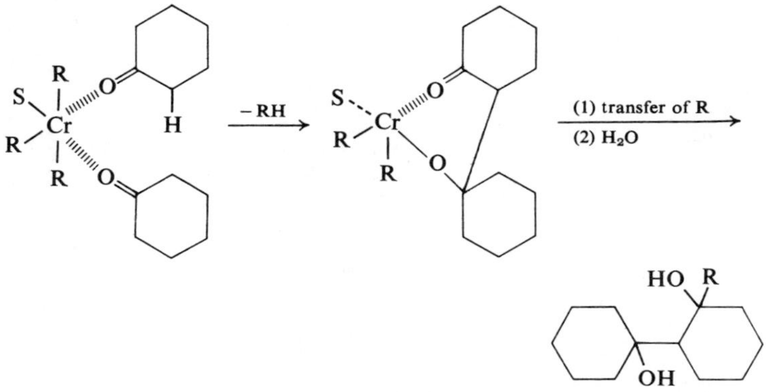

(3-149)

iii. Esters and Aldehydes. Tris(phenyl)chromium reacts with methyl benzoate to give triphenyl carbinol [383] **219**, Scheme 3-26. Tris(allyl)-chromium on the other hand, does not appear to react with ethyl acetate [382]. Both tris(phenyl)chromium and the tris(ethyl)chromium system react with isobutyraldehyde to give the secondary alcohols **220**, and, by hydrogen transfer, the corresponding ketones **221**, Scheme 3-26 [384].

iv. β-Diketones. The only recorded instance of the reaction of a tris-(organo)chromium compound and a β-diketone is that of $(C_6H_5)_3Cr(THF)_3$ and acetylacetone to give, quantitatively, benzene and tris(acetylacetonato)-chromium(III) [378]. This reaction should be of value in the characterization of σ-bonded organochromium compounds (see Chapter 2, Section F,2,b,iii).

v. Olefins and Diolefins. Metlesics, Wheatley, and Zeiss have found that whereas tris(phenyl)Cr(THF)$_3$ is relatively inert toward simple olefins (e.g., cyclohexene, isoprene, and stilbene), it reacts smoothly with bicyclo[2.2.1]-heptadiene-2,5 and maleic anhydride to give, **222–224**, Scheme 3-26. In these products, one or two aryl groups have added to the C=C. This difference in the reactivity of simple and activated (or strained) olefins illustrates that the coordinative capacity of the olefin is a determining factor in its reaction with an organochromium compound. In the present reactions the diene and anhydride may displace tetrahydrofuran [from $R_3Cr(THF)_n$] to give π-(olefin)chromium complexes. Phenyl group transfer within these complexes and subsequent hydrogen transfer (hydrolysis step) produce the observed products [385].

c. *Thermal Decomposition*

σ-Bonded tris(organo)chromium compounds $[R_3Cr(THF)_n]$ show marked differences in their thermal stabilities and, though the only crystalline compounds isolated to date are those where R = aryl or allyl, the apparent stability sequence for the present group of compounds is aryl ≫ allyl ≫ benzyl ≫ alkyl ≫ vinyl. This sequence indicates that the stabilities of a tris(organo)chromium compound is not determined uniquely by the resistance of the C–Cr bond to homolysis (i.e., to give a free radical). Some of the factors influencing the stability of organochromium compounds have been enumerated earlier in this section (see F,1) and their roles will be illustrated in the ensuing discussion. The stabilities of tris(organo)chromium compounds and the pathways by which they undergo fragmentation are determined mainly by the nature of the organic groups and of the ligands associated with the chromium center. Thus, tris(aryl) compounds are the most stable of the known tris(organo)chromium compounds and they may undergo fragmentation either by homolysis of the C–Cr bond (to give the arene and the biaryl) or by an irreversible rearrangement (σ- to π-rearrangement) to give bis(arene)chromium π-complexes. Tris(allyl)- and (vinyl)chromium compounds undergo fragmentation by reductive dimerization. Tris(alkyl)-chromium compounds undergo fragmentation by homolytic and metal-hydride elimination processes. The thermal decomposition of these three different classes of compounds are discussed separately in the ensuing pages.

i. Tris(aryl)chromium Compounds, σ- to π-Rearrangement. One of the most fascinating reactions of certain σ-bonded tris(aryl)chromium compounds is their rearrangement to bis(arene)chromium π-complexes. Thus, Zeiss and his colleagues have found that when the solvating species are removed from pure crystalline $(C_6H_5)_3Cr(THF)_3$, either thermally or by displacement with diethyl ether, the σ-bonded compound undergoes an irreversible change to a pyrophoric paramagnetic black solid. Hydrolysis of the latter, in the presence of air, gives finally a mixture of π-bis(arene)chromium(I) complexes, **225–227**, isolable as their tetraphenylboron salts [(a) in Eq. (3-150), $X = B(C_6H_5)^{\ominus}$] [377,386–389]. The same black intermediate and final π-complex mixture could be obtained from the controlled interaction of phenylmagnesium bromide and $CrCl_3$ in diethyl ether solution [390–392] and subsequent hydrolysis in air [(b) in Eq. (3-150)]. The history of the reaction of phenylmagnesium bromide and $CrCl_3$ in diethyl ether, culminating in the identification of the products as π-bis(arene)chromium compounds, has been reviewed [377,389].

Analogous mixtures of alkyl-substituted π-bis(arene)chromium compounds may be obtained either by the rearrangement of the solvated σ-bonded tris(3-, or $4\text{-}CH_3C_6H_4$)chromium compounds [393,394] or by interaction of the appropriate tolyl Grignard reagent and $CrCl_3$ in diethyl ether solution [393,395]. Only one π-bis(arene)chromium complex is formed, however, by the reaction of benzylmagnesium chloride and $CrCl_3(THF)_3$ in diethyl ether [315,396,397].

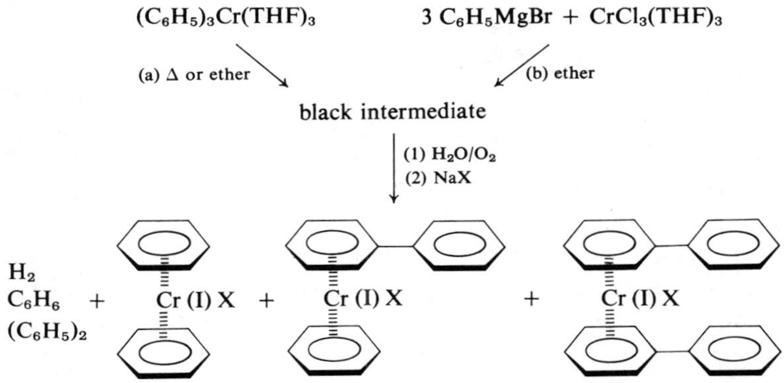

(3-150)

225 226 227

The feature common to these σ- to π-rearrangements is that the organic groups originally bonded to chromium have either coupled or acquired hydrogen. The rationalization of the processes involved in the σ- to π-rearrangement presents a particular challenge in spite of a considerable amount of experimental information available. The relevant facts, together with some of the current ideas, are summarized in the following pages.

α. *Tris(benzyl)chromium.* The salient feature of the rearrangement of benzylchromium is that two of the original benzyl groups have undergone ortho coupling to give 2-benzyltoluene both free (in solution) and in the π-complex, Eq. (3-151), X = $B(C_6H_5)_4^{\ominus}$. The π-complex may have either

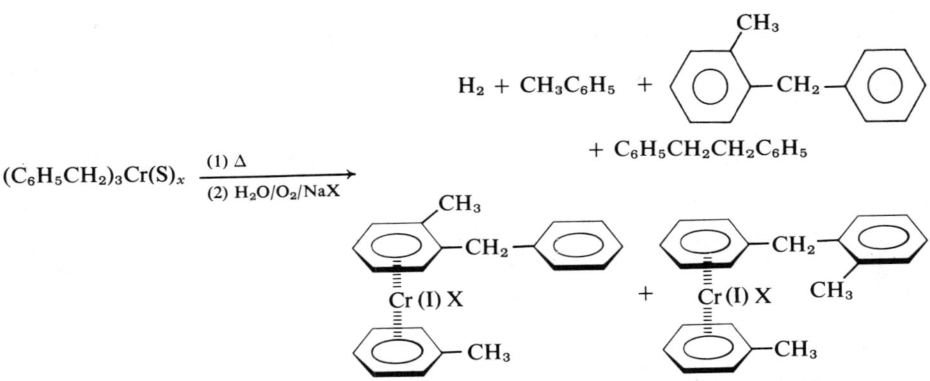

228 229

(3-151)

of the isomeric structures **228** or **229** [Eq. (3-151)] and, for convenience only one formulation i.e., (**228**) will be used in the ensuing discussion. When deuterium oxide is used in place of water in the final hydrolysis step, deuterium is incorporated in the toluene and 2-benzyltoluene present as the free hydrocarbons and as arene units in the π-complex [397]. This incorporation of deuterium into both arene parts of the π-complex can only be associated with the σ- to π-rearrangement process since: (1) σ-bonded organochromium-(III) or (II) compounds do not undergo hydrogen–deuterium exchange in the presence of D_2O [Chapter 2, Section F] or D_2 gas [354], and (2) π-bis-(arene)chromium compounds only undergo appreciable H/D exchange in strongly alkaline media (Chapter 3, Section B,1,b). Furthermore, the isotopic compositions of the toluene and 2-benzyltoluene isolated on pyrolysis of the π-bis(arene)complex are similar, which suggests that both future arene parts are already present in the precursor to the π-complex. The transformation outlined in Scheme 3-27 could account for the σ- to π-rearrangement of tris(benzyl)chromium [397]. The key step involves the ortho coupling of two *cis*-benzyl units, with concomitant transfer of hydrogen to the chromium center to give the hydridochromium(III) compound **230**. A reversible hydrogen transfer, or tautomerism, analogous to that observed in π-(naphthalene)RuL_2 [398] converts the organochromium(III) compound into the half-sandwich organochromium(I) complexes **231** and **232**. These, in analogy with the known σ-(benzyl)- and π-(benzyl)molybdenum compounds [σ-$(C_6H_5CH_2)Mo(CO)_3(C_5H_5)$ and π-$(C_6H_5CH_2)Mo(CO)_2(C_5H_5)$] [399,400] could, in fact, have the alternative π-(benzyl)-π-(arene)chromium(I) structures **233** and **234**, respectively. Reaction of any of these chromium(I) complexes (i.e., **231**–**234**) with water could now give π-(toluene)-π-(2-benzyltoluene)chromium(I) **228**. The incorporation of deuterium into both arene parts of the π-complex, observed when deuterium oxide is used in place of water in the final hydrolysis step, may be accounted for by the interaction of any of the intermediates with D_2O (or the D_2 gas generated).

β. *Tris(aryl)chromium Compounds.* The rearrangement of σ-bonded tris(aryl)chromium compounds has been extensively investigated and the accumulated results (given in Table 3.6) indicate that substitution in the aryl ring has an effect upon the ease of the σ- to π-rearrangement and may even inhibit it. Thus, tris(4- or 3-methylphenyl)chromium rearrange more readily than the parent tris(phenyl)chromium [394]. The blue (2-methylphenyl)chromium, which has not been isolated in the crystalline state, gives only traces of π-complex; this could be due to the interaction of the hydrogens of the *ortho*-methyl group with the chromium center.

The tris(3- or 4-halogen-substituted)arylchromium compounds are stable in diethyl ether solution; though on more drastic thermal treatment, they

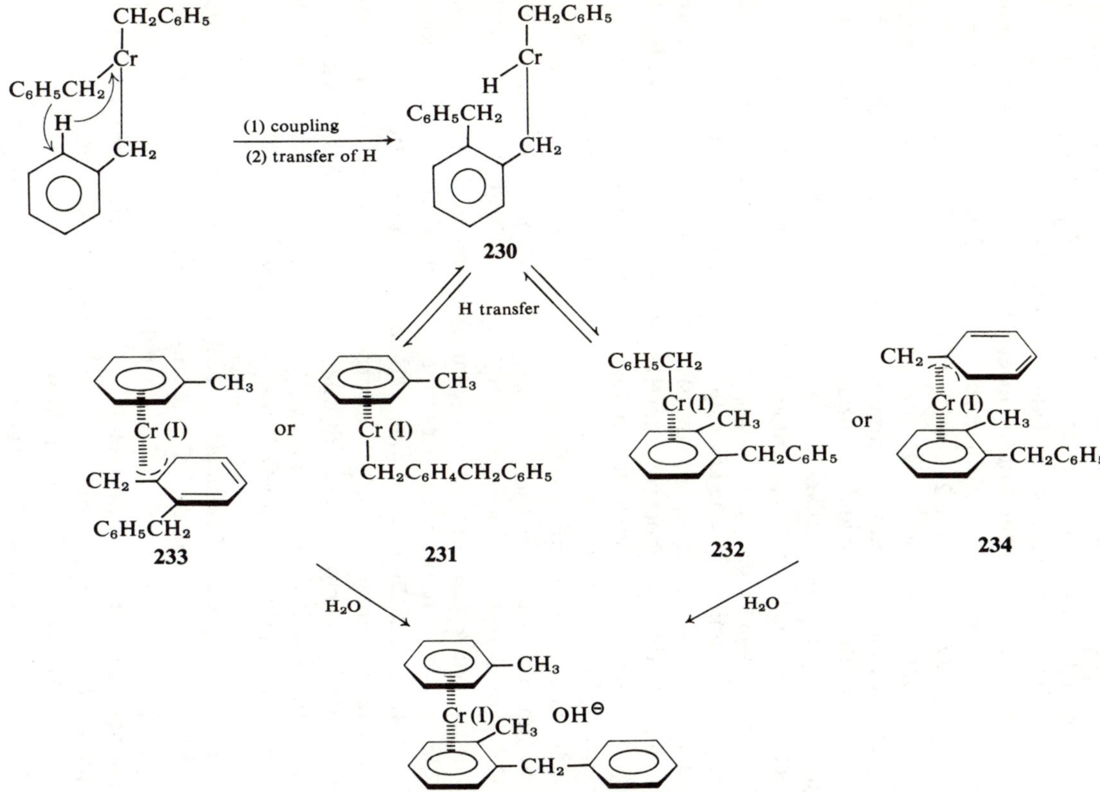

Scheme 3-27. σ- to π-rearrangement of benzylchromium system.

TABLE 3.6

"*Fragmentation*" *of σ-Bonded Tris(aryl)chromium Compounds*

ARYL	CONDITIONS	PRODUCTS	REF.
C_6H_5	Δ or ether	π-Complex mixture; arene components, benzene, and biphenyl	[386–388]
4-$C_6H_5C_6H_4$	Ether	π-Complex mixture; arene components, biphenyl, and 4,4′,4″,4″-quaterphenyl	[394]
3-$CH_3C_6H_4$	Ether	π-Complex mixture; arene components, toluene and 3,3′-bitolyl	[394,395]
4-$CH_3C_6H_4$	Ether	π-Complex mixture; arene components, toluene and 4,4′-bitolyl	[393–395]
2-$CH_3C_6H_4$[a]	Ether	Traces of π-complex[a]	[395,401]
4-(F,Br,Cl)C_6H_4	Ether	Stable	[394]
	Δ	No details; halogen-free π-complex mixtures	
3-ClC_6H_4	Ether	Stable	[394]
	Δ	No details; halogen-free π-complex mixtures	
2-ClC_6H_4	20°C/THF	No π-complex; decomposition products: biphenyl and triphenylene	[357]
2-$CH_3OC_6H_4$	Reflux Dioxane	No π-complex; decomposition products [2-$(CH_3OC_6H_4)_2Cr]_n$ $CH_3OC_6H_5$, 2,2′-$(CH_3OC_6H_4)_2$, [(2-$CH_3OC_6H_4)_2CH]_2$	[402]
3-$CH_3OC_6H_4$	Ether	π-Complex mixture; arene components anisole 3,3′-bianisyl	[395]
4-$CH_3OC_6H_4$	Ether	No π-complex	[395]
	Δ	Traces of π-complex; decomposition products: anisole 4,4′-bianisyl	
2,4,6-$(CH_3)_3C_6H_2$	Warm ether	No π-complex; compound unchanged	[391,401,403]
2,4,6-$(CH_3)_3C_6H_2$	80°C *in vacuo*	No π-complex; decomposition products: mesitylene and bimesityl	[401,404]
2-$(CH_3)_2NC_6H_4$	Warm ether	No π-complex; compound unchanged	[343]
2-$(CH_3CH_2)_2NCH_2C_6H_4$	Warm ether/octane	No π-complex; compound unchanged	[346]
2-(2′-$C_5H_4N)C_6H_4$	Warm ether	No π-complex; compound unchanged	[342]

[a] Not isolated crystalline; the direct reaction of the *o*-tolyl-Grignard and $CrCl_3$ $(THF)_3$ in diethyl ether also gave only traces of π-complex [395].

give a complex mixture of arenechromium π-complexes containing halogen-free arene components [394]. In contrast, the ortho isomer $(2\text{-ClC}_6H_4)_3Cr$, which has not been isolated crystalline, is said to decompose at room temperature to give various organic products but no bis(arene)chromium π-complex [357].

With methoxyl substituents, only the tris$(3\text{-CH}_3OC_6H_4)$Cr undergoes the σ- to π-rearrangement to give substantial yields of a mixture of bis(arene)-chromium π-complexes [395]. The ortho isomer undergoes thermal decomposition in a variety of solvents to give consistently the polymeric bis(2-methoxyphenyl)chromium(II) but no π-complex [402]. When the decomposition is carried out in dioxane, the organic products contain tetrakis(2-methoxyphenyl)ethane [**235**, Eq. (3-152), An $= 2\text{-CH}_3OC_6H_4$].

$$An_3Cr(THF)_n \xrightarrow[\text{dioxane}]{\Delta} [An_2Cr]_n + CH_3OC_6H_5 + AnAn + (An)_2CH{-}CH(An)_2 \qquad (3\text{-}152)$$

235

The latter compound is formed presumably by cleavage of the anisyl–Cr bond to give anisyl radicals that subsequently attack and cleave coordinated dioxane. The unusual stability of the diamagnetic bis(anisyl)chromium(II) compound can be attributed to the fact that it is a polymeric substance containing Cr–Cr bonds and "bridging anisyl groups." Examination of molecular models shows that the methoxyl group of an anisyl unit bonded to one chromium may coordinate to an adjacent chromium center, as

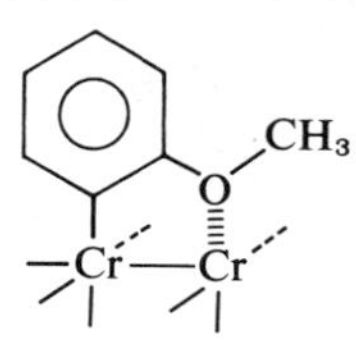

236

illustrated in **236**. Tris(*p*-anisyl)chromium is stable in diethyl ether and under more forcing conditions gives only traces of π-complex mixture [395].

The presence of two *ortho*-alkyl groups in the aryl unit bonded to chromium completely inhibits the σ- to π-rearrangement. Thus, neither tris(2,4,6-trimethylphenyl)chromium(III) nor bis(2,4,6-trimethylphenyl)chromium(II) rearrange to π-complex [391,401,403]; though they do undergo thermal decomposition to give mesitylene and bimesityl [401,404]. The σ- to π-rearrangement is also inhibited by the presence of certain ortho-chelating groups in the aryl unit bonded to chromium. Thus, the compounds (**237**–**239**) may be recrystallized from warm diethyl ether, and apparently do not rearrange to π-complex [342,343,346].

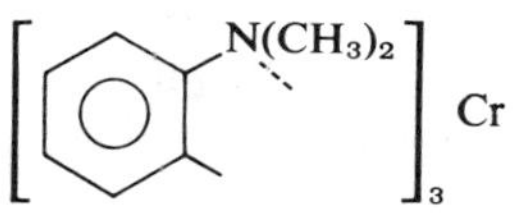

237

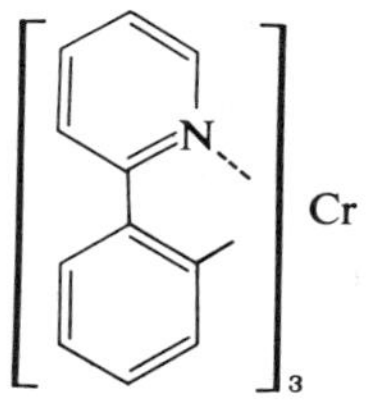

238

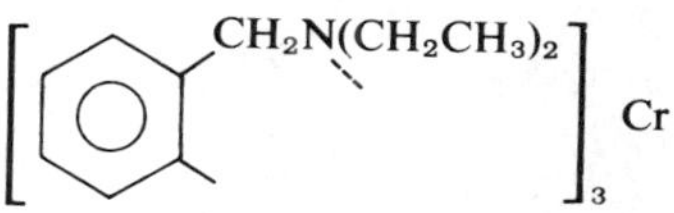

239

The limited information available does not permit any definitive decision to be made concerning the role of substituents on the ease of the σ- and π-rearrangement of tris(aryl)chromium compounds. It may be that the inductive effect of the substituent influences the electron density in the C_{aryl}–Cr bond. This in turn could determine whether the arylchromium compound undergoes simple homolytic thermal decomposition (to give the arene and biaryl) or undergoes the σ- to π-rearrangement. Two points do, however, emerge from the data in Table 3.6, namely, it seems that a free ortho position in the aryl ring bonded to chromium and a free coordination site on the chromium center are prerequisites for the σ- to π-rearrangement.

As mentioned earlier [Eq. (3-150)], when tris(aryl)chromium(III) compounds undergo σ- to π-rearrangement, they give a mixture of three bis-(arene)chromium π-complexes (containing biaryl and arene units complexed to Cr). An understanding of the mechanism of the rearrangement, therefore, involves knowing the particular type of biaryl (or biaryls) formed in the case of substituted aryl species and the origin of the hydrogen required for the transformation of the σ-bonded aryl group to an arene unit.

The results, summarized in Table 3.6, show that the symmetrical biaryl is formed consistently in the σ- to π-rearrangement of substituted arylchromium compounds, i.e., the coupling of the aryl groups to give both the free biaryl and that complexed to chromium, involves the carbon atoms originally bonded to chromium. This excludes any mechanism that involves a benzyne-type intermediate or the random attack of one aryl group by a second aryl radical.

Studies with the phenyl [405,406], 3- and 4-methylphenyl [395] systems showed that the hydrogen required for arene formation was acquired at the final hydrolysis step. Thus, treatment of the "black intermediate" from any of the three systems with D_2O, in place of H_2O, resulted in the incorporation of substantial quantities of deuterium into the arene unit complexed to chromium and only minor quantities into the biaryl unit. Mass spectroscopic analysis of the deuterioarenes (deuteriotoluenes) showed that up to two deuterium atoms were incorporated in an arene unit. This suggests that a σ-bonded arylchromium compound is the precursor to the π-complex. Confirmation of this comes from the observation that carbonation of the "black intermediate" from $(C_6H_5)_3Cr$ gave π-(benzoic acid)-π-(biphenyl)-chromium [15]. Furthermore, treatment of the "black intermediate" from tris(3- and 4-methylphenyl)chromium with CH_3I led to the formation of *meta*- and *para*-xylene both free and in the bis(arene)chromium π-complex mixture [395]. The foregoing results argue strongly against the suggestion that the precursor to the bis(arene)chromium π-complex is some species containing aryl radicals complexed to the chromium center [407,408], while

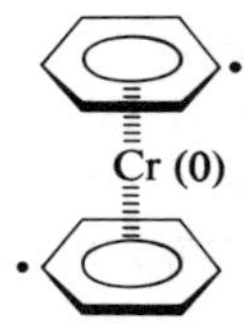

they support the contention that the precursor to the π-complex is a σ-bonded organochromium(I)-π-arene complex, e.g., (**240**). The latter may react with water to give directly the bis-π-(arene)chromium(I) [Eq. (3-153)] or alternatively with methyl iodide to give the π-(methylarene)-π-(arene)chromium(I) iodide.

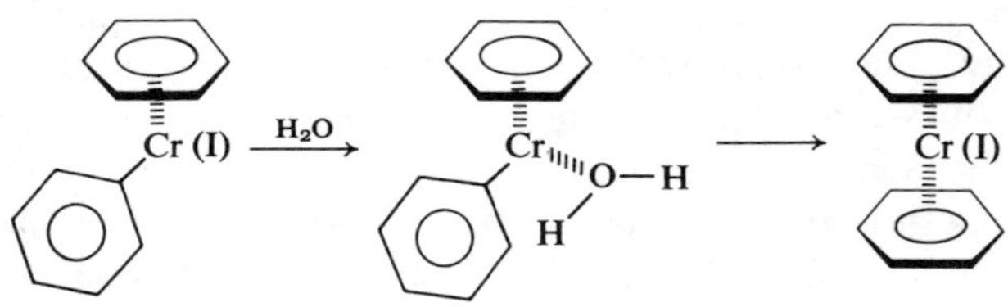

(3-153)

240

As regards the overall rearrangement process, Hähle and Stolze have found that $(C_6H_5)_3Cr(THF)_3$ loses, reversibly, one molecule of tetrahydrofuran to give $(C_6H_5)_3Cr(THF)_2$. The latter undergoes rearrangement, on treatment with diethyl ether, to give *in solution* benzene, biphenyl, and π-bis(biphenyl)chromium(0) Eq. (3-154), and *as precipitate* a black paramagnetic solid. Hydrolysis of the black solid now gives π-bis(benzene)-

$$(C_6H_5)_3Cr\,(THF)_3 \rightleftharpoons (C_6H_5)_3Cr(THF)_2 \longrightarrow$$

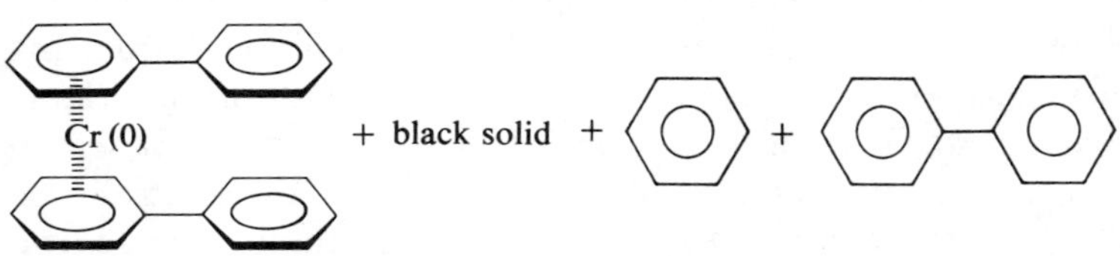

(3-154)

chromium and π-(benzene)-π-(biphenyl)chromium, isolated finally as the chromium(I) salts together with the products shown in Eq. (3-155) [388,409]. Since, however, the black solid contains reducing species (liberation of H_2

from H_2O and D_2 from D_2O), it is not possible to say what the valency state of the chromium is in the π-bis(arene) complexes at the moment of their formation. Analogous results were obtained in the rearrangement of tris-(4-phenylphenyl)chromium and tris(3- and 4-methylphenyl)chromium compounds [394]. It is clear, therefore, that the bis(biaryl)chromium(0) π-complex is formed before the hydrolysis stage and the two other bis(arene)chromium π-complexes result from hydrolysis.

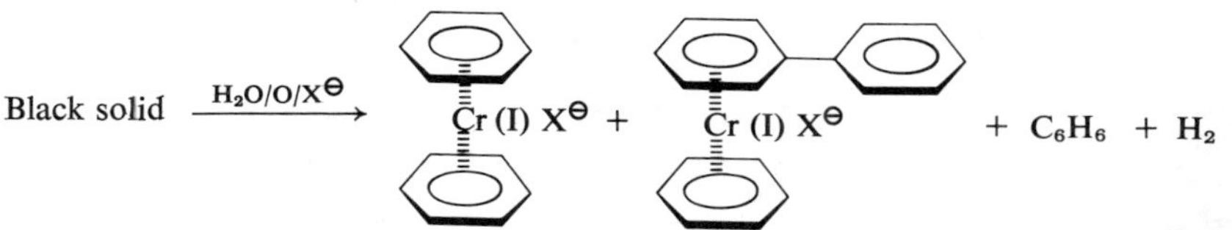

(3-155)

The accumulated results may be interpreted in terms of Scheme 3-28. Thus, assuming that the rearrangement is unimolecular, the reversible loss of tetrahydrofuran, from $(C_6H_5)_3Cr(THF)_3$ may be accompanied by a change from octahedral (**241**) to trigonal bipyramidal coordination **242**. Rearrangement of **242** could give the half-sandwich compound **243** by direct attack of one phenyl group on a second phenyl group bonded to chromium (i.e., by reductive coupling). The half-sandwich compound **243** could disproportionate to the π-bis(biaryl)chromium(0) compound (**244**) and the bis(aryl)chromium(II) species **245**. This would account for the formation of **244** prior to the hydrolysis of the black intermediate. Alternatively, the half-sandwich compound **243** could react with D_2O to give the π-(monodeuteriobenzene)-π-(biphenyl)chromium(I) complex **246**.

The bis(aryl)chromium(II) compound (**245**) could proceed either by hydrogen abstraction from solvent to the half-sandwich compound **247** or, by hydrogen abstraction, from an adjacent σ-bonded phenyl group to the "benzyne-π-arene chromium(II) complex" formulated as **248**. Reaction of the compounds with D_2O and D_2 would give respectively π-(monodeuteriobenzene)-π-(benzene)chromium(I) and π-(dideuteriobenzene)-π-(benzene)-chromium(0). Introduction of deuterium into the biphenyl portion of the π-complex could arise by tautomerism of **243** to the π-(benzene)-σ-(biphenyl)-chromium(I) compound **249** via a hydridochromium(III) compound.

ii. Tris(allyl)- and Tris(vinyl)chromium Compounds. Kurras has found that tris(allyl)chromium(III) undergoes thermal decomposition to give 1,5-hexadiene and tetra(allyl)dichromium(II) [410]; the same transformation can be effected photochemically [411,412]. The mechanism of the reaction has not been investigated and it is not known whether the biallyl is formed by the head-to-head, head-to-tail, or tail-to-tail coupling of two allyl groups, and there is no report of the formation of propylene during the reaction. However,

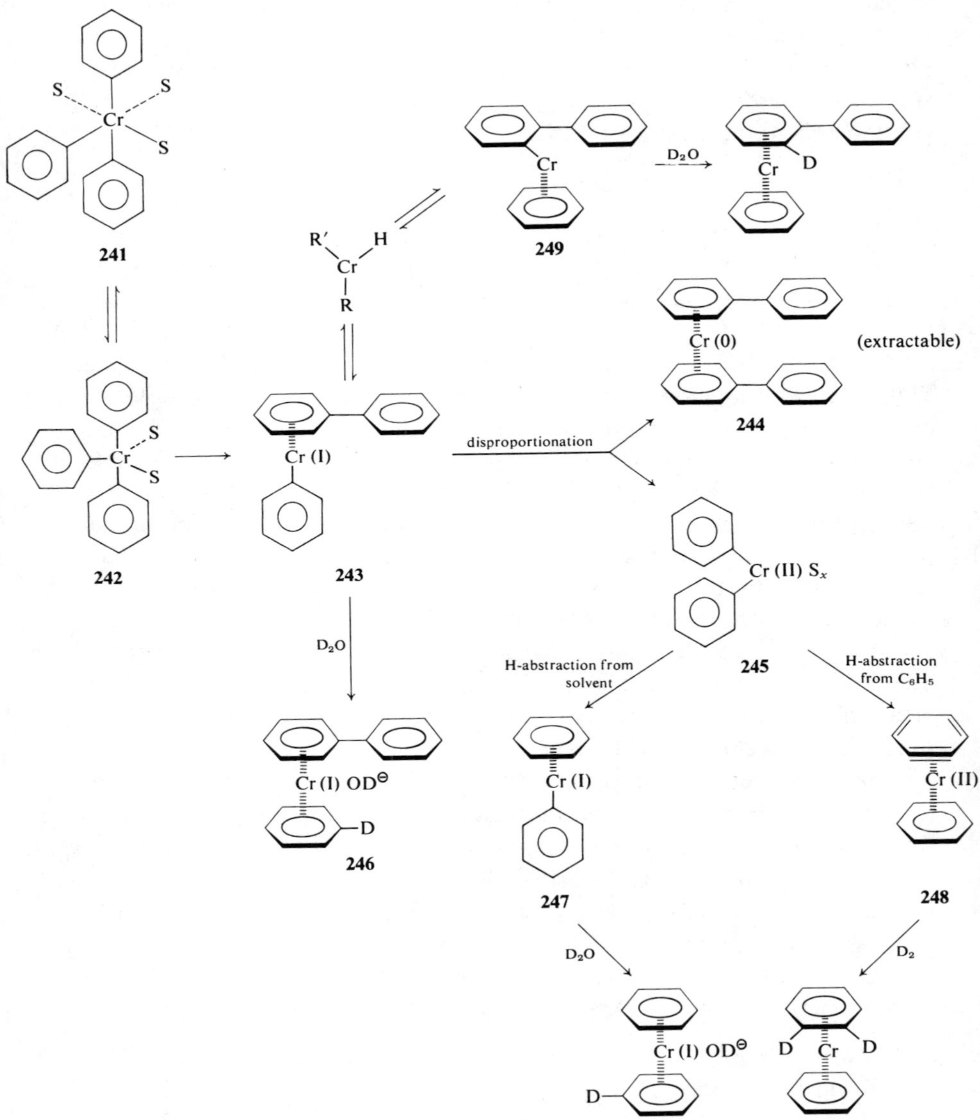

Scheme 3-28. σ- to π-rearrangement of $(C_6H_5)_3Cr(THF)_3$, $R = C_6H_5$, $R' = C_6H_5C_6H_4$.

since the end-product is a bridged dimeric chromium(II) compound [413], it is possible that the reaction proceeds via allyl-bridge formation followed by the reductive elimination of two allyl groups and the concomitant formation of the Cr–Cr bond, as outlined in Eq. (3-156).

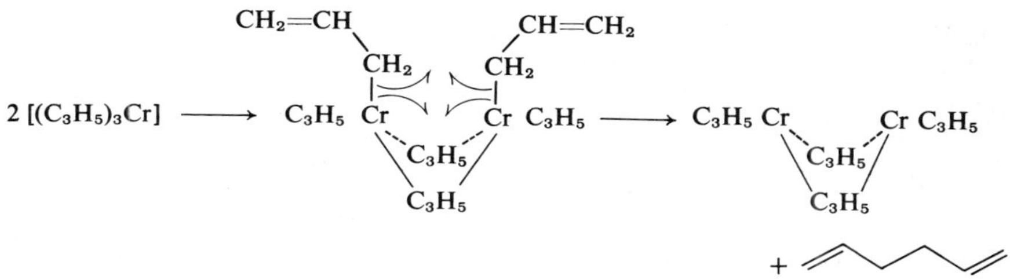

(3-156)

There are no crystalline vinylchromium compounds known; however, vinyl- and substituted vinylchromium systems have been prepared by the interaction of the appropriate organomagnesium halide and $CrCl_3(THF)_3$ [311,354,359]. These systems are stable at low temperatures but undergo fragmentation, when their solutions are allowed to warm to room temperature, to give butadiene or the substituted butadienes as major product. The reactions have not been studied in any detail; however, the presence of ethylene and acetylene in the products from tris(vinyl)chromium [311] suggests that homolysis of the carbon–chromium bond occurs. The radicals may then either couple (butadiene) or they may abstract hydrogen from a second vinyl group to give acetylene and ethylene. In the case of tris-β-(α-phenylstyryl)chromium $\{[(C_6H_5)_2C{=}CH]_3Cr\}$ there are no hydrogen atoms on the carbon beta to the chromium, therefore the dimer should be (and is) the main fragmentation product [354].

iii. Tris(alkyl)chromium Systems. The early studies on alkylchromium compounds suggested that though these compounds might be formed by the interaction of an alkylmagnesium halide and $CrCl_3$ they were exceedingly unstable. Their fragmentation was variously described [414–417] as a homolytic process leading to alkane and 1-alkene and colloidal or "active chromium."

Zeiss and his collaborators showed, however, that the interaction of methyl- and ethylmagnesium halides and $CrCl_3(THF)_3$ in tetrahydrofuran led to the formation of homogeneous red-brown solutions of alkylchromium compounds [348,361]. Other tris(alkyl)chromium compounds have been prepared since [418–420] and their common feature is that though they are stable in solution at −70° to −80°C for prolonged periods of time, they all undergo total fragmentation on warming to room temperature. With the exception of some bulky tris(alkyl)chromium(III) compounds, e.g., the unsolvated tris(norcamphyl)chromium(III) [421], no solvated σ-bonded tris(alkyl)-chromium(III) compound has been isolated in a crystalline state. All the known alkylchromium systems, however, react with $HgCl_2$ and D_2O to give the organomercurial and the deuterioalkane (free from any alkene) and this together with the ESR spectra of their solutions [422,423] suggest that these "systems" do consist of octahedral σ-bonded alkylchromium(III) compounds. The nature of the ligands associated with the chromium center is not known and could be either tetrahydrofuran or $ClMgX(THF)_n$ (i.e., compounds with Cr–Cl–Mg bridges).

Zeiss, Light, and Sneeden have studied the fragmentation reactions of several alkylchromium systems and in particular those of certain specifically deuterated systems. The apparent stability sequence for the alkylchromium systems is CH_3 > *n*-alkyl > *sec*-alkyl > *tert*-alkyl. Alkylchromium systems undergo fragmentation on warming to room temperature, to give after deuteriolysis, unequal proportions of nondeuterated alkane and 1-, and 2-alkenes, H_2, HD, and D_2 [418–420]. A closer study of the precise products formed at different temperatures has shown that at low temperatures these consist of alkanes and 1-alkenes and at higher temperatures they consist of alkanes, 1- and 2-alkenes, and a hydridochromium species. The overall fragmentation of these systems involves, therefore, the superposition of at least three processes; the first leads to the formation of the alkane and the terminal olefin (1-alkene), the second to the formation of the alkane, 1-alkene and a hydridochromium compound and the third leads to the formation of the 2-alkene (presumably an isomerization process). An insight into the reactions involved in these fragmentation processes was obtained from a study [424] of the thermal decomposition of the specifically deuterated systems $(C_6H_5CH_2CH_2CD_2CH_2)_3Cr(THF)_n$ and $(C_6H_5CH_2CH_2CH_2CD_2)_3$-$Cr(THF)_n$ [424,425]. This shows that (1) the deuterium on the carbon atoms alpha and beta to the chromium is transferred to the metal (formation of HD on hydrolysis); (2) the deuterium on the carbon atom beta to the chromium is acquired by one of the alkyl groups originally bonded to the metal (formation of 1,2,2-trideuterioalkane); and (3) the alkyl groups originally bonded to chromium acquire hydrogen (formation of dideuterioalkane).

The initial (low-temperature) fragmentation process can be formulated analogously to that of the monoalkylchromium compounds. Thus, one of the alkyl groups bonded to chromium can either abstract hydrogen from coordinated solvent to give the dideuterioalkane, e.g., Eq. (3-157) ($R = C_6H_5CH_2CH_2$), or it can abstract deuterium from the β-position of an

$$(RCD_2CH_2)_3Cr(THF)_n \xrightarrow{\text{solvent}} RCD_2CH_3 + (RCD_2CH_2)_2Cr(THF)_n \qquad (3\text{-}157)$$

adjacent alkyl group to give trideuterioalkane and monodeuterioalkene. The latter reaction may occur either in a mononuclear compound, to give an organochromium(I) compound **250**, Eq. (3-158) ($R = C_6H_5CH_2CH_2$), or in a dinuclear complex, to give the bis(organo)chromium(II) compound **251** Eq. (3-159) ($R = C_6H_5CH_2CH_2$). At higher temperatures, either **250** or **251**

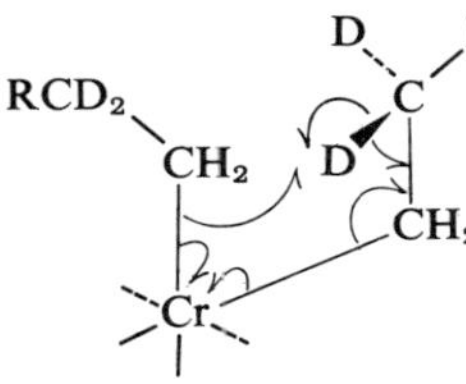

$$\longrightarrow RCD_2CH_2D + RCD{=}CH_2 + \underset{\mathbf{250}}{RCD_2CH_2Cr(I)(THF)_n} \qquad (3\text{-}158)$$

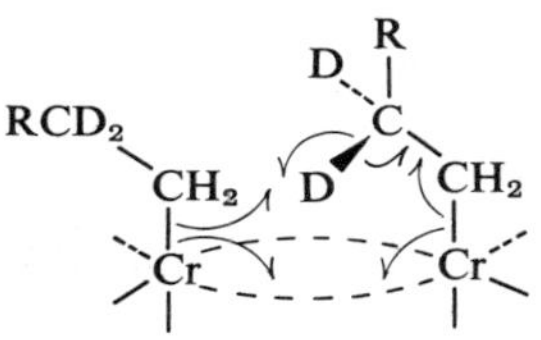

$$\longrightarrow RCD_2CH_2D + RCD{=}CH_2 + 2\,\underset{\mathbf{251}}{(RCD_2CH_2)_2Cr(II)(THF)_n} \qquad (3\text{-}159)$$

may undergo β-metal deuteride elimination to give the 1-alkene and the catalytically active deuteriochromium species, and the latter subsequently isomerizes the 1-alkene to the 2-alkene (see olefin isomerization).

The formation of a deuteriochromium species, from the 1,1-dideuterio-alkylchromium compound [424] can be interpreted in terms of α-metal-deuteride elimination to give a transient carbenoid complex that subsequently rearranges to the alkene, Eq. (3-160) [$R = C_6H_5(CH_2)_2$].

$$(RCH_2CD_2)_nCr \longrightarrow [RCH_2CD\colon \longrightarrow RCH{=}CDH] + DCr(CD_2CH_2R)_{n-1} \qquad (3\text{-}160)$$

Zeiss and Light have shown that the tris(CD_3)Cr system undergoes fragmentation by α-metal-deuteride elimination and suggest that a methyl group when bonded to chromium may undergo tautomerism: $CD_3Cr \rightleftharpoons D\text{-}CrCD_2$ [362]. Pu and Yamamoto have recently interpreted their results concerning the H-D exchange in the system $D_2/CH_3Rh(PPh_3)_3$ in terms of an analogous tautomerism [426].

Evidence from other alkyl transition metal systems suggests that α-metal-hydride elimination may indeed be quite common. Thus, Overberger and his colleagues had concluded, in 1965, from their results on the polymerization of α-deuteriostyrene by Ziegler–Natta catalysts, that α-metal-hydride elimination was a terminating step [427].

Also, Wilkinson and his colleagues found that the reaction of $(CH_3)_3$-$SiCH_2MgCl$ with $NbCl_5$ did not give the penta(alkyl)niobium compound but the carbyne-bridged dimer [428].

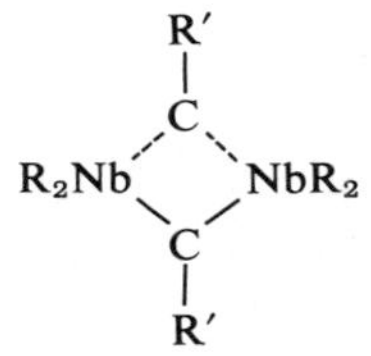

$R = (CH_3)_3SiCH_2$, $R' = (CH_3)_3Si$

In this case, however, it is also possible that the "α-hydrogen" is lost in a process whereby it is abstracted by an alkyl group bonded originally to the same or an adjacent niobium atom. Finally, α-metal-hydride elimination would explain the scrambling of the deuterium observed by Chatt and his colleagues during their studies on the fragmentation of the specifically deuterated ethylplatinum complex [429]:

$$CH_3CD_2PtBr(PEt_3)_2 \longrightarrow [CH_3CD: \longrightarrow CH_2{=}CDH] + DPtBr(PEt_3)_2$$

The accumulated evidence proves that the fragmentation of alkylchromium systems involves the major processes of (1) homolysis of the carbon–chromium bond with a concomitant hydrogen abstraction from either coordinated solvent or from the β-position of a second alkyl group bonded to chromium; (2) the four-center β-metal-hydride elimination; and (3) α-metal-hydride elimination.

There are doubtless other processes involved in these fragmentation reactions (see under isomerization), however, the present results show that it is false to attempt to constrain the metal-hydride elimination process to one involving only the hydrogen atoms on the carbon beta to the metal center; rather, the metal-hydride elimination process will involve the hydrogen atom (or atoms), which is (are) contiguous to the metal center and these may be on carbon atoms α, β, γ, or even δ to the metal center. In the latter case, the actual elimination of the alkyl residue may well proceed via the oxidative addition of the C–H bond to the metal center and reductive elimination of an alicyclic residue.

d. Olefin Isomerization

Tris(aryl)chromium compounds and (alkyl)chromium(III) systems, per se do not isomerize olefins or dienes. Zeiss and Sneeden found, however, that the thermal decomposition products from both types of compounds, i.e., "the black intermediate," from tris(phenyl)chromium (see σ- to π-rearrangement), and the chromium-based fragmentation product(s) from the tris(alkyl)chromium systems did isomerize terminal olefins and dienes to the corresponding 2-alkenes and bis-Δ^2dienes [418–420,424,430–433]. In the case of the terminal olefins, the *trans*-2-alkene was obtained as the major product and in the case of dienes, a mixture of *cis,cis- cis,trans-* and *trans,trans*-bis-Δ^2dienes were obtained. In all cases the saturated alkane was found as minor product. It has been established earlier that alkylchromium systems undergo fragmentation to give hydridochromium species [418–420,424] and it has been suggested that the black intermediate [from tris(aryl)chromium compound] is also a hydridochromium species [397]. The foregoing results on the isomerization of olefins can be explained in terms of the reversible cis addition and elimination of Hcr (cr = chromium and associated ligands) to complexed olefin or diene.

Addition of the chromium center to the terminal (α) carbon would give an *n*(alkyl)cr compound and elimination of Hcr would regenerate the terminal olefin (a) in Eq. (3-161). Addition of the chromium center to the β-carbon atom would give a *sec*-(alkyl)cr compound (**252**) and elimination of Hcr can now give either the terminal olefin [(b) in Eq. (3-161)] or the 2-alkene [(c) and (d)] in Eq. (3-161). Cis elimination of Hcr from **252** (to give the 2-alkene)

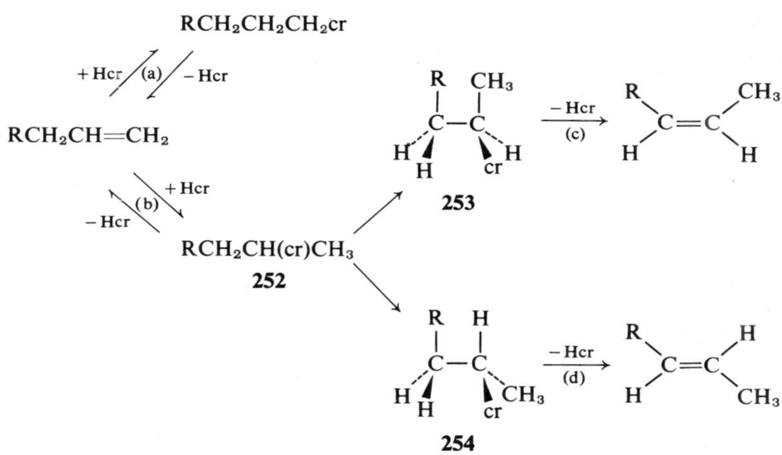

(3-161)

constrains the molecule to adopt either the conformation **253** or **254** [Eq. (3-161)]. The repulsion between the substituents on neighboring carbon atoms is least in **254** and will therefore make it the preferred conformation for the elimination step, which in turn will favor the formation of the *trans*-2-alkene, as in (d), Eq. (3-161). There is very little known concerning the initial step of this isomerization process, i.e., the addition of Hcr to the olefin and what role, if any, it plays in the selectiveness observed in the isomerization (i.e., the formation of the 2-alkene). The addition may proceed either by the coordination of the olefin to the metal followed by the transfer of H and cr to the complexed olefin:

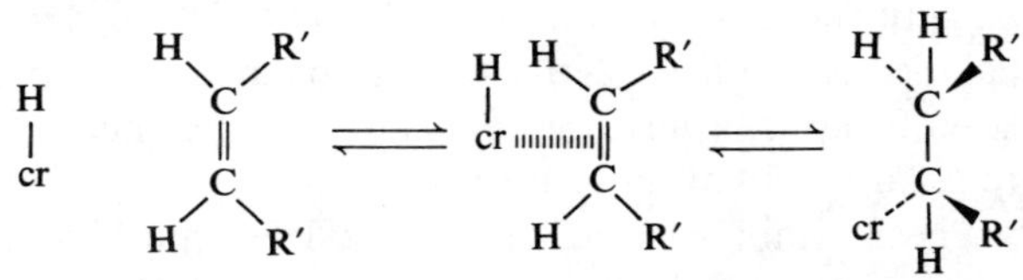

or it may proceed by the direct dipolar addition of Hcr to olefin without a preequilibrium step:

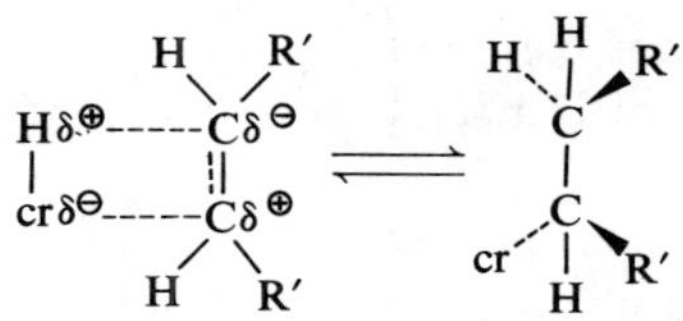

In the former case, a free coordination site on chromium is required and the selective formation of the 2-alkene reflects the better ability of a terminal olefin to coordinate to a metal center as compared to an internal olefin.

In the latter case, no free coordination site on chromium is required and the selective formation of the 2-alkene reflects the difference in the steric encumberment of terminal and internal olefins and also differences in their polarization and polarizability.

The presence of minor amounts of alkane (derived from olefin substrate) in these isomerization reactions may be accounted for in terms of the formation of a mixed bis(alkyl)chromium compound **255**, Eq. (3-162). Fragmenta-

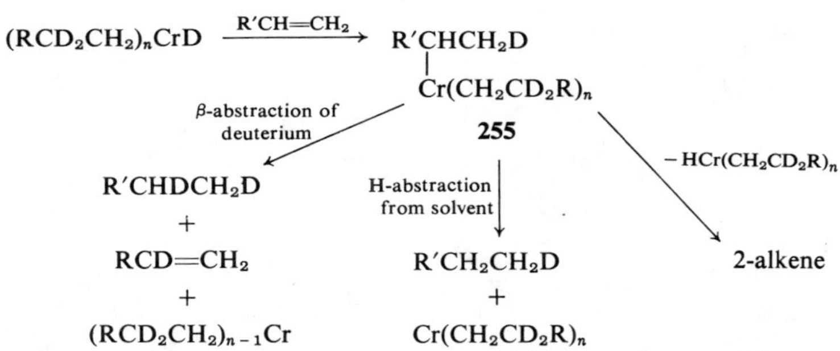

(3-162)

tion of **255** may proceed not only by β-metal-hydride elimination (to olefin) but also by abstraction of hydrogen from solvent or a second alkyl group, to give the alkane, Eq. (3-162). Experiments with tris(2,2-dideuterio-4-phenylbutyl)chromium [433] confirmed that part of the hydrogen required for alkane formation (from olefin substrate) does indeed come from the carbon atom beta to the chromium center.

Clearly the mechanism of olefin isomerization is intimately associated with the fragmentation pathways available to alkylchromium compounds. Though the majority of the observed facts in both domains can be rationalized in terms of the processes outlined in the foregoing pages, there are others that cannot. Thus, in the 1-alkene, formed in the fragmentation of tris-(1,1-dideuterio-4-phenylbutyl)chromium, there is transfer of deuterium over C_3, C_2, and C_1, without concomitant isomerization of the double bond [424]. Frankel, Selke, and Glass observed a similar phenomenon in the system $D_2/(C_6H_6)Cr(CO)_3$/methyllinoleate (a 1,4-diene), namely, the rapid introduction of deuterium, without isomerization and/or reduction of the diene [204]. Therefore, besides the (Hcr and Dcr) addition-elimination reactions (which would account for the scrambling of D over C_1 and C_2), another hydrogen transfer process involving the hydrogen on the γ-carbon atom must also be occurring. This transfer could proceed via a π-allylic complex derived from the olefin:

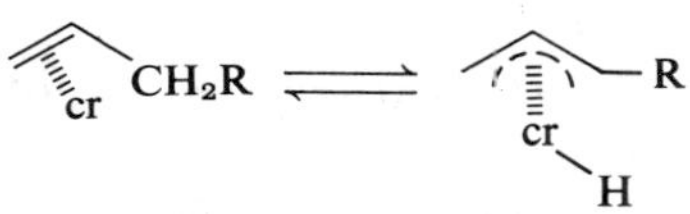

or a metallocycle derived from the alkylchromium compound:

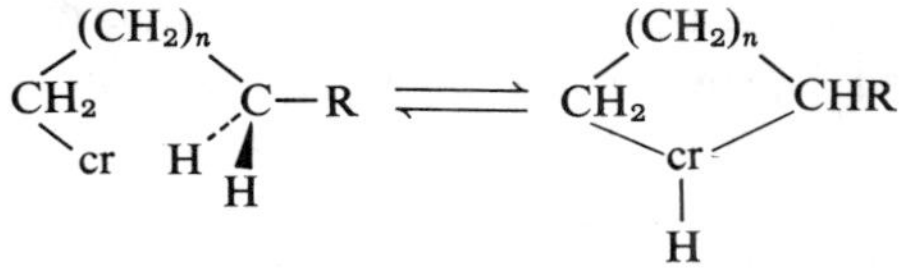

e. Catalytic Reactions

i. Hydrogenation. There are many reports of the formation of "hydrogenation catalysts" from the interaction of hydrogen with reacting solutions of aryl- and alkylmagnesium halides and $CrCl_3$ [434–436]. The precise nature of the active chromium catalyst has never been established; however, Weichselfelder and Thiede [434] have suggested that it could be CrH_3. The chromium-based fragmentation products from the tris(alkyl)chromium systems are also catalysts for the hydrogenation of olefins [383] though again the nature of the active species is not known.

Tris(phenyl)chromium reacts with hydrogen to give, by hydrogenolysis of the C–Cr bond, benzene and this, in turn, is hydrogenated to cyclohexane [437,438].

ii. Polymerization. It has been reported that butadiene is polymerized to a mixture of 1,2- and *trans*-1,4-polybutadiene by a reacting solution of an organomagnesium halide RMgX (R = CH_3, $CH_2{=}CH$, CH_3CH_2, C_6H_5)

and $CrCl_3$ [439]. However, since tris(phenyl)Cr(THF)$_3$ may be recrystallized from a solution of butadiene in tetrahydrofuran [383], it is unlikely that the organochromium(III) compound is the active catalyst in these polymerizations.

Allylchromium compounds, on the other hand, are catalysts for the polymerization, and oligomerization, of butadiene [338,440] and other unsaturated substrates (e.g., ethylene [338], methylmethacrylate, and acrylonitrile [441]). W. Oberkirch found that tris(allyl)chromium, in benzene solution, polymerized butadiene to 1,2-polybutadiene, and, in heptane solution polymerized ethylene to a polyethylene ($M = 1.88 \times 10^6$) [338]. Bis(allyl)chromium iodide, on the other hand, trimerized butadiene to a mixture of all *trans*- and *trans, trans, cis*-cyclododecatriene [338]. Dolgoplosk and his associates have studied, in detail [e.g., 442–445], the polymerization of butadiene by allyl- and crotylchromium compounds. The overall results are contained in a review [440] and may be thus summarized:

(1) The nature of the polymer obtained depends, critically, upon the ligands associated with the chromium center, thus in the absence of halide ion or oxygen the product is a 1,2-polybutadiene, in the presence of halide it is a 1,4-*cis*-polybutadiene and in the presence of oxygen a 1,4-*trans*-polybutadiene.

(2) The polymerization proceeds via π-allylic intermediates, formed by the transfer of an allyl group from chromium to complexed butadiene, Eq.

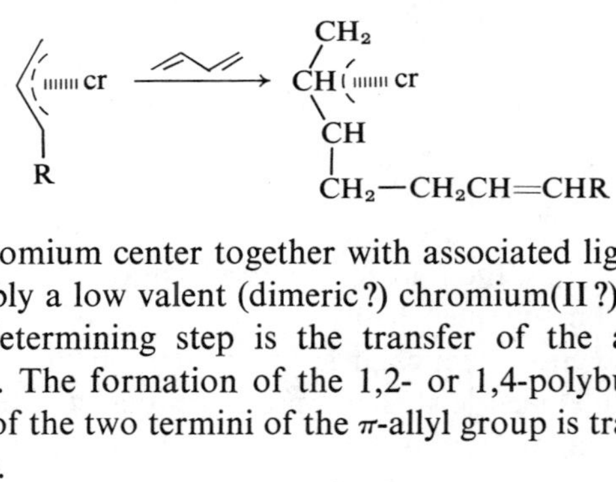

(3-163)

(3-163) (cr = chromium center together with associated ligands). The active catalyst is probably a low valent (dimeric?) chromium(II?) compound.

(3) The rate-determining step is the transfer of the allylic residue to complexed diene. The formation of the 1,2- or 1,4-polybutadiene is determined by which of the two termini of the π-allyl group is transferred to complexed butadiene.

(4) The stereoregulation of the polymerization is effected by the overall configuration of the chromium-π-allyl complex. This, in turn, is presumed to be determined by the mode in which the entering butadiene molecule is

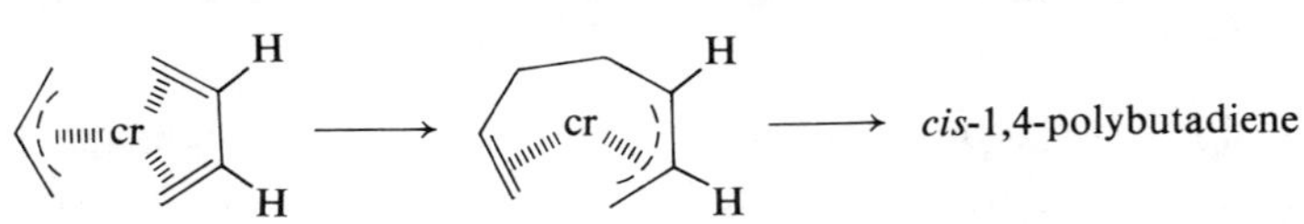

(3-164)

complexed to the chromium center. Coordination of both double bonds of a *cisoid*-butadiene to the chromium gives an intermediate **256** in which the hydrogens of the future CH═CH are disposed "cis" to each other, Eq. (3-164) (cr = chromium and associated ligands). Coordination of one double bond of a transoid butadiene gives an intermediate **257** in which the hydrogens of the future CH═CH are trans to each other, Eq. (3-165).

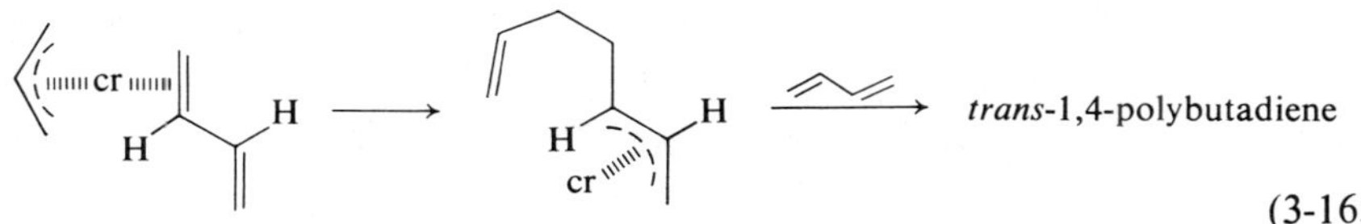

257

(3-165)

An alternative explanation would be that the stereochemistry of the final polymer is determined by the monomeric chromium(III) ⇌ dimeric chromium(II) equilibrium in the π-(allyl)chromium compound. Thus, the monomer could complex selectively with the *transoid*-butadiene (via one double bond) while the dimer could complex selectively with the *cisoid*-butadiene (via one double bond at each chromium center).

(5) Termination of the polymerization can occur either by ligand exchange, or more likely, by metal-hydride elimination, Eq. (3-166) (R = polymer chain). The hydridochromium compound can react anew with butadiene to regenerate an active π-(allyl)chromium compound, Eq. (3-166).

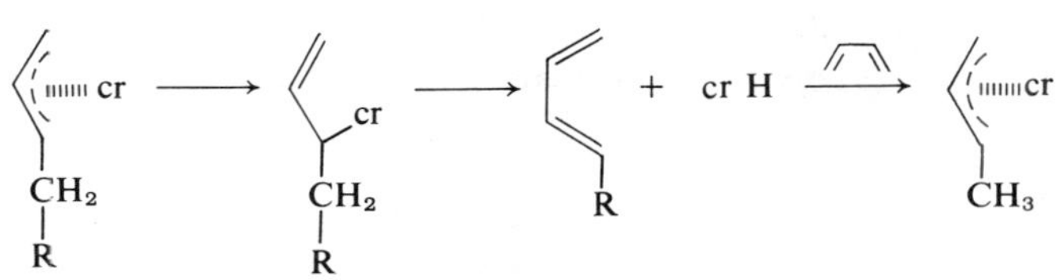

(3-166)

Ballard and Medinger have studied the polymerization of methyl methacrylate by tris(allyl)chromium(III) [412,441]. They conclude that the active species is a six-coordinate chromium(III) complex formed by the transfer of one of the allyl groups from chromium to complexed ester. The process is envisaged as involving an initial coordination of ester (via the C═C) to chromium to give the σ-π-allyl complex **258**, Eq. (3-167). Coordination of a second molecule of ester induces an allyl group transfer to give the active complex **259**. Coordination of successive molecules of ester leads to polymer formation as in **260**. Termination is presumed to involve the transfer of one of the remaining allyl groups, bonded to chromium, to the ester group of the growing chain to give the inactive chromium alkoxide **261**, Eq. (3-167) [R = $CH_2{=}C(CH_3)COOCH_3$].

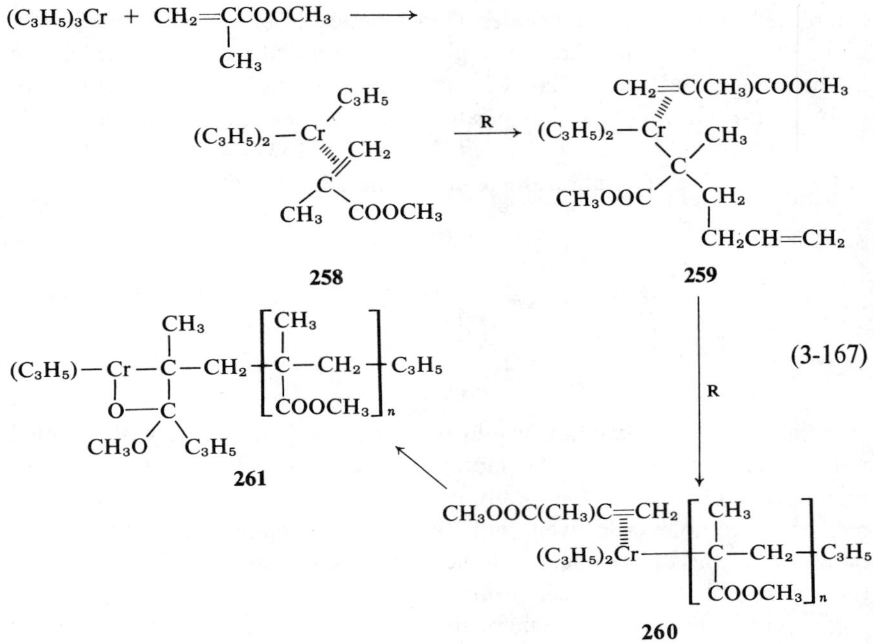

(3-167)

Arakawa has described the polymerization of ethylene by a catalyst derived from bis(organo)chromium halides and $AlBr_3$ [337]. Studies with tris(alkyl)chromium systems have shown that, though under normal temperatures and pressures they isomerized olefins and dienes (see the foregoing Section d), they did not polymerize olefins or nonconjugated dienes. Butadiene, however, is in part reduced (butene-2) and in part polymerized [432].

4. Reactions of σ-Bonded Organochromium(II) Compounds

The information concerning the chemistry of σ-bonded mono- and bis-(organo)chromium(II) compounds is sparse, and this is due largely to the extreme difficulty of preparing such compounds (see Chapter 1, Table 1.5).

Zeiss and Sneeden have reported that bis(2-methoxyphenyl)chromium(II) is inert toward various unsaturated substrates (e.g., olefins, dienes, and 2-butyne); it does, however, react with $HgCl_2$ (to give 2-methoxyphenyl mercuric chloride), with D_2O to give (2-deuterioanisole), with 2-(2′-pyridyl)-imidazole {to give tris[2-(2′-pyridyl)imidazolato]chromium(III)} and with 2,2′-bipyridyl [to give bis(2-methoxyphenyl)-bis(2,2′-bipyridyl)chromium(II)]. This last compound is readily oxidized by various reagents (e.g., O_2, allylic and benzylic halides) to give the corresponding cationic chromium(III) compound, and it will polymerize acrylonitrile [340].

G. Lithium and Sodium Poly(organo)chromate(III) and (II) Complexes

Hein and his colleagues have prepared several monomeric lithium or sodium poly(organo)chromate(III) complexes and monomeric or dimeric lithium poly(organo)chromate(II) complexes (Table 3.7 and also Table 1.6, Chapter 1, Section G). These compounds are all relatively stable thermally, but they are highly reactive toward oxygen, water, and other protic solvents. This reactivity makes the study of their chemistry an exceedingly delicate task, the more so since in some instances the products are also highly reactive organometallic compounds or complexes. Therefore, it is not surprising that the chemistry of the present group of complexes has not been extensively investigated. The bulk of the information available concerns the reactions of the hexa(aryl)chromate(III) complexes, and some of these reactions (e.g., fragmentation reactions and reactions with H_2O, $HgCl_2$, and I_2) have already been mentioned in Chapter 1, Section G, and Chapter 2, Section F.

1. Reactions of Lithium or Sodium Hexa(organo)chromates(III)

a. Fragmentation Reactions

The hexa(organo)chromate(III) complexes undergo various fragmentation reactions (i.e., reactions in which one or more of the σ-bonded ligands are lost) either when treated with an organic solvent, or when treated with $CrCl_3(THF)_3$. These fragmentation reactions are of two types: those in which the valency state of the chromium center does not change and those

TABLE 3.7

Lithium Poly(organo)chromate Complexes

	CHROMIUM(II)	
CHROMIUM(III)	MONOMERIC	DIMERIC
$Li_3[CrR_6](S)_n$ octahedral	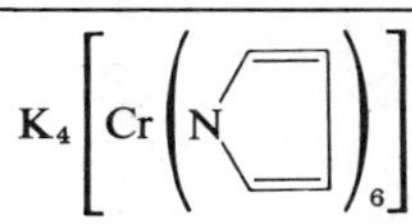 octahedral?	
$Li_2[CrR_5](S)_n$ trigonal bipyramidal	$M_3[CrR_5]$ not known	
$Li[CrR_4 2DME]_2$[a] octahedral?	$Li_2[CrR_4 2\cdot THF](THF)_2$[b] octahedral?	$[Li_2CrR_4]_2$ square pyramidal around Cr
	$M[CrR_3]$ not known	$[LiCrR_3]_2$ tetrahedral around Cr?

[a] DME = 1,2-dimethoxyethane.
[b] THF = tetrahydrofuran.

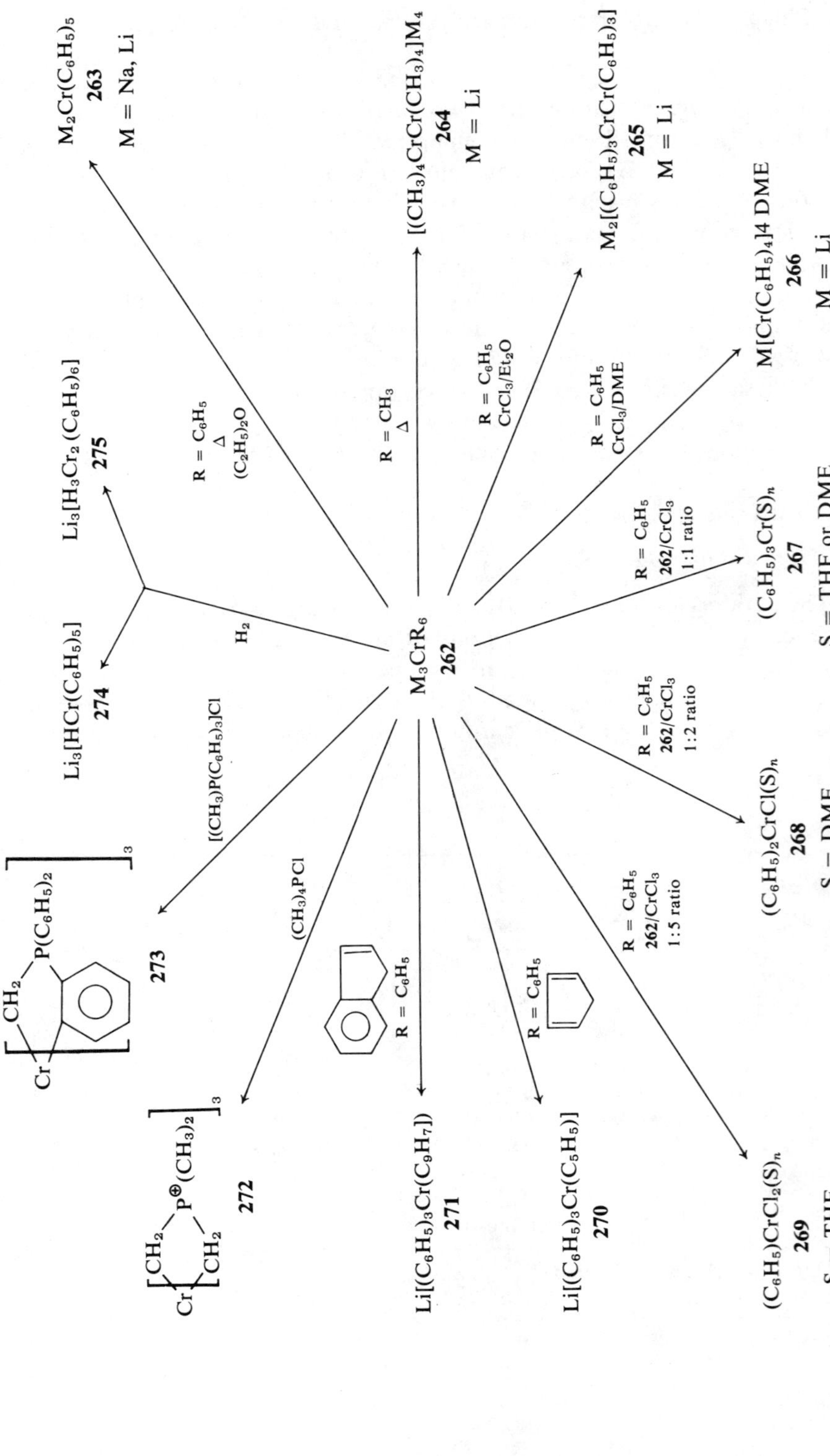

Scheme 3-29. Some reactions of hexa(organo)chromate(III) complexes.

in which it does (see Scheme 3-29). When the hexa(aryl) complex **262** (R = C_6H_5) is treated with 1,2-dimethoxyethane, it loses phenyllithium to give, without any change in the valency of the chromium, the penta(phenyl) complex **263**, Scheme 3-29 [446]. On the other hand, when the hexa(alkyl) compounds (e.g., **262**, R = CH_3) are treated with ether/dioxane, they slowly undergo fragmentation to give the tetra(methyl)chromium(II) dimers (e.g., **264**, R = CH_3) [447]. The fragmentation of the hexa(aryl)complex **262** (R = C_6H_5) to a dimeric chromate(II) complex can be achieved by reaction with $CrCl_3$ in diethyl ether. But in this case the product is the tris(aryl)-chromium(II) dimer **265** (R = C_6H_5) [448]. This reaction probably proceeds via the formation of the tetra(aryl)chromate(III) complex, and it is this compound that undergoes reductive dimerization to **265**, as outlined in

$$4\ Li_3CrR_6 + 2\ CrCl_3 \xrightarrow{\text{ether}} 6\ Li[CrR_4] + 6\ LiCl \longrightarrow \underset{\mathbf{265}}{3\ Li_2[R_3CrCrR_3]} + 3\ R\text{–}R \qquad (3\text{-}168)$$

Eq. (3-168). Some support for this reaction sequence is found in the observation that when the same reaction is carried out in 1,2-dimethoxyethane (DME) as solvent, the product is the DME solvate of the tetra(aryl)chromate-(III) **266** [449]. This tetra(aryl)chromate(III) complex may also be obtained by the reaction of $Na_2Cr(C_6H_5)_5$ with $CrCl_3$ [449] and its stability (as compared with the ether solvate) may be due to the stronger donor character of DME, but it may also be due to the fact that DME can act as a bidentate donor ligand, thereby stabilizing an octahedral configuration within the complex.

b. Ligand Replacement Reactions

Solvated tris(phenyl)chromium(III) **267**, bis(phenyl)chlorochromium **268** and mono(phenyl)dichlorochromium **269** are formed in a metathetical reaction between the hexa(aryl)chromate(III) and the appropriate quantity of $CrCl_3$ [449,450].

Lithium hexa(phenyl)chromate **262** (R = C_6H_5) reacts with several "acidic hydrocarbons" (e.g., cyclopentadiene, indene, fluorene, or pyrrole), and the precise products formed depend upon the relative ratios of the reagents. Thus, with equimolar ratios, one of the chromium-to-phenyl bonds is cleaved to give the pentaphenyl complex $Li_2Cr(C_6H_5)_5$ [451,452]. Excess cyclopentadiene or indene reacts with the hexa(phenyl)chromate to give new complexes **270** and **271** in which three aryl units are replaced by a cyclopentadienyl or indenyl unit [452]; excess cyclopentadiene also reacts with to give, analogously [453]:

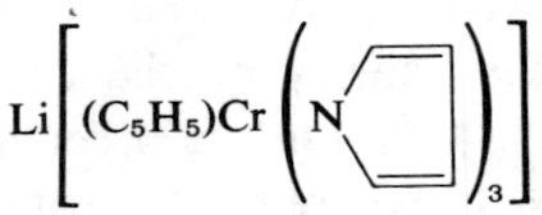

Excess pyrrole reacts with the hexa(phenyl)chromate to give, by replacement of all the C_6H_5 groups [453]:

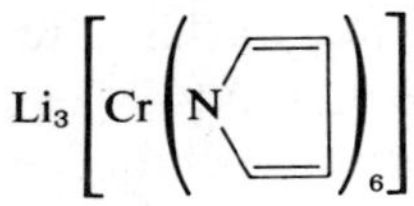

Lithium hexa(phenyl)chromate(III) reacts with methylphosphonium halides to give a new class of zwitterionic hexa(organo)chromate(III) complex, i.e., compounds **272** and **273**. The formation of **273** involves the abstraction of one of the *ortho*-hydrogen atoms of aryl ring bonded to phosphorus [454].

Hein and his colleagues have found that lithium hexa(phenyl)chromate reacts rapidly with one molecule of hydrogen to give, by hydrogenolysis of one Cr–aryl bond, the crystalline hydride $Li_3[HCr(C_6H_5)_5]$ (**274**). This compound reacts slowly with more hydrogen to give a hydrido complex formulated as $Li_3[H_3Cr_2(C_6H_5)_6]$ (**275**) [455].

2. Catalytic Reactions of Lithium(organo)chromate Complexes

There are no reports of the utilization of the hydridochromium complexes **274** and **275** for the hydrogenation or isomerization of simple olefins. It has been reported that the lithium hexa(phenyl)chromate complexes in combination with neighboring group halides (e.g., TiI_4) will effect the polymerization of butadiene [456].

H. Reactions of σ-Bonded Tetrakis(organo)chromium(IV) Compounds

The scant data available suggest that the known tetrakis(organo)chromium(IV) compounds (see Chapter 1, Table 1.7) are relatively inert chemically. Wilkinson and his colleagues report that these compounds do not react with primary amines, alkyl and aryl tertiary phosphines, CS_2, ethylene and liquid alkenes, phenyl acetylene, CO, succinimide, and triphenylmethyl fluoroborate [457,458]. This chemical inertness suggests that in the tetrahedral chromium compounds either there are no sites available for the coordination of the substrates (and therefore also for β-metal-hydride elimination!) or that the steric encumberment around the chromium center is such as to impede the approach of substrate. There are no reports concerning the substitution reactions at the carbon bonded to chromium (e.g., reaction with $HgCl_2$); such reactions, in contrast to hydrolysis and oxidation reactions, should not require free coordination sites on chromium.

The tetrakis(organo)chromium(IV) compounds vary in their reactivity toward oxygen and nitric oxide and the apparent sequence of their reactivity is methyl > neopentyl > trimethylsilylmethyl > neophyl > 1-camphyl > 1-norbornyl > tritylmethyl [457–459].

There are no published reports on the systematic study of the fragmentation of the tetrakis(alkyl)chromium(IV) compounds, though Kruse mentions that $[(CH_3)_3C]_4Cr$ decomposes in heptane solution to give 6.5 parts of isobutane to 1 part of isobutylene (the activation energy for this process being 29 kcal/mole at temperatures of 55°–80°C). Furthermore, it is reported that the compound with eight β-hydrogens $(CH_3CH_2CH_2CH_2)_4Cr(IV)$ is stable for several hours at 0°C (though in this case the fragmentation products are not given) [460]. These scant facts suggest that the main pathway for the fragmentation of these compounds probably does not involve a β-metal-hydride elimination process but may involve homolytic and/or radical chain processes. The reason why β-metal-hydride elimination does not occur to an appreciable extent in the case of $[(CH_3)_3C]_4Cr(IV)$ might be the steric encumberment around the chromium, but, in view of the stability of the tetra-*n*-butyl compound, it seems more probable that there are no sites available in the tetrahedral complex for β-metal-hydride elimination. The differences in the stabilities of organochromium(III) and organochromium-(IV) compounds could then be ascribed to differences in their major fragmentation pathways occasioned by differences in geometries around the chromium center.

Wilkinson has reported on the redox potential for systems

$$R_4Cr\ (IV) \underset{-e}{\overset{+e}{\rightleftharpoons}} [R_4Cr\ (III)]^{\ominus}$$

$$\begin{aligned} R = {} & (CH_3)_3SiCH_2 \\ & (C_6H_5)C(CH_3)_2CH_2 \\ & (C_6H_5)_3CCH_2 \end{aligned}$$

and the half-wave potentials (*vs.* saturated calomel) are in the range −1.3–1.97 V. Spectroscopic studies on the trimethylsilylmethyl system suggested that the tetrahedral environment around the chromium center was retained in both the Cr(IV) and the Cr(III) compounds [457,458].

The tetrakis(organo)chromium(IV)compounds, by themselves, do not react with ethylene or alkenes [457, 458], however, when they are modified chemically (e.g. by reaction with a suitable compound), they do polymerize ethylene [461].

References

1. G. E. Coates, M. L. H. Green, and K. Wade, "Organometallic Compounds," Vol. 2. Methuen, London, 1968.
2. F. A. Cotton and G. Wilkinson, "Advanced Inorganic Chemistry," 3rd ed. Wiley (Interscience), New York, 1972.
3. J. S. Basi, D. C. Bradley, and M. Chisholm, *J. Chem. Soc., A* p. 1433 (1971).
4. H. P. Fritz, and E. O. Fischer, *Z. Naturforsch. B* **12**, 67 (1957).
5. N. Hagihara and H. Yamazaki, *J. Amer. Chem. Soc.* **81**, 3160 (1959).
6. F. Hein and K. Kartte, *Z. Anorg. Allg. Chem.* **307**, 22 (1960).
7. E. O. Fischer and J. Seeholzer, *Z. Anorg. Allg. Chem.* **312**, 244 (1961).
8. W. Kleinwächter, *Monatsber. Deut. Akad. Wiss. Berlin* **3**, 640 (1961).
9. E. O. Fischer and K. Oefele, *Chem. Ber.* **90**, 2532 (1957).
10. D. N. Kursanov, V. N. Setkina, and B. G. Gribov, *J. Organometal. Chem.* **37**, C35 (1972).
11. C. Elschenbroich, F. Gerson, and J. Heinzer, *Z. Naturforsch. B* **27**, 312 (1972).
12. E. O. Fischer and H. Brunner, *Z. Naturforsch. B* **16**, 406 (1961).
13. E. O. Fischer and H. Brunner, *Chem. Ber.* **95**, 1999 (1962).
14. E. O. Fischer and H. Brunner, *Chem. Ber.* **98**, 175 (1965).
15. T. F. Burger and H. H. Zeiss, *Chem. Ind. (London)* p. 183 (1962).
16. C. Elschenbroich, *J. Organometal. Chem.* **14**, 157 (1968).
17. C. Elschenbroich, *J. Organometal. Chem.* **22**, 677 (1970).
18. A. N. Nesmeyanov, L. P. Yur'eva, and S. N. Levchenko, *Proc. Acad. Sci. USSR Chem. Sect.*, **190**, 24 (1970).
19. G. G. Petukhov, B. I. Rabosin, and A. N. Artemov, *J. Gen. Chem. USSR* **42**, 2685 (1972).
20. G. A. Razuvaev, G. A. Domrachev, O. N. Suvorova, and L. G. Abakumova, *J. Organometal. Chem.* **32**, 115 (1971).
21. G. A. Razuvaev, G. A. Domrachev, and L. G. Abakumova, *Bull. Acad. Sci. USSR Div. Chem. Sci.*, **20**, p. 1974 (1971).
22. E. O. Fischer and W. Hafner, *Z. Naturforsch. B* **10**, 665 (1955).
23. B. G. Gribov, D. D. Mozzhukhin, I. A. Suskina, and B. A. Salamatin, *Proc. Acad. Sci. USSR Chem. Sect.*, **196**, 77 (1971).
24. F. Hein and K. Kartte, *Monatsber. Deut. Akad. Wiss. Berlin* **2**, 185 (1960).
25. G. A. Razuvaev, G. A. Domrachev, and O. N. Druzhkov, *J. Gen. Chem. USSR* **33**, 3011 (1963).
26. G. A. Razuvaev and G. A. Domrachev, *Tetrahedron* **19**, 341 (1963).
27. B. G. Gribov, D. D. Mozzhukhin, B. I. Kozyrkin, and A. S. Strizhkova, *J. Gen. Chem. USSR* **42**, 2521 (1972).
28. H. P. Schröer and A. A. Vlcek, *Z. Anorg. Allg. Chem.* **334**, 205 (1964).
29. A. A. Vlcek, *Z. Anorg. Allg. Chem.* **304**, 109 (1960).
30. L. N. Vertyulina, G. A. Domrachev, I. A. Korshunov, and G. A. Razuvaev, *J. Gen. Chem. USSR* **33**, 277 (1963).
31. I. A. Korshunov, L. N. Vertyulina, and G. A. Domrachev, *J. Gen. Chem. USSR* **32**, 10 (1962).
32. I. A. Suskina, B. G. Gribov, R. A. Idrisova, L. I. Denisovich, and S. P. Gubin, *Bull. Acad. Sci. USSR Div. Chem. Sci.*, **20**, 360 (1971).
33. C. Furlani and E. O. Fischer, *Z. Elektrochem.* **61**, 481 (1957).
34. N. S. Hush, *Z. Elektrochem.* **61**, 734 (1957).

35. D. D. Mozzhukin, B. G. Gribov, G. A. Tychin, A. S. Strizhkova, and M. L. Khidekel, *Bull. Acad. Sci. USSR Div. Chem. Sci.*, p. 163 (1967).
36. D. W. Walker and E. L. Czenkusch, U.S. Pat. 3,157,712 (1964); *Chem. Abstr.* **62**, 2846 (1965).
37. Y. Tajima, K. Tani, and S. Yuguchi, *Polym. Lett.* **3**, 529 (1965).
38. F. J. Karol, Ger. Offen. 2,024,765 (1970); *Chem. Abstr.* **74**, 42867 (1971).
39. H. Nagase, Jap. Pat. 68/05892 (1968); *Chem. Abstr.* **69**, 19756 (1968).
40. Y. Tazima and S. Yuguchi, *Bull. Chem. Soc. Jap.* **39**, 2534 (1966).
41. M. Tsutsui and T. Koyano, *J. Polym. Sci. Part A* **5**, 683 (1967).
42. K. M. Roch and J. Saunders, *J. Polym. Sci.* **38**, 554 (1959).
43. J. R. Olechowski and R. F. Clark, Ger. Offen. 1,801,723 (1969); *Chem. Abstr.* **71**, 90935 (1969).
44. E. W. Gluesenkamp, W. R. Richard, and J. F. K. Wilshire, U.S. Pat. 3,328,380 (1967); *Chem. Abstr.* **67**, 64852 (1967).
45. G. A. Domrachev and G. A. Razuvaev, *Vysokomol. Soedin.* **4**, 1822 (1962).
46. H. Yamazaki and N. Hagihara, *Nippon Kagaku Zasshi* **81**, 819 (1960); *Chem. Abstr.* **56**, 1588 (1962).
47. H. Yamazaki and N. Hagihara, *Nippon Kagaku Zasshi* **81**, 822 (1960); *Chem. Abstr.* **56**, 1589 (1962).
48. H. Yamazaki, M. Yamaguchi, and N. Hagihara, *Mem. Inst. Sci. Ind. Res., Osaka Univ.* **20**, 107 (1963); *Chem. Abstr.* **60**, 13320 (1964).
49. R. Nakanishi and N. Nishimura, Japan. 19,228, (1964); *Chem. Abstr.* **62**, P7894 (1965).
50. H. Yamazaki and N. Hagihara, *Mem. Inst. Sci. Ind. Res., Osaka Univ.* **22**, 163 (1965); *Chem. Abstr.* **63**, 15819 (1965).
51. H. Yamazaki and N. Hagihara, *Nippon Kagaku Zasshi* **81**, 821 (1960); *Chem. Abstr.* **56**, 1589 (1962).
52. N. Hagihara, *Ann. N.Y. Acad. Sci.* **125**, 98 (1965).
53. V. M. Fomin, Yu. A. Alexandrov, and V. A. Umilin, *J. Organometal. Chem.* **61**, 267 (1973).
54. H. Yamazaki, Y. Matsumoto, and N. Hagihara, *Mem. Inst. Sci. Ind. Res., Osaka Univ.* **21**, 131 (1964); *Chem. Abstr.* **64**, 8410 (1966).
55. K. W. Barnett and D. W. Slocum, *J. Organometal. Chem.* **44**, 1 (1972).
56. G. Wilkinson, F. A. Cotton, and J. M. Birmingham, *J. Inorg. Nucl. Chem.* **2**, 95 (1956).
57. M. E. Switzer and M. R. Rettig, *Chem. Commun.* p. 687 (1972).
58. E. O. Fischer and K. Ulm, *Chem. Ber.* **94**, 2413 (1961).
59. S. A. Keppie and M. F. Lappert, *J. Chem. Soc., A* p. 3216 (1971).
60. E. O. Fischer and W. Hafner, *Z. Naturforsch B* **10**, 140 (1955).
61. E. O. Fischer and K. Ulm, *Z. Naturforsch B.* **16**, 757 (1961).
62. E. O. Fischer, K. Ulm, and P. Kuzel, *Z. Anorg. Allg. Chem.* **319**, 253 (1963).
63. E. O. Fischer and K. Ulm, *Chem. Ber.* **95**, 692 (1962).
64. T. S. Piper and G. Wilkinson, *J. Inorg. Nucl. Chem.* **3**, 104 (1956).
65. G. W. Watt and L. J. Baye, *J. Inorg. Nucl. Chem.* **26**, 2099 (1964).
66. H. Werner and J. H. Richards, *J. Amer. Chem. Soc.* **90**, 4976 (1968).
67. F. J. Karol, G. L. Karapinka, C. Wu, A. W. Dow, R. N. Johnson, and W. L. Carrick, *J. Polym. Sci. Part A-1* **10**, 2621 (1972).
68. D. W. Walker and E. L. Czenkusch, U.S. Pat. 3,163,682 (1964); *Chem. Abstr.* **62**, 6325 (1965).
69. G. L. Karapinka Ger. Offen. 1,808,388 (1970); *Chem. Abstr.* **72**, 80105 (1970).

70. R. E. Dessy, P. M. Weissman, and R. L. Pohl, *J. Amer. Chem. Soc.* **88,** 5117 (1966).
71. R. E. Dessy, R. L. Pohl, and R. B. King, *J. Amer. Chem. Soc.* **88,** 5121 (1966).
71a. R. D. Adams, D. E. Collins and F. A. Cotton, *J. Amer. Chem. Soc.* **96,** 749 (1974).
72. E. O. Fischer and K. Plesske, *Chem. Ber.* **94,** 93 (1961).
73. E. O. Fischer, H. P. Kögler, and P. Kuzel, *Chem. Ber.* **93,** 3006 (1960).
74. H. P. Fritz, H. Keller, and E. O. Fischer, *Naturwissenschaften* **15,** 518 (1961).
75. R. B. King, *J. Amer. Chem. Soc.* **85,** 1587 (1963).
76. A. R. Manning and D. J. Thornhill, *J. Chem. Soc., A* p. 637 (1971).
76a. P. Hackett, P. S. O'Neill, and A. R. Manning, *J. Chem. Soc., Dalton Trans.* p. 1625 (1974).
77. K. W. Barnett, D. L. Beach, and T. G. Pollman, *Inorg. Nucl. Chem. Lett.* **9,** 131 (1973).
78. R. B. King and F. G. A. Stone, *Inorg. Syn.* **7,** 99 (1963).
79. R. D. Fischer and K. Noack, *J. Organometal. Chem.* **16,** 125 (1969).
80. J. M. Burlitch and A. Ferrari, *Inorg. Chem.* **9,** 563 (1970).
81. E. N. Gladyshev, N. S. Vyazankin, Yu. A. Sorokin, V. I. Ermoleav, G. N. Bortnikov, and G. A. Razuvaev, *Proc. Acad. Sci. USSR Chem. Sect.,* **194,** 729 (1970).
82. A. N. Nesmeyanov, K. N. Anisimov, V. E. Kolobova, and M. Ya. Zakharova, *Proc. Acad. Sci. USSR Chem. Sect.,* **156,** 532 (1964).
83. H. R. H. Patil and W. A. G. Graham, *Inorg. Chem.* **5,** 1401 (1966).
84. D. J. Cardin, S. A. Keppie, and M. F. Lappert, *Inorg. Nucl. Chem. Lett.* **4,** 365 (1968).
85. D. J. Cardin, S. A. Keppie, and M. F. Lappert, *J. Chem. Soc., A* p. 2594 (1970).
86. A. P. Hagen, C. R. Higgins, and P. J. Russo, *Inorg. Chem.* **10,** 1657 (1971).
87. P. J. Russo and A. P. Hagen, *Inorg. Nucl. Chem. Lett.* **5,** 885 (1969).
88. Yu. T. Struchkov, K. N. Anisimov, O. P. Osipova, N. E. Kolobova, and A. N. Nesmeyanov, *Proc. Acad. Sci. USSR Chem. Sect.,* **172,** 15 (1967).
89. R. J. Haines, R. S. Nyholm, and M. H. B. Stiddard, *J. Chem. Soc., A* p. 46 (1968).
90. A. Miyake, H. Kondo, and M. Aoyama, *Angew. Chem. Int. Ed. Engl.* **8,** 520 (1969).
91. P. Hackett and A. R. Manning, *J. Chem. Soc., Dalton Trans.* p. 2434 (1972).
92. A. Miyake and H. Kondo, *Angew. Chem. Int. Ed. Engl.* **7,** 631 (1968).
93. A. Miyake and H. Kondo, *Angew. Chem. Int. Ed. Engl.* **7,** 880 (1968).
94. T. Kondo and A. Miyake, Jap. Pat. 70/21,283 (1970); *Chem. Abstr.* **73,** 66143 (1970).
95. T. Kondo and A. Miyake, Jap. Pat. 70/21,484 (1970); *Chem. Abstr.* **73,** 87393 (1970)
96. H. Kondo and A. Miyake, Jap. Pat. 69/10,245 (1969); *Chem. Abstr.* **71,** 60659 (1969).
97. H. Brunner, *J. Organometal. Chem.* **16,** 119 (1969).
98. R. B. King, *Inorg. Chem.* **6,** 30 (1967).
99. G. R. Knox and C. Robertson, C. Robertson's PhD Thesis, University Strathclyde Glasgow U.K. (1966).
100. M. Herberhold and H. Alt, *J. Organometal. Chem.* **42,** 407 (1972).
101. H. Brunner, *Angew. Chem., Int. Ed. Engl.* **10,** 249 (1971).
102. E. O. Fischer, M. von Foerster, C. G. Krieter, and K. E. Schwarzhans, *J. Organometal. Chem.* **7,** 113 (1967).
103. E. O. Fischer and H. J. Beck, *Chem. Ber.* **104,** 3101 (1971).
104. H. Brunner, *Chem. Ber.* **102,** 305 (1969).
105. C. Krüger, E. G. Rochow, and U. Wannaga, *Chem. Ber.* **96,** 2132 (1963).

106. B. F. G. Johnson and J. A. McCleverty, *Progr. Inorg. Chem.* **7**, 277 (1966).
107. T. S. Piper and G. Wilkinson, *J. Inorg. Nucl. Chem.* **2**, 38 (1956).
108. M. A. Jennings and A. Wojcicki, *J. Organometal. Chem.* **14**, 231 (1968).
109. S. R. Su, J. A. Hanna, and A. Wojcicki, *J. Organometal. Chem.* **21**, P21 (1970).
110. E. O. Fischer and H. Strametz, *J. Organometal. Chem.* **10**, 323 (1967).
111. E. O. Fischer and P. Kuzel, *Z. Anorg. Allg. Chem.* **317**, 226 (1962).
112. N. Flitcroft, *J. Organometal. Chem.* **15**, 254 (1968).
113. R. B. King and M. B. Bisnette, *Inorg. Chem.* **3**, 791 (1964).
114. R. B. King, *J. Amer. Chem. Soc.* **85**, 2527 (1963).
115. M. Ahmad, R. Bruce, and G. R. Knox, *Z. Naturforsch. B* **21**, 289 (1966).
116. E. O. Fischer and S. Breitschaft, *Chem. Ber.* **99**, 2905 (1966).
117. E. O. Fischer and S. Breitschaft, *Chem. Ber.* **96**, 2451 (1963).
118. E. O. Fischer and S. Breitschaft, *Chem. Ber.* **99**, 2213 (1966).
119. E. O. Fischer and J. Müller, *Z. Naturforsch. B* **18**, 1137 (1963).
120. R. B. King and M. S. Bisnette, *Inorg. Chem.* **3**, 785 (1964).
121. C. Furlani, A. Furlani, and L. Sestili, *J. Electroanal. Chem.* **9**, 140 (1965).
122. H. H. Zeiss, P. J. Wheatley, and H. J. S. Winkler, "Benzenoid-Metal Complexes." Ronald Press, New York, 1966.
123. B. Nicholls and M. C. Whiting, *J. Chem. Soc., London* p. 551 (1959).
124. J. D. Holmes, D. A. K. Jones, and R. Pettit, *J. Organometal. Chem.* **4**, 324 (1965).
125. W. Strohmeier and H. Hellmann, *Chem. Ber.* **98**, 1598 (1965).
126. W. S. Trahanovsky and D. K. Wells, *J. Amer. Chem. Soc.* **91**, 5870 (1969).
127. D. K. Wells and W. S. Trahanovsky, *J. Amer. Chem. Soc.* **91**, 5871 (1969).
128. R. S. Bly and R. C. Strickland, *J. Amer. Chem. Soc.* **92**, 7459 (1970).
129. D. K. Wells and W. S. Trahanovsky, *J. Amer. Chem. Soc.* **92**, 7461 (1970).
130. D. A. Brown, *J. Chem. Soc., London* p. 4389 (1963).
131. D. A. Brown and J. R. Raju, *J. Chem. Soc., A* p. 40 (1966).
132. D. A. Brown and F. J. Hughes, *Inorg. Chim. Acta* **1**, 448 (1967).
133. D. A. Brown and R. M. Rawlinson, *J. Chem. Soc., A* p. 1534 (1969).
134. S. P. Gubin and V. S. Khandkarova, *J. Organometal. Chem.* **22**, 449 (1970).
135. W. Strohmeier and H. Hellmann, *Chem. Ber.* **96**, 2859 (1963).
136. V. S. Khandkarova, S. P. Gubin, and B. A. Kvasov, *J. Organometal. Chem.* **23**, 509 (1970).
137. W. Strohmeier and F. J. Müller, *Chem. Ber.* **100**, 2812 (1967).
138. J. F. Guttenberger and W. Strohmeier, *Chem. Ber.* **100**, 2807 (1967).
139. C. U. Pittmann, Jr., P. L. Grube, O. E. Ayers, S. P. McManus, M. D. Rausch, and G. A. Moser, *J Polym. Sci., Part A-1*, **10**, 379 (1972).
140. E. O. Fischer, K. Oefele, H. Essler, W. Fröhlich, J. P. Mortensen, and W. Semmlinger, *Chem. Ber.* **91**, 2763 (1958).
141. A. Z. Rubezhov and S. P. Gubin, *Advan. Organometal. Chem.* **10**, 347 (1972).
142. H. Werner, *Top. Curr. Chem.* **28**, 141 (1972).
143. H. Behrens, K. Meyer, and A. Müller, *Z. Naturforsch. B* **20**, 74 (1965).
144. T. A. Magee, C. N. Matthews, T. S. Wang, and J. H. Wotiz, *J. Amer. Chem. Soc.* **83**, 3200 (1961).
145. W. Strohmeier and H. Mittnacht, *Z. Phys. Chem.* **29**, 339 (1961).
146. W. Strohmeier and E. H. Staricco, *Z. Phys. Chem.* **38**, 315 (1963).
147. R. J. Angelici and L. Busetto, *Inorg. Chem.* **7**, 1935 (1968).
148. E. O. Fischer and P. Kuzel, *Z. Naturforsch. B* **16**, 475 (1961).
149. F. Zingales, A. Chiesi, and F. Basolo, *J. Amer. Chem. Soc.* **88**, 2707 (1966).
150. D. Sellmann and G. Maisel, *Z. Naturforsch. B* **27**, 718 (1972).

151. D. Sellmann and G. Maisel, *Z. Naturforsch. B* **27**, 465 (1972).
152. G. Jaouen and R. Dabard, *J. Organometal. Chem.* **61**, C36 (1973).
153. G. Jaouen, *Tetrahedron Lett.* p. 5159 (1973).
154. G. Jaouen, A. Meyer, and G. Simonneaux, *Tetrahedron Lett.* p. 5163 (1973).
155. W. Jetz and W. A. G. Graham, *Inorg. Chem.* **10**, 4 (1971).
156. V. N. Setkina, N. K. Baranetskaya, K. N. Anisimov, and D. N. Kursanov, *Bull. Acad. Sci. USSR Div. Chem. Sci.*, p. 1772 (1964).
157. D. N. Kursanov, V. N. Setkina, N. K. Baranetskaya, and K. N. Anisimov, *Bull. Acad. Sci. USSR Div. Chem. Sci.*, p. 1528 (1968).
158. D. N. Kursanov, V. N. Setkina, N. K. Baranetskaya, E. I. Fedin, K. N. Anisinov, and V. M. Urinyuk, *Proc. Acad. Sci. USSR Chem. Sect.*, **183**, 1118 (1968).
159. V. N. Setkina, N. K. Baranetskaya, K. N. Anisimov, and D. N. Kursanov, *Bull. Acad. Sci. USSR Div. Chem. Sci.*, p. 429 (1970).
160. M. Ashraf, *Can. J. Chem.* **50**, 118 (1972).
161. D. N. Kursanov, V. N. Setkina, N. K. Baranetskaya, E. I. Fedin, and K. N. Anisimov, *Bull. Acad. Sci. USSR Div. Chem. Sci.*, **20**, p. 1727 (1971).
162. V. N. Setkina, N. V. Kislyakova, N. E. Kolobova, and D. N. Kursanov, *Bull. Acad. Sci. USSR Div. Chem. Sci.*, **20**, p. 751 (1971).
163. W. S. Trahanovsky and R. J. Card, *J. Amer. Chem. Soc.* **94**, 2897 (1972).
164. D. N. Kursanov, V. N. Setkina, N. K. Baranetskaya, Z. I. Zdanovic, and K. N. Anisimov, *Proc. Acad. Sci. USSR Chem. Sect.*, **190**, 127 (1970).
165. V. N. Setkina, Z. I. Zdanovich, N. K. Baranetskaya, A. I. Yurtanov, K. N. Anisimov, and D. N. Kursanov, *J. Gen. Chem. USSR* **41**, 1345 (1971).
166. D. N. Kursanov, V. N. Setkina, N. K. Baranetskaya, E. I. Fedin, E. I. Mysov, and K. N. Anisimov, *Proc. Acad. Sci. USSR Chem. Sect.*, **202**, 75 (1972).
167. C. P. Lillya and R. A. Sahatjian, *Inorg. Chem.* **11**, 889 (1972).
168. B. V. Lokshin, V. I. Zdanovich, N. K. Baranetskaya, V. N. Setkina, and D. N. Kursanov, *J. Organometal. Chem.* **37**, 331 (1972).
169. D. N. Kursanov, V. N. Setkina, P. V. Petrovskii, V. I. Zdanovich, N. K. Baranetskaya, and I. D. Rubin, *J. Organometal. Chem.* **37**, 339 (1972).
170. A. Davison, W. McFarlane, L. Pratt, and G. W. Wilkinson, *J. Chem. Soc., London* p. 3653 (1962).
171. V. N. Setkina, N. K. Baranetskaya, A. G. Ginzburg, V. I. Zdanovich, M. N. Nefedowa, and D. N. Kursanov, *J. Organometal. Chem.* **61**, 287 (1973).
172. A. N. Nesmeyanov, N. E. Kolobova, K. N. Anisimov, and Yu. V. Makarov, *Bull. Acad. Sci. USSR Div. Chem. Sci.*, p. 2538 (1968).
173. H. J. Beck, Dissertation, Technische Universität, Munich (1970).
174. H. J. Beck, E. O. Fischer, and C. G. Kreiter, *J. Organometal. Chem.* **26**, C41 (1971).
175. G. E. Herberich and E. O. Fischer, *Chem. Ber.* **95**, 2803 (1962).
176. R. Riemschneider, O. Becker, and K. Franz, *Monatsh. Chem.* **90**, 571 (1959).
177. W. R. Jackson and W. B. Jennings, *J. Chem. Soc., B* p. 1221 (1969).
178. G. K. I. Magomedov, V. G. Syrkin, and A. S. Frenkel, *J. Gen. Chem. USSR* **42**, 2443 (1972).
179. G. A. Razuvaev, G. G. Petuchov, A. N. Artemov, and N. I. Sirotkin, *J. Organometal. Chem.* **37**, 313 (1972).
180. G. K.-I. Magomedov, V. G. Syrkin, A. S. Frenkel, A. V. Medvedeva, and L. V. Morozova, *J. Gen. Chem. USSR* **43**, 803 (1973).
181. G. Jaouen and R. Dabard, *J. Organometal. Chem.* **21**, P43 (1970).
182. L. Tchissambou, G. Jaouen, and R. Dabard, *C.R. Acad. Sci.* **274**, 806 (1972).

183. S. P. Gubin, V. S. Khandkarova, and A. Z. Kreindlin, *J. Organometal. Chem.* **64**, 229 (1974).
184. R. S. Bly, R. C. Strickland, R. T. Swindell, and R. L. Veazy, *J. Amer. Chem. Soc.* **92**, 3722 (1970).
185. R. S. Bly and R. L. Veazy, *J. Amer. Chem. Soc.* **91**, 4221 (1969).
186. A. Ceccon, *J. Organometal. Chem.* **29**, C19 (1971).
187. G. Jaouen and G. Simonneaux, *J. Organometal. Chem.* **61**, C39 (1973).
188. H. Falk, K. Schlögl, and W. Steyrer, *Monatsh. Chem.* **97**, 1029 (1966).
189. W. R. Jackson and T. R. B. Mitchell, *J. Chem. Soc., B* p. 1228 (1969).
190. G. Jaouen and R. Dabard, *Tetrahedron Lett.* p. 1015 (1971).
191. G. Jaouen, B. Caro, and J. Y. Le Bihan, *C.R. Acad. Sci.* **274**, 902 (1972).
192. J. Besançon, A. Card, Y. Dusausoy, and J. Tirouflet, *C.R. Acad. Sci.* **274**, 545 (1972).
193. A. Meyer and R. Dabard, *J. Organometal. Chem.* **36**, C38 (1972).
194. J. Besançon, J. Tirouflet, A. Card, and Y. Dusausoy, *J. Organometal. Chem.* **59**, 267 (1973).
195. G. R. Knox, D. G. Leppard, P. L. Pauson, and W. E. Watts, *J. Organometal. Chem.* **34**, 347 (1972).
196. M. D. Rausch, G. A. Moser, E. J. Zaiko, and A. L. Lipman, Jr., *J. Organometal. Chem.* **23**, 185 (1970).
197. C. U. Pittman, Jr., R. L. Voges, and J. Elder, *Macromolecules* **3**, 302 (1971).
198. M. F. Farona and J. F. White, *J. Amer. Chem. Soc.* **93**, 2826 (1971).
199. G. L. Karapinka, Ger. Offen. 1,949,574 (1970); *Chem. Abstr.* **73**, 67029 (1970).
200. D. P. Tate, J. M. Augl, W. M. Ritchey, B. L. Ross, and J. G. Grasselli, *J. Amer. Chem. Soc.* **86**, 3261 (1964).
201. W. Strohmeier and C. Barbeau, *Z. Naturforsch. B* **19**, 262 (1964).
202. M. Cais, E. N. Frankel, and A. Rejoan, *Tetrahedron Lett.* p. 1919 (1968).
203. E. N. Frankel, E. Selke, and C. A. Glass, *J. Amer. Chem. Soc.* **90**, 2446 (1968).
204. E. N. Frankel, E. Selke, and C. A. Glass, *J. Org. Chem.* **34**, 3936 (1969).
205. E. N. Frankel and R. O. Butterfield, *J. Org. Chem.* **34**, 3930 (1969).
206. M. Cais and A. Rejoan, *Inorg. Chim. Acta* **4**, 509 (1970).
207. E. N. Frankel and F. L. Little, *J. Amer. Chem. Soc.* **46**, 256 (1969).
208. E. N. Frankel, *J. Amer. Oil Chem. Soc.* **47**, 11 (1970).
209. E. N. Frankel, *J. Amer. Oil Chem. Soc.* **47**, 33 (1970).
210. E. N. Frankel, F. L. Thomas, and J. C. Cowan, *J. Amer. Oil Chem. Soc.* **47**, 497 (1970).
211. M. Cais and D. Fraenkel, *Proc. Int. Conf. Organometal. Chem., 5th, 1971* Abstracts Vol. II, p. 620 (1971).
212. E. N. Frankel, *J. Catal.* **24**, 358 (1972).
213. M. Cais and E. N. Frankel, U.S. Pat. 3,632,614 (1972); *Chem. Abstr.* **76**, 72051 (1972).
214. E. N. Frankel, *J. Org. Chem.* **37**, 1549 (1972).
215. D. A. Brown, J. P. Hargaden, C. M. McMullin, N. Cogan, and H. Sloan, *J. Chem. Soc., London* p. 4914 (1963).
216. J. Nasielski, P. Kirsch, and L. Wilputte-Steinert, *J. Organometal. Chem.* **27**, C13 (1971).
217. M. Wrighton and M. A. Schroeder, *J. Amer. Chem. Soc.* **95**, 5764 (1973).
218. J. D. Munro and P. L. Pauson, *J. Chem. Soc., London* p. 3475 (1961).
219. J. A. Connor and E. J. Rasburn, *J. Organometal. Chem.* **24**, 441 (1970).

220. P. L. Pauson, G. H. Smith, and J. H. Valentine, *J. Chem. Soc., C* p. 1057 (1967).
221. P. L. Pauson and K. H. Todd, *J. Chem. Soc., C* p. 2315 (1970).
222. P. L. Pauson and K. H. Todd, *J. Chem. Soc., C* p. 2638 (1970).
223. E. W. Abel, M. A. Bennett, R. Burton, and G. Wilkinson, *J. Chem. Soc., London* p. 4559 (1958).
224. A. Pidcock and B. W. Taylor, *J. Chem. Soc., A* p. 877 (1967).
225. E. Deckelmann and H. Werner, *Helv. Chim. Acta* **52**, 892 (1969).
226. H. Werner, K. Deckelmann, and U. Schönenberger, *Helv. Chim. Acta* **53**, 2002 (1970).
227. W. P. Anderson, W. G. Blenderman, and K. A. Drews, *J. Organometal. Chem.* **42**, 139 (1972).
228. F. A. Cotton and F. Zingales, *J. Amer. Chem. Soc.* **83**, 351 (1961).
229. J. A. Connor, E. M. Jones, G. K. McEwen, M. K. Lloyd, and J. A. McCleverty, *J. Chem. Soc., Dalton Trans.* p. 1246 (1972).
230. C. E. Jones and K. J. Coskran, *Inorg. Chem.* **10**, 55 (1971).
231. R. B. King, *Inorg. Chem.* **2**, 936 (1963).
232. R. B. King and A. Efraty, *Inorg. Chim. Acta* **4**, 319 (1970).
233. W. E. Stanclift and D. G. Hendricker, *Inorg. Chem.* **7**, 1242 (1968).
234. F. Zingales and F. Canziani, *Gazz. Chim. Ital.* **92**, 343 (1962).
235. F. Zingales, F. Canziani, and R. Ugo, *Gazz. Chim. Ital.* **92**, 761 (1962).
236. H. Behrens, E. Lindner, and G. Lehnert, *J. Organometal. Chem.* **22**, 439 (1970).
237. J. A. S. Howell, B. F. G. Johnson, and J. Lewis, *J. Organometal. Chem.* **42**, C54 (1972).
238. M. I. Foreman, G. R. Knox, P. L. Pauson, K. H. Todd, and W. E. Watts, *Chem. Commun.* p. 843 (1970).
239. M. I. Foreman, G. R. Knox, P. L. Pauson, K. H. Todd, and W. E. Watts, *J. Chem. Soc., Perkin Trans.* p. 1141 (1972).
240. K. W. Egger, *J. Amer. Chem. Soc.* **89**, 3688 (1969).
241. P. L. Pauson, G. H. Smith, and J. H. Valentine, *J. Chem. Soc., C* p. 1061 (1967).
242. J. D. Munro and P. L. Pauson, *J. Chem. Soc., London* p. 3484 (1961).
243. J. D. Munro and P. L. Pauson, *J. Chem. Soc. London* p. 3479 (1961).
244. P. L. Pauson and K. H. Todd, *J. Chem. Soc., C* p. 2636 (1970).
245. E. O. Fischer, *Pure Appl. Chem.* **24**, 407 (1970).
246. M. Ryang, *Organometal. Chem. Rev. Sect. A* **5**, 67 (1970).
247. E. W. Abel and F. G. A. Stone, *Quart. Rev., Chem. Soc.* **24**, 498 (1970).
248. F. A. Cotton and C. M. Lukehart, *Progr. Inorg. Chem.* **16**, 487 (1972).
249. D. J. Cardin, B. Cetinkaya, and M. F. Lappert, *Chem. Rev.* **72**, 545 (1972).
250. E. O. Fischer, *Pure Appl. Chem.* **30**, 353 (1972).
251. H. Werner and H. Rascher, *Helv. Chim. Acta* **51**, 1765 (1968).
252. H. Werner and H. Rascher, *Inorg. Chim. Acta* **2**, 181 (1968).
253. E. O. Fischer, H. Fischer, and H. Werner, *Angew. Chem., Int. Ed. Engl.* **11**, 644 (1972).
254. E. O. Fischer and R. Aumann, *Chem. Ber.* **102**, 1495 (1969).
255. F. R. Kreissl, C. G. Kreiter, and E. O. Fischer, *Angew. Chem., Int. Ed. Engl.* **11**, 643 (1972).
256. F. R. Kreissl, E. O. Fischer, C. G. Kreiter, and H. Fischer, *Chem. Ber.* **106**, 1262 (1973).
257. U. Schubert and E. O. Fischer, *J. Organometal. Chem.* **51**, C11 (1973).
258. U. Schubert and E. O. Fischer, *Chem. Ber.* **106**, 3882 (1973).
259. E. O. Fischer and L. Knauss, *Chem. Ber.* **102**, 223 (1969).

260. E. O. Fischer, E. Louis, W. Bathelt, E. Moser, and J. Müller, *J. Organometal. Chem.* **14**, P9 (1968).
261. E. O. Fischer and A. Maasböl, *J. Organometal. Chem.* **12**, P15 (1968).
262. E. O. Fischer, B. Heckl, K. H. Dötz, J. Müller, and H. Werner, *J. Organometal. Chem.* **16**, P29 (1969).
263. E. O. Fischer, B. Heckl, and H. Werner, *J. Organometal. Chem.* **28**, 359 (1971).
264. J. A. Connor and P. D. Rose, *J. Organometal. Chem.* **46**, 329 (1972).
265. E. O. Fischer and K. H. Dötz, *Chem. Ber.* **103**, 1273 (1970).
266. K. H. Dötz and E. O. Fischer, *Chem. Ber.* **105**, 1356 (1972).
267. E. O. Fischer and K. H. Dötz, *Chem. Ber.* **105**, 3966 (1972).
268. M. D. Cooke and E. O. Fischer, *J. Organometal. Chem.* **56**, 279 (1973).
269. J. A. Connor and P. D. Rose, *J. Organometal. Chem.* **24**, C45 (1970).
270. J. A. Connor, P. D. Rose, and R. M. Turner, *J. Organometal. Chem.* **55**, 111 (1973).
271. J. A. Connor, J. P. Day, and R. M. Turner, *Chem. Commun.* 578 (1973).
272. E. O. Fischer and K. H. Dötz, *J. Organometal. Chem.* **35**, C4 (1972).
273. J. A. Connor and J. P. Lloyd, *J. Chem. Soc., Perkin Trans. 1* p. 17 (1973).
274. C. P. Casey and T. J. Burkhardt, *J. Amer. Chem. Soc.* **94**, 6543 (1972).
275. E. O. Fischer and M. Leupold, *Chem. Ber.* **105**, 599 (1972).
276. J. A. Connor and J. P. Lloyd, *J. Chem. Soc., A* p. 3237 (1970).
277. J. A. Connor and E. M. Jones, *J. Organometal. Chem.* **60**, 77 (1973).
278. E. O. Fischer, K. Weiss, and K. Burger, *Chem. Ber.* **106**, 1581 (1973).
279. E. Moser and E. O. Fischer, *J. Organometal. Chem.* **12**, P1 (1968).
280. U. Klabunde and E. O. Fischer, *J. Amer. Chem. Soc.* **89**, 7141 (1967).
281. J. A. Connor and E. O. Fischer, *J. Chem. Soc., A* p. 578 (1969).
282. J. A. Connor and P. D. Rose, *Tetrahedron Lett.* p. 3623 (1970).
283. J. A. Connor and E. O. Fischer, *Chem. Commun.* p. 1024 (1967).
284. J. A. Connor and E. M. Jones, *J. Chem. Soc., A* p. 3368 (1971).
285. B. Heckl, H. Werner, and E. O. Fischer, *Angew. Chem., Int. Ed. Engl.* **7**, 817 (1968).
286. H. Werner, E. O. Fischer, B. Heckl, and C. G. Kreiter, *J. Organometal. Chem.* **28**, 367 (1971).
287. F. R. Kreissl, E. O. Fischer, C. G. Kreiter, and K. Weiss, *Angew. Chem., Int. Ed. Engl.* **12**, 563 (1973).
288. E. O. Fischer and S. Fontana, *J. Organometal. Chem.* **40**, 367 (1972).
289. E. O. Fischer and R. Aumann, *Chem. Ber.* **101**, 963 (1968).
290. L. Knauss and E. O. Fischer, *Chem. Ber.* **103**, 3744 (1970).
291. E. O. Fischer and L. Knauss, *Chem. Ber.* **103**, 1262 (1970).
292. L. Knauss and E. O. Fischer, *J. Organometal. Chem.* **31**, C68 (1971).
293. R. Aumann and E. O. Fischer, *Chem. Ber.* **101**, 954 (1968).
294. G. Huttner and S. Lange, *Chem. Ber.* **103**, 3149 (1970).
295. K. Weiss and E. O. Fischer, *Chem. Ber.* **106**, 1277 (1973).
296. E. O. Fischer, M. Leupold, C. G. Kreiter, and J. Müller, *Chem. Ber.* **105**, 150 (1972).
297. B. D. Dombek and R. J. Angelici, *J. Amer. Chem. Soc.* **95**, 7516 (1973).
298. E. O. Fischer and V. Kiener, *Angew. Chem., Int. Ed. Engl.* **6**, 961 (1967).
299. E. O. Fischer, G. Kreis, F. R. Kreissl, C. G. Kreiter, and J. Müller, *Chem. Ber.* **106**, 3910 (1973).
300. E. O. Fischer and G. Kreis, *Chem. Ber.* **106**, 2310 (1973).
301. E. O. Fischer, K. R. Schmid, W. Kalbfuss, and C. G. Kreiter, *Chem. Ber.* **106**, 3893 (1973).
302. L. Knauss and E. O. Fischer, *J. Organometal. Chem.* **31**, C71 (1971).

303. E. O. Fischer, G. Kreis, C. G. Kreiter, J. Müller, G. Huttner, and H. Lorenz, *Angew. Chem., Int. Ed. Engl.* **12**, 564 (1973).
304. C. G. Kreiter, *Angew. Chem., Int. Ed. Engl.* **7**, 390 (1968).
305. C. P. Casey, R. A. Boggs, and R. L. Anderson, *J. Amer. Chem. Soc.* **94**, 8947 (1972).
306. J. A. Connor and E. M. Jones, *Chem. Commun.* p. 570 (1971).
307. J. A. Connor and E. M. Jones, *J. Chem. Soc., Dalton Trans.* p. 2119 (1973).
308. P. S. Braterman and R. J. Cross, *Chem. Soc. Rev.* **2**, 271 (1973).
309. M. C. Baird, *J. Organometal. Chem.* **64**, 289 (1974).
310. W. Seidel and K. Fischer, *Z. Chem.* **12**, 147 (1972).
311. H. P. Throndsen, W. Metlesics, and H. Zeiss, *J. Organometal. Chem.* **5**, 176 (1966).
312. K. Nishimura, H. Kuribayashi, A. Yamamoto, and S. Ikeda, *J. Organometal. Chem.* **37**, 317 (1972).
313. J. J. Daly, R. P. A. Sneeden, and H. H. Zeiss, *J. Amer. Chem. Soc.* **88**, 4287 (1966).
314. E. Kurras, Dissertation, University of Jena (1959).
315. F. Glockling, R. P. A. Sneeden, and H. Ziess, *J. Organometal. Chem.* **2**, 109 (1964).
316. R. P. A. Sneeden and H. P. Throndsen, *J. Organometal. Chem.* **6**, 542 (1966).
317. E. Kurras, *Monatsber. Deut. Akad. Wiss. Berlin* **5**, 378 (1963).
318. R. P. A. Sneeden and H. H. Zeiss, *J. Organometal. Chem.* **13**, 369 (1968).
319. R. P. A. Sneeden and H. H. Zeiss, *J. Organometal. Chem.* **16**, 449 (1969).
320. K. W. Egger and A. T. Cocks, *Helv. Chim. Acta* **56**, 1516 (1973).
321. T. Arakawa, *Kogyo Kagaku Zasshi* **72**, 1733 (1969).
322. F. A. L. Anet and E. Leblanc, *J. Amer. Chem. Soc.* **79**, 2649 (1957).
323. J. K. Kochi and D. D. Davis, *J. Amer. Chem. Soc.* **86**, 5264 (1964).
324. R. G. Coombes, M. D. Johnson, M. L. Tobe, N. Winterton, and L. Y. Wong, *Chem. Commun.* p. 251 (1965).
325. R. G. Coombes, M. D. Johnson, and N. Winterton, *J. Chem. Soc., London* p. 7029 (1965).
326. R. G. Coombes and M. D. Johnson, *J. Chem. Soc., A* p. 1805 (1966).
327. R. G. Coombes and M. D. Johnson, *J. Chem. Soc., London* p. 177 (1966).
328. F. A. L. Anet, *Can. J. Chem.* **37**, 58 (1959).
329. D. Dodd and M. D. Johnson, *J. Chem. Soc., A* p. 34 (1968).
330. W. Schmidt, J. H. Swinehart, and H. Taube, *J. Amer. Chem. Soc.* **93**, 1117 (1971).
331. J. K. Kochi and D. Buchanan, *J. Amer. Chem. Soc.* **87**, 853 (1965).
332. R. G. Coombes, M. D. Johnson, and D. Vamplew, *J. Chem. Soc., A* p. 2297 (1968).
333. J. K. Kochi and F. F. Rust, *J. Amer. Chem. Soc.* **83**, 2017 (1961).
334. L. H. Slaugh and J. H. Raley, *Tetrahedron* **20**, 1005 (1964).
335. J. C. Chang and J. H. Espenson, *Chem. Commun.* p. 233 (1974).
336. H. E. Lunk and E. A. Youngman, *J. Polym. Sci., Part A* **3**, 2983 (1965).
337. T. Arakawa, *Kogyo Kagaku Zasshi* **70**, 1742 (1967).
338. W. Oberkirch, Dissertation, University of Aachen (1963).
339. F. Hein and K. Schmiedeknecht, *Z. Anorg. Allg. Chem.* **352**, 138 (1967).
340. R. P. A. Sneeden and H. H. Zeiss, *J. Organometal. Chem.* **47**, 125 (1973).
341. H. Müller, *Z. Chem.* **9**, 311 (1969).
342. K. Madeja, E. Hüsing, and N. Ahrens, *Z. Chem.* **7**, 22 (1967).
343. F. Hein and D. Tille, *Z. Anorg. Allg. Chem.* **239**, 72 (1964).
344. A. Tzschach and H. Nindel, *J. Organometal. Chem.* **24**, 159 (1970).
345. R. P. A. Sneeden and H. H. Zeiss, unpublished results.
346. G. Bähr and H. Zohm, *Angew. Chem.* **75**, 94 (1963).

347. H. H. Zeiss and W. Herwig, *J. Amer. Chem. Soc.* **80**, 2913 (1958).
348. W. Herwig, W. Metlesics, and H. H. Zeiss, *J. Amer. Chem. Soc.* **81**, 6203 (1959).
349. R. P. A. Sneeden and H. H. Zeiss, *J. Organometal. Chem.* **20**, 153 (1969).
350. R. P. A. Sneeden and H. H. Zeiss, *J. Organometal. Chem.* **40**, 163 (1972).
351. I. Hashimoto, M. Ryang, and S. Tsutsumi, *J. Org. Chem.* **33**, 3955 (1968).
352. W. Metlesics and H. H. Zeiss, *J. Amer. Chem. Soc.* **81**, 4117 (1959).
353. J. L. Boston, D. W. A. Sharp, and G. Wilkinson, *J. Chem. Soc., London p.* 3488 (1962).
354. G. M. Whitesides and W. J. Ehmann, *J. Amer. Chem. Soc.* **92**, 5625 (1970).
355. Y. T. Huang, H. I. Tai, S. L. Ch'en, K. Y. Hou, and T. N. Ni, *Hua Hsueh Hsueh Pao* **31**, 149 (1965); *Chem. Abstr.* **63**, 5543 (1965).
356. R. P. A. Sneeden and H. H. Zeiss, *J. Organometal. Chem.* **29**, C31 (1971).
357. M. Sato and Y. Ishida, *Nippon Kagaku Zasshi* **91**, 173 (1970); *Chem. Abstr.* **73**, 25624b (1970).
358. M. Tsutsui and H. Zeiss, *J. Amer. Chem. Soc.* **82**, 6255 (1960).
359. M. Sato and Y. Ishida, *Bull. Chem. Soc. Jap.* **41**, 730 (1968).
360. R. P. A. Sneeden and H. H. Zeiss, *J. Organometal. Chem.* **28**, 259 (1971).
361. M. Tsutsui and H. H. Zeiss, *J. Amer. Chem. Soc.* **81**, 6090 (1959).
362. J. R. C. Light and H. H. Zeiss, *J. Organometal. Chem.* **21**, 391 (1970).
363. M. Michman and H. H. Zeiss, *J. Organometal. Chem.* **13**, P23 (1968).
364. M. Michman and H. H. Zeiss, *J. Organometal. Chem.* **25**, 167 (1970).
365. M. Michman and H. H. Zeiss, *J. Organometal. Chem.* **25**, 161 (1970).
366. J. R. C. Light and H. H. Zeiss, *J. Organometal. Chem.* **21** 517 (1970).
367. M. Michman and H. H. Zeiss, *J. Organometal. Chem.* **15**, 139 (1968).
368. H. H. Zeiss, *Bull. Soc. Chim. Fr.* p. 1500 (1963).
369. G. M. Whitesides and W. J. Ehmann, *J. Amer. Chem. Soc.* **90**, 804 (1968).
370. G. M. Whitesides and W. J. Ehmann, *J. Amer. Chem. Soc.* **91**, 3800 (1969).
371. T. Blackmore, M. I. Bruce, F. G. A. Stone, R. E. Davis, and A. Garza, *Chem. Commun.* p. 852 (1971).
372. E. Müller, E. Langer, H. Jäckle, H. Muhm, W. Hoppe, R. Graziani, A. Gieren, and F. Brandl, *Z. Naturforsch. B* **26**, 305 (1971).
373. H Benn, G. Wilke, and D. Henneberg, *Angew. Chem., Int. Ed. Engl.* **12**, 1001 (1973).
374. A. Mortreux and M. Blanchard, *Bull. Soc. Chim. Fr.* p. 1641 (1972).
375. J. W. Kang, K. Moseley, and P. M. Maitlis, *J. Amer. Chem. Soc.* **91**, 5970 (1969).
376. A. Job and A. Cassal, *C.R. Acad. Sci.* **183**, 58 (1926).
377. H. Zeiss, "Organometallic Chemistry," Amer. Chem. Soc. Monogr. No. 147. Van Nostrand-Reinhold, Princeton, New Jersey, 1960.
378. R. P. A. Sneeden, T. F. Burger, and H. H. Zeiss, *J. Organometal. Chem.* **4**, 397 (1965).
379. M Ryang, K. Yoshida, H. Yokoo, and S. Tsutsumi, *Bull. Chem. Soc. Jap.* **38**, 636 (1965).
380. M. Ryang, K. Yoshida, and S. Tsutsumi, *Technol. Rep. Osaka Univ.* **16**, 205 (1966).
381. Beilstein, "Handbuch der organischen Chemie," 4th ed., Vol. VI, p. 5924. Springer-Verlag, Berlin and New York, 1967.
382. D. G. H. Ballard, W. H. Janes, and T. Medinger, *J. Chem. Soc., B* p. 1168 (1968).
383. R. P. A. Sneeden and H. H. Zeiss, unpublished results.
384. T. F. Burger and H. H. Zeiss, unpublished results.
385. W. Metlesics, P. J. Wheatley, and H. H. Zeiss, *J. Amer. Chem. Soc.* **84**, 2327 (1962).
386. W. Herwig and H. H. Zeiss, *J. Amer. Chem. Soc.* **79**, 6561 (1957).

387. W. Herwig and H. H. Zeiss, *J. Amer. Chem. Soc.* **81**, 4798 (1959).
388. J. Hähle and G. Stolze, *Z. Naturforsch B* **19**, 1081 (1964).
389. H. H. Zeiss, P. J. Wheatley, and H. J. S. Winkler, "Benzenoid-Metal Complexes." Ronald Press, New York, 1966.
390. F. Hein, *Ber. Deut. Chem. Ges.* **52**, 195 (1919).
391. M. Tsutsui and H. H. Zeiss, *J. Amer. Chem. Soc.* **81**, 1367 (1959).
392. H. H. Zeiss and M. Tsutsui, *J. Amer. Chem. Soc.* **79**, 3062 (1957).
393. R. P. A. Sneeden and H. H. Zeiss, *J. Organometal. Chem.* **4**, 355 (1965).
394. G. Stolze and J. Hähle, *J. Organometal. Chem.* **5**, 545 (1966).
395. R. P. A. Sneeden and H. H. Zeiss, unpublished results.
396. R. P. A. Sneeden, F. Glockling, and H. H. Zeiss, *J. Organometal. Chem.* **6**, 149 (1966).
397. H. H. Zeiss and R. P. A. Sneeden, *Angew. Chem., Int. Ed. Engl.* **6**, 435 (1967).
398. J. Chatt and J. M. Davidson, *J. Chem. Soc., London* p. 843 (1965).
399. F. A. Cotton and M. D. LaPrade, *J. Amer. Chem. Soc.* **90**, 5418 (1968).
400. R. B. King and A. Fronzaglia, *J. Amer. Chem. Soc.* **88**, 709 (1966).
401. G. Stolze, *J. Organometal. Chem.* **6**, 383 (1966).
402. J. J. Daly, F. Sanz, R. P. A. Sneeden, and H. H. Zeiss, *J. Chem. Soc., Perkin Trans*, **2**, p. 1614 (1972).
403. M. Tsutsui and H. H. Zeiss, *J. Amer. Chem. Soc.* **82**, 6255 (1960).
404. M. Tsutsui and M. N. Levy, *Z. Naturforsch. B* **21**, 823 (1966).
405. H. H. Zeiss and W. Herwig, *J. Amer. Chem. Soc.* **78**, 5959 (1956).
406. H. H. Zeiss and W. Herwig, *Justus Liebigs Ann. Chem.* **606**, 209 (1957).
407. M. Tsutsui, *Ann. N.Y. Acad. Sci.* **93**, 133 (1961).
408. M. Tsutsui, *Z. Chem.* **3**, 215 (1963).
409. J. Hähle and G. Stolze, *J. Organometal. Chem.* **8**, 311 (1967).
410. E. Kurras and P. Klimsch, *Monatsber. Deut. Akad. Wiss. Berlin* **6**, 735 (1964).
411. P. Klimsch and E. Kurras, East Ger. Pat. 45,708 (1966).
412. D. G. H. Ballard and T. Medinger, *J. Chem. Soc., B* p. 1176 (1968).
413. T. Aoki, A. Furusaki, Y. Tomiee, K. Ono, and K. Tanaka, *Bull. Chem. Soc. Jap.* **42**, 545 (1969).
414. N. V. Kondyrev and D. A. Fomin, *Russ. J. Phys. Chem.* **47**, 190 (1915).
415. A. Job and R. Reich, *C.R. Acad. Sci.* **177**, 1439 (1923).
416. H. Gilman, R. G. Jones, and L. A. Woods, *J. Amer. Chem. Soc.* **76**, 3615 (1954).
417. E. O. Fischer and J. Müller, *Z. Naturforsch. B* **18**, 413 (1963).
418. R. P. A. Sneeden and H. H. Zeiss, *J. Organometal. Chem.* **13**, 369 (1968).
419. R. P. A. Sneeden and H. H. Zeiss, *J. Organometal. Chem.* **16**, 449 (1969).
420. R. P. A. Sneeden and H. H. Zeiss, *J. Organometal. Chem.* **26**, 101 (1971).
421. B. K. Bower and H. G. Tennent, *J. Amer. Chem. Soc.* **94**, 2512 (1972).
422. W. Mowat, A. J. Shortland, N. J. Hill, and G. Wilkinson, *J. Chem. Soc., Dalton Trans.* p. 770 (1973).
423. T. Tsuda and J. K. Kochi, *Bull. Chem. Soc. Jap.* **45**, 648 (1972).
424. R. P. A. Sneeden and H. H. Zeiss, *Angew. Chem. Int., Ed. Engl.* **7**, 951 (1968).
425. R. P. A. Sneeden and H. H. Zeiss, *J. Label. Compounds* **5**, 54 (1969).
426. L. S. Pu and A. Yamamoto, *Chem. Commun.* p. 9 (1974).
427. C. G. Overberger and P. A. Jarovitzky, *Polym. Prepr., Amer. Chem. Soc., Div. Polym. Chem.* **6**, 234 (1965).
428. F. Hug, W. Mowat, A. C. Skapski, and G. Wilkinson, *Chem. Commun.* p. 1477 (1971).

429. J. Chatt, R. S. Coffey, A. Gough, and D. T. Thompson, *J. Chem. Soc., A* p. 190 (1968).
430. R. P. A. Sneeden and H. H. Zeiss, *J. Organometal. Chem.* **13**, 377 (1968).
431. R. P. A. Sneeden and H. H. Zeiss, *J. Organometal. Chem.* **22**, 713 (1970).
432. R. P. A. Sneeden and H. H. Zeiss, *J. Organometal. Chem.* **19**, 93 (1969).
433. R. P. A. Sneeden and H. H. Zeiss, *J. Organometal. Chem.* **27**, 89 (1971).
434. T. Weichselfelder and B. Thiede, *Justus Liebigs Ann. Chem.* **447**, 64 (1926).
435. G. Dupont and P. Piganiol, *Bull. Soc. Chim. Fr.* p. 322 (1939).
436. Y. Takegami and T. Fujimaki, *Kogyo Kagaku Zasshi* **64**, 287 (1961).
437. K. Clauss and H. Bestian, *Justus Liebigs Ann. Chem.* **654**, 8 (1962).
438. F. Hein, H. Müller, and R. Weiss, *Handb. Int. Congr. Pure Appl. Chem., 17th 1959* p. 21 (1959).
439. M. Tsutsui and J. Ariyoshi, *J. Polym. Sci., Part A* **3**, 1729 (1965).
440. B. A. Dolgoplosk and E. I. Tinyakova, *Bull. Acad. Sci. USSR Div Chem. Sci.*, p. 291 (1970).
441. D. G. Ballard, W. H. Janes, and T. Medinger, *J. Chem. Soc., B* p. 1168 (1968).
442. O. K. Sharayev, A. V. Alferov, Ye. I. Tinyakova, and B. A. Dolgoplosk, *Vysokomol. Soedin., Ser. A* **9**, 633 (1967).
443. I. A. Oreshkin, I. Ya. Ostrovskaya, V. A. Yakovlev, E. I. Tinyakova, and B. A. Dolgoplosk, *Proc. Acad. Sci. USSR Chem. Sect.*, **173**, 402 (1967).
444. B. A. Dolgoplosk, I. A. Oreshkin, E. I. Tinyakova, and V. A. Yakovlev, *Bull. Acad. Sci. USSR Div. Chem. Sci.*, p. 2059 (1967).
445. I A. Oreshkin, G. M. Chernenko, E. I. Tinyakova, and B. A. Dolgoplosk, *Proc. Acad. Sci. USSR Chem. Sect.*, **169**, 790 (1966).
446. F. Hein and K. Schmiedeknecht, *J. Organometal. Chem.* **5**, 454 (1966).
447. E. Kurras and J. Otto, *J. Organometal. Chem.* **4**, 114 (1965).
448. F. Hein and K. Schmiedeknecht, *J. Organometal. Chem.* **6**, 45 (1966).
449. F. Hein and K. Schmiedeknecht, *J. Organometal. Chem.* **8**, 503 (1967).
450. F. Hein and K. Schmiedeknecht, *Z. Anorg. Allg. Chem.* **352**, 138 (1967).
451. F. Hein and B. Heyn, *Monatsber. Deut. Akad. Wiss. Berlin* **4**, 223 (1962).
452. F. Hein and B. Heyn, *Monatsber. Deut. Akad. Wiss. Berlin* **4**, 220 (1962).
453. D. Tille, *Z. Anorg. Allg. Chem.* **384**, 136 (1971).
454. E. Kurras, U. Rosenthal, H. Mennenga, and G. Oehme, *Angew. Chem., Int. Ed. Engl.* **12**, 854 (1973).
455. F. Hein and R. Weiss, *Naturwissenschaften* **46**, 321 (1959).
456. K. Nuetzel, Belg. Pat. 630,721 (1963); *Chem. Abstr.* **60**, 14713 (1964).
457. W. Mowat, A. Shortland, G. Yagupsky, N. J. Hill, M. Yagupsky, and G. Wilkinson, *J. Chem. Soc., Dalton Trans.* p. 533 (1972).
458. W. Mowat, A. J. Shortland, N. J. Hill, and G. Wilkinson, *J. Chem. Soc., Dalton Trans.* p. 770 (1973).
459. B. K. Bower and H. G. Tennent, *J. Amer. Chem. Soc.* **94**, 2512 (1972).
460. W. Kruse, *J. Organometal. Chem.* **42**, C39 (1972).
461. B. K. Bower and W. P. Long, U.S. Pat. 3,666,743 (1972).

SUBJECT INDEX

Chromium compounds are indexed under the name of the ligand and, when relevant, the oxidation states of the compounds are indicated by Roman numerals. The abbreviation Cp has been used in the index for cyclopentadienyl. The generic term "phosphine" has been used throughout the book to denote organophosphorus compounds.

B

Q

R

S

U

V

X

A 5
B 6
C 7
D 8
E 9
F 0
G 1
H 2
I 3
J 4